PHYSICS in NUCLEAR MEDICINE

PHYSICS
in NUCLEAR
MEDICINE

FOURTH EDITION

Simon R. Cherry, PhD
Professor, Departments of Biomedical Engineering and Radiology
Director, Center for Molecular and Genomic Imaging
University of California—Davis
Davis, California

James A. Sorenson, PhD
Emeritus Professor of Medical Physics
Department of Medical Physics
University of Wisconsin—Madison
Madison, Wisconsin

Michael E. Phelps, PhD
Norton Simon Professor
Chief, Division of Nuclear Medicine
Chair, Department of Molecular and Medical Pharmacology
Director, Crump Institute for Molecular Imaging
David Geffen School of Medicine
University of California—Los Angeles
Los Angeles, California

SAUNDERS

ELSEVIER

SAUNDERS

1600 John F. Kennedy Blvd.
Ste 1800
Philadelphia, PA 19103-2899

Physics in Nuclear Medicine   ISBN: 978-1-4160-5198-5

Library of Congress Cataloging-in-Publication Data

Cherry, Simon R.
  Physics in nuclear medicine / Simon R. Cherry, James A. Sorenson, Michael E. Phelps.
  —4th ed.
       p. ; cm.
  Includes bibliographical references and index.
  ISBN 978-1-4160-5198-5 (hardback : alk. paper)
  1. Medical physics. 2. Nuclear medicine. I. Sorenson, James A., 1938- II. Phelps,
Michael E. III. Title.
  [DNLM: 1. Health Physics. 2. Nuclear Medicine. WN 110]
  R895.S58 2012
  610.1′53—dc23

                                                                2011021330

Senior Content Strategist: Don Scholz
Content Development Specialist: Lisa Barnes
Publishing Services Manager: Anne Altepeter
Senior Project Manager: Janaki Srinivasan Kumar
Project Manager: Cindy Thoms
Design Direction: Ellen Zanolle

Printed in China

Last digit is the print number: 9 8 7 6 5 4

Preface

Physics and instrumentation affect all of the subspecialty areas of nuclear medicine. Because of their fundamental importance, they usually are taught as a separate course in nuclear medicine training programs. This book is intended for use in such programs by physicians, technologists, and scientists who desire to become specialists in nuclear medicine and molecular imaging, as well as a reference source for physicians, scientists, and engineers in related fields.

Although there have been substantial and remarkable changes in nuclear medicine, the goal of this book remains the same as it was for the first edition in 1980: to provide an introductory text for such courses, covering the physics and instrumentation of nuclear medicine in sufficient depth to be of permanent value to the trainee or student, but not at such depth as to be of interest only to the physics or instrumentation specialist. The fourth edition includes many recent advances, particularly in single-photon emission computed tomography (SPECT) and positron emission tomography (PET) imaging. As well, a new chapter is included on hybrid imaging techniques that combine the exceptional functional and physiologic imaging capabilities of SPECT and PET with the anatomically detailed techniques of computed tomography (CT) and magnetic resonance imaging (MRI). An introduction to CT scanning is also included in the new chapter.

The fourth edition also marks the first use of color. We hope that this not only adds cosmetic appeal but also improves the clarity of our illustrations.

The organization of this text proceeds from basic principles to more practical aspects. After an introduction to nuclear medicine (Chapter 1), we provide a review of atomic and nuclear physics (Chapter 2) and basic principles of radioactivity and radioactive decay (Chapters 3 and 4). Radionuclide production methods are discussed in Chapter 5, followed by radiation interactions in Chapter 6. Basic principles of radiation detectors (Chapter 7), radiation-counting electronics (Chapter 8), and statistics (Chapter 9) are provided next.

Following the first nine chapters, we move on to detailed discussions of nuclear medicine systems and applications. Pulse-height spectrometry, which plays an important role in many nuclear medicine procedures, is described in Chapter 10, followed by general problems in nuclear radiation counting in Chapter 11. Chapter 12 is devoted to specific types of nuclear radiation-counting instruments, for both in vivo and in vitro measurements.

Chapters 13 through 20 cover topics in radionuclide imaging, beginning with a description of the principles and performance characteristics of gamma cameras (Chapters 13 and 14), which are still the workhorse of many nuclear medicine laboratories. We then discuss general concepts of image quality in nuclear medicine (Chapter 15), followed by an introduction to the basic concepts of reconstruction tomography (Chapter 16).

The instrumentation for and practical implementation of reconstruction tomography are discussed for SPECT in Chapter 17 and for PET in Chapter 18. Hybrid imaging systems, as well as the basic principles of CT scanning, are covered in Chapter 19. Chapter 20 provides a summary of digital image processing techniques, which are important for all systems and applications.

The imaging section of this text focuses primarily on instruments and techniques that now enjoy or appear to have the potential for achieving clinical acceptance. However, nuclear medicine imaging has become increasingly important in the research environment. Therefore we have included some systems that are used for small-animal or other research purposes in these chapters.

We then move on to basic concepts and some applications of tracer kinetic modeling (Chapter 21). Tracer kinetic modeling and its applications embody two of the most important strengths of nuclear medicine techniques: the ability to perform studies with minute (tracer) quantities of labeled molecules and the ability to extract quantitative biologic data from these studies. We describe the main assumptions and mathematical models used and present several examples of the application of these models for calculating physiologic, metabolic, and biochemical parameters

The final two chapters address radiation dose and safety issues. Internal radiation dosimetry is presented in Chapter 22, and the final chapter presents an introduction to the problems of radiation safety and health physics (Chapter 23). We did not deal with more general problems in radiation biology, believing this topic to be of sufficient importance to warrant its own special treatment, as has been done already in several excellent books on the subject.

Additional reading for more detailed information is suggested at the end of each chapter. We also have included sample problems with solutions to illustrate certain quantitative relationships and to demonstrate standard calculations that are required in the practice of nuclear medicine. Systeme Internationale (SI) units are used throughout the text; however, traditional units still appear in a few places in the book, because these units remain in use in day-to-day practice in many laboratories. Appendix A provides a summary of conversion factors between SI and traditional units.

Appendixes B, C, and D present tables of basic properties of elements and radionuclides, and of attenuation properties of some materials of basic relevance to nuclear medicine. Appendix E provides a summary of radiation dose estimates for a number of nuclear medicine procedures. Although much of this information now is available on the Internet, we believe that users of this text will find it useful to have a summary of the indicated quantities and parameters conveniently available.

Appendixes F and G provide more detailed discussions of Fourier transforms and convolutions, both of which are essential components of modern nuclear medicine imaging, especially reconstruction tomography. This is the only part of the book that makes extensive use of calculus.

The fourth edition includes extensive revisions, and we are grateful to our many colleagues and friends who have assisted us with information, data, and figures. Particular gratitude is extended to Hendrik Pretorius, Donald Yapp, Jarek Glodo, Paul Kinahan, David Townsend, Richard Carson, Stephen Mather, and Freek Beekman. We also wish to thank readers who reported errors and inconsistencies in the third edition and brought these to our attention. In particular, we recognize the contributions of Andrew Goertzen, Tim Turkington, Mark Madsen, Ing-Tsung Hsiao, Jyh Cheng Chen, Scott Metzler, Andrew Maidment, Lionel Zuckier, Jerrold Bushberg, Zongjian Cao, Marvin Friedman, and Fred Fahey. This feedback from our readers is critical in ensuring the highest level of accuracy in the text. Naturally, any mistakes that remain in this new edition are entirely our responsibility.

We are grateful to Susie Helton (editorial assistance), and Robert Burnett and Simon Dvorak (graphics), at the University of California–Davis for their dedication to this project. We also appreciate the patience and efforts of the editorial staff at Elsevier, especially Lisa Barnes, Cindy Thoms, and Don Scholz. Finally, we thank our many colleagues who have used this book over the years and who have provided constructive feedback and suggestions for improvements that have helped to shape each new edition.

Simon R. Cherry, James A. Sorenson, and Michael E. Phelps

Contents

Animations, Calculators, and Graphing Tools

(Available online at expertconsult.com.)

ANIMATIONS

1. Emission of a characteristic x ray (Figure 2-4)
2. Emission of an Auger electron (Figure 2-5)
3. Internal conversion involving K-shell electron (Figure 3-5)
4. Positron emission and annihilation (Figure 3-7)
5. Positive ion cyclotron (Figure 5-3)
6. Ionization of an atom (Figure 6-1A)
7. Bremsstrahlung production (Figure 6-1B)
8. Photoelectric effect (Figure 6-11)
9. Compton scattering (Figure 6-12)
10. Pair production (Figure 6-14)
11. Basic principles of a gas-filled chamber (Figure 7-1)
12. Basic principles of a photomultiplier tube (Figure 7-13)
13. Scintillation detector (Figure 7-16)
14. Pulse-height spectrum (Figure 8-9 and Figure 10-2)
15. Gamma camera (Figure 13-1)
16. Sinogram formation and SPECT (Figure 16-4)
17. Backprojection (Figure 16-5)

CALCULATORS

1. Decay of activity (Equations 4-7 and 4-10)
2. Image-frame decay correction (Equations 4-15 and 4-16)
3. Carrier-free specific activity (Equations 4-21 to 4-23)
4. Cyclotron particle energy (Equation 5-12)
5. Compton scatter kinematics (Equations 6-11 and 6-12)
6. Photon absorption and transmission (Equation 6-22)
7. Effective atomic number (Equations 7-2 and 7-3)
8. Propagation of errors for sums and differences (Equation 9-12)
9. Solid angle calculation for a circular detector (Equation 11-7)
10. Activity conversions (Appendix A)

GRAPHING TOOLS

1. Bateman equation (Equation 4-25)
2. Dead time models (Equations 11-16 and 11-18)
3. Resolution and sensitivity of a parallel-hole collimator (Equations 14-6 and 14-7)
4. Resolution and sensitivity of a pinhole collimator (Equations 14-15 to 14-18)

PHYSICS in NUCLEAR MEDICINE

What Is Nuclear Medicine?

A. FUNDAMENTAL CONCEPTS

The science and clinical practice of nuclear medicine involve the administration of trace amounts of compounds labeled with radioactivity (radionuclides) that are used to provide diagnostic information in a wide range of disease states. Although radionuclides also have some therapeutic uses, with similar underlying physics principles, this book focuses on the diagnostic uses of radionuclides in modern medicine.

In its most basic form, a nuclear medicine study involves injecting a compound, which is labeled with a gamma-ray-emitting or positron-emitting radionuclide, into the body. The radiolabeled compound is called a *radiopharmaceutical*, or more commonly, a *tracer* or *radiotracer*. When the radionuclide decays, gamma rays or high-energy photons are emitted. The energy of these gamma rays or photons is such that a significant number can exit the body without being scattered or attenuated. An external, position-sensitive gamma-ray "camera" can detect the gamma rays or photons and form an image of the distribution of the radionuclide, and hence the compound (including radiolabeled products of reactions of that compound) to which it was attached.

There are two broad classes of nuclear medicine imaging: *single photon imaging* [which includes single photon emission computed tomography (SPECT)] and *positron imaging* [positron emission tomography (PET)]. Single photon imaging uses radionuclides that decay by gamma-ray emission. A *planar* image is obtained by taking a picture of the radionuclide distribution in the patient from one particular angle. This results in an image with little depth information, but which can still be diagnostically useful (e.g., in bone

scans, where there is not much tracer uptake in the tissue lying above and below the bones). For the tomographic mode of single photon imaging (SPECT), data are collected from many angles around the patient. This allows cross-sectional images of the distribution of the radionuclide to be reconstructed, thus providing the depth information missing from planar imaging.

Positron imaging makes use of radionuclides that decay by positron emission. The emitted positron has a very short lifetime and, following annihilation with an electron, simultaneously produces two high-energy photons that subsequently are detected by an imaging camera. Once again, tomographic images are formed by collecting data from many angles around the patient, resulting in PET images.

B. THE POWER OF NUCLEAR MEDICINE

The power of nuclear medicine lies in its ability to provide exquisitely sensitive measures of a wide range of biologic processes in the body. Other medical imaging modalities such as magnetic resonance imaging (MRI), x-ray imaging, and x-ray computed tomography (CT) provide outstanding anatomic images but are limited in their ability to provide biologic information. For example, magnetic resonance methods generally have a lower limit of detection in the millimolar concentration range ($\approx 6 \times 10^{17}$ molecules per mL tissue), whereas nuclear medicine studies routinely detect radiolabeled substances in the nanomolar ($\approx 6 \times 10^{11}$ molecules per mL tissue) or picomolar ($\approx 6 \times 10^{8}$ molecules per mL tissue) range. This sensitivity advantage, together with the ever-growing selection of radiolabeled compounds, allows nuclear medicine studies to be targeted to the very specific biologic processes underlying disease. Examples of the diverse biologic processes that can be measured by nuclear medicine techniques include tissue perfusion, glucose metabolism, the somatostatin receptor status of tumors, the density of dopamine receptors in the brain, and gene expression.

Because radiation detectors can easily detect very tiny amounts of radioactivity, and because radiochemists are able to label compounds with very high specific activity (a large fraction of the injected molecules are labeled with a radioactive atom), it is possible to form high-quality images even with nanomolar or picomolar concentrations of compounds. Thus trace amounts of a compound, typically many orders of magnitude below the millimolar to micromolar concentrations that generally are required for pharmacologic effects, can be injected and followed safely over time without perturbing the biologic system. Like CT, there is a small radiation dose associated with performing nuclear medicine studies, with specific doses to the different organs depending on the radionuclide, as well as the spatial and temporal distribution of the particular radiolabeled compound that is being studied. The safe dose for human studies is established through careful dosimetry for every new radiopharmaceutical that is approved for human use.

C. HISTORICAL OVERVIEW

As with the development of any field of science or medicine, the history of nuclear medicine is a complex topic, involving contributions from a large number of scientists, engineers, and physicians. A complete overview is well beyond the scope of this book; however, a few highlights serve to place the development of nuclear medicine in its appropriate historical context.

The origins of nuclear medicine[1] can be traced back to the last years of the 19th century and the discovery of radioactivity by Henri Becquerel (1896) and of radium by Marie Curie (1898). These developments came close on the heels of the discovery of x rays in 1895 by Wilhelm Roentgen. Both x rays and radium sources were quickly adopted for medical applications and were used to make shadow images in which the radiation was transmitted through the body and onto photographic plates. This allowed physicians to see "inside" the human body noninvasively for the first time and was particularly useful for the imaging of bone. X rays soon became the method of choice for producing "radiographs" because images could be obtained more quickly and with better contrast than those provided by radium or other naturally occurring radionuclides that were available at that time. Although the field of diagnostic x-ray imaging rapidly gained acceptance, nuclear medicine had to await further developments.

The biologic foundations for nuclear medicine were laid down between 1910 and 1945. In 1913, Georg de Hevesy developed the principles of the tracer approach[2] and was the first to apply them to a biologic system in 1923, studying the absorption and translocation of radioactive lead nitrate in plants.[3] The first human study employing radioactive tracers was probably that of Blumgart and Weiss (1927),[4] who injected an aqueous solution of radon intravenously and measured the transit time of the blood from one arm to the other using a cloud chamber as the radiation detector. In the 1930s, with the invention of the cyclotron by Lawrence (Fig. 1-1),[5] it became possible to artificially produce new radionuclides, thereby extending the range of biologic processes that could be studied. Once again, de Hevesy was at the forefront of using these new radionuclides to study biologic processes in plants and in red blood cells. Finally, at the end of the Second World War, the nuclear reactor facilities that were developed as part of the Manhattan Project started to be used for the production of radioactive isotopes in quantities sufficient for medical applications.

The 1950s saw the development of technology that allowed one to obtain images of the distribution of radionuclides in the human body rather than just counting at a few measurement points. Major milestones included the development of the rectilinear scanner in 1951 by Benedict Cassen[6] (Fig. 1-2) and the Anger camera, the forerunner of all modern nuclear medicine single-photon imaging systems, developed in 1958 by Hal Anger (Fig. 1-3).[7] In 1951, the use of positron emitters and the advantageous imaging properties of these radionuclides also were described by Wrenn and coworkers.[8]

Until the early 1960s, the fledgling field of nuclear medicine primarily used ^{131}I in the study and diagnosis of thyroid disorders and an assortment of other radionuclides that were individually suitable for only a few specific organs. The use of ^{99m}Tc for imaging in

FIGURE 1-1 Ernest O. Lawrence standing next to the cyclotron he invented at Berkeley, California. (*From Myers WG, Wagner HN: Nuclear medicine: How it began. Hosp Pract 9:103-113, 1974.*)

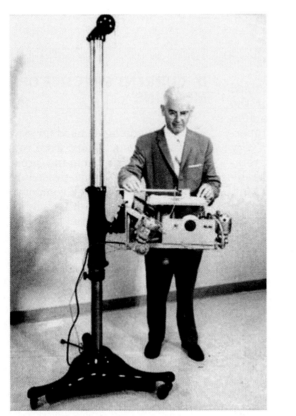

FIGURE 1-2 *Left*, Benedict Cassen with his rectilinear scanner (1951), a simple scintillation counter (see Chapter 7) that scans back and forth across the patient. *Right*, Thyroid scans from an early rectilinear scanner following administration of ^{131}I. The output of the scintillation counter controlled the movement of an ink pen to produce the first nuclear medicine images. (Left, *Courtesy William H. Blahd, MD; with permission of Radiology Centennial, Inc.* Right, *From Cassen B, Curtis L, Reed C, Libby R: Instrumentation for ^{131}I use in medical studies. Nucleonics 9:46-50, 1951.*)

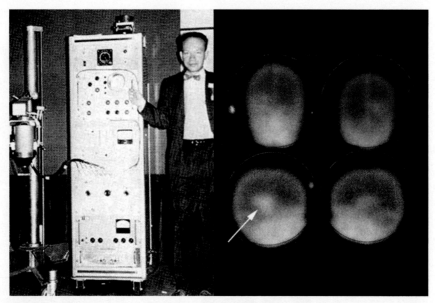

FIGURE 1-3   *Left*, Hal Anger with the first gamma camera in 1958. *Right*, ^{99m}Tc-pertechnetate brain scan of a patient with glioma at Vanderbilt University Hospital (1971). Each image represents a different view of the head. The glioma is indicated by an *arrow* in one of the views. In the 1960s, this was the only noninvasive test that could provide images showing pathologic conditions inside the human brain. These studies played a major role in establishing nuclear medicine as an integral part of the diagnostic services in hospitals. (Left, *From Myers WG: The Anger scintillation camera becomes of age.* J Nucl Med 20:565-567, 1979. Right, *Courtesy Dennis D. Patton, MD, University of Arizona, Tucson, Arizona.*)

1964 by Paul Harper and colleagues[9] changed this and was a major turning point for the development of nuclear medicine. The gamma rays emitted by ^{99m}Tc had very good properties for imaging. It also proved to be very flexible for labeling a wide variety of compounds that could be used to study virtually every organ in the body. Equally important, it could be produced in a relatively long-lived generator form, allowing hospitals to have a readily available supply of the radionuclide. Today, ^{99m}Tc is the most widely used radionuclide in nuclear medicine.

The final important development was the mathematics to reconstruct tomographic images from a set of angular views around the patient. This revolutionized the whole field of medical imaging (leading to CT, PET, SPECT and MRI) because it replaced the two-dimensional representation of the three-dimensional radioactivity distribution, with a true three-dimensional representation. This allowed the development of PET by Phelps and colleagues[10] and SPECT by Kuhl and colleagues[11] during the 1970s and marked the start of the modern era of nuclear medicine.

D. CURRENT PRACTICE OF NUCLEAR MEDICINE

Nuclear medicine is used for a wide variety of diagnostic tests. There were roughly 100 different diagnostic imaging procedures available in 2006.* These procedures use many different radiolabeled compounds, cover all the major organ systems in the body, and provide many different measures of biologic function. Table 1-1 lists some of the more common clinical procedures.

As of 2008, more than 30 million nuclear medicine imaging procedures were performed on a global basis.† There are more than 20,000 nuclear medicine cameras capable of imaging gamma-ray-emitting radionuclides installed in hospitals across the world. Even many small hospitals have their own nuclear medicine clinic. There also were more than 3,000 PET scanners installed in the world performing on the order of 4 million procedures

---

*Data courtesy Society of Nuclear Medicine, Reston, Virginia.

†Data courtesy Siemens Molecular Imaging, Hoffman Estates, Illinois.

TABLE 1-1
SELECTED CLINICAL NUCLEAR MEDICINE PROCEDURES

Radiopharmaceutical	Imaging	Measurement	Examples of Clinical Use
^{99m}Tc-MDP	Planar	Bone metabolism	Metastatic spread of cancer, osteomyelitis vs. cellulitis
^{99m}Tc-sestamibi (Cardiolite) ^{99m}Tc-tetrofosmin (Myoview) ^{201}Tl-thallous chloride	SPECT or planar	Myocardial perfusion	Coronary artery disease
^{99m}Tc-MAG3 ^{99m}Tc-DTPA	Planar	Renal function	Kidney disease
^{99m}Tc-HMPAO (Ceretec)	SPECT	Cerebral blood flow	Neurologic disorders
^{99m}Tc-ECD	SPECT	Cerebral blood flow	Neurologic disorders
^{123}I-sodium iodide	Planar	Thyroid function	Thyroid disorders
^{131}I-sodium iodide			Thyroid cancer
^{67}Ga-gallium citrate	Planar	Sequestered in tumors	Tumor localization
^{99m}Tc-macroaggregated albumin and ^{133}Xe gas	Planar	Lung perfusion/ ventilation	Pulmonary embolism
^{111}In-labeled white blood cells	Planar	Sites of infection	Detection of inflammation
^{18}F-fluorodeoxyglucose	PET	Glucose metabolism	Cancer, neurological disorders, and myocardial diseases
^{82}Rb-rubidium chloride	PET	Myocardial perfusion	Coronary artery disease

MDP, methylene diphosphonate; MAG3, mercapto-acetyl-triglycine; DTPA, diethylenetriaminepenta-acetic acid; HMPAO, hexamethylpropyleneamine oxime; ECD, ethyl-cysteine-dimer; SPECT, single photon emission computed tomography; PET, positron emission tomography.

annually. The short half-lives of the most commonly used positron-emitting radionuclides require an onsite accelerator or delivery of PET radiopharmaceuticals from regional radiopharmacies. To meet this need, there is now a PET radiopharmacy within 100 miles of approximately 90% of the hospital beds in the United States. The growth of clinical PET has been driven by the utility of a metabolic tracer, ^{18}F-fluorodeoxyglucose, which has widespread applications in cancer, heart disease, and neurologic disorders.

One major paradigm shift that has occurred since the turn of the millennium has been toward multimodality instrumentation. Virtually all PET scanners, and a rapidly growing number of SPECT systems, are now integrated with a CT scanner in combined PET/CT and SPECT/CT configurations. These systems enable the facile correlation of structure (CT) and function (PET or SPECT), yielding better diagnostic insight in many clinical situations. The combination of nuclear medicine scanners with MRI systems also is under investigation, and as of 2011, first commercial PET/MRI systems were being delivered.

In addition to its clinical role, PET (and to a certain extent, SPECT) continues to play a major role in the biomedical research community. PET has become an established and powerful research tool for quantitatively and noninvasively measuring the rates of biologic processes, both in the healthy and diseased state. In this research environment, the radiolabeled compounds and clinical nuclear medicine assays of the future are being developed. In preclinical, translational and clinical research, nuclear medicine has been at the forefront in developing new diagnostic opportunities in the field of molecular medicine, created by the merger of biology and medicine. A rapid growth is now occurring in the number and diversity of PET and SPECT molecular imaging tracers targeted to specific proteins and molecular pathways implicated in disease. These nuclear medicine technologies also have been embraced by the pharmaceutical and biotechnology industries to aid in drug development and validation.

E. THE ROLE OF PHYSICS IN NUCLEAR MEDICINE

....

Although the physics underlying nuclear medicine is not changing, the technology for producing radioactive tracers and for obtaining images of those tracer distributions most certainly is. We can expect to continue seeing major improvements in nuclear medicine technology, which will come from combining advances in detector and accelerator physics, electronics, signal processing, and computer technology with the underlying physics of nuclear medicine. Methods for accurately quantifying the concentrations of radiolabeled tracers in structures of interest, measuring biologic processes, and then relaying this information to the physician in a clinically meaningful and biologically relevant format are also an important challenge for the future. Refinement in the models used in dosimetry will allow better characterization of radiation exposure and make nuclear medicine even safer than it already is. Physics therefore continues to play an important and continuing role in providing high-quality, cost-effective, quantitative, reliable, and safe biologic assays in living humans.

REFERENCES

1. Mould RF: *A Century of X-Rays and Radioactivity in Medicine*, Bristol, 1993, Institute of Physics.
2. de Hevesy G: Radioelements as tracers in physics and chemistry. *Chem News* 108:166, 1913.
3. de Hevesy G: The absorption and translocation of lead by plants: A contribution to the application of the method of radioactive indicators in the investigation of the change of substance in plants. *Biochem J* 17:439-445, 1923.
4. Blumgart HL, Weiss S: Studies on the velocity of blood flow. *J Clin Invest* 4:15-31, 1927.
5. Lawrence EO, Livingston MS: The production of high-speed light ions without the use of high voltages. *Phys Rev* 40:19-30, 1932.
6. Cassen B, Curtis L, Reed C, Libby R: Instrumentation for ^{131}I use in medical studies. *Nucleonics* 9:46-50, 1951.
7. Anger HO: Scintillation camera. *Rev Sci Instr* 29:27-33, 1958.
8. Wrenn FR, Good ML, Handler P: The use of positron-emitting radioisotopes for the localization of brain tumors. *Science* 113:525-527, 1951.
9. Harper PV, Beck R, Charleston D, Lathrop KA: Optimization of a scanning method using technetium-99m. *Nucleonics* 22:50-54, 1964.
10. Phelps ME, Hoffman EJ, Mullani NA, Ter Pogossian MM: Application of annihilation coincidence detection of transaxial reconstruction tomography. *J Nucl Med* 16:210-215, 1975.
11. Kuhl DE, Edwards RQ, Ricci AR, et al: The Mark IV system for radionuclide computed tomography of the brain. *Radiology* 121:405-413, 1976.

BIBLIOGRAPHY

For further details on the history of nuclear medicine, we recommend the following:

Myers WG, Wagner HN: Nuclear medicine: How it began. *Hosp Pract* 9(3):103-113, 1974.
Nutt R: The history of positron emission tomography. *Mol Imaging Biol* 4:11-26, 2002.
Thomas AMK, editor: *The Invisible Light: One Hundred Years of Medical Radiology*, Oxford, England, 1995, Blackwell Scientific.
Webb S: From the Watching of Shadows: The Origins of Radiological Tomography, Bristol, England, 1990, Adam Hilger.

Recommended texts that cover clinical nuclear medicine in detail are the following:

Ell P, Gambhir S, editors: *Nuclear Medicine in Clinical Diagnosis and Treatment*, ed 3, Edinburgh, Scotland, 2004, Churchill Livingstone.
Sandler MP, Coleman RE, Patton JA, et al, editors: *Diagnostic Nuclear Medicine*, ed 4, Baltimore, 2002, Williams & Wilkins.
Schiepers C, editor: *Diagnostic Nuclear Medicine*, ed 2, New York, 2006, Springer.
Von Schulthess GK, editor: *Molecular Anatomic Imaging: PET-CT and SPECT-CT Integrated Modality Imaging*, ed 2, Philadelphia, 2006, Lippincott, Williams and Wilkins.

Basic Atomic and Nuclear Physics

Radioactivity is a process involving events in individual atoms and nuclei. Before discussing radioactivity, therefore, it is worthwhile to review some of the basic concepts of atomic and nuclear physics.

A. QUANTITIES AND UNITS

1. Types of Quantities and Units

Physical properties and processes are described in terms of *quantities* such as time and energy. These quantities are measured in *units* such as seconds and joules. Thus a quantity describes *what* is measured, whereas a unit describes *how much*.

Physical quantities are characterized as fundamental or derived. A *base* quantity is one that "stands alone"; that is, no reference is made to other quantities for its definition. Usually, base quantities and their units are defined with reference to standards kept at national or international laboratories. Time (s or sec), distance (m), and mass (kg) are examples of base quantities. *Derived* quantities are defined in terms of combinations of base quantities. Energy ($kg \cdot m^2/sec^2$) is an example of a derived quantity.

The international scientific community has agreed to adopt so-called System International (SI) units as the standard for scientific communication. This system is based on seven base quantities in metric units, with all other quantities and units derived by appropriate definitions from them. The four quantities of mass, length, time and electrical charge are most relevant to nuclear medicine. The use of specially defined quantities (e.g., "atmospheres" of barometric pressure) is specifically discouraged. It is hoped that this will improve scientific communication, as well as eliminate some of the more irrational units (e.g., feet and pounds). A useful discussion of the SI system, including definitions and values of various units, can be found in reference 1.

SI units or their metric subunits (e.g., centimeters and grams) are the standard for this text; however, in some instances traditional or other non-SI units are given as well (in parentheses). This is done because some traditional units still are used in the day-to-day practice of nuclear medicine (e.g., units of activity and absorbed dose). In other instances, SI units are unreasonably large (or small) for describing the processes of interest and specially defined units are more convenient and widely used. This is particularly true for units of mass and energy, as discussed in the following section.

2. Mass and Energy Units

Events occurring at the atomic level, such as radioactive decay, involve amounts of mass and energy that are very small when described in SI or other conventional units. Therefore they often are described in terms of specially defined units that are more convenient for the atomic scale.

The basic unit of mass is the *unified atomic mass unit,* abbreviated u. One u is defined as being equal to exactly $\frac{1}{12}$ the mass of an unbound ^{12}C atom* at rest and in its ground state. The conversion from SI mass units to unified atomic mass units is[1]

$$1\,u = 1.66054 \times 10^{-27}\ kg \qquad (2\text{-}1)$$

---

*Atomic notation is discussed in Section D.2.

The universal mass unit often is called a *Dalton* (Da) when expressing the masses of large biomolecules. The units are equivalent (i.e., 1 Da = 1 u). Either unit is convenient for expressing atomic or molecular masses, because a hydrogen atom has a mass of approximately 1 u or 1 Da.

The basic unit of energy is the *electron volt* (eV). One eV is defined as the amount of energy acquired by an electron when it is accelerated through an electrical potential of 1 V. Basic multiples are the kiloelectron volt (keV) (1 keV = 1000 eV) and the megaelectron volt (MeV) (1 MeV = 1000 keV = 1,000,000 eV). The conversion from SI energy units to the electron volt is

$$1 \text{ eV} = 1.6022 \times 10^{-19} \text{ kg} \cdot \text{m}^2/\text{sec}^2 \quad (2\text{-}2)$$

Mass m and energy E are related to each other by Einstein's equation $E = mc^2$, in which c is the velocity of light (approximately 3×10^8 m/sec in vacuum). According to this equation, 1 u of mass is equivalent to 931.5 MeV of energy.

Relationships between various units of mass and energy are summarized in Appendix A. Universal mass units and electron volts are very small, yet, as we shall see, they are quite appropriate to the atomic scale.

B. RADIATION

The term *radiation* refers to "energy in transit." In nuclear medicine, we are interested principally in the following two specific forms of radiation:

1. Particulate radiation, consisting of atomic or subatomic particles (electrons, protons, etc.) that carry energy in the form of kinetic energy of mass in motion.

2. Electromagnetic radiation, in which energy is carried by oscillating electrical and magnetic fields traveling through space at the speed of light.

Radioactive decay processes, discussed in Chapter 3, result in the emission of radiation in both of these forms.

The wavelength, λ, and frequency, ν, of the oscillating fields of electromagnetic radiation are related by:

$$\lambda \times \nu = c \quad (2\text{-}3)$$

where c is the velocity of light.

Most of the more familiar types of electromagnetic radiation (e.g., visible light and radio waves) exhibit "wavelike" behavior in their interactions with matter (e.g., diffraction patterns and transmission and detection of radio signals). In some cases, however, electromagnetic radiation behaves as discrete "packets" of energy, called *photons* (also called *quanta*). This is particularly true for interactions involving individual atoms. Photons have no mass or electrical charge and also travel at the velocity of light. These characteristics distinguish them from the forms of particulate radiation mentioned earlier. The energy of the photon E, in kiloelectron volts, and the wavelength of its associated electromagnetic field λ (in nanometers) are related by

$$E(\text{keV}) = 1.24/\lambda(\text{nm}) \quad (2\text{-}4)$$

Figure 2-1 illustrates the photon energies for different regions of the electromagnetic spectrum. Note that x rays and γ rays occupy the highest-energy, shortest-wavelength end of the spectrum; x-ray and γ-ray photons have energies in the keV-MeV range, whereas visible light photons, for example, have

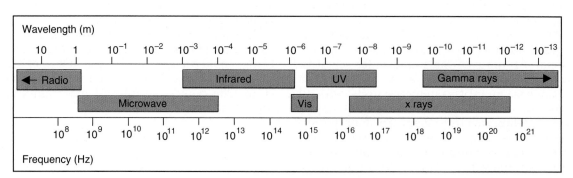

FIGURE 2-1 Schematic representation of the different regions of the electromagnetic spectrum. Vis, visible light; UV, ultraviolet light.

energies of only a few electron volts. As a consequence of their high energies and short wavelengths, x rays and γ rays interact with matter quite differently from other, more familiar types of electromagnetic radiation. These interactions are discussed in detail in Chapter 6.

C. ATOMS

1. Composition and Structure

All matter is composed of atoms. An atom is the smallest unit into which a chemical element can be broken down without losing its chemical identity. Atoms combine to form molecules and chemical compounds, which in turn combine to form larger, macroscopic structures.

The existence of atoms was first postulated on philosophical grounds by Ionian scholars in the 5th century BC. The concept was formalized into scientific theory early in the 19th century, owing largely to the work of the chemist, John Dalton, and his contemporaries. The exact structure of atoms was not known, but at that time they were believed to be indivisible. Later in the century (1869), Mendeleev produced the first *periodic table,* an ordering of the chemical elements according to the weights of their atoms and arrangement in a grid according to their chemical properties. For a time it was believed that completion of the periodic table would represent the final step in understanding the structure of matter.

Events of the late 19th and early 20th centuries, beginning with the discovery of x rays by Roentgen (1895) and radioactivity by Becquerel (1896), revealed that atoms had a substructure of their own. In 1910, Rutherford presented experimental evidence indicating that atoms consisted of a massive, compact, positively charged core, or *nucleus,* surrounded by a diffuse cloud of relatively light, negatively charged *electrons.* This model came to be known as the *nuclear atom.* The number of positive charges in the nucleus is called the *atomic number* of the nucleus (Z). In the electrically neutral atom, the number of orbital electrons is sufficient to balance exactly the number of positive charges, Z, in the nucleus. The chemical properties of an atom are determined by orbital electrons; therefore the atomic number Z determines the *chemical element* to which the atom belongs. A listing of chemical elements and their atomic numbers is given in Appendix B.

According to classical theory, orbiting electrons should slowly lose energy and spiral into the nucleus, resulting in atomic "collapse." This obviously is not what happens. The simple nuclear model therefore needed further refinement. This was provided by Niels Bohr in 1913, who presented a model that has come to be known as the *Bohr atom.* In the Bohr atom there is a set of stable electron orbits, or "shells," in which electrons can exist indefinitely without loss of energy. The diameters of these shells are determined by *quantum numbers,* which can have only integer values ($n = 1, 2, 3, \ldots$). The innermost shell ($n = 1$) is called the K shell, the next the L shell ($n = 2$), followed by the M shell ($n = 3$), N shell ($n = 4$), and so forth.

Each shell actually comprises a set of orbits, called *substates,* which differ slightly from one another. Each shell has $2n - 1$ substates, in which n is the quantum number of the shell. Thus the K shell has only one substate; the L shell has three substates, labeled L_I, L_{II}, L_{III}; and so forth. Figure 2-2 is a schematic representation of the K, L, M, and N shells of an atom.

The Bohr model of the atom was further refined with the statement of the *Pauli Exclusion Principle* in 1925. According to this principle, no two orbital electrons in an atom can move with exactly the same motion. Because of different possible electron "spin" orientations, more than one electron can exist in each substate; however, the number of electrons that can exist in any one shell or its substates is limited. For a shell with quantum number n, the maximum number of electrons allowed is $2n^2$. Thus the K shell ($n = 1$) is limited to two electrons, the L shell ($n = 2$) to eight electrons, and so forth.

The Bohr model is actually an oversimplification. According to modern theories, the orbital electrons do not move in precise circular orbits but rather in imprecisely defined "regions of space" around the nucleus, sometimes actually passing through the nucleus; however, the Bohr model is quite adequate for the purposes of this text.

2. Electron Binding Energies and Energy Levels

In the most stable configuration, orbital electrons occupy the innermost shells of an atom, where they are most "tightly bound" to the nucleus. For example, in carbon, which has a total of six electrons, two electrons (the maximum number allowed) occupy the K

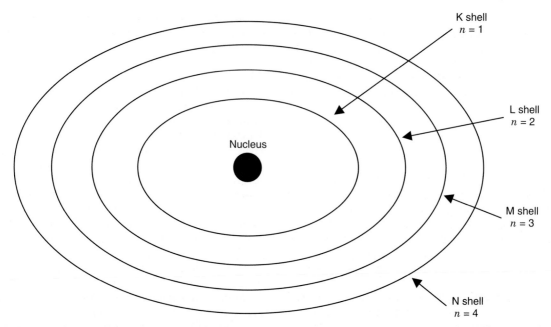

FIGURE 2-2 Schematic representation of the Bohr model of the atom; n is the quantum number of the shell. Each shell has multiple substates, as described in the text.

shell, and the four remaining electrons are found in the L shell. Electrons can be moved to higher shells or completely removed from the atom, but doing so requires an energy input to overcome the forces of attraction that "bind" the electron to the nucleus. The energy may be provided, for example, by a particle or a photon striking the atom.

The energy required to completely remove an electron from a given shell in an atom is called the *binding energy* of that shell. It is symbolized by the notation K_B for the K shell,* L_B for the L shell (L_{IB}, L_{IIB}, L_{IIIB} for the L shell substates), and so forth. Binding energy is greatest for the innermost shell, that is, $K_B > L_B > M_B$. Binding energy also increases with the positive charge (atomic number Z) of the nucleus, because a greater positive charge exerts a greater force of attraction on an electron. Therefore binding energies are greatest for the heaviest elements. Values of K-shell binding energies for the elements are listed in Appendix B.

The energy required to move an electron from an inner to an outer shell is exactly equal to the difference in binding energies between the two shells. Thus the energy required to move an electron from the K shell to the L shell in an atom is $K_B - L_B$ (with slight differences for different L shell substates).

Binding energies and energy differences are sometimes displayed on an *energy-level diagram*. Figure 2-3 shows such a diagram for the K and L shells of the element iodine. The top line represents an electron completely separated from the parent atom ("unbound" or "free" electron). The bottom line represents the most tightly bound electrons, that is, the K shell. Above this are lines representing substates of the L shell. (The M shell and other outer shell lines are just above the L shell lines.) The distance from the K shell to the top level represents the K-shell binding energy for iodine (33.2 keV). To move a K-shell electron to the L shell requires approximately $33 - 5 = 28$ keV of energy.

3. Atomic Emissions

When an electron is removed from one of the inner shells of an atom, an electron from an outer shell promptly moves in to fill the vacancy and energy is released in the process. The energy released when an electron drops from an outer to an inner shell is exactly equal to the difference in binding energies between the two shells. The energy may appear as a photon of electromagnetic radiation (Fig. 2-4). Electron binding energy differences have exact characteristic values for different elements; therefore the photon emissions are called *characteristic radiation* or *characteristic x rays*. The notation used to

---

*Sometimes the notation K_{ab} also is used.

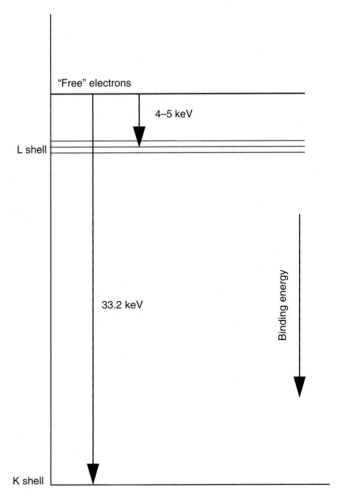

FIGURE 2-3 Electron energy-level diagram for an iodine atom. *Vertical axis* represents the energy required to remove orbital electrons from different shells (binding energy). Removing an electron from the atom, or going from an inner (e.g., K) to an outer (e.g., L) shell, requires an energy input, whereas an electron moving from an outer to an inner shell results in the emission of energy from the atom.

FIGURE 2-4 Emission of characteristic x rays occurs when orbital electrons move from an outer shell to fill an inner-shell vacancy. (K_α x-ray emission is illustrated.)

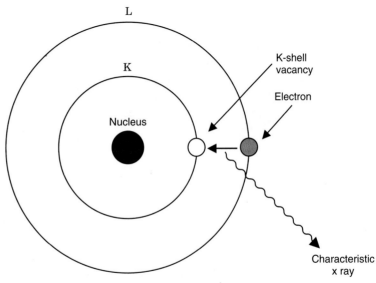

identify characteristic x rays from various electron transitions is summarized in Table 2-1. Note that some transitions are not allowed, owing to the selection rules of quantum mechanics.

As an alternative to characteristic x-ray emission, the atom may undergo a process known as the *Auger* (pronounced oh-zhaý) *effect*. In the Auger effect, an electron from an outer shell again fills the vacancy, but the energy released in the process is transferred to another orbital electron. This electron then is emitted from the atom instead of characteristic radiation. The process is shown schematically in Figure 2-5. The emitted electron is called an *Auger electron*.

The kinetic energy of an Auger electron is equal to the difference between the binding energy of the shell containing the original vacancy and the sum of the binding energies of the two shells having vacancies at the end. Thus the kinetic energy of the Auger electron emitted in Figure 2-5 is $K_B - 2L_B$ (ignoring small differences in L-substate energies).

Two orbital vacancies exist after the Auger effect occurs. These are filled by electrons from the other outer shells, resulting in the emission of additional characteristic x rays or Auger electrons.

The number of vacancies that result in emission of characteristic x rays versus Auger electrons is determined by probability values that depend on the specific element and orbital shell involved. The probability that a vacancy will yield characteristic x rays is called the *fluorescent yield,* symbolized by ω_K for the K shell, ω_L for the L shell, and so forth. Figure 2-6 is a graph of ω_K versus Z. Both characteristic x rays and Auger electrons are emitted by all elements, but heavy elements are more likely to emit x rays (large ω), whereas light elements are more likely to emit electrons (small ω).

The notation used to identify the shells involved in Auger electron emission is e_{abc}, in which a identifies the shell with the original vacancy, b the shell from which the electron dropped to fill the vacancy, and c the shell from which the Auger electron was emitted.

TABLE 2-1
SOME NOTATION USED FOR CHARACTERISTIC X RAYS

Shell with Vacancy	Shell from Which Filled	Notation
K	L_I	Not allowed
K	L_{II}	$K_{\alpha 2}$
K	L_{III}	$K_{\alpha 1}$
K	M_I	Not allowed
K	M_{II}	$K_{\beta 3}$
K	M_{III}	$K_{\beta 1}$
K	N_I	Not allowed
K	N_{II}, N_{III}	$K_{\beta 2}$
L_{II}	M_{IV}	$L_{\beta 1}$
L_{III}	M_{IV}	$L_{\alpha 2}$
L_{III}	M_V	$L_{\alpha 1}$

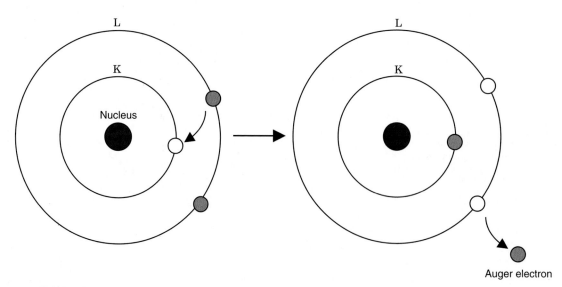

FIGURE 2-5   Emission of an Auger electron as an alternative to x-ray emission. No x ray is emitted.

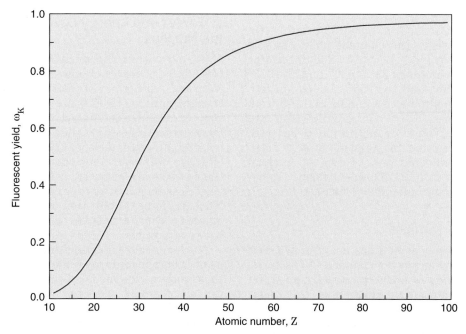

FIGURE 2-6　Fluorescent yield, ω_K, or probability that an orbital electron shell vacancy will yield characteristic x rays rather than Auger electrons, versus atomic number Z of the atom. (*Data from Hubbell JH, Trehan PN, Singh N, et al: A review, bibliography, and tabulation of K, L, and higher atomic shell x-ray fluorescence yields.* J Phys Chem Ref Data 23:339-364, 1994.)

Thus the electron emitted in Figure 2-5 is a *KLL* Auger electron, symbolized by e_{KLL}. In the notation e_{Kxx}, the symbol x is inclusive, referring to all Auger electrons produced from initial K-shell vacancies.

D. THE NUCLEUS

1. Composition

The atomic nucleus is composed of *protons* and *neutrons*. Collectively, these particles are known as *nucleons*. The properties of nucleons and electrons are summarized in Table 2-2.

TABLE 2-2
BASIC PROPERTIES OF NUCLEONS AND ELECTRONS[1]

Particle	Charge*	Mass	
		u	**MeV**
Proton	+1	1.007276	938.272
Neutron	0	1.008665	939.565
Electron	−1	0.000549	0.511

*One unit of charge is equivalent to 1.602×10^{-19} coulombs.

Nucleons are much more massive than electrons (by nearly a factor of 2000). Conversely, nuclear diameters are very small in comparison with atomic diameters (10^{-13} vs. 10^{-8} cm). Thus it can be deduced that the density of nuclear matter is very high ($\sim 10^{14}$ g/cm^3) and that the rest of the atom (electron cloud) is mostly empty space.

2. Terminology and Notation

An atomic nucleus is characterized by the number of neutrons and protons it contains. The number of protons determines the *atomic number* of the atom, Z. As mentioned earlier, this also determines the number of orbital electrons in the electrically neutral atom and therefore the *chemical element* to which the atom belongs.

The total number of nucleons is the *mass number* of the nucleus, A. The difference, A − Z, is the *neutron number*, N. The mass number A is approximately equal to, but not the same as, the *atomic weight* (AW) used in chemistry. The latter is the average weight of an atom of an element in its natural abundance (see Appendix B).

The notation now used to summarize atomic and nuclear composition is $^A_Z X_N$, in which X represents the chemical element to which the atom belongs. For example, an atom composed of 53 protons, 78 neutrons (and thus 131 nucleons), and 53 orbital electrons represents the element iodine and is symbolized by $^{131}_{53}I_{78}$. Because all iodine atoms have atomic number 53, the "I" and the "53" are redundant and the "53" can be omitted. The neutron number, 78, can be inferred from the difference, 131 – 53, so this also can be omitted. Therefore a shortened but still complete notation for this atom is ^{131}I. An acceptable alternative in terms of medical terminology is I-131. Obsolete forms (sometimes found in older texts) include I^{131}, $_{131}I$, and I_{131}.

3. Nuclear Families

Nuclear species sometimes are grouped into families having certain common characteristics. A *nuclide* is characterized by an exact nuclear composition, including the mass number A, atomic number Z, and arrangement of nucleons within the nucleus. To be classified as a nuclide, the species must have a "measurably long" existence, which for current technology means a lifetime greater than about 10^{-12} sec. For example, ^{12}C, ^{16}O, and ^{131}I are nuclides.

Figure 2-7 summarizes the notation used for identifying a particular nuclear species, as well as the terminology used for nuclear families. Nuclides that have the same atomic number Z are called *isotopes*. Thus ^{125}I, ^{127}I, and ^{131}I are isotopes of the element iodine. Nuclides with the same mass number A are *isobars* (e.g., ^{131}I, ^{131}Xe, and ^{131}Cs). Nuclides with the same neutron number N are *isotones* (e.g., $^{131}_{53}I_{78}$, $^{132}_{54}Xe_{78}$, and $^{133}_{55}Cs_{78}$). A mnemonic device for remembering these relationships is that iso**p**es have the same number of protons, iso**n**es the same number of neutrons, and isob**a**rs the same mass number (A).

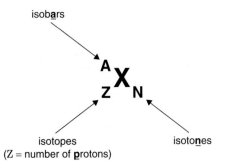

FIGURE 2-7 Notation and terminology for nuclear families.

4. Forces and Energy Levels within the Nucleus

Nucleons within the nucleus are subject to two kinds of forces. Repulsive *coulombic* or *electrical forces* exist between positively charged protons. These are counteracted by very strong forces of attraction, called *nuclear forces* (sometimes also called *exchange forces*), between any two nucleons. Nuclear forces are effective only over very short distances, and their effects are seen only when nucleons are very close together, as they are in the nucleus. Nuclear forces hold the nucleus together against the repulsive coulombic forces between protons.

Nucleons move about within the nucleus in a very complicated way under the influence of these forces. One model of the nucleus, called the *shell model,* portrays the nucleons as moving in "orbits" about one another in a manner similar to that of orbital electrons moving about the nucleus in the Bohr atom. Only a limited number of motions are allowed, and these are determined by a set of nuclear quantum numbers.

The most stable arrangement of nucleons is called the *ground state.* Other arrangements of the nucleons fall into the following two categories:

1. *Excited states* are arrangements that are so unstable that they have only a transient existence before transforming into some other state.
2. *Metastable states* also are unstable, but they have relatively long lifetimes before transforming into another state. These also are called isomeric states.

The dividing line for lifetimes between excited and metastable states is approximately 10^{-12} sec. This is not a long time according to everyday standards, but it is "relatively long" by nuclear standards. (The prefix *meta* derives from the Greek word for "almost.") Some metastable states are quite long-lived; that is, they have average lifetimes of several hours. Because of this, metastable states are considered to have separate identities and are themselves classified as nuclides. Two nuclides that differ from one another in that one is a metastable state of the other are called *isomers.*

In nuclear notation, excited states are identified by an asterisk ($^{A}X^*$) and metastable states by the letter m (^{Am}X or X-Am).[†] Thus

---

[†]The notation $^{A}X^m$ is sometimes used in Europe (e.g., $^{99}Tc^m$).

^{99m}Tc (or Tc-99m) represents a metastable state of ^{99}Tc, and ^{99m}Tc and ^{99}Tc are isomers.

Nuclear transitions between different nucleon arrangements involve discrete and exact amounts of energy, as do the rearrangements of orbital electrons in the Bohr atom. A *nuclear energy-level diagram* is used to identify the various excited and metastable states of a nuclide and the energy relationships among them. Figure 2-8 shows a partial diagram for ^{131}Xe.* The bottom line represents the ground state, and other lines represent excited or metastable states. Metastable states usually are indicated by somewhat heavier lines. The vertical distances between lines are proportional to the energy differences between levels. A transition from a lower to a higher state requires an energy input of some sort, such as a photon or particle striking the nucleus. Transitions from higher to lower states result in the release of energy, which is given to emitted particles or photons.

5. Nuclear Emissions

Nuclear transformations can result in the emission of particles (primarily electrons or α particles) or photons of electromagnetic radiation. This is discussed in detail in Chapter 3.

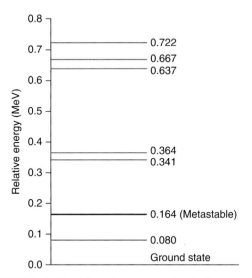

FIGURE 2-8 Partial nuclear energy-level diagram for the ^{131}Xe nucleus. The *vertical axis* represents energy differences between nuclear states (or "arrangements" of nucleons). Going up the scale requires energy input. Coming down the scale results in the emission of nuclear energy. Heavier lines indicate metastable states.

---

*Actually, these are the excited and metastable states formed during radioactive decay by β⁻ emission of ^{131}I (see Chapter 3, Section D, and Appendix C).

Photons of nuclear origin are called γ *rays* (*gamma rays*). The energy difference between the states involved in the nuclear transition determines the γ-ray energy. For example, in Figure 2-8 a transition from the level marked 0.364 MeV to the ground state would produce a 0.364-MeV γ ray. A transition from the 0.364-MeV level to the 0.080-MeV level would produce a 0.284-MeV γ ray.

As an alternative to emitting a γ ray, the nucleus may transfer the energy to an orbital electron and emit the electron instead of a photon. This process, which is similar to the Auger effect in x-ray emission (see Section C.3, earlier in this chapter), is called *internal conversion*. It is discussed in detail in Chapter 3, Section E.

6. Nuclear Binding Energy

When the mass of an atom is compared with the sum of the masses of its individual components (protons, neutrons, and electrons), it always is found to be less by some amount, Δm. This mass deficiency, expressed in energy units, is called the *binding energy* E_B of the atom:

$$E_B = \Delta mc^2 \qquad (2\text{-}5)$$

For example, consider an atom of ^{12}C. This atom is composed of six protons, six electrons, and six neutrons, and its mass is precisely 12 u (by definition of the universal mass unit u). The sum of the masses of its components is

electrons	6×0.000549 u	=0.003294 u
protons	6×1.007276 u	=6.043656 u
neutrons	6×1.008665 u	=6.051990 u
total		12.098940 u

Thus Δm = 0.098940 u. Because 1 u = 931.5 MeV, the binding energy of a ^{12}C atom is 0.098940 × 931.5 MeV = 92.16 MeV.

The binding energy is the minimum amount of energy required to overcome the forces holding the atom together to separate it completely into its individual components. Some of this represents the binding energy of orbital electrons, that is, the energy required to strip the orbital electrons away from the nucleus; however, comparison of the total binding energy of a ^{12}C atom with the K-shell binding energy of carbon (see Appendix B) indicates that most of this energy is *nuclear binding energy*, that is, the energy required to separate the nucleons.

Nuclear processes that result in the release of energy (e.g., γ-ray emission) always *increase* the binding energy of the nucleus. Thus a nucleus emitting a 1-MeV γ ray would be found to weigh *less* (by the mass equivalent of 1 MeV) after the γ ray was emitted than before. In essence, mass is converted to energy in the process.

7. Characteristics of Stable Nuclei

Not all combinations of protons and neutrons produce stable nuclei. Some are unstable, even in their ground states. An unstable nucleus emits particles or photons to transform itself into a more stable nucleus. This is the process of *radioactive disintegration* or *radioactive decay,* discussed in Chapter 3. A survey of the general characteristics of naturally occurring *stable nuclides* provides clues to the factors that contribute to nuclear instability and thus to radioactive decay.

Figure 2-9 is a plot of the nuclides found in nature, according to their neutron and proton numbers. For example, the nuclide $^{12}_{6}C$ is represented by a dot at the point Z = 6, N = 6. Most of the naturally occurring nuclides are stable; however, 17 very long-lived but unstable (radioactive) nuclides that still are present from the creation of the elements also are shown.

A first observation is that there are favored neutron-to-proton ratios among the naturally occurring nuclides. They are clustered around an imaginary line called the *line of stability.* For light elements, the line corresponds to N ≈ Z, that is, approximately equal numbers of protons and neutrons. For heavy elements, it corresponds to N ≈ 1.5 Z, that is, approximately 50% more neutrons than protons. The line of stability ends at ^{209}Bi (Z = 83, N = 126). All heavier nuclides are unstable.

In general, there is a tendency toward instability in atomic systems composed of large numbers of identical particles confined in a small volume. This explains the instability of very heavy nuclei. It also explains why, for light elements, stability is favored by more or less equal numbers of neutrons and protons rather than grossly unequal numbers. A moderate excess of neutrons is favored among heavier elements because neutrons provide only exchange forces (attraction), whereas protons provide both exchange forces and coulombic forces (repulsion). Exchange forces are effective over very short distances and thus affect only "close neighbors" in the nucleus, whereas the repulsive coulombic forces are effective over much greater distances. Thus

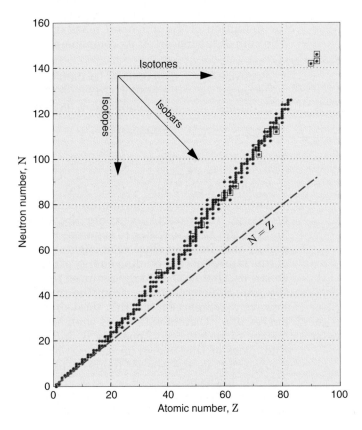

FIGURE 2-9 Neutron number (N) versus atomic number (Z) for nuclides found in nature. The *boxed data points* identify very long-lived, naturally occurring unstable (radioactive) nuclides. The remainder are stable. The nuclides found in nature are clustered around an imaginary line called the *line of stability.* N ≈ Z for light elements; N ≈ 1.5 Z for heavy elements.

an excess of neutrons is required in heavy nuclei to overcome the long-range repulsive coulombic forces between a large number of protons.

Nuclides that are not close to the line of stability are likely to be unstable. Unstable nuclides lying above the line of stability are said to be "proton deficient," whereas those lying below the line are "neutron deficient." Unstable nuclides generally undergo radioactive decay processes that transform them into nuclides lying closer to the line of stability, as discussed in Chapter 3.

Figure 2-9 demonstrates that there often are many stable isotopes of an element. Isotopes fall on vertical lines in the diagram. For example, there are ten stable isotopes of tin (Sn, $Z = 50$)*. There may also be several stable isotones. These fall along horizontal lines. In relatively few cases, however, is there more than one stable isobar (isobars fall along descending 45-degree lines on the graph).

---

*Although most element symbols are simply one- or two-letter abbreviations of their (English) names, ten symbols derive from Latin or Greek names of metals known for more than 2 millennia: antimony (stibium, Sb), copper (cuprum, Cu), gold (aurum, Au), iron (ferrum, Fe), lead (plumbum, Pb), mercury (hydrargyrum, Hg), potassium (kalium, K), silver (argentum, Ag), sodium (natrium, Na), and tin (stannum, Sn). The symbol for tungsten, W, derives from the German "wolfram," the name it was first given in medieval times.

This reflects the existence of several modes of "isobaric" radioactive decay that permit nuclides to transform along isobaric lines until the most stable isobar is reached. This is discussed in detail in Chapter 3.

One also notes among the stable nuclides a tendency to favor even numbers. For example, there are 165 stable nuclides with both even numbers of protons and even numbers of neutrons. Examples are $^4_2$He and $^{12}_6$C. There are 109 "even-odd" stable nuclides, with even numbers of protons and odd numbers of neutrons or vice versa. Examples are $^9_4$Be and $^{11}_5$B. However, there are only four stable "odd-odd" nuclides: $^2_1$H, $^6_3$Li, $^{10}_5$B, and $^{14}_7$N. The stability of even numbers reflects the tendency of nuclei to achieve stable arrangements by the "pairing up" of nucleons in the nucleus.

Another measure of relative nuclear stability is nuclear binding energy, because this represents the amount of energy required to break the nucleus up into its separate components. Obviously, the greater the number of nucleons, the greater the total binding energy. Therefore a more meaningful parameter is the *binding energy per nucleon*, E_B/A. Higher values of E_B/A are indicators of greater nuclear stability.

Figure 2-10 is a graph of E_B/A versus A for the stable nuclides. Binding energy is greatest (≈ 8 MeV per nucleon) for nuclides of mass number A ≈ 60. It decreases slowly with increasing A, indicating the tendency toward

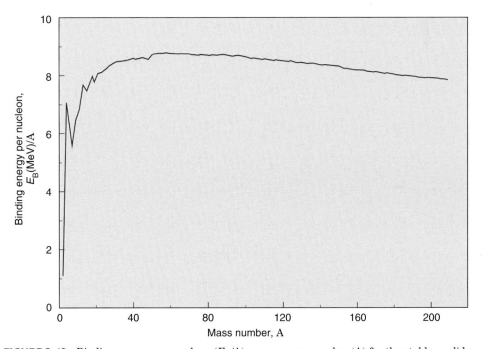

FIGURE 2-10 Binding energy per nucleon (E_B/A) versus mass number (A) for the stable nuclides.

instability for very heavy nuclides. Finally, there are a few peaks in the curve representing very stable light nuclides, including $^{4}_{2}$He, $^{12}_{6}$C, and $^{16}_{8}$O. Note that these are all even-even nuclides.

REFERENCES

1. National Institute of Standards and Technology (NIST): Fundamental Physics Constants. Available at http://physics.nist.gov/cuu/Constants/index.html [accessed July 4, 2011].

BIBLIOGRAPHY

Fundamental quantities of physics and mathematics, as well as constants and conversion factors, can be found in reference 1.

Recommended texts for in-depth discussions of topics in atomic and nuclear physics are the following:

Evans RD: *The Atomic Nucleus*, New York, 1972, McGraw-Hill.

Jelley NA: *Fundamentals of Nuclear Physics*, New York, 1990, Cambridge University Press.

Yang F, Hamilton JH: *Modern Atomic and Nuclear Physics*, New York, 1996, McGraw-Hill.

Modes of Radioactive Decay

Radioactive decay is a process in which an unstable nucleus transforms into a more stable one by emitting particles, photons, or both, releasing energy in the process. Atomic electrons may become involved in some types of radioactive decay, but it is basically a *nuclear* process caused by *nuclear* instability. In this chapter we discuss the general characteristics of various modes of radioactive decay and their general importance in nuclear medicine.

A. GENERAL CONCEPTS

It is common terminology to call an unstable radioactive nucleus the *parent* and the more stable product nucleus the *daughter*. In many cases, the daughter also is radioactive and undergoes further radioactive decay. Radioactive decay is *spontaneous* in that the exact moment at which a given nucleus will decay cannot be predicted, nor is it affected to any significant extent by events occurring outside the nucleus.

Radioactive decay results in the conversion of mass into energy. If all the products of a particular decay event were gathered together and weighed, they would be found to weigh less than the original radioactive atom. Usually, the energy arises from the conversion of nuclear mass, but in some decay modes, electron mass is converted into energy as well. The total mass-energy conversion amount is called the *transition energy*, sometimes designated Q.* Most of this energy is imparted as kinetic energy to emitted particles or

converted to photons, with a small (usually insignificant) portion given as kinetic energy to the recoiling nucleus. Thus radioactive decay results not only in the transformation of one nuclear species into another but also in the transformation of mass into energy.

Each radioactive nuclide has a set of characteristic properties. These properties include the mode of radioactive decay and type of emissions, the transition energy, and the average lifetime of a nucleus of the radionuclide before it undergoes radioactive decay. Because these basic properties are characteristic of the nuclide, it is common to refer to a radioactive species, such as ^{131}I, as a *radionuclide*. The term *radioisotope* also is used but, strictly speaking, should be used only when specifically identifying a member of an isotopic family as radioactive; for example, ^{131}I is a radioisotope of iodine.

B. CHEMISTRY AND RADIOACTIVITY

Radioactive decay is a process involving primarily the nucleus, whereas chemical reactions involve primarily the outermost orbital electrons of the atom. Thus the fact that an atom has a radioactive nucleus does not affect its chemical behavior and, conversely, the chemical state of an atom does not affect its radioactive characteristics. For example, an atom of the radionuclide ^{131}I exhibits the same chemical behavior as an atom of ^{127}I, the naturally occurring stable nuclide, and ^{131}I has the same radioactive characteristics whether it exists as iodide ion (I^-) or incorporated into a

---

*Some texts and applications consider only nuclear mass, rather than the mass of the entire atom (i.e., atomic mass), in the definition of transition energy. As will be seen, the use of atomic mass is more appropriate for the analysis of radioactive decay because both nuclear and nonnuclear mass are converted into energy in some decay

modes. As well, energy originating from either source can contribute to usable radiation or to radiation dose to the patient. For a detailed discussion of the two methods for defining transition energy, see Evans RD: *The Atomic Nucleus.* New York, 1972, McGraw-Hill, pp 117-133.

large protein molecule as a radioactive label. Independence of radioactive and chemical properties is of great significance in tracer studies with radioactivity—a radioactive *tracer* behaves in chemical and physiologic processes exactly the same as its stable, naturally occurring counterpart, and, further, the radioactive properties of the tracer do not change as it enters into chemical or physiologic processes.

There are two minor exceptions to these generalizations. The first is that chemical behavior can be affected by differences in atomic *mass*. Because there are always mass differences between the radioactive and the stable members of an isotopic family (e.g., ^{131}I is heavier than ^{127}I), there may also be chemical differences. This is called the *isotope effect*. Note that this is a *mass* effect and has nothing to do with the fact that one of the isotopes is radioactive. The chemical differences are small unless the relative mass differences are large, for example, ^{3}H versus ^{1}H. Although the isotope effect is important in some experiments, such as measurements of chemical bond strengths, it is, fortunately, of no practical consequence in nuclear medicine.

A second exception is that the average lifetimes of radionuclides that decay by processes involving orbital electrons (e.g., internal conversion, Section E, and electron capture, Section F) can be changed very slightly by altering the chemical (orbital electron) state of the atom. The differences are so small that they cannot be detected except in elaborate nuclear physics experiments and again are of no practical consequence in nuclear medicine.

C. DECAY BY β⁻ EMISSION

Radioactive decay by β⁻ emission is a process in which, essentially, a neutron in the nucleus is transformed into a proton and an electron. Schematically, the process is

$$n \rightarrow p^+ + e^- + \nu + \text{energy} \qquad (3\text{-}1)$$

The electron (e⁻) and the neutrino (ν) are ejected from the nucleus and carry away the energy released in the process as kinetic energy. The electron is called a β⁻ *particle*. The neutrino is a "particle" having no mass or electrical charge.* It undergoes virtually no interactions with matter and therefore is essentially undetectable. Its only practical consequence is that it carries away some of the energy released in the decay process.

Decay by β⁻ emission may be represented in standard nuclear notation as

$$_{Z}^{A}X \xrightarrow{\ \beta^-\ } {_{Z+1}^{A}}Y \qquad (3\text{-}2)$$

The parent radionuclide (X) and daughter product (Y) represent different chemical elements because atomic number increases by one. Thus β⁻ decay results in a *transmutation* of elements. Mass number A does not change because the total number of nucleons in the nucleus does not change. This is therefore an *isobaric* decay mode, that is, the parent and daughter are isobars (see Chapter 2, Section D.3).

Radioactive decay processes often are represented by a *decay scheme diagram*. Figure 3-1 shows such a diagram for ^{14}C, a radionuclide that decays solely by β⁻ emission. The line representing ^{14}C (the parent) is drawn above and to the left of the line representing ^{14}N (the daughter). Decay is "to the right" because atomic number *increases* by one (reading Z values from left to right). The vertical distance between the lines is proportional to the total amount of energy released, that is, the transition energy for the decay process ($Q = 0.156$ MeV for ^{14}C).

---

*Actually, in β⁻ emission an antineutrino, $\bar{\nu}$, is emitted, whereas in β⁺ emission and EC, a neutrino, ν, is emitted. For simplicity, no distinction is made in this text. Also, evidence from high-energy physics experiments suggests that neutrinos may indeed have a very small mass, but an exact value has not yet been assigned.

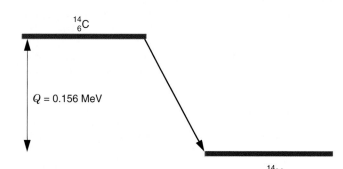

FIGURE 3-1 Decay scheme diagram for ^{14}C, a β⁻ emitter. Q is the transition energy.

The energy released in β^- decay is shared between the β^- particle and the neutrino. This sharing of energy is more or less random from one decay to the next. Figure 3-2 shows the distribution, or *spectrum*, of β^--particle energies resulting from the decay of ^{14}C. The maximum possible β^--particle energy (i.e., the transition energy for the decay process) is denoted by E_β^{max} (0.156 MeV for ^{14}C). From the graph it is apparent that the β^- particle usually receives something less than half of the available energy. Only rarely does the β^- particle carry away all the energy ($E_\beta = E_\beta^{max}$).

The *average energy* of the β^- particle is denoted by $\bar{E}_\beta$. This varies from one radionuclide to the next but has a characteristic value for any given radionuclide. Typically, $\bar{E}_\beta \approx (1/3)E_\beta^{max}$. For ^{14}C, $\bar{E}_\beta = 0.0497$ MeV $(0.32\,E_\beta^{max})$.

Beta particles present special detection and measurement problems for nuclear medicine applications. These arise from the fact that they can penetrate only relatively small thicknesses of solid materials (see Chapter 6, Section B.2). For example, the thickness is at most only a few millimeters in soft tissues. Therefore it is difficult to detect β^- particles originating from inside the body with a detector that is located outside the body. For this reason, radionuclides emitting only β^- particles rarely are used when measurement in vivo is required. Special types of detector systems also are needed to detect β^- particles

because they will not penetrate even relatively thin layers of metal or other outside protective materials that are required on some types of detectors. The implications of this are discussed in Chapter 7.

The properties of various radionuclides of medical interest are presented in Appendix C. Radionuclides decaying solely by β^- emission listed there include ^{3}H, ^{14}C, and ^{32}P.

D. DECAY BY (β^-, γ) EMISSION

In some cases, decay by β^- emission results in a daughter nucleus that is in an excited or metastable state rather than in the ground state. If an excited state is formed, the daughter nucleus promptly decays to a more stable nuclear arrangement by the emission of a γ ray (see Chapter 2, Section D.5). This sequential decay process is called (β^-, γ) *decay*. In standard nuclear notation, it may be represented as

$$\,^A_Z X \xrightarrow{\;\beta^-\;} \,^A_{Z+1}Y^* \xrightarrow{\;\gamma\;} \,^A_{Z+1}Y \qquad (3\text{-}3)$$

Note that γ emission does not result in a transmutation of elements.

An example of (β^-, γ) decay is the radionuclide ^{133}Xe, which decays by β^- emission to one of three different excited states of ^{133}Cs.

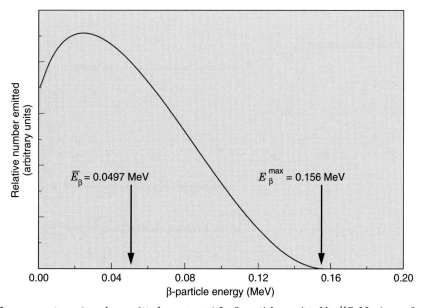

FIGURE 3-2 Energy spectrum (number emitted vs. energy) for β particles emitted by ^{14}C. Maximum β^--particle energy is Q, the transition energy (see Fig. 3-1). Average energy $\bar{E}_\beta$ is 0.0497 MeV, approximately ($\frac{1}{3}$) E_β^{max}. (*Data courtesy Dr. Jongwha Chang, Korea Atomic Energy Research Institute.*)

Figure 3-3 is a decay scheme for this radionuclide. The daughter nucleus decays to the ground state or to another, less energetic excited state by emitting a γ ray. If it is to another excited state, additional γ rays may be emitted before the ground state is finally reached. Thus in (β⁻, γ) decay more than one γ ray may be emitted before the daughter nucleus reaches the ground state (e.g., β_2 followed by γ_1 and γ_2 in ^{133}Xe decay).

The number of nuclei decaying through the different excited states is determined by probability values that are characteristic of the particular radionuclide. For example, in ^{133}Xe decay (Fig. 3-3), 99.3% of the decay events are by β_3 decay to the 0.081-MeV excited state, followed by emission of the 0.081-MeV γ ray or conversion electrons (Section E). Only a very small number of the other β particles and γ rays of other energies are emitted. The data presented in Appendix C include the relative number of emissions of different energies for each radionuclide listed.

In contrast to β⁻ particles, which are emitted with a continuous distribution of energies (up to E_β^{max}), γ rays are emitted with a precise and discrete series of energy values. The spectrum of emitted radiation energies is therefore a series of discrete lines at energies that are characteristic of the radionuclide rather than a continuous distribution of energies (Fig. 3-4). In (β⁻, γ) decay, the transition energy between the parent radionuclide and the ground state of the daughter has a fixed characteristic value. The distribution of this energy among the β⁻ particle, the neutrino, and the γ rays may vary from one nuclear decay to the next, but the sum of their energies in any decay event is always equal to the transition energy.

Because γ rays are much more penetrating than β⁻ particles, they do not present some of the measurement problems associated with β⁻ particles that were mentioned earlier, and they are suitable for a wider variety of applications in nuclear medicine. Some radionuclides of medical interest listed in Appendix C that undergo (β⁻, γ) decay include ^{131}I, ^{133}Xe, and ^{137}Cs.

E. ISOMERIC TRANSITION AND INTERNAL CONVERSION

The daughter nucleus of a radioactive parent may be formed in a "long-lived" metastable or isomeric state, as opposed to an excited state. The decay of the metastable or isomeric state by the emission of a γ ray is called an *isomeric transition* (see Chapter 2, Section D.4). Except for their average lifetimes, there are no differences in decay by γ emission of metastable or excited states.

An alternative to γ-ray emission is *internal conversion*. This can occur for any excited state, but is especially common for metastable states. In this process, the nucleus decays by

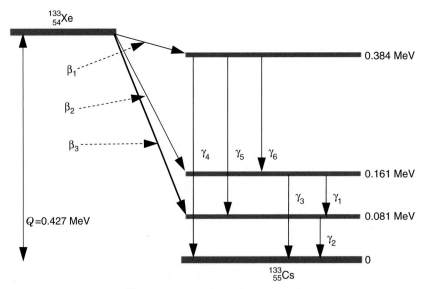

FIGURE 3-3 Decay scheme diagram for ^{133}Xe, a (β⁻, γ) emitter. More than one γ ray may be emitted per disintegrating nucleus. The *heavy line* (for β_3) indicates most-probable decay mode.

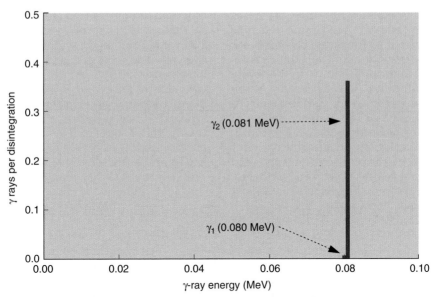

FIGURE 3-4 Emission spectrum for 0.080- and 0.081-MeV γ rays emitted in the decay of ^{133}Xe (γ_1 and γ_2 in Fig. 3-3; higher-energy emissions omitted). Compare with Figure 3-2 for β⁻ particles.

transferring energy to an orbital electron, which is ejected instead of the γ ray. It is as if the γ ray were "internally absorbed" by collision with an orbital electron (Fig. 3-5). The ejected electron is called a *conversion electron*. These electrons usually originate from one of the inner shells (K or L), provided that the γ-ray energy is sufficient to overcome the binding energy of that shell. The energy excess above the binding energy is imparted to the conversion electron as kinetic energy. The orbital vacancy created by internal conversion subsequently is filled by an outer-shell electron, accompanied by emission of characteristic x rays or Auger electrons (see Chapter 2, Section C.3).

Whether a γ ray or a conversion electron is emitted is determined by probabilities that have characteristic values for different radionuclides. These probabilities are expressed in terms of the ratio of conversion electrons emitted to γ rays emitted (e/γ) and denoted by α (or α_K = e/γ for K-shell conversion electrons, and so on) in detailed charts and tables of nuclear properties.

Internal conversion, like β⁻ decay, results in the emission of electrons. The important differences are that (1) in β⁻ decay the electron originates from the nucleus, whereas in internal conversion it originates from an electron orbit; and (2) β⁻ particles are emitted with a continuous spectrum of energies, whereas conversion electrons have a discrete series of energies determined by the differences between the γ-ray energy and orbital electron-binding energies.

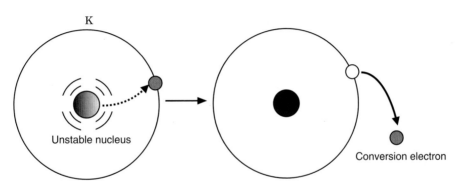

FIGURE 3-5 Schematic representation of internal conversion involving a K-shell electron. An unstable nucleus transfers its energy to the electron rather than emitting a γ ray. Kinetic energy of conversion electron is γ ray energy minus electron-binding energy ($E_\gamma - K_B$).

Metastable radionuclides are of great importance in nuclear medicine. Because of their relatively long lifetimes, it sometimes is possible to separate them from their radioactive parent and thus obtain a relatively "pure" source of γ rays. The separation of the metastable daughter from its radioactive parent is accomplished by chemical means in a radionuclide "generator" (see Chapter 5, Section C). Metastable nuclides always emit a certain number of conversion electrons, and thus they are not really "pure" γ-ray emitters. Because conversion electrons are almost totally absorbed within the tissue where they are emitted (Chapter 6, Section B.2), they can cause substantial radiation dose to the patient, particularly when the conversion ratio, e/γ, is large. However, the ratio of photons to electrons emitted by metastable nuclides usually is greater than for (β⁻,γ) emitters, and this is a definite advantage for studies requiring detection of γ rays from internally administered radioactivity.

A metastable nuclide of medical interest listed in Appendix C is ^{99m}Tc. Technetium-99m is currently by far the most popular radionuclide for nuclear medicine imaging studies.

F. ELECTRON CAPTURE AND (EC, γ) DECAY

Electron capture (EC) decay looks like, and in fact is sometimes called, "inverse β⁻ decay." An orbital electron is "captured" by the nucleus and combines with a proton to form a neutron:

$$p^+ + e^- \rightarrow n + \nu + \text{energy} \qquad (3\text{-}4)$$

The neutrino is emitted from the nucleus and carries away some of the transition energy. The remaining energy appears in the form of characteristic x rays and Auger electrons, which are emitted by the daughter product when the resulting orbital electron vacancy is filled. Usually, the electron is captured from orbits that are closest to the nucleus, that is, the K and L shells. The notation EC(K) is used to indicate capture of a K-shell electron, EC(L) an L-shell electron, and so forth.

EC decay may be represented as:

$$_Z^A X \xrightarrow{\text{EC}} {}_{Z-1}^A Y \qquad (3\text{-}5)$$

Note that like β⁻ decay it is an isobaric decay mode leading to a transmutation of elements.

The characteristic x rays emitted by the daughter product after EC may be suitable for external measurement if they are sufficiently energetic to penetrate a few centimeters of body tissues. There is no precise energy cutoff point, but 25 keV is probably a reasonable value, at least for shallow organs such as the thyroid. For elements with Z of 50 or more, the energy of K-x rays exceeds 25 keV. The K-x rays of lighter elements and all L-x rays are of lower energy and generally are not suitable for external measurements. These lower-energy radiations introduce measurement problems similar to those encountered with particles.

EC decay results frequently in a daughter nucleus that is in an excited or metastable state. Thus γ rays (or conversion electrons) may also be emitted. This is called (EC, γ) decay. Figure 3-6 shows a decay scheme for ^{125}I, an (EC, γ) radionuclide finding application

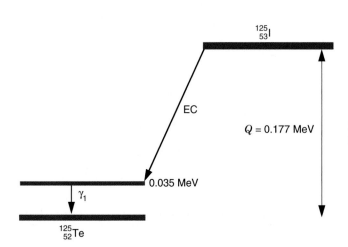

FIGURE 3-6 Decay scheme diagram for ^{125}I, an (EC, γ) emitter.

in radioimmunoassay studies. Note that EC decay is "to the left" because EC *decreases* the atomic number by one. Medically important EC and (EC, γ) radionuclides listed in Appendix C include ^{57}Co, ^{67}Ga, ^{111}In, ^{123}I, ^{125}I, and ^{201}Tl.

G. POSITRON (β^+) AND (β^+, γ) DECAY

In radioactive decay by positron emission, a proton in the nucleus is transformed into a neutron and a positively charged electron. The positively charged electron—or *positron* (β^+)—and a neutrino are ejected from the nucleus. Schematically, the process is:

$$p^+ \rightarrow n + e^+ + \nu + \text{energy} \qquad (3\text{-}6)$$

A positron is the antiparticle of an ordinary electron. After ejection from the nucleus, it loses its kinetic energy in collisions with atoms of the surrounding matter and comes to rest, usually within a few millimeters of the site of its origin in body tissues. More accurately, the positron and an electron momentarily form an "atom" called *positronium*, which has the positron as its "nucleus" and a lifetime of approximately 10^{-10} sec. The positron then combines with the negative electron in an *annihilation reaction*, in which their masses are converted into energy (see Fig. 3-7). The mass-energy equivalent of each particle is 0.511 MeV. This energy appears in the form of two 0.511-MeV *annihilation photons,** which leave the site of the annihilation event in nearly exact opposite directions (180 degrees apart).

The "back-to-back" emission of annihilation photons is required for conservation of momentum for a stationary electron-positron pair. However, because both particles actually are moving, the annihilation photons may be emitted in directions slightly off from the ideal by perhaps a few tenths of a degree. The effects of this on the ability to localize positron-emitting radionuclides for imaging purpose are discussed in Chapter 18, Section A.4.

Energy "bookkeeping" is somewhat more complicated for β^+ decay than for some of the previously discussed decay modes. There is a minimum transition energy requirement of 1.022 MeV before β^+ decay can occur. This requirement may be understood by evaluating the difference between the atomic mass of the parent and the daughter atom (including the orbital electrons). In β^+ decay, a positron is ejected from the nucleus, and because β^+ decay reduces the atomic number by one, the daughter atom also has an excess electron that it releases to reach its ground state. Thus two particles are emitted from the atom during β^+ decay, and because the rest-mass energy of an electron or a positron is 511 keV, a total transition energy of 1.022 MeV is required. Note that no such requirement is present for β^- decay, because the daughter atom must take up an electron from the environment to become neutral, thereby compensating for the electron released during β^- decay.

In β^+ decay, the excess transition energy above 1.022 MeV is shared between the positron (kinetic energy) and the neutrino. The positron energy spectrum is similar to that observed for β^- particles (see Fig. 3-2). The average β^+ energy also is denoted by $\bar{E}_\beta$ and again is approximately $\bar{E}_\beta \approx (1/3)E_\beta^{\text{max}}$, in which E_β^{max} is the transition energy minus 1.022 MeV.

In standard notation, β^+ decay is represented as

$$^A_Z\text{X} \xrightarrow{\ \beta^+\ } ^{\ A}_{Z-1}\text{Y} \qquad (3\text{-}7)$$

It is another isobaric decay mode, with a transmutation of elements. Figure 3-8 shows

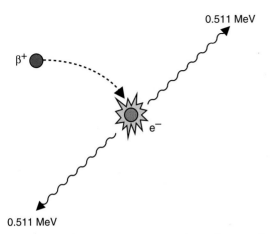

0.511 MeV

β^+

e^-

0.511 MeV

FIGURE 3-7 Schematic representation of mutual-annihilation reaction between a positron (β^+) and an ordinary electron. A pair of 0.511-MeV annihilation photons are emitted "back-to-back" at 180 degrees to each other.

---

*Although the photons produced when the positron and an electron undergo annihilation are not of nuclear origin, they sometimes are called *annihilation γ rays.* This terminology may be used in some instances in this book.

a decay scheme for ^{15}O, a β^+ emitter of medical interest. Decay is "to the left" because atomic number *decreases* by one. The vertical line represents the minimum transition energy requirement for β^+ decay (1.022 MeV). The remaining energy (1.7 MeV) is E_β^{max}. With some radionuclides, β^+ emission may leave the daughter nucleus in an excited state, and thus additional γ rays may also be emitted [(β^+, γ) decay].

Positron emitters are useful in nuclear medicine because two photons are generated per nuclear decay event. Furthermore, the precise directional relationship between the annihilation photons permits the use of novel "coincidence-counting" techniques (see Chapter 18). Medically important pure β^+ radionuclides listed in Appendix C include ^{13}N and ^{15}O.

H. COMPETITIVE β^+ AND EC DECAY

Positron emission and EC have the same effect on the parent nucleus. Both are isobaric decay modes that decrease atomic number by one. They are alternative means for reaching the same endpoint (see Equations 3-5 and 3-7, and Figs. 3-6 and 3-8). Among the radioactive nuclides, one finds that β^+ decay occurs more frequently among lighter elements, whereas EC is more frequent among heavier elements, because in heavy elements orbital electrons tend to be closer to the nucleus and are more easily captured.

There also are radionuclides that can decay by either mode. An example is ^{18}F, the decay scheme for which is shown in Figure 3-9. For this radionuclide, 3% of the nuclei decay by EC and 97% decay by β^+ emission. Radionuclides of medical interest that undergo competitive (β^+, EC) decay listed in Appendix C include ^{11}C and ^{18}F.

I. DECAY BY α EMISSION AND BY NUCLEAR FISSION

Radionuclides that decay by α-particle emission or by nuclear fission are of relatively little importance for direct usage as tracers in nuclear medicine but are described here for the sake of completeness. Both of these decay modes occur primarily among very heavy elements that are of little interest as physiologic tracers. As well, they are highly energetic and tend to be associated with relatively large radiation doses (see Table 22-1).

In decay by α-particle emission, the nucleus ejects an α particle, which consists of two neutrons and two protons (essentially a $^4_2$He nucleus). In standard notation this is represented as:

$$^A_Z X \xrightarrow{\alpha} {}^{A-4}_{Z-2} Y \qquad (3\text{-}8)$$

The α particle is emitted with kinetic energy usually between 4 and 8 MeV. Although quite energetic, α particles have *very* short ranges

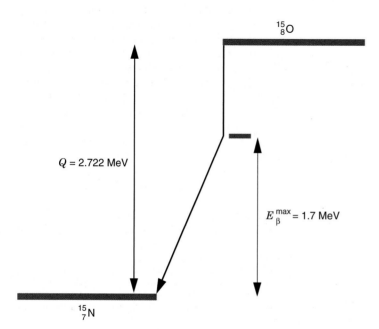

FIGURE 3-8 Decay scheme diagram for ^{15}O, a β^+ emitter. E_β^{max} is Q, the transition energy, minus 1.022 MeV, the minimum transition energy for β^+ decay.

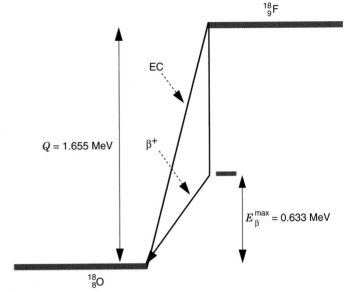

FIGURE 3-9 Decay scheme diagram for ^{18}F, which decays by both electron capture and β^+ emission competitively.

$^{18}_{9}$F

EC

$Q = 1.655$ MeV

β^+

$E_\beta^{max} = 0.633$ MeV

$^{18}_{8}$O

in solid materials, for example, approximately 0.03 mm in body tissues. Thus they present very difficult detection and measurement problems.

Decay by α-particle emission results in a transmutation of elements, but it is not isobaric. Atomic mass is decreased by 4; therefore this process is common among very heavy elements that must lose mass to achieve nuclear stability. Heavy, naturally occurring radionuclides such as ^{238}U and its daughter

products undergo a series of decays involving α-particle and β^--particle emission to transform into lighter, more stable nuclides. Figure 3-10 illustrates the "decay series" of ^{238}U $\rightarrow$ ^{206}Pb. The radionuclide ^{226}Ra in this series is of some medical interest, having been used at one time in encapsulated form for implantation into tumors for radiation therapy. The ubiquitous, naturally occurring ^{222}Rn also is produced in this series. Note that there are "branching points" in the series where either

FIGURE 3-10 Illustration of series decay, starting from ^{238}U and ending with stable ^{206}Pb. (*Adapted from Hendee WR: Medical Radiation Physics. Chicago, 1970, Year Book Publishers Inc., p 501.*)

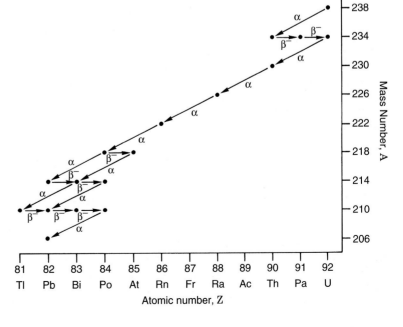

α or β⁻ emission may occur. Only every fourth atomic number value appears in this series because α emission results in atomic number differences of four units. The $^{238}U \rightarrow ^{206}Pb$ series is called the "4n + 2" series. Two others are $^{235}U \rightarrow ^{207}Pb$ (4n + 3) and $^{232}Th \rightarrow ^{208}Pb$ (4n). These three series are found in nature because in each case the parent is a very long-lived radionuclide (half-lives ~ 10^8 to 10^{10} yr) and small amounts remain from the creation of the elements. The fourth series, 4n + 1, is not found naturally because all its members have much shorter lifetimes and have disappeared from nature.

An (α, γ) radionuclide of interest in nuclear medicine is ^{241}Am. It is used in encapsulated form as a source of 60-keV γ rays for instrument calibration and testing.

Nuclear fission is the spontaneous fragmentation of a very heavy nucleus into two lighter nuclei. In the process a few (two or three) *fission neutrons* also are ejected. The distribution of nuclear mass between the two product nuclei varies from one decay to the next. Typically it is split in approximately a 60:40 ratio. The energy released is very large, often amounting to hundreds of MeV per nuclear fission, and is imparted primarily as kinetic energy to the recoiling nuclear fragments (*fission fragments*) and the ejected neutrons. Nuclear fission is the source of energy from nuclear reactors. More precisely, the kinetic energy of the emitted particles is converted into heat in the surrounding medium, where it is used to create steam for driving turbines and other uses. The fission process is of interest in nuclear medicine because the fission fragment nuclei usually are radioactive and, if chemically separable from the other products, can be used as medical tracers. Also, the neutrons are used to produce radioactive materials by neutron activation, as discussed in Chapter 5, Section A.3. The parent fission nuclides themselves are of no use as tracers in nuclear medicine.

J. DECAY MODES AND THE LINE OF STABILITY

In Chapter 2, Section D.7, it was noted that on a graph of neutron versus proton numbers the stable nuclides tend to be clustered about an imaginary line called the *line of stability* (see Fig. 2-9). Nuclides lying off the line of stability generally are radioactive. The type of radioactive decay that occurs usually is such as to move the nucleus closer to this line. A radionuclide that is proton deficient (above the line) usually decays β⁻ emission, because this transforms a neutron into a proton, moving the nucleus closer to the line of stability. A neutron-deficient radionuclide (below the line) usually decays by EC or β⁺ emission, because these modes transform a proton into a neutron. Heavy nuclides frequently decay by α emission or by fission, because these are modes that reduce mass number.

It also is worth noting that β⁻, β⁺, and EC decay all can transform an "odd-odd" nucleus into an "even-even" nucleus. As noted in Chapter 2, Section D.7 even-even nuclei are relatively stable because of pairing of alike particles within the nucleus. There are in fact a few odd-odd nuclides lying on or near the line of stability that can decay either by β⁻ emission or by EC and β⁺ emission. An example is ^{40}K (89% β⁻, 11% EC or β⁺). In this example, the instability created by odd numbers of protons and neutrons is sufficient to cause decay in both directions *away* from the line of stability; however, this is the exception rather than the rule.

K. SOURCES OF INFORMATION ON RADIONUCLIDES

There are several sources of information providing useful summaries of the properties of radionuclides. One is a chart of the nuclides, a portion of which is shown in Figure 3-11. Every stable or radioactive nuclide is assigned a square on the diagram. Isotopes occupy horizontal rows and isotones occupy vertical columns. Isobars fall along descending 45-degree lines. Basic properties of each nuclide are listed in the boxes. Also shown in Figure 3-11 is a diagram indicating the transformations that occur for various decay modes. A chart of the nuclides is particularly useful for tracing through a radioactive series.

Perhaps the most useful sources of data for radionuclides of interest in nuclear medicine are the Medical Internal Radiation Dosimetry (MIRD) publications, compiled by the MIRD Committee of the Society of Nuclear Medicine.[1] Decay data for some of the radionuclides commonly encountered in nuclear medicine are presented in Appendix C. Also presented are basic data for internal dosimetry, which will be discussed in Chapter 22.

Radioactive transformations

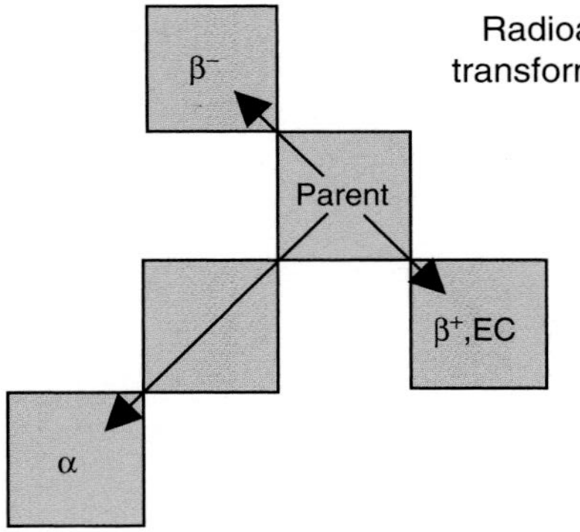

Decay scheme (relative positions of daughter nuclides):
- **β⁻** — upper-left of Parent
- **Parent** — center
- **β⁺, EC** — upper-right of Parent
- **α** — lower-right of Parent

Chart of the Nuclides (extract)

Half-lives of nuclides (vertical axis = atomic number Z; horizontal axis = neutron number N). Stable nuclides indicated in **bold**; values for stable nuclides indicate percent natural abundance. Where two half-lives are listed, the metastable-state half-life is given on the left.

Z \ N	42	43	44	45	46	47	48	49	50	51	52	53	54	55	56	57
Ag (47)						Ag94 0.42 s	Ag95 2.0 s	Ag96 5.1 s	Ag97 19 s	Ag98 47 s	Ag99 11 s / 2.1 m	Ag100 2.3 m / 2.0 m	Ag101 3.1 s / 11.1 m	Ag102 7.8 m / 13 m	Ag103 5.7 s / 66 m	Ag104 33 m / 69 m
Pd (46)							Pd94 9 s	Pd95 13.4 s	Pd96 2.03 m	Pd97 3.1 m	Pd98 18 m	Pd99 21.4 m	Pd100 3.6 d	Pd101 8.4 h	**Pd102** 1.02	Pd103 16.97 d
Rh (45)								Rh94 26 s / 1.2 m	Rh95 1.96 m / 5 m	Rh96 1.5 m / 9.9 m	Rh97 46 m / 31 m	Rh98 3.5 m / 8.6 m	Rh99 4.7 h / 16.1 d	Rh100 4.7 h / 20.8 h	Rh101 4.34 d / 3.3 y	Rh102 206 d / 2.9 y
Ru (44)					Ru90 11 s	Ru91 9 s	Ru92 3.7 m	Ru93 10.8 s / 1 m	Ru94 52 m	Ru95 1.64 h	**Ru96** 5.54	Ru97 2.89 d	**Ru98** 1.86	**Ru99** 12.7	**Ru100** 12.6	**Ru101** 17.1
Tc (43)					Tc89 13 s / 13 s	Tc90 49.2 s / 8.3 s	Tc91 3.3 m / 3.1 m	Tc92 4.2 m	Tc93 43 m / 2.8 h	Tc94 52 m / 4.9 h	Tc95 61 d / 20 h	Tc96 52 m / 4.3 d	Tc97 90 d / 2.6E6 y	Tc98 4.2E6 y	Tc99 6 h / 2.1E5 y	Tc100 15.8 s
Mo (42)			Mo86 20 s	Mo87 14 s	Mo88 8.0 m	Mo89 0.2 s / 2.2 m	Mo90 5.7 h	Mo91 1.1 m / 15.5 m	**Mo92** 14.84	Mo93 6.9 h / 3.5E3 y	**Mo94** 9.25	**Mo95** 15.92	**Mo96** 16.68	**Mo97** 9.55	**Mo98** 24.13	Mo99 65.9 h
Nb (41)	Nb83 4.1 s	Nb84 12 s	Nb85 21 s	Nb86 56 s / 1.5 m	Nb87 3.7 m / 2.6 m	Nb88 7.7 m / 14.4 m	Nb89 1.1 h / 2 h	Nb90 14.6 h / 19 s	Nb91 62 d / 7E2 y	Nb92 10 d / 3.7E7 y	**Nb93** 6.1 y / 100	Nb94 6.2 m / 2.4E4 y	Nb95 3.6 d / 35 d	Nb96 23.4 h	Nb97 54 s / 74 m	Nb98 51 m / 2.8 s

FIGURE 3-11 Portion of a chart of the nuclides. *Vertical axis* = atomic number; *horizontal axis* = neutron number. Also listed are half-lives of radioactive nuclides (see Chapter 4, Section B.2). Stable nuclides are indicated in **bold** font. Values listed for these nuclides indicate their percent natural abundance. Half-lives of metastable states are listed on the left, where applicable.

REFERENCE

1. Eckerman KF, Endo A: MIRD: Radionuclide Data and Decay Schemes, New York, 2008, Society of Nuclear Medicine.

BIBLIOGRAPHY

A comprehensive source of radionuclide data can be found at the National Nuclear Data Center [accessed July 6, 2011]. Available at http://www.nndc.bnl.gov/.

Decay of Radioactivity

Radioactive decay is a spontaneous process; that is, there is no way to predict with certainty the exact moment at which an unstable nucleus will undergo its radioactive transformation into another, more stable nucleus. Mathematically, radioactive decay is described in terms of probabilities and average decay rates. In this chapter we discuss these mathematical aspects of radioactive decay.

A. ACTIVITY

1. The Decay Constant

If one has a sample containing N radioactive atoms of a certain radionuclide, the average decay rate, $\Delta N/\Delta t$, for that sample is given by:

$$\Delta N/\Delta t = -\lambda N \qquad (4\text{-}1)$$

where λ is the *decay constant* for the radionuclide. The decay constant has a characteristic value for each radionuclide. It is the fraction of the atoms in a sample of that radionuclide undergoing radioactive decay per unit of time during a period that is so short that only a small fraction decay during that interval. Alternatively, it is the probability that any individual atom will undergo decay during the same period. The units of λ are (time)$^{-1}$. Thus 0.01 sec^{-1} means that, on the average, 1% of the atoms undergo radioactive decay each second. In Equation 4-1 the minus sign indicates that $\Delta N/\Delta t$ is negative; that is, N is decreasing with time.

Equation 4-1 is valid only as an estimate of the *average* rate of decay for a radioactive sample. From one moment to the next, the actual decay rate may differ from that predicted by Equation 4-1. These *statistical fluctuations* in decay rate are described in Chapter 9.

Some radionuclides can undergo more than one type of radioactive decay (e.g., ^{18}F: 97% β^+, 3% electron capture). For such types of "branching" decay, one can define a value of λ for each of the possible decay modes, for example, λ_1, λ_2, λ_3, and so on, where λ_1 is the fraction decaying per unit time by decay mode 1, λ_2 by decay mode 2, and so on. The total decay constant for the radionuclide is the sum of the branching decay constants:

$$\lambda = \lambda_1 + \lambda_2 + \lambda_3 + \cdots \qquad (4\text{-}2)$$

The fraction of nuclei decaying by a specific decay mode is called the *branching ratio* (B.R.). For the ith decay mode, it is given by:

$$\text{B.R.} = \lambda_i/\lambda \qquad (4\text{-}3)$$

2. Definition and Units of Activity

The quantity $\Delta N/\Delta t$, the average decay rate, is the *activity* of the sample. It has dimensions of disintegrations per second (dps) or disintegrations per minute (dpm) and is essentially a measure of "how radioactive" the sample is. The Systeme International (SI) unit of activity is the *becquerel* (Bq). A sample has an activity of 1 Bq if it is decaying at an average rate of 1 sec^{-1} (1 dps). Thus:

$$A(\text{Bq}) = |\Delta N/\Delta t| = \lambda N \qquad (4\text{-}4)$$

where λ is in units of sec^{-1}. The absolute value is used to indicate that activity is a "positive" quantity, as compared with the change in number of radioactive atoms in Equation 4-1, which is a negative quantity. Commonly used multiples of the becquerel are the kilobecquerel (1 kBq = 10^3 sec^{-1}), the megabecquerel (1 MBq = 10^6 sec^{-1}), and the gigabecquerel (1 GBq = 10^9 sec^{-1}).

The traditional unit for activity is the *curie* (Ci), which is defined as 3.7×10^{10} dps (2.22×10^{12} dpm). Subunits and multiples of the curie are the millicurie (1 mCi = 10^{-3} Ci), the microcurie (1 μCi = 10^{-3} mCi = 10^{-6} Ci), the nanocurie (1 nCi = 10^{-9} Ci), and the kilocurie (1 kCi = 1000 Ci). Equation 4-1 may be modified for these units of activity:

$$A(\text{Ci}) = \lambda N/(3.7 \times 10^{10}) \qquad (4\text{-}5)$$

The curie was defined originally as the activity of 1 g of ^{226}Ra; however, this value "changed" from time to time as more accurate measurements of the ^{226}Ra decay rate were obtained. For this reason, the ^{226}Ra standard was abandoned in favor of a fixed value of 3.7×10^{10} dps. This is not too different from the currently accepted value for ^{226}Ra (3.656×10^{10} dps/g).

SI units are the "official language" for nuclear medicine and are used in this text; however, because traditional units of activity still are used in day-to-day practice in many laboratories, we sometimes also indicate activities in these units as well. Conversion factors between traditional and SI units are provided in Appendix A.

The amounts of activity used for nuclear medicine studies typically are in the MBq-GBq range (10s of μCi to 10s of mCi). Occasionally, 10s of gigabecquerels (curie quantities) may be acquired for long-term supplies. External-beam radiation sources (e.g., ^{60}Co therapy units) use source strengths of 1000s of GBq [1000 GBq = 1 terraBq (TBq) = 10^{12} Bq]. At the other extreme, the most sensitive measuring systems used in nuclear medicine can detect activities at the level of a few becquerels (nanocuries).

B. EXPONENTIAL DECAY

1. The Decay Factor

With the passage of time, the number N of radioactive atoms in a sample decreases. Therefore the activity A of the sample also decreases (see Equation 4-4). Figure 4-1 is used to illustrate radioactive decay with the passage of time.

Suppose one starts with a sample containing $N(0)$ = 1000 atoms[*] of a radionuclide having a decay constant λ = 0.1 sec^{-1}. During the first 1-sec time interval, the approximate

---

[*]$N(t)$ is symbolic notation for the number of atoms present as a function of time t. $N(0)$ is the number N at a specific time t = 0, that is, at the starting point.

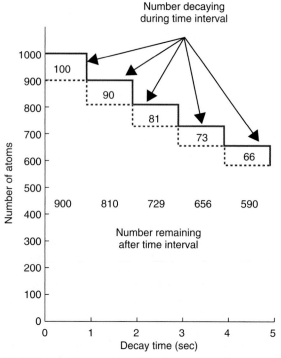

FIGURE 4-1 Decay of a radioactive sample during successive 1-sec increments of time, starting with 1000 atoms, for λ = 0.1 sec^{-1}. Both the number of atoms remaining and activity (decay rate) decrease with time. Note that the values shown are approximations, because they do not account precisely for the changing number of atoms present during the decay intervals (see Section D).

number of atoms decaying is $0.1 \times 1000 = 100$ atoms (see Equation 4-1). The activity is therefore 100 Bq, and after 1 sec there are 900 radioactive atoms remaining. During the next second, the activity is $0.1 \times 900 = 90$ Bq, and after 2 sec, 810 radioactive atoms remain. During the next second the activity is 81 Bq, and after 3 sec 729 radioactive atoms remain. Thus both the activity and the number of radioactive atoms remaining in the sample are decreasing continuously with time. A graph of either of these quantities is a curve that gradually approaches zero.

An exact mathematical expression for $N(t)$ can be derived using methods of calculus.[*] The result is:

$$N(t) = N(0)\, e^{-\lambda t} \qquad (4\text{-}6)$$

---

[*]The derivation is as follows:

$$dN/dt = -\lambda N \qquad (4\text{-}6a)$$

$$dN/N = -\lambda dt \qquad (4\text{-}6b)$$

$$\int dN/N = -\int \lambda\, dt \qquad (4\text{-}6c)$$

from which follows Equation 4-6.

Thus $N(t)$, the number of atoms remaining after a time t, is equal to $N(0)$, the number of atoms at time $t = 0$, multiplied by the factor $e^{-\lambda t}$. This factor $e^{-\lambda t}$, the fraction of radioactive atoms remaining after a time t, is called the *decay factor* (DF). It is a *number* equal to e—the base of natural logarithms (2.718 …)—raised to the power $-\lambda t$. For given values of λ and t, the decay factor can be determined by various methods as described in Section C later in this chapter. Note that because activity A is proportional to the number of atoms N (see Equation 4-4), the decay factor also applies to activity versus time:

$$A(t) = A(0)\, e^{-\lambda t} \qquad (4\text{-}7)$$

The decay factor $e^{-\lambda t}$ is an *exponential function* of time t. Exponential decay is characterized by the disappearance of a *constant fraction* of activity or number of atoms present per unit time interval. For example if $\lambda = 0.1$ sec^{-1}, the fraction is 10% per second. Graphs of $e^{-\lambda t}$ versus time t for $\lambda = 0.1$ sec^{-1} are shown in Figure 4-2. On a *linear* plot, it is a curve gradually approaching zero; on a *semilogarithmic* plot, it is a straight line. It should be noted that there are other processes besides radioactive decay that can be described by exponential functions. Examples are the

absorption of x- and λ-ray beams (see Chapter 6, Section D) and the clearance of certain tracers from organs by physiologic processes (see Chapter 22, Section B.2).

When the exponent in the decay factor is "small," that is, $\lambda t \lesssim 0.1$, the decay factor may be approximated by $e^{-\lambda t} \approx 1 - \lambda t$. This form may be used as an approximation in Equations 4-6 and 4-7.

2. Half-Life

As indicated in the preceding section, radioactive decay is characterized by the disappearance of a constant fraction of the activity present in the sample during a given time interval. The *half-life* ($T_{1/2}$) of a radionuclide is the time required for it to decay to 50% of its initial activity level. The half-life and decay constant of a radionuclide are related as*

$$T_{1/2} = \ln 2/\lambda \qquad (4\text{-}8)$$

$$\lambda = \ln 2/T_{1/2} \qquad (4\text{-}9)$$

_____

*The relationships are derived as follows:

$$1/2 = e^{-\lambda T_{1/2}} \qquad (4\text{-}8a)$$

$$2 = e^{\lambda T_{1/2}} \qquad (4\text{-}8b)$$

$$\ln 2 = \lambda T_{1/2} \qquad (4\text{-}8c)$$

from which follow Equations 4-8 and 4-9.

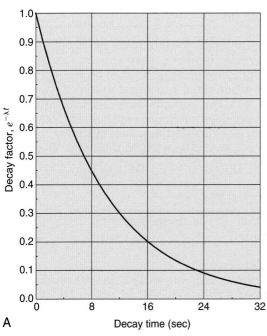

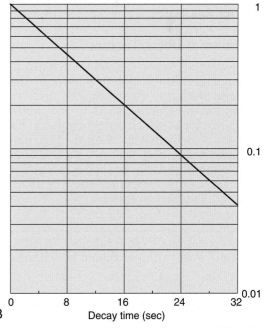

FIGURE 4-2 Decay factor versus time shown on linear (A) and semilogarithmic (B) plots, for radionuclide with $\lambda = 0.1$ sec^{-1}.

where $\ln 2 \approx 0.693$. Usually, tables or charts of radionuclides list the half-life of the radionuclide rather than its decay constant. Thus it often is more convenient to write the decay factor in terms of half-life rather than decay constant:

$$DF = e^{-\ln 2 \times t/T_{1/2}} \qquad (4\text{-}10)$$

3. Average Lifetime

The actual lifetimes of individual radioactive atoms in a sample range anywhere from "very short" to "very long." Some atoms decay almost immediately, whereas a few do not decay for a relatively long time (see Fig. 4-2). The *average lifetime* τ of the atoms in a sample has a value that is characteristic of the nuclide and is related to the decay constant λ by*

$$\tau = 1/\lambda \qquad (4\text{-}11)$$

Combining Equations 4-9 and 4-11, one obtains

$$\tau = T_{1/2}/\ln 2 \qquad (4\text{-}12)$$

The average lifetime for the atoms of a radionuclide is therefore longer than its half-life, by a factor $1/\ln 2$ (≈ 1.44). The concept of average lifetime is of importance in radiation dosimetry calculations (see Chapter 22).

C. METHODS FOR DETERMINING DECAY FACTORS

1. Tables of Decay Factors

It is essential that an individual working with radionuclides know how to determine decay factors. Perhaps the simplest and most straightforward approach is to use tables of decay factors, which are available from vendors of radiopharmaceuticals, instrument manufacturers, and so forth. An example of such a table for ^{99m}Tc is shown in Table 4-1. Such tables are generated easily with computer spreadsheet programs.

EXAMPLE 4-1

A vial containing ^{99m}Tc is labeled "75 kBq/mL at 8 am." What volume should be withdrawn

---

*The equation from which Equation 4-11 is derived is:

$$\tau = \int_0^\infty t e^{-\lambda t} dt \bigg/ \int_0^\infty e^{-\lambda t} dt \qquad (4\text{-}11a)$$

TABLE 4-1
DECAY FACTORS FOR ^{99m}Tc

Hours	Minutes			
	0	**15**	**30**	**45**
0	1.000	0.972	0.944	0.917
1	0.891	0.866	0.841	0.817
2	0.794	0.771	0.749	0.727
3	0.707	0.687	0.667	0.648
4	0.630	0.612	0.595	0.578
5	0.561	0.545	0.530	0.515
6	0.500	0.486	0.472	0.459
7	0.445	0.433	0.420	0.408
8	0.397	0.385	0.375	0.364
9	0.354	0.343	0.334	0.324
10	0.315	0.306	0.297	0.289
11	0.281	0.273	0.264	0.257
12	0.250	0.243	0.236	0.229

at 4 pm on the same day to prepare an injection of 50 kBq for a patient?

Answer

From Table 4-1 the DF for ^{99m}Tc after 8 hours is found to be 0.397. Therefore the concentration of activity in the vial is 0.397×75 kBq/mL = 29.8 kBq/mL. The volume required for 50 kBq is 50 kBq divided by 29.8 kBq/mL = 1.68 mL.

Tables of decay factors cover only limited periods; however, they can be extended by employing principles based on the properties of exponential functions, specifically $e^{a + b} = e^a \times e^b$. For example, suppose that the desired time t does not appear in the table but that it can be expressed as a sum of times, $t = t_1 + t_2 + \cdots$, that do appear in the table. Then

$$DF(t_1 + t_2 + \cdots) = DF(t_1) \times DF(t_2) \cdots \quad (4\text{-}13)$$

EXAMPLE 4-2

What is the decay factor for ^{99m}Tc after 16 hours?

Answer

Express 16 hours as 6 hours + 10 hours. Then, from Table 4-1, DF(16 hr) = DF(10 hr) × DF(6 hr) = $0.315 \times 0.5 = 0.1575$. Other combinations of times totaling 16 hours provide the same result.

Occasionally, radionuclides are shipped in *precalibrated* quantities. A precalibrated shipment is one for which the activity calibration is given for some *future* time. To determine its present activity, it is therefore necessary to calculate the decay factor for a time preceding the calibration time, that is, a "negative" value of time. One can make use of tables of decay factors by employing another of the properties of exponential functions, specifically $e^{-x} = 1/e^x$. Thus:

$$DF(-t) = 1/DF(t) \qquad (4\text{-}14)$$

EXAMPLE 4-3

A vial containing ^{99m}Tc is labeled "50 kBq at 3 pm." What is the activity at 8 am on the same day?

Answer

The decay time is $t = -7$ hours. From Table 4-1, $DF(7 \text{ hr}) = 0.445$. Thus $DF(-7 \text{ hr}) = 1/0.445 = 2.247$. The activity at 8 am is therefore 2.247×50 kBq $= 112.4$ kBq.

2. Pocket Calculators

Many pocket calculators have capabilities for calculating exponential functions. First compute the exponent, $x = \ln 2 \times (t/T_{1/2})$, then press the appropriate keys to obtain e^{-x}. For precalibrated shipments, use e^{+x}.

3. Universal Decay Curve

Exponential functions are straight lines on a semilogarithmic plot (see Fig. 4-2). This useful property allows one to construct a "universal decay curve" by plotting the number of half-lives elapsed on the horizontal (linear) axis and the decay factor on the vertical (logarithmic) axis. A straight line can be drawn by connecting any two points on the curve. These could be, for example, $(t = 0, DF = 1)$, $(t = T_{1/2}, DF = 0.5)$, $(t = 2T_{1/2}, DF = 0.25)$, and so on. The graph can be used for any radionuclide provided that the elapsed time is expressed in terms of the number of radionuclide half-lives elapsed. An example of a universal decay curve is shown in Figure 4-3.

EXAMPLE 4-4

Use the decay curve in Figure 4-3 to determine the decay factor for ^{99m}Tc after 8 hours.

Answer

The half-life of ^{99m}Tc is 6 hours. Therefore the elapsed time is $8/6 = 1.33$ half-lives. From Figure 4-3, the decay factor is approximately

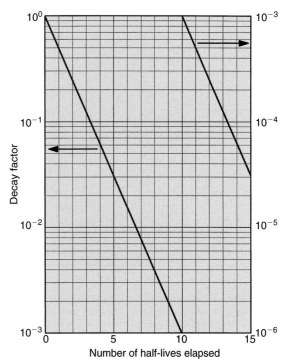

FIGURE 4-3   Universal decay curve.

0.40. (Compare this result with the value used in Example 4-1.)

D. IMAGE-FRAME DECAY CORRECTIONS

In some applications, data are acquired during periods that are not short in comparison with the half-life of the radionuclide. An example is the measurement of glucose metabolism using deoxyglucose labeled with fluorine-18 (see Chapter 21, Section E.5). In such measurements, it often is necessary to correct for decay that occurs during each measurement period while data collection is in progress. Because data are acquired in a series of image frames, these sometimes are called *image-frame decay corrections*.

The concept for these corrections is illustrated in Figure 4-4, showing the decay curve for an image frame starting at time t and ending at a time Δt later. The number of counts acquired during the image frame is proportional to the area a_d, shown with darker shading. The counts that would be recorded in the absence of decay are proportional to the area a_0, which includes both the darker and lighter shaded areas. Using the appropriate mathematical integrals, the *effective decay*

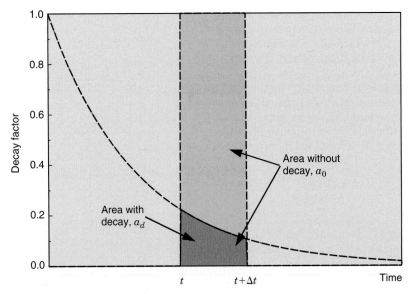

FIGURE 4-4   Basic concept for calculating the decay factor for an image frame starting at time t with duration Δt. The counts recorded with decay are proportional to the darker shaded area, a_d. The counts that would be recorded in the absence of decay are proportional to the total shaded area, a_0. The effective decay factor is the ratio a_d/a_0.

factor for a radionuclide with half-life $T_{1/2}$ for the indicated measurement interval is given by:

$$\mathrm{DF}_{\mathrm{eff}}\ (t, \Delta t) = a_d / a_0$$
$$= e^{-(\ln 2 \times t / T_{1/2})} \times [(1 - e^{-x}) / x]$$
$$= \mathrm{DF}(t) \times [(1 - e^{-x}) / x] \qquad (4\text{-}15)$$

where

$$x = \ln 2 \times \Delta t / T_{1/2} \qquad (4\text{-}16)$$

To correct the recorded counts back to what would have been recorded in the absence of decay, one would multiply the counts recorded during the interval $(t, t + \Delta t)$ by the inverse of $\mathrm{DF}_{\mathrm{eff}}$.

The effective decay factor in Equation 4-15 is composed of two parts. The first term is just the standard decay factor (Equation 4-10) at the start of the image frame, $\mathrm{DF}(t)$. The second term is a factor that depends on the parameter x, which in turn depends on the duration of the frame, Δt, relative to the half-life of the radionuclide (Equation 4-16). This term accounts for decay that occurs while data are being acquired during the image frame. Note again that the correction in Equation 4-15 uses $t = 0$ as the reference point, not the start of the individual image frame for which the correction is being calculated. To compute the decay occurring during the image frame itself, only the second term should be used.

In a quantitative study, the data for each image frame would be corrected according to the appropriate values for t and Δt and for the half-life $T_{1/2}$. For computational simplicity and efficiency, various approximations can be used when the parameter x in Equation 4-16 is small. For example, the following approximation is accurate to within 1% when $x < 0.25$:

$$\mathrm{DF}_{\mathrm{eff}}\ (t, \Delta t) \approx \mathrm{DF}(t) \times [1 - (x/2)] \qquad (4\text{-}17)$$

where x again is defined as in Equation 4-16.

Another approach is to use the standard DF (see Equation 4-10) for the midpoint of the frame:

$$\mathrm{DF}_{\mathrm{eff}}\ (t, \Delta t) \approx \mathrm{DF}\ [t + (\Delta t / 2)] \qquad (4\text{-}18)$$

This approximation is accurate to within 1% for $x < 0.5$.

Yet another possibility is to use the average of the standard decay factors for the beginning and end of the frame:

$$\mathrm{DF}_{\mathrm{eff}}\ (t, \Delta t) \approx \frac{[\mathrm{DF}(t) + \mathrm{DF}(t + \Delta t)]}{2} \qquad (4\text{-}19)$$

This approximation is accurate to within 1% for $x < 0.35$.

EXAMPLE 4-5

What are the effective decay factor and decay correction factor for the counts recorded in an image frame starting 30 sec and ending 45 sec after injection in a study performed with $^{15}\mathrm{O}$? Compare the results obtained with Equation 4-15 and the approximation given by Equation 4-17. Assume that the data are to be corrected to $t = 0$, the time of injection.

Answer

From Appendix C, the half-life of ^{15}O is 122 sec. The decay factor at the beginning of the image frame, $t = 30$ sec, is

$$DF(30 \text{ sec}) = e^{-\ln 2 \times 30 \text{ sec}/122 \text{ sec}}$$

$$\approx e^{-0.170}$$

$$\approx 0.843$$

The duration of the image frame is $\Delta t = 15$ sec. The parameter x (Equation 4-16) is given by

$$x = \ln 2 \times (\Delta t/T_{1/2})$$

$$= \ln 2 \times (15 \text{ sec}/122 \text{ sec})$$

$$\approx 0.0852$$

Thus, decay during the image frame is given by

$$(1 - e^{-x})/x = (1 - e^{-0.0852})/0.0852$$

$$\approx 0.0817/0.0852$$

$$\approx 0.959$$

Taking the product of the two decay factors gives

$$DF_{eff} \approx 0.843 \times 0.959$$

$$\approx 0.808$$

The decay correction factor to apply to the counts recorded in this frame is

$$CF \approx 1/0.808$$

$$\approx 1.237$$

Using the approximation given by Equation 4-17 yields

$$DF_{eff} \approx 0.843 \times [1 - (0.0852/2)]$$

$$\approx 0.843 \times 0.957$$

$$\approx 0.807$$

which differs from the exact result obtained with Equation 4-15 by only approximately 0.1%.

E. SPECIFIC ACTIVITY

A radioactive sample may contain stable isotopes of the element represented by the radionuclide of interest. For example, a given ^{131}I sample may also contain the stable isotope

^{127}I. When stable isotopes of the radionuclide of interest are present in the sample, they are called *carrier*, and the sample is said to be *with carrier*. A sample that does not contain stable isotopes of the element represented by the radionuclide is called *carrier-free*.* Radionuclides may be produced carrier-free or with carrier, depending on the production method (see Chapter 5).

The ratio of radioisotope activity to total mass of the element present is called the *specific activity* of the sample. Specific activity has units of becquerels per gram, megabecquerels per gram, and so forth. The highest possible specific activity of a radionuclide is its *carrier-free specific activity* (CFSA). This value can be calculated in a straightforward manner from the basic properties of the radionuclide.

Suppose a carrier-free sample contains 1 g of a radionuclide ^AX, having a half-life $T_{1/2}$ (sec). The atomic weight of the radionuclide is approximately equal to A, its mass number (see Chapter 2, Section D.2). A sample containing A g of the radionuclide has approximately 6.023×10^{23} atoms (Avogadro's number); therefore a 1-g sample has $N \approx 6.023 \times 10^{23}/A$ atoms. The decay rate of the sample is $\Delta N/\Delta t$ (dps) $= \lambda N = 0.693N/T_{1/2}$. Therefore the activity per gram is:

$$A(\text{Bq/g}) \approx \ln 2 \times 6.023 \times 10^{23}/(A \times T_{1/2})$$
$$(4\text{-}20)$$

Because the sample contains 1 g of the radioisotope, this is also its specific activity in becquerels per gram. When the equation is normalized for the half-life in *days* (1 day = 86,400 sec), the result is

$$CFSA (\text{Bq/g}) \approx 4.8 \times 10^{18}/(A \times T_{1/2})$$
$$(4\text{-}21)$$

where $T_{1/2}$ is given in days. With appropriate normalization, Equation 4-21 also applies for specific activity in kBq/mg, GBq/g, and so on.

In radiochemistry applications, specific activities sometimes are specified in becquerels per mole of labeled compound. Because 1 mole of compound contains A g of radionuclide, this quantity is

$$CFSA (\text{Bq/mole}) = CFSA (\text{Bq/g}) \times A (\text{g/mole})$$

$$\approx 4.8 \times 10^{18}/T_{1/2} \qquad (4\text{-}22)$$

where $T_{1/2}$ again is in days.

---

*Because it is virtually impossible to prepare a sample with absolutely no other atoms of the radioactive element, the terminology *without carrier* sometimes is used as well.

In traditional units, the equations for CFSA are

$$CFSA\ (Ci\,/\,g) \approx 1.3 \times 10^8 /(A \times T_{1/2})$$
$$CFSA\ (Ci\,/\,mole) \approx 1.3 \times 10^8 /T_{1/2} \quad (4\text{-}23)$$

where $T_{1/2}$ again is in days.

EXAMPLE 4-6

What are the CFSAs of ^{131}I and ^{99m}Tc?

Answer

For ^{131}I, A = 131 and $T_{1/2}$ = 8 days. Using Equation 4-21,

$$CFSA\ (^{131}I) \approx \frac{(4.8 \times 10^{18})}{(1.31 \times 10^2 \times 8)}$$
$$\approx 4.6 \times 10^{15}\ Bq\,/\,g$$

For ^{99m}Tc, A = 99 and $T_{1/2}$ = 6 hours = 0.25 days. Thus,

$$CFSA\ (^{99m}Tc) \approx \frac{(4.8 \times 10^{18})}{(0.99 \times 10^2 \times 0.25)}$$
$$\approx 1.94 \times 10^{17}\ Bq\,/\,g$$

In traditional units (Equation 4-23), the answers are

$$CFSA\ (^{131}I) \approx \frac{(1.3 \times 10^8)}{(1.31 \times 10^2 \times 8)}$$
$$\approx 1.24 \times 10^5\ Ci\,/\,g$$

$$CFSA\ (^{99m}Tc) \approx \frac{(1.3 \times 10^8)}{(0.99 \times 10^2 \times 0.25)}$$
$$\approx 5.3 \times 10^6\ Ci\,/\,g$$

As shown by Example 4-6, CFSAs for radionuclides having half-lives of hours, days, or even weeks are very high. Most of the radionuclides used in nuclear medicine are in this category.

In most instances, a high specific activity is desirable because then a moderate amount of activity contains only a very small mass of the element represented by the radioisotope and can be administered to a patient without causing a pharmacologic response to that element. This is an essential requirement of a "tracer study." For example, a capsule containing 0.4 MBq (~10 µCi) of carrier-free ^{131}I contains only approximately 10^{-10} g of elemental iodine (mass = activity/specific activity), which is well below the amount necessary to cause any "iodine reaction." Even radioisotopes of highly toxic elements, such as arsenic, have been given to patients in a carrier-free state. It is not possible to obtain carrier-free ^{99m}Tc because it cannot be separated from its daughter product, ^{99}Tc, a very long-lived and essentially stable isotope of technetium. Nevertheless, the mass of technetium in most ^{99m}Tc preparations is very small and has no physiologic effect when administered to a patient.

Not all production methods result in carrier-free radionuclides. Also, in some cases carrier may be added to promote certain chemical reactions in radiochemistry procedures. When a preparation is supplied with carrier, usually the packaging material indicates specific activity. If the radioactivity exists as a label attached to some complex molecule, such as a protein molecule, the specific activity may be expressed in terms of the activity per unit mass of labeled substance, such as MBq/g of protein. Methods of calculating the specific activities of radionuclides produced in a non–carrier-free state are discussed in Chapter 5.

On rare occasions, radioactive preparations that are not carrier-free or that are attached as labels to complex molecules may present problems if the carrier or labeled molecule is toxic or has undesired pharmacologic effects. Two examples in the past were reactor-produced ^{42}K in K$^+$ solution (intravenous K$^+$ injections may cause cardiac arrhythmia) and ^{131}I-labeled serum albumin (serum albumin could cause undesirably high protein levels when injected into intrathecal spaces for cerebrospinal fluid studies). In situations such as these, the amount of material that can be administered safely to a patient may be limited by the amount of carrier or unlabeled molecule present rather than by the amount of radioactivity and associated radiation hazards.

F. DECAY OF A MIXED RADIONUCLIDE SAMPLE

The equations and methods presented in Sections B and C apply only to samples containing a single radionuclide species. When a sample contains a mixture of *unrelated* species (i.e., no parent-daughter relationships), the total activity A_t is just the sum of the individual activities of the various species:

$$A_t(t) = A_1(0)e^{-0.693t/T_{1/2,1}} + A_2(0)e^{-0.693t/T_{1/2,2}} + \cdots$$
$$(4\text{-}24)$$

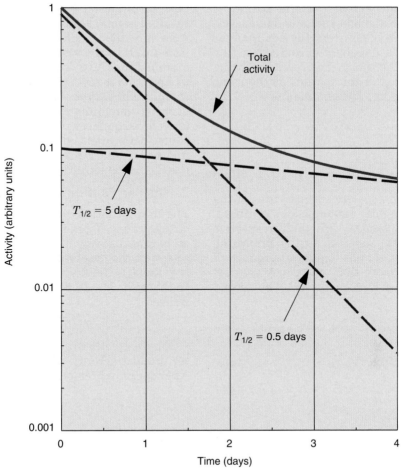

FIGURE 4-5 Activity versus time for a mixed sample of two unrelated radionuclides. The sample contains initially (at $t = 0$) 0.9 units of activity with a half-life of 0.5 days and 0.1 units of activity with a half-life of 5 days.

where $A_1(0)$ is the initial activity of the first species and $T_{1/2,1}$ is its half-life, and so forth.

Figure 4-5 shows total activity versus time for a sample containing two unrelated radionuclides. A characteristic of such a curve is that it *always* eventually follows the slope of the curve for the radionuclide having the longest half-life. Once the final slope has been established, it can be extrapolated as a straight line on a semilogarithmic graph back to time zero. This curve can then be subtracted from the total curve to give the net curve for the other radionuclides present. If more than two radionuclide species are present, the "curve-stripping" operation can be repeated for the next-longest-lived species and so forth.

Curve stripping can be used to determine the relative amounts of various radionuclides present in a mixed sample and their half-lives. It is especially useful for detecting and quantifying long-lived contaminants in radioactive preparations (e.g., ^{99}Mo in ^{99m}Tc).

G. PARENT-DAUGHTER DECAY

1. The Bateman Equations

A more complicated situation occurs when a sample contains radionuclides having parent-daughter relationships (Fig. 4-6). The

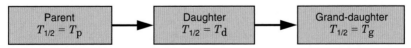

FIGURE 4-6 Schematic representation of series decay. Activities of the parent (p), daughter (d), and grand-daughter (g) are described by the Bateman equations.

equation for the activity of the parent is simply that for a single radionuclide species (see Equation 4-7); however, the equation for the activity of a daughter is complicated by the fact that the daughter product is being formed (by decay of the parent) at the same time it is decaying. The equation is

$$A_d(t) = \left\{ \left[A_p(0) \frac{\lambda_d}{\lambda_d - \lambda_p} \times \left(e^{-\lambda_p t} - e^{-\lambda_d t} \right) \right] \times \text{B.R.} \right\}$$
$$+ A_d(0) e^{-\lambda_d t}$$
$$(4\text{-}25)$$

where $A_p(t)$ and $A_d(t)$ are the activities of the parent and daughter radionuclides at time t, respectively, λ_p and λ_d are their respective decay constants, and B.R. is the branching ratio for decay to the daughter product of interest when more than one decay channel is available (see Equation 4-3).* The second

---

*The differential equations from which Equation 4-25 is derived are

$$dN_p/dt = -\lambda_p N_p \qquad (4\text{-}25a)$$

$$dN_d/dt = -\lambda_d N_d + \lambda_p N_p \qquad (4\text{-}25b)$$

These equations provide

$$N_d(t) = N_p(0) \frac{\lambda_p}{\lambda_d - \lambda_p} \times (e^{-\lambda_p t} - e^{-\lambda_d t}) + N_d(0) e^{-\lambda_d t} \qquad (4\text{-}25c)$$

Multiplying Equation 4-25c by λ_d and substituting $A_d = \lambda_d N_d$, $A_p = \lambda_p N_p$, one obtains Equation 4-25.

term in Equation 4-25, $A_d(0) e^{-\lambda_d t}$, is just the residual daughter-product activity remaining from any that might have been present at time $t = 0$. In the rest of this discussion, it is assumed that $A_d(0) = 0$, and only the first term in Equation 4-25 is considered.

Equation 4-25 is the *Bateman equation* for a parent-daughter mixture. Bateman equations for sequences of three or more radionuclides in a sequential decay scheme are found in other texts.[1] Equation 4-25 is analyzed for three general situations.†

2. Secular Equilibrium

The first situation applies when the half-life of the parent, T_p, is so long that the decrease of parent activity is negligible during the course of the observation period. An example is ^{226}Ra ($T_p = 1620$ yr) $\rightarrow$ ^{222}Rn ($T_d = 4.8$ days). In this case, $\lambda_p \approx 0$; thus Equation 4-25 can be written

$$A_d(t) \approx A_p(0)(1 - e^{-\lambda_d t}) \times \text{B.R.} \qquad (4\text{-}26)$$

Figure 4-7 illustrates the buildup of daughter product activity versus time for B.R. = 1. After one daughter-product half-life, $e^{-\lambda_d t} = 1/2$

---

†A fourth (but unlikely) situation occurs when $\lambda_p = \lambda_d = \lambda$, that is when parent and daughter have the same half-life. In this case, it can be shown that Equation 4-25 reduces to

$$A_d(t) = A_p(0) t e^{-\lambda t} + A_d(0) e^{-\lambda t} \qquad (4\text{-}25d)$$

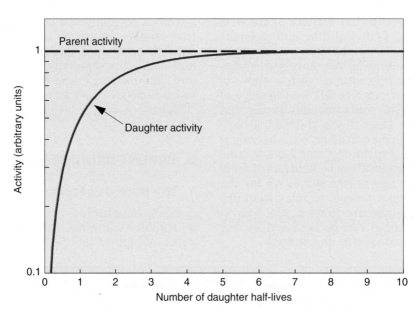

FIGURE 4-7   Buildup of daughter activity when $T_d \ll T_p \approx \infty$, branching ratio = 1. Eventually, secular equilibrium is achieved.

and $A_d \approx (1/2)A_p$. After two half-lives, $A_d \approx (3/4)A_p$, and so forth. After a "very long" time, ($\sim 5 \times T_d$), $e^{-\lambda_d t} \approx 0$, and the activity of the daughter equals that of the parent. When this occurs ($A_d \approx A_p \times$ B.R.), the parent and daughter are said to be in *secular equilibrium*.

3. Transient Equilibrium

The second situation occurs when the parent half-life is longer than the daughter half-life but is not "infinite." An example of this case is ^{99}Mo ($T_{1/2} = 66$ hr) $\rightarrow$ ^{99m}Tc ($T_{1/2} = 6$ hr). When there is a significant decrease in parent activity during the course of the observation period, one can no longer assume $\lambda_p \approx 0$, and Equation 4-25 cannot be simplified. Figure 4-8 shows the buildup and decay of daughter-product activity for a hypothetical parent-daughter pair with $T_p = 10T_d$ and B.R. = 1. The daughter-product activity increases and eventually exceeds that of the parent, reaches a maximum value, and then decreases and follows the decay of the parent. When this stage of "parallel" decay rates has been reached—that is, parent and daughter activities are decreasing but the *ratio* of parent-to-daughter activities are constant—the parent and daughter are said to be in *transient equilibrium*. The ratio of daughter-to-parent activity in transient equilibrium is

$$A_d/A_p = [T_p/(T_p - T_d)] \times \text{B.R.} \qquad (4\text{-}27)$$

The time at which maximum daughter activity is available is determined using the methods of calculus* with the result

$$t_{\max} = [1.44T_pT_d/(T_p - T_d)]\ln(T_p/T_d) \qquad (4\text{-}28)$$

where T_p and T_d are the half-lives of the parent and daughter, respectively.

Figure 4-8 is similar to that for ^{99}Mo ($T_p = 66$ hr) $\rightarrow$ ^{99m}Tc ($T_d = 6$ hr); however, the time-activity curve for ^{99m}Tc is somewhat lower because only a fraction (B.R. = 0.876) of the parent ^{99}Mo atoms decay to ^{99m}Tc (see Fig. 5-7). The remainder bypass the ^{99m}Tc metastable state and decay directly to the ground state of ^{99}Tc. Thus the ^{99m}Tc activity is given by Equation 4-25 multiplied by 0.876 and the ratio of ^{99m}Tc/^{99}Mo activity in transient equilibrium by Equation 4-27 multiplied by the same factor; however, $t_{\max}$ remains as given by Equation 4-28.

4. No Equilibrium

When the daughter half-life is longer than the parent half-life, there is no equilibrium between them. An example of this combination is ^{131m}Te ($T_{1/2} = 30$ hr) $\rightarrow$ ^{131}I ($T_{1/2} = 8$ days). Figure 4-9 shows the buildup and

---

*Set $dA_d/dt = 0$ and solve for $t_{\max}$.

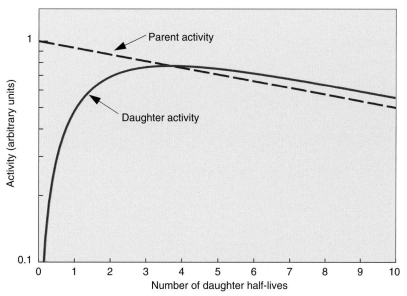

FIGURE 4-8 Buildup and decay of activity for $T_p = 10 \, T_d$, branching ratio = 1. Eventually, transient equilibrium is achieved when the parent and daughter decay curves are parallel.

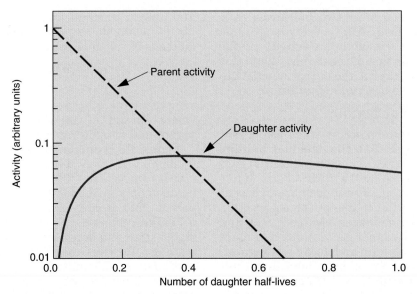

FIGURE 4-9 Buildup and decay of activity for $T_p = 0.1\ T_d$, branching ratio = 1. There is no equilibrium relationship established between the parent and daughter decay curves.

decay of the daughter product activity for a hypothetical parent-daughter pair with $T_p = 0.1T_d$. It increases, reaches a maximum (Equation 4-28 still applies for t_{max}), and then decreases. Eventually, when the parent activity is essentially zero, the remaining daughter activity decays with its own characteristic half-life.

REFERENCE

1. Evans RD: *The Atomic Nucleus*, New York, 1972, McGraw-Hill, pp 477-499.

Radionuclide and Radiopharmaceutical Production

Most of the naturally occurring radionuclides are very long-lived (e.g., ^{40}K, $T_{1/2} \sim 10^9$ years), represent very heavy elements (e.g., uranium and radium) that are unimportant in metabolic or physiologic processes, or both. Some of the first applications of radioactivity for medical tracer studies in the 1920s and 1930s made use of natural radionuclides; however, because of their generally unfavorable characteristics indicated here, they have found virtually no use in medical diagnosis since that time. The radionuclides used in modern nuclear medicine all are of the manufactured or "artificial" variety. They are made by bombarding nuclei of stable atoms with subnuclear particles (such as neutrons and protons) so as to cause nuclear reactions that convert a stable nucleus into an unstable (radioactive) one. This chapter describes the methods used to produce radionuclides for nuclear medicine as well as some considerations in the labeling of biologically relevant compounds to form radiopharmaceuticals.

A. REACTOR-PRODUCED RADIONUCLIDES

1. Reactor Principles

Nuclear reactors have for many years provided large quantities of radionuclides for nuclear medicine. Because of their long and continuing importance for this application, a brief description of their basic principles is presented.

The "core" of a nuclear reactor contains a quantity of fissionable material, typically natural uranium (^{235}U and ^{238}U) enriched in ^{235}U content. Uranium-235 undergoes spontaneous nuclear fission ($T_{1/2} \sim 7 \times 10^8$ years),

splitting into two lighter *nuclear fragments* and emitting two or three *fission neutrons* in the process (see Chapter 3, Section I). Spontaneous fission of ^{235}U is not a significant source of neutrons or energy in of itself; however, the fission neutrons emitted stimulate additional fission events when they bombard ^{235}U and ^{238}U nuclei. The most important reaction is

$$^{235}U + n \rightarrow {}^{236}U^* \qquad (5-1)$$

The ^{236}U* nucleus is highly unstable and promptly undergoes nuclear fission, releasing additional fission neutrons. In the nuclear reactor, the objective is to have the fission neutrons emitted in each spontaneous or stimulated fission event stimulate, on the average, one additional fission event. This establishes a controlled, self-sustaining *nuclear chain reaction.*

Figure 5-1 is a schematic representation of a nuclear reactor core. "Fuel cells" containing fissionable material (e.g., uranium) are surrounded by a *moderator* material. The purpose of the moderator is to slow down the rather energetic fission neutrons. Slow neutrons (also called *thermal neutrons*) are more efficient initiators of additional fission events. Commonly used moderators are "heavy water" [containing deuterium (D_2O)] and graphite. *Control rods* are positioned to either expose or shield the fuel cells from one another. The control rods contain materials that are strong neutron absorbers but that do not themselves undergo nuclear fission (e.g., cadmium or boron). The fuel cells and control rods are positioned carefully so as to establish the critical conditions for a controlled chain reaction. If the control rods were removed (or

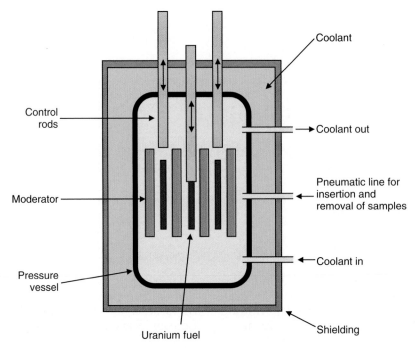

FIGURE 5-1 Schematic representation of a nuclear reactor.

incorrectly positioned), conditions would exist wherein each fission event would stimulate more than one additional nuclear fission. This could lead to a runaway reaction and to a possible "meltdown" of the reactor core. (This sequence occurs in a very rapid time scale in nuclear explosives. Fortunately, the critical conditions of a nuclear explosion cannot be achieved in a nuclear reactor.) Insertion of additional control rods results in excess absorption of neutrons and terminates the chain reaction. This procedure is used to shut down the reactor.

Each nuclear fission event results in the release of a substantial amount of energy (200-300 MeV per fission fragment), most of which is dissipated ultimately as thermal energy. This energy can be used as a thermal power source in reactors. Some radionuclides are produced directly in the fission process and can be subsequently extracted by chemical separation from the fission fragments.

A second method for producing radionuclides uses the large neutron flux in the reactor to activate samples situated around the reactor core. Pneumatic lines are used for the insertion and removal of samples. The method of choice largely depends on yield of the desired radionuclide, whether suitable sample materials are available for neutron activation, the desired specific activity, and cost considerations.

2. Fission Fragments

The fission process that takes place in a reactor can lead to useful quantities of medically important radionuclides such as ^{99}Mo, the parent material in the ^{99m}Tc generator (see Section C). As described earlier, ^{236}U* promptly decays by splitting into two fragments. A typical fission reaction (Fig. 5-2A) is

$$^{235}_{92}\text{U} + \text{n} \rightarrow \, ^{236}_{92}\text{U}^* \rightarrow \, ^{144}_{56}\text{Ba} + \, ^{89}_{36}\text{Kr} + 3\text{n} \quad (5\text{-}2)$$

More than 100 nuclides representing 20 different elements are found among the fission products of ^{236}U*. The mass distribution of the fission fragments is shown in Figure 5-2B. It can be seen that fission of ^{236}U* generally leads to one fragment with a mass number in the range of 85 to 105 and the other fragment with a mass number in the range of 130 to 150. It also is apparent that fission rarely results in fragments with nearly equal masses.

The fission products always have an excess of neutrons and hence undergo further radioactive decay by β^- emission, until a stable nuclide is reached. If one of the radioactive intermediates has a sufficiently long half-life, it can be extracted from the fission products and used as a medical radionuclide. For example,

$$^{99}_{39}\text{Y} \xrightarrow{\beta^- (1.5\,\text{s})} \, ^{99}_{40}\text{Zr} \xrightarrow{\beta^- (21\,\text{s})} \, ^{99}_{41}\text{Nb} \xrightarrow{\beta^- (15\,\text{s})} \, ^{99}_{42}\text{Mo}$$
$$(5\text{-}3)$$

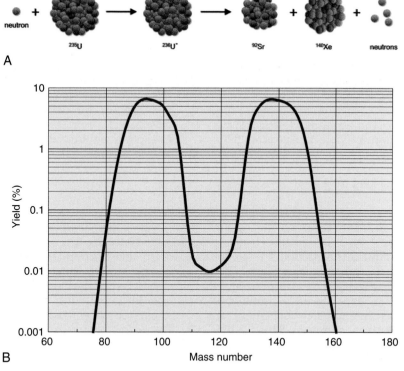

FIGURE 5-2 *A,* Example of production of fission fragments produced when neutrons interact with $^{236}U^*$. *B,* Mass distribution of fragments following fission of $^{236}U^*$.

The half-life of ^{99}Mo is 65.9 hours, which is sufficiently long to allow it to be chemically separated from other fission fragments. Molybdenum-99 plays an important role in nuclear medicine as the parent radionuclide in the ^{99}Mo-^{99m}Tc generator (see Section C). Technetium-99m is the most common radionuclide used in clinical nuclear medicine procedures today. Fission has also been used to produce ^{131}I and ^{133}Xe for nuclear medicine studies.

Radionuclides produced by the fission process have the following general characteristics:

1. Fission products always have an excess of neutrons, because N/Z is substantially higher for ^{235}U than it is for nuclei falling in the mass range of the fission fragments, even after the fission products have expelled a few neutrons (see Fig. 2-9). These radionuclides therefore tend to decay by β⁻ emission.

2. Fission products may be carrier free (no stable isotope of the element of interest is produced), and therefore radionuclides can be produced with high specific activity by chemical separation. (Sometimes other isotopes of the element of

interest are also produced in the fission fragments. For example, high-specific-activity ^{131}I cannot be produced through fission because of significant contamination from ^{127}I and ^{129}I.)

3. The lack of specificity of the fission process is a drawback that results in a relatively low yield of the radionuclide of interest among a large amount of other radionuclides.

3. Neutron Activation

Neutrons carry no net electrical charge. Thus they are neither attracted nor repelled by atomic nuclei. When neutrons (e.g., from a nuclear reactor core) strike a target, some of the neutrons are "captured" by nuclei of the target atoms. A target nucleus may be converted into a radioactive product nucleus as a result. Such an event is called *neutron activation.* Two types of reactions commonly occur.

In an *(n,γ) reaction* a target nucleus, $^A_Z X$, captures a neutron and is converted into a product nucleus, $^{A+1}_Z X^*$, which is formed in an excited state. The product nucleus immediately undergoes de-excitation to its ground state by emitting a *prompt γ ray*. The reaction is represented schematically as

$$_Z^A X(n, \gamma)_Z^{A+1} X \qquad (5\text{-}4)$$

The target and product nuclei of this reaction represent different isotopes of the same chemical element.

A second type of reaction is the *(n,p) reaction*. In this case, the target nucleus captures a neutron and promptly ejects a proton. This reaction is represented as

$$_Z^A X(n, p)_{Z-1}^A Y \qquad (5\text{-}5)$$

Note that the target and product nuclei for an (n,p) reaction do not represent the same chemical element.

In these examples, the products ($_Z^{A+1}X$ or $_{Z-1}^A Y$) usually are radioactive species. The quantity of radioactivity that is produced by neutron activation depends on a number of factors, including the intensity of the neutron flux and the neutron energies. This is discussed in detail in Section D. Production methods for biomedically important radionuclides produced by neutron activation are summarized in Table 5-1.

Radionuclides produced by neutron activation have the following general characteristics:

1. Because neutrons are added to the nucleus, the products of neutron activation generally lie above the line of stability (see Fig. 2-9). Therefore they tend to decay by β⁻ emission.

2. The most common production mode is by the (n,γ) reaction, and the products of this reaction are not carrier free because they are the same chemical element as the bombarded target material. It is possible to produce carrier-free products in a reactor by using the (n,p) reaction (e.g., ^{32}P from ^{32}S) or by activating a short-lived intermediate product, such as ^{131}I from ^{131}Te using the reaction

$$^{130}Te(n, \gamma)^{131}Te \xrightarrow{\beta^-} {}^{131}I \qquad (5\text{-}6)$$

3. Even in intense neutron fluxes, only a very small fraction of the target nuclei actually are activated, typically $1:10^6$ to 10^9 (see Section D). Thus an (n,γ) product may have very low specific activity because of the overwhelming presence of a large amount of unactivated stable carrier (target material).

There are a few examples of the production of electron capture (EC) decay or β⁺-emitting radionuclides with a nuclear reactor, for example, ^{51}Cr by (n,γ) activation of ^{50}Cr. They may also be produced by using more complicated production techniques. An example is the production of ^{18}F (β⁺, $T_{1/2}$ = 110 min). The target material is lithium carbonate (Li_2CO_3). The first step is the reaction

TABLE 5-1
NEUTRON-ACTIVATED RADIONUCLIDES OF IMPORTANCE IN BIOLOGY AND MEDICINE

Radionuclide	Decay Mode	Production Reaction	Natural Abundance of Target Isotope (%)*	$\sigma_c(b)$†
^{14}C	β⁻	$^{14}N(n,p)^{14}C$	99.6	1.81
^{24}Na	(β⁻,γ)	$^{23}Na(n,\gamma)^{24}Na$	100	0.53
^{32}P	β⁻	$^{31}P(n,\gamma)^{32}P$	100	0.19
		$^{32}S(n,p)^{32}P$	95.0	0.1
^{35}S	β⁻	$^{35}Cl(n,p)^{35}S$	75.8	0.4
^{42}K	(β⁻,γ)	$^{41}K(n,\gamma)^{42}K$	6.7	1.2
^{51}Cr	(EC,γ)	$^{50}Cr(n,\gamma)^{51}Cr$	4.3	17
^{59}Fe	(β⁻,γ)	$^{58}Fe(n,\gamma)^{59}Fe$	0.3	1.1
^{75}Se	(EC,γ)	$^{74}Se(n, \gamma)^{75}Se$	0.9	30
^{125}I	(EC,γ)	$^{124}Xe(n, \gamma)^{125}Xe \xrightarrow{EC} {}^{125}I$	0.1	110
^{131}I	(β⁻,γ)	$^{130}Te(n, \gamma)^{131}Te \xrightarrow{\beta^-} {}^{131}I$	33.8	0.24

*Values from Browne E, Firestone RB: Table of Radioactive Isotopes. New York, 1986, John Wiley.[1]

†Thermal neutron capture cross-section, in barns (b) (see "Activation Cross-Sections"). Values from Wang Y: Handbook of Radioactive Nuclides, Cleveland, Chemical Rubber Company, 1969.[2]

EC, Electron capture.

$$^6\text{Li}\,(n,\gamma)^7\text{Li} \qquad (5\text{-}7)$$

Lithium-7 is very unstable and promptly disintegrates:

$$^7_3\text{Li} \rightarrow {}^4_2\text{He} + {}^3_1\text{H} + \text{energy} \qquad (5\text{-}8)$$

Some of the energetic recoiling tritium nuclei (^3_1H) bombard stable ^{16}O nuclei, causing the reaction

$$^{16}_{8}\text{O}\left(^3_1\text{H}, n\right)^{18}_{9}\text{F} \qquad (5\text{-}9)$$

Useful quantities of ^{18}F can be produced in this way. One problem is removal from the product (by chemical means) of the rather substantial quantity of radioactive tritium that is formed in the reaction. More satisfactory methods for producing ^{18}F involve the use of charged particle accelerators, as discussed in Section B.

B. ACCELERATOR-PRODUCED RADIONUCLIDES

1. Charged-Particle Accelerators

Charged-particle accelerators are used to accelerate electrically charged particles, such as protons, deuterons (^2_1H nuclei), and α particles (^4_2He nuclei), to very high energies. When directed onto a target material, these particles may cause nuclear reactions that result in the formation of radionuclides in a manner similar to neutron activation in a reactor. A major difference is that the particles must have very high energies, typically 10-20 MeV, to penetrate the repulsive coulomb forces surrounding the nucleus.

Two types of nuclear reactions are commonly used to produce radionuclides using a charged-particle accelerator. In a *(p,n) reaction*, the target nucleus captures a proton and promptly releases a neutron. This reaction is represented as

$$^A_Z\text{X}(p, n)^A_{Z+1}\text{Y} \qquad (5\text{-}10)$$

This reaction can be considered the inverse of the (n,p) reaction that uses neutrons as the bombarding particle and was discussed in Section A.3.

A second common reaction is the *(d,n) reaction* in which the accelerated particle is a deuteron (d). The target nucleus captures a deuteron from the beam and immediately releases a neutron. This reaction is represented as

$$^A_Z\text{X}(d, n)^{A+1}_{Z+1}\text{Y} \qquad (5\text{-}11)$$

and results in a change of both the element (atomic number) and the mass number. In some cases, more than one neutron may be promptly released from the target nucleus after the bombarding particle has been captured. For example, a (p,2n) reaction involves the release of two neutrons following proton capture and a (d,3n) reaction involves the release of three neutrons following deuteron capture. Some accelerators also use alpha-particles to bombard a target and produce radionuclides. Indium-111 can be produced in this way using the reaction $^{109}\text{Ag}(\alpha,2n)^{111}\text{In}$.

Van de Graaff accelerators, linear accelerators, cyclotrons, and variations of cyclotrons have been used to accelerate charged particles. The cyclotron is the most widely used form of particle accelerator for production of medically important radionuclides.[3] Many larger institutions have their own compact *biomedical cyclotrons* for onsite production of the shorter-lived, positron-emitting radionuclides. The principles and design of cyclotrons dedicated to production of radionuclides for nuclear medicine are described briefly.

2. Cyclotron Principles

A cyclotron consists of a pair of hollow, semi-circular metal electrodes (called *dees* because of their shape), positioned between the poles of a large electromagnet (Fig. 5-3). The dees are separated from one another by a narrow gap. Near the center of the dees is an ion source, S, (typically an electrical arc device in a gas) that is used to generate the charged particles. All these components are contained in a vacuum tank at $\sim10^{-3}$ Pa($\sim10^{-8}$ atm).

During operation, particles are generated in bursts by the ion source, and a high-frequency alternating current (AC) voltage generated by a high-frequency oscillator (typically 30 kV, 25-30 MHz) is applied across the dees. The particles are injected into the gap and immediately are accelerated toward one of the dees by the electrical field generated by the applied AC voltage. Inside the dee there is no electrical field, but because the particles are in a magnetic field, they follow a curved, circular path around to the opposite side of the dee. The AC voltage frequency is such that the particles arrive at the gap just as the voltage across the dees reaches its maximum

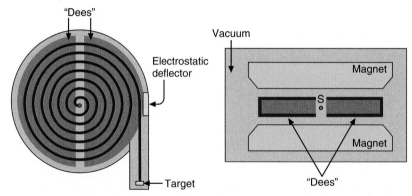

FIGURE 5-3 Schematic representation of a positive ion cyclotron: top (*left*) and side (*right*) views. The accelerating voltage is applied by a high-frequency oscillator to the two "dees." S is a source of positive ions.

value (30 kV) in the opposite direction. The particles are accelerated across the gap, gaining about 30 keV of energy in the process, and then continue on a circular path within the opposite dee.

Each time the particles cross the gap they gain energy, so the orbital radius continuously increases and the particles follow an outwardly spiraling path. The increasing speed of the particles exactly compensates for the increasing distance traveled per half orbit, and they continue to arrive back at the gap exactly in phase with the AC voltage. This condition applies so long as the charge-to-mass ratio of the accelerated particles remains constant. Because of their large relativistic mass increase, even at relatively low energies (~100 keV), it is not practical to accelerate electrons in a cyclotron. Protons can be accelerated to 20-30 MeV, and heavier particles can be accelerated to even higher energies (in proportion to their rest mass), before relativistic mass changes become limiting.*

Higher particle energies can be achieved in a variation of the cyclotron called the *synchrocyclotron* or *synchrotron*, in which the AC voltage frequency changes as the particles spiral outward and gain energy. These machines are used in high-energy nuclear physics research.

The energy of particles accelerated in a cyclotron is given by

$$E \text{ (MeV)} \approx 4.8 \times 10^{-3} (H \times R \times Z)^2 / A \quad (5\text{-}12)$$

in which H is the magnetic field strength in tesla, R is the radius of the particle orbit in centimeters, and Z and A are the atomic number (charge) and mass number of the accelerated particles, respectively. The energies that can be achieved are limited by the magnetic field strength and the dee size. In a typical biomedical cyclotron with magnetic field strength of 1.5 tesla and a dee diameter of 76 cm, protons (Z = 1, A = 1) and α particles (Z = 2, A = 4) can be accelerated to approximately 15 MeV and deuterons (Z = 1, A = 2) to approximately 8 MeV.

When the particles reach the maximum orbital radius allowed within the cyclotron dees, they may be directed onto a target placed directly in the orbiting beam path (internal beam irradiation). More commonly, the beam is extracted from the cyclotron and directed onto an external target (external-beam radiation). Typical beam currents at the target are in the range of 50-100 μA. For cyclotrons using positively charged particles (positive ion cyclotron), the beam is electrostatically deflected by a negatively charged plate and directed to the target (Fig. 5-3). Unfortunately electrostatic deflectors are relatively inefficient, as much as 30% of the beam current being lost during extraction. This "lost" beam activates the internal parts of the cyclotron, thus making servicing and maintenance of the cyclotron difficult.

In a negative-ion cyclotron, negatively charged ions (e.g. H⁻, a proton plus two electrons) are generated and then accelerated in the same manner as the positive ions in a positive-ion cyclotron (but in the opposite direction because of the different polarity). When the negatively charged ions reach the outermost orbit within the dee electrodes,

---

*Even at low energies, protons, deuterons, and α particles gain some mass when accelerated in a cyclotron. Magnetic "field shaping" is used in the cyclotron to compensate for this effect.

they are passed through a thin (5-25 μm) carbon foil, which strips off the electrons and converts the charge on the particle from negative to positive. The interaction of the magnetic beam with this positive ion bends its direction of motion outward and onto the target (Fig. 5-4). The negative-ion cyclotron has a beam extraction efficiency close to 100% and can therefore be described as a "cold" machine that requires minimal levels of shielding. Furthermore, two beams can be extracted simultaneously by positioning a carbon-stripping foil part way into the path of the beam, such that only a portion of the beam is extracted to a target. The remainder of the beam is allowed to continue to orbit and then is extracted with a second stripping foil onto a different target (Fig. 5-4). This allows two different radionuclides to be prepared simultaneously. One disadvantage of negative-ion cyclotrons is the requirement for a much higher vacuum (typically 10^{-5} Pa compared with 10^{-3} Pa for positive ion machines) because of the unstable nature of the H^- ion, the most commonly used particle in negative ion cyclotrons.

3. Cyclotron-Produced Radionuclides

Cyclotrons are used to produce a variety of radionuclides for nuclear medicine, some of which are listed in Table 5-2. General characteristics of cyclotron-produced radionuclides include the following:

1. Positive charge is added to the nucleus in most activation processes. Therefore, the products lie below the line of stability (see Fig. 2-9) and tend to decay by EC or β^+ emission.

2. Addition of positive charge to the nucleus changes its atomic number. Therefore cyclotron-activation products usually are carrier free.

3. Cyclotrons generally produce smaller quantities of radioactivity than are obtained from nuclear reactors. In part this results from generally smaller activation cross-sections for charged particles as compared with neutron irradiation (see Section D) and in part from lower beam intensities obtained in cyclotrons as compared with nuclear reactors.

Cyclotron products are attractive for nuclear medicine imaging studies because of the high photon/particle emission ratios that are obtained in β^+ and EC decay. Of special interest are the short-lived positron emitters ^{11}C ($T_{1/2} = 20.4$ min), ^{13}N ($T_{1/2} = 9.97$ min), and ^{15}O ($T_{1/2} = 2.03$ min). These radionuclides represent elements that are important constituents of all biologic substances, and they can be used to label a wide variety of biologically relevant tracers. Because of their very short lifetimes, these positron-emitting radionuclides must be prepared on site with a dedicated biomedical cyclotron. The high cost of owning and operating such machines has impeded their widespread use. Nevertheless, because of the importance of several positron emitter–labeled radiopharmaceuticals, there are now many hundreds of cyclotrons worldwide producing short-lived positron-emitting isotopes for nuclear medicine imaging studies. A typical biomedical cyclotron is shown in Figure 5-5.

Fluorine-18 ($T_{1/2} = 110$ min) is another important positron-emitting radionuclide.

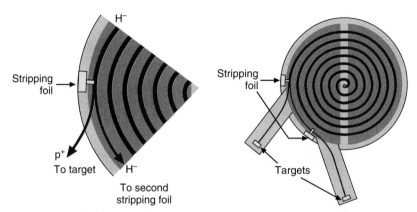

FIGURE 5-4 *Left,* Schematic representation of a negative-ion cyclotron. The carbon stripping foils remove two electrons from negative hydrogen (H^-) ions, converting them into protons (p^+) that bend in the opposite direction in the applied magnetic field. *Right,* The first stripping foil intersects only part of the beam, allowing two beams to be extracted simultaneously.

TABLE 5-2
SOME CYCLOTRON-PRODUCED RADIONUCLIDES USED IN NUCLEAR MEDICINE

Product	Decay Mode	Common Production Reaction	Natural Abundance of Target Isotope* (%)	Energy Threshold (MeV)[†]
^{11}C	β^+, EC	^{14}N(p,α)^{11}C	99.6	3.1
		^{10}B(d,n)^{11}C	19.9	0
^{13}N	β^+	^{16}O(p,α)^{13}N	99.8	5.5
		^{12}C(d,n)^{13}N	98.9	0.35
^{15}O	β^+	^{14}N(d,n)^{15}O	99.6	0
		^{15}N(p,n)^{15}O	0.37	—
^{18}F	β^+, EC	^{18}O(p,n)^{18}F	0.20	2.57
		^{20}Ne(d,α)^{18}F	90.5	0
^{67}Ga	(EC,γ)	^{68}Zn(p,2n)^{67}Ga	18.8	5.96
^{111}In	(EC,γ)	^{109}Ag(α,2n)^{111}In	48.2	—
		^{111}Cd(p,n)^{111}In	12.8	—
^{123}I	(EC,γ)	^{122}Te(d,n)^{123}I	2.6	—
		^{124}Te(p,3n)^{123}I	4.8	—
^{201}Tl	(EC,γ)	^{201}Hg(d,2n)^{201}Tl	13.2	—

*Values from Browne E, Firestone RB: *Table of Radioactive Isotopes*. New York, 1986, John Wiley.[1]
[†]Values from Helus F, Colombetti LG: *Radionuclides Production*, Vols I, II. Boca Raton, 1983, CRC Press.[4]
EC, electron capture.

FIGURE 5-5 Photograph of a negative-ion biomedical cyclotron. *Left,* Cyclotron within concrete shield. *Right,* The cyclotron itself. (*Courtesy Siemens Molecular Imaging Inc., Knoxville, TN.*)

One of its main applications is in the labeling of a glucose analog, ^{18}F-fluorodeoxyglucose (FDG), which provides a measure of the metabolic rate for glucose in the cells of the body. The longer half-life of the ^{18}F label allows FDG to be produced in regional distribution centers and shipped to hospitals tens or even hundreds of miles away. FDG is the most widely used positron-emitting radiopharmaceutical with a wide range of clinical applications in the heart, and brain and especially in cancer imaging. (See Chapter 18, Section F.)

C. RADIONUCLIDE GENERATORS

A radionuclide generator consists of a parent-daughter radionuclide pair contained in an apparatus that permits separation and extraction of the daughter from the parent. The daughter product activity is replenished continuously by decay of the parent and may be extracted repeatedly.

Table 5-3 lists some radionuclide generators of interest to nuclear medicine. They are an important source of metastable

TABLE 5-3

SOME RADIONUCLIDE GENERATORS USED IN NUCLEAR MEDICINE

Daughter*	Decay Mode	$T_{1/2}$	Parent	$T_{1/2}$
^{62}Cu	β^+,EC	9.7 min	^{62}Zn	9.3 hr
^{68}Ga	β^+,EC	68 min	^{68}Ge	271 d
^{82}Rb	β^+,EC	1.3 min	^{82}Sr	25 d
^{87m}Sr	IT	2.8 hr	^{87}Y	80 hr
^{99m}Tc	IT	6 hr	^{99}Mo	66 hr
^{113m}In	IT	100 min	^{113}Sn	120 d

*Generator product.

EC, electron capture; IT, isomeric transition.

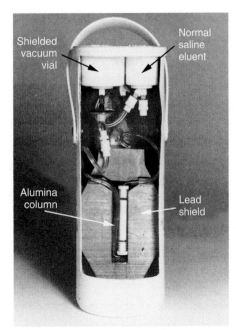

FIGURE 5-6 Cut-away view of a ^{99}Mo-^{99m}Tc generator. (*Adapted from* A Guide to Radiopharmaceutical Quality Control. *Billerica, MA, 1985, Du Pont Company.*)

radionuclides. The most important generator is the ^{99}Mo-^{99m}Tc system, because of the widespread use of ^{99m}Tc for radionuclide imaging. Technetium-99m emits γ rays (140 keV) that are very favorable for use with a gamma camera (Chapter 13). It has a reasonable half-life (6 hours), delivers a relatively low radiation dose per emitted γ ray (Chapter 22), and can be used to label a wide variety of imaging agents. More than 1850 TBq (50,000 Ci) of ^{99}Mo per week are required to meet the worldwide requirements for nuclear medicine procedures.

A ^{99}Mo-^{99m}Tc generator is shown in Figure 5-6. The parent ^{99}Mo activity in the form of molybdate ion, MoO_4^{2-} is bound to an alumina (Al_2O_3) column. The daughter ^{99m}Tc activity, produced in the form of ^{99m}TcO_4^-$ (pertechnetate), is not as strongly bound to alumina and is eluted from the column with 5 to 25 mL of normal saline. Technetium-99m activity builds up again after an elution and maximum activity is available about 24 hours later (Equation 4-28); however, usable quantities are available 3 to 6 hours later. Commercially prepared generators are sterilized, well shielded, and largely automated in operation. Typically they are used for approximately 1 week and then discarded because of natural decay of the ^{99}Mo parent.

Decay of the ^{99}Mo-^{99m}Tc parent-daughter pair is an example of transient equilibrium (see Chapter 4, Section G.3). Equation 4-25 and Figure 4-8 describe the buildup and decay of activity for such a pair. Under idealized conditions, and a branching ratio of 0.876, the ratio of ^{99m}Tc/^{99}Mo activity in a generator in a state of transient equilibrium (see Equation 4-27) would be approximately 0.96, and the time to maximum activity following an elution

(Equation 4-28) would be approximately 23 hours.

However, these equations do not accurately predict the amount of ^{99m}Tc actually obtained in individual elutions, because most generators do not yield 100% of the available activity. Typical generator elution efficiencies are 80% to 90%, depending on the size and type of generator, volume of eluant, and so on. Furthermore, the efficiency can vary from one elution to the next. In practice, efficiency variations of ±10% or more can occur in successive elutions of the same generator.[5] These may be caused by chemical changes in the column, including some that are caused by the intense radiation levels. Failure to keep a "dry" column in a dry state also can substantially degrade elution efficiency. These issues, as well as other complexities of ^{99}Mo-^{99m}Tc generators, are discussed in detail in references 5 and 6.

If 90% of the ^{99m}Tc activity in a generator is removed during an elution, the activity obtained would be 10% less than predicted from Equation 4-25 and Figure 4-8. Furthermore, the 10% residual ^{99m}Tc activity left in the generator becomes "$A_d(0)$" in Equation 4-25 for the next elution interval. This activity provides a "jump start" for regrowth of ^{99m}Tc in the generator, thereby shortening the time to maximum activity after an elution from that predicted by Equation 4-28.

Figure 5-7A, shows the available ^{99m}Tc activity, relative to parent ^{99}Mo activity, for a generator that is eluted with 90% efficiency at 24-hour intervals, starting at t = 0 hours. Under these conditions, the activity obtained is approximately 77% of the parent ^{99}Mo activity in the generator at the time of elution, and the time to maximum activity after an elution is shortened to approximately 21 hours.

If a generator is eluted at irregular intervals, the situation becomes more complicated, because the residual ^{99m}Tc activity left in the generator varies from one elution to the next. In this situation, the ^{99m}Tc activity in generator can be predicted using Equation 4-25, using the ideal versus actual yield to estimate the amount of residual ^{99m}Tc for $A_d(0)$

for the next elution interval. Figure 5-7B, shows the results of such a calculation for elutions at 0, 24, 30, 48, and 96 hours, each done with 90% elution efficiency.

In a practical environment, it is useful to keep records comparing generator yields to those predicted from the idealized equations. This can be helpful for identifying "low-yield" generators, as well as possible problems that may develop in an individual generator. A simplified equation that can be used to predict yields for elutions performed at regular 24-hour or other similarly "long" intervals is

$$Y_2 = \frac{Y_1 \times \left(e^{-\lambda_p \Delta t_2} - e^{-\lambda_d \Delta t_2}\right)}{\left[1 - e^{-(\lambda_d - \lambda_p)\Delta t_1}\right]} \qquad (5\text{-}13)$$

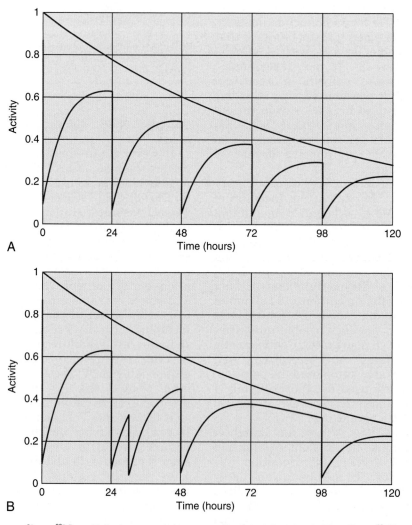

FIGURE 5-7 Orange lines: ^{99}Mo activity in a generator, normalized to 1.0 at t = 0. Blue lines: ^{99m}Tc activity available for elution, assuming 90% elution efficiency. A, Generator eluted at regular 24-hour intervals. B, Generator eluted at irregular intervals. ^{99m}Tc activities also are expressed relative to the ^{99}Mo activity in the generator, and assume consistent 90% elution efficiency from one elution to the next.

Here, Y_2 is the predicted yield of an elution (in units of activity), Y_1 is the actual yield of the immediately preceding elution, Δt_2 is the time since that elution, Δt_1 is the elution interval between that elution and the one immediately preceding it (i.e., prior to the elution yielding Y_1), and λ_p and λ_d are the decay constants of ^{99}Mo (~0.0105 hr^{-1}) and ^{99m}Tc (~0.115 hr^{-1}), respectively. This equation assumes that the elution efficiency is constant from one elution to the next and that there is insignificant carryover of residual ^{99m}Tc activity in the column at the time of the next elution. The latter condition is reasonably satisfied for 24-hr or similarly long elution intervals that allow for virtually complete decay of any ^{99m}Tc left over from previous elutions.

Molybdenum-99 activity is obtained by separation from fission fragments produced in a target containing uranium or by (n,γ) activation of stable molybdenum (23.8% ^{98}Mo). The former, sometimes called *fission moly,* has significantly higher specific activity and is the production method of choice for large quantities. The reaction by which it is produced sometimes is called an (n,f) reaction, indicating neutron irradiation causing fission. The production of ^{99}Mo is described in detail in reference 7.

Fission moly is produced by inserting a target (typically shaped as a pin, cylinder, or plate) containing natural uranium, enriched with ^{235}U, via an access port into the reactor core. The target is encapsulated in aluminum or stainless steel. Fission neutrons from the reactor core induce fission reactions in the target, as shown in Equation 5-1. Molybdenum-99 is one of the more abundant fission products (6.1% of fission products), but a wide variety of others are produced as well (see Fig. 5-2B).

After a suitable period of irradiation (typically 5-7 days), the uranium target is removed, allowed to cool, and dissolved either using an acid or alkaline dissolution process. The ^{99}Mo is then extracted by chemical means. Special care is required to assure that the many other radioactive fission products do not contaminate the desired ^{99}Mo product. As well, a large fraction of the original ^{235}U remains in the solution and must be stored as long-term radioactive waste.

The amount of stable molybdenum produced by the (n,f) reaction is small, as compared with its concentration in a target used for neutron activation of ^{98}Mo. Therefore the specific activity of ^{99}Mo in "fission moly" is

much higher, and can be loaded into generators containing much smaller quantities of the alumina column.

The volume of alumina required in a ^{99}Mo-^{99m}Tc generator is determined essentially by the amount of stable ^{99}Mo carrier that is present. Therefore "fission moly" generators require much smaller volumes of alumina per unit of ^{99}Mo activity. They can be eluted with very small volumes of normal saline (~5 mL), which is useful in some dynamic imaging studies requiring bolus injections of very small volumes of high activity (740 MBq, 20 mCi) of ^{99m}Tc.

One problem with ^{99m}Tc generators is ^{99}Mo "breakthrough," that is, partial elution of the ^{99}Mo parent along with ^{99m}Tc from the generator. From the standpoint of patient radiation safety, the amount of ^{99}Mo should be kept to a minimum. Maximum amounts, according to Nuclear Regulatory Commission regulations, are 0.15 Bq ^{99}Mo per kBq ^{99m}Tc (0.15 μCi ^{99}Mo per mCi ^{99m}Tc). It is possible to assay ^{99}Mo activity in the presence of much larger ^{99m}Tc activity using NaI(Tl) counting systems by surrounding the sample with approximately 3 mm of lead, which is an efficient absorber of the 140 keV γ rays of ^{99m}Tc but relatively transparent to the 740-780-keV γ rays of ^{99}Mo. Thus small quantities of ^{99}Mo can be detected in the presence of much larger amounts of ^{99m}Tc. Some dose calibrators are provided with a lead-lined container called a "moly shield" specifically for this purpose. Other radioactive contaminants also are occasionally found in ^{99}Mo-^{99m}Tc generator eluate.

A second major concern is breakthrough of aluminum ion, which interferes with labeling processes and also can cause clumping of red blood cells and possible microemboli. Maximum permissible levels are 10 μg aluminum per mL of ^{99m}Tc solution. Chemical test kits are available from generator manufacturers to test for the presence of aluminum ion.

D. EQUATIONS FOR RADIONUCLIDE PRODUCTION

1. Activation Cross-Sections

The amount of activity produced when a sample is irradiated in a particle beam depends on the intensity of the particle beam, the number of target nuclei in the sample, and the probability that a bombarding particle will interact with a target nucleus. The probability of interaction is determined by the *activation cross-section*. The activation cross-section is the effective "target area" presented by a target nucleus to a bombarding particle. It has dimensions of area and is symbolized by σ. The Systeme International units for σ are m^2. The traditional and more commonly used unit is the *barn* (1 b = 10^{-28} m^2) or *millibarn* (1 mb = 10^{-3} b = 10^{-31} m^2).

Activation cross-sections for a particular nucleus depend on the type of bombarding particle, the particular reaction involved, and the energy of the bombarding particles. Figure 5-8 shows the activation cross-section for the production of ^{18}F from the $^{18}O(p,n)^{18}F$ reaction. Note that the cross-section is a strong function of the energy of the bombarding proton beam, and that for the reaction shown there is a threshold energy of 2.57 MeV below which production of ^{18}F is not possible. The threshold energies for several other cyclotron-produced radionuclides are given in Table 5-2.

Because of their importance in radionuclide production by nuclear reactors, activation cross-sections for thermal neutrons have been measured in some detail. These are called *neutron-capture cross-sections*,

symbolized by σ_c. Some values of σ_c of interest for radionuclide production in nuclear medicine are listed in Table 5-1.

2. Activation Rates

Suppose a sample containing n target nuclei per cm^3, each having an activation cross-section σ, is irradiated in a beam having a *flux density* ϕ (particles/cm^2 sec) (Fig. 5-9). It is assumed that the sample thickness Δx (cm) is sufficiently thin that ϕ does not change much as the beam passes through it. The total number of targets, per cm^2 of beam area, is $n \Delta x$. They present a total area $n \phi \Delta x$ per cm^2 of beam area. The reduction of beam flux in passing through the target thickness Δx is therefore

$$\Delta\phi/\phi = n\ \sigma\ \Delta x \qquad (5\text{-}14)$$

The number of particles removed from the beam (i.e., the number of nuclei activated) per cm^2 of beam area per second is

$$\Delta\phi = n\ \sigma\ \phi\ \Delta x \qquad (5\text{-}15)$$

Each atom of target material has mass AW/ (6.023 × 10^{23}) g, in which AW is its atomic weight and 6.023 × 10^{23} is Avogadro's number. The total mass m of target material per cm^2 in the beam is therefore

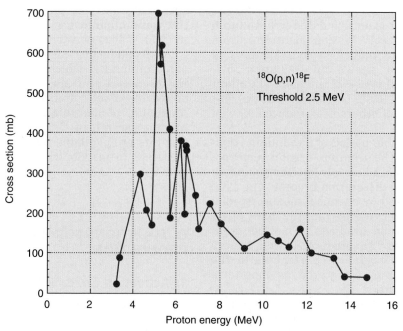

FIGURE 5-8   Activation cross-section versus particle energy for the reaction $^{18}O(p,n)^{18}F$. The energy threshold for this reaction is ~2.5 MeV. [*From Ruth TJ, Wolf AP: Absolute cross sections for the production of ^{18}F via the $^{18}O(p,n)^{18}F$ reaction.* Radiochim Acta *26:21-24, 1979.*]

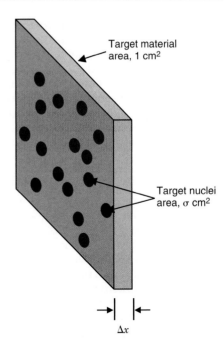

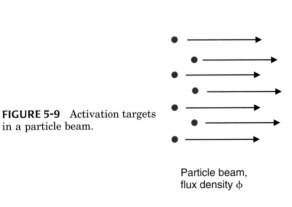

FIGURE 5-9 Activation targets in a particle beam.

Particle beam, flux density ϕ

$$m \approx n \times \Delta x \times AW/(6.023 \times 10^{23}) \quad (5\text{-}16)$$

and the activation rate R per unit mass of target material is thus

$$R \approx \Delta\phi/m \quad (5\text{-}17)$$

$$R \approx \frac{(6.023 \times 10^{22}) \times \sigma \times \phi}{AW} \text{ (activations/g} \cdot \text{sec)} \quad (5\text{-}18)$$

Equation 5-18 can be used to calculate the rate at which target nuclei are activated in a particle beam per gram of target material in the beam.

EXAMPLE 5-1

What is the activation rate per gram of sodium for the reaction ^{23}Na(n,γ)^{24}Na in a reactor thermal neutron flux density of 10^{13} neutrons/cm^2 · sec?

Answer
From Table 5-1, the thermal neutron capture cross-section for ^{23}Na is $\sigma_c = 0.53$ b. The atomic weight of sodium is approximately 23. Therefore (Equation 5-18)

$$R \approx (6.023 \times 10^{23}) \times (0.53 \times 10^{-24}) \times 10^{13}/23$$

$$\approx 1.38 \times 10^{11} \text{ activations/g} \cdot \text{sec}$$

Equation 5-18 and Example 5-1 describe situations in which the isotope represented by

the target nucleus is 100% abundant in the irradiated sample (e.g., naturally occurring sodium is 100% ^{23}Na). When the target is not 100% abundant, then the activation rate *per gram of irradiated element* is decreased by the percentage abundance of the isotope of interest in the irradiated material.

EXAMPLE 5-2

Potassium 42 is produced by the reaction ^{41}K(n,γ)^{42}K. Naturally occurring potassium contains 6.8% ^{41}K and 93.2% ^{39}K. What is the activation rate of ^{42}K per gram of K in a reactor with thermal neutron flux density 10^{13} neutrons/cm^2 · sec?

Answer
From Table 5-1, the neutron capture cross-section of ^{41}K is 1.2 b. The atomic weight of ^{41}K is approximately 41. Thus (Equation 5-18)

$$R \approx (6.023 \times 10^{23}) \times (1.2 \times 10^{-24}) \times 10^{13}/41$$

$$\approx 1.76 \times 10^{11} \text{ activations/g}(^{41}\text{K}) \cdot \text{sec}$$

The activation rate per gram of potassium is 6.8% of this, that is,

$$R \approx 0.068 \times (1.76 \times 10^{11})$$

$$\approx 1.20 \times 10^{10} \text{ activations/g(K)} \cdot \text{sec}$$

Activation rates are less than predicted by Equation 5-18 when the target thickness is such that there is significant attenuation of particle beam intensity as it passes through the target (i.e., some parts of the target are irradiated by a weaker flux density). Also, when "thick" targets are irradiated by charged-particle beams, the particles lose energy and activation cross-sections change as the beam penetrates the target. The equations for these conditions are beyond the scope of this book and are discussed in reference 8.

3. Buildup and Decay of Activity

When a sample is irradiated in a particle beam, the buildup and decay of product radioactivity is exactly analogous to a special case of parent-daughter radioactive decay discussed in Chapter 4, Section G.2. The irradiating beam acts as an inexhaustible, long-lived "parent," generating "daughter" nuclei at a constant rate. Thus, as shown in Figure 4-7, the product activity starts from zero and increases with irradiation time, gradually approaching a saturation level at which its disintegration rate equals its production rate. The saturation level can be determined from Equation 5-18. The saturation *disintegration rate* per gram is just equal to R, the *activation rate* per gram, so the saturation specific activity A_s is

$$A_s \text{ (Bq/g)} = R \qquad (5\text{-}19)$$

which, when combined with Equation 5-18, yields

$$A_s \text{ (Bq/g)} \approx 0.6023 \times \sigma \times \phi / \text{AW} \qquad (5\text{-}20)$$

where σ is the activation cross-section in barns, ϕ is the flux in units of particles per $cm^2 \cdot sec$, and AW is the atomic weight of the target material. The final equation for specific activity, A, versus irradiation time is

$$A(t) \text{ (Bq/g)} = A_s (1 - e^{-\lambda t}) \qquad (5\text{-}21)$$

where λ is the decay constant of the product (compare with Equation 4-26). The specific activity of the target reaches 50% of the saturation level after irradiating for one daughter product half-life, 75% after two half-lives, and so on (see Fig. 4-7). No matter how long the irradiation, the sample-specific activity cannot exceed the saturation level. Therefore it is unproductive to irradiate a target for longer than approximately three or four times the product half-life.

EXAMPLE 5-3

What is the saturation specific activity for the ^{42}K production problem described in Example 5-2? Compare this with the carrier-free

specific activity (CFSA) of ^{42}K (the half-life of ^{42}K is 12.4 hours).

Answer
Applying Equation 5-20 with $\sigma = 1.2$ b, $\phi = 10^{13}$, and AW ≈ 41,

$$A_s = 0.6023 \times 1.2 \times 10^{13} / 41$$
$$= 1.76 \times 10^{11} \text{ (Bq } ^{42}K/g \ ^{41}K)$$

If natural potassium is used, only 6.8% is ^{41}K. Therefore the saturation specific activity is

$$A_s = (1.76 \times 10^{11}) \times 0.068$$
$$= 1.20 \times 10^{10} \text{ Bq } ^{42}K/g \text{ K}$$

The CFSA of ^{42}K ($T_{1/2} \sim 0.5$ days) is (Equation 4-21)

$$\text{CFSA} \approx (4.8 \times 10^{18})/(41 \times 0.5)$$
$$\approx (2.3 \times 10^{17}) \text{ Bq } ^{42}K/g \ ^{42}K$$

Example 5-3 illustrates the relatively low specific activity that typically is obtained by (n,γ) activation procedures in a nuclear reactor.

A parameter that is related directly to the saturation activity in an activation problem is the *production rate, A'.* This is the rate at which radioactivity is produced during an irradiation, disregarding the simultaneous decay of radioactivity that occurs during the irradiation. It is the slope of the production curve at time $t = 0$ (before any of the generated activity has had opportunity to decay). The production rate can be shown by methods of differential calculus to be equal to

$$A' \text{ (Bq/g} \cdot \text{hr)} = \ln 2 \times A_s \text{ (Bq/g)} / T_{1/2} \text{(hr)} \quad (5\text{-}22)$$

where $T_{1/2}$ is the half-life of the product.

Reactor production capabilities may be defined in terms of either saturation levels or production rates. If the irradiation time t is "short" in comparison with the product half-life, one can approximate the activity produced from the production rate according to

$$A \text{ (Bq/g)} \approx A' \times t \qquad (5\text{-}23)$$

$$\approx \ln 2 \times A_s \times t / T_{1/2} \qquad (5\text{-}24)$$

where t and $T_{1/2}$ must be in the same units.

EXAMPLE 5-4

What is the production rate of ^{42}K for the problem described in Example 5-2, and what specific activity would be available after an irradiation period of 3 hours? (The half-life of ^{42}K is 12.4 hours.)

Answer
From Example 5-3, $A_s = 1.20 \times 10^{10}$ Bq ^{42}K/g K. Therefore (Equation 5-22)

$$A' = 0.693 \times (1.20 \times 10^{10})/12.4$$
$$\approx 6.7 \times 10^8 \text{ Bq } ^{42}\text{K/g K} \cdot \text{hr}$$

After 3 hours, which is "short" in comparison with the half-life of ^{42}K, the specific activity of the target is (Equation 5-23)

$$A(\text{Bq/g}) \approx (6.7 \times 10^8) \times 3$$
$$\approx 2.0 \times 10^9 \text{ Bq } ^{42}\text{K/g K}$$

E. RADIONUCLIDES FOR NUCLEAR MEDICINE
....

1. General Considerations

In elemental form, radionuclides themselves generally have a relatively small range of biologically interesting properties. For example, ^{131}I as an iodide ion (I^-) is useful for studying the uptake of elemental iodine in the thyroid or in metastatic thyroid cancer or for delivering a concentrated radiation dose to thyroid tissues for therapeutic purposes; however, elemental iodine has no other generally interesting properties for medical usage. For this reason, most studies in nuclear medicine employ *radiopharmaceuticals,* in which the radionuclide is attached as a label to a compound that has useful biomedical properties.

For most applications, the radiopharmaceutical is injected into the patient, and the emissions are detected using external imaging or counting systems. Certain practical requirements must be met for a radionuclide to be a useful label. A portion of the Chart of Nuclides was shown in Figure 3-11. A complete chart contains hundreds of radionuclides that could conceivably be used for some biomedical application, either in elemental form or as a radiopharmaceutical label. However, the number of radionuclides actually used is much smaller because of various practical considerations, as discussed in the following section. A listing of some of the more commonly used

radionuclides for nuclear medicine procedures is presented in Table 5-4.

2. Specific Considerations

The *type and energy of emissions* from the radionuclide determine the availability of useful photons or γ rays for counting or imaging. For external detection of a radionuclide inside the body, photons or γ rays in the 50-600 keV energy range are suitable. Very-low-energy photons and γ rays (<50 keV), or particulate radiation, have a high likelihood of interacting in the body and will not in general escape for external detection. The presence of such low energy or particulate emissions increases the radiation dose to the patient. An example of this is ^{131}I, which decays by (β^-,γ) emitting a β^- particle, followed by γ rays at 364 (82%), 637 (6.5%), 284 (5.8%), or 80 keV (2.6%). The γ rays are in an appropriate range for external detection; however, the β^- particle contributes additional dose as compared with radionuclides that decay by (EC,γ).

The *physical half-life* of the radionuclide should be within the range of seconds to days (preferably minutes to hours) for clinical applications. If the half-life is too short, there is insufficient time for preparation of the radiopharmaceutical and injection into the patient. An example of this is the positron emitter ^{15}O ($T_{1/2} = 122$ sec). This limits ^{15}O-labeled radiopharmaceuticals to simple compounds such as $H_2{}^{15}$O and C^{15}O. If the half-life were longer, a much wider range of compounds could be labeled with ^{15}O. Other radionuclides have half-lives that are too long for practical purposes. Most of the radiation is emitted outside of the examination time, which can result in a high radiation dose to the patient in relation to the number of decays detected during the study. Long-lived radionuclides also can cause problems in terms of storage and disposal. An example of a very long-lived radionuclide that is not used in human studies because of half-life considerations is ^{22}Na ($T_{1/2} = 2.6$ yr).

The *specific activity* of the radionuclide largely determines the mass of a compound that is introduced for a given radiation dose. Because nuclear medicine relies on the use of subpharmacologic tracer doses that do not perturb the biologic system under study, the mass should be low and the specific activity high. At low specific activities, only a small fraction of the molecules in the sample are radioactive and therefore signal producing, whereas the rest of the molecules add to the

TABLE 5-4

PHYSICAL PROPERTIES OF RADIONUCLIDES USED IN NUCLEAR MEDICINE STUDIES

Radionuclide	Decay Mode	Principal Photon Emissions	Half-Life	Primary Use
^{11}C	β^+	511 keV	20.4 min	Imaging
^{13}N	β^+	511 keV	9.97 min	Imaging
^{15}O	β^+	511 keV	2.03 min	Imaging
^{18}F	β^+	511 keV	110 min	Imaging
^{32}P	β^-	—	14.3 d	Therapy
^{67}Ga	EC	93, 185, 300 keV	3.26 d	Imaging
^{82}Rb	β^+	511 keV	1.25 min	Imaging
^{89}Sr	β^-	—	50.5 d	Therapy
^{99m}Tc	IT	140 keV	6.02 hr	Imaging
^{111}In	EC	172, 247 keV	2.83 d	Imaging
^{123}I	EC	159 keV	13.2 hr	Imaging
^{125}I	EC	27-30 keV x rays	60.1 d	In vitro assays
^{131}I	β^-	364 keV	8.04 d	Therapy/ imaging
^{153}Sm	β^-	41, 103 keV	46.7 hr	Therapy
^{186}Re	β^-	137 keV	3.8 d	Therapy
^{201}Tl	EC	68-80 keV x rays	3.04 d	Imaging

EC, electron capture; IT, isomeric transition.

mass of the compound being introduced, without producing signal. Theoretically, the attainable specific activity of a radionuclide is inversely proportional to its half-life, although in practice, many other factors (e.g., the abundance of stable isotopes in air and glassware) can determine the actual specific activity of the injected labeled compound, as described in Section F.1.

The *radionuclidic purity* is defined as the fraction of the total radioactivity in a sample that is in the form of the desired radionuclide. Radionuclidic contaminants arise in the production of radionuclides and can be significant in some situations. The effect of these contaminants is to increase the radiation dose to the patient. They may also increase detector dead time, and if the energy of the emissions falls within the acceptance window of the detector system, contaminants may result in incorrect counting rate or pixel intensities in images. Of concern in radionuclide generator systems is contamination with the long-lived parent radionuclide. In the case of the ^{99}Mo-^{99m}Tc generator, the radionuclidic purity of the ^{99m}Tc must be higher than 99.985%, as discussed in Section C.

The *chemical properties* of the radionuclide also are an important factor. Radionuclides of elements that can easily produce useful *precursors* (chemical forms that react readily to form a wide range of labeled products) and that can undergo a wide range of chemical syntheses are preferred (e.g., ^{123}I, ^{18}F, and ^{11}C). Radionuclides of elements that are easily incorporated into biomolecules, without significantly changing their biochemical properties, also are attractive. Examples are ^{11}C, ^{13}N, ^{15}O, elements that are found naturally in many biomolecules. Metals such as ^{99m}Tc and ^{67}Ga also are widely used as labels in nuclear medicine, because of the desirable imaging properties of the radionuclide. To incorporate such elements into biologically relevant molecules is challenging but can be achieved by *chelation* and other techniques that seek to "hide" or shield the metal atom from the biologically active sites of the molecule.

Finally, the *cost and complexity* of preparing a radionuclide must be considered. Sufficient quantities of radionuclide for radiopharmaceutical labeling and subsequent patient injection must be produced at a cost (both materials and labor) consistent with today's health care market.

F. RADIOPHARMACEUTICALS FOR CLINICAL APPLICATIONS

As noted earlier, radionuclides almost always are attached as labels to compounds of biomedical interest for nuclear medicine applications. Because of the practical considerations discussed in the preceding section, the number of different radionuclides routinely used in nuclear medicine is relatively small, perhaps fewer than a dozen even in large hospitals. On the other hand, the number of labeled compounds is much larger and continuously growing, owing to very active research in radiochemistry and radiopharmaceutical preparation. The following sections summarize the properties of some radiopharmaceuticals that enjoy widespread usage at this time. More detailed discussions are found in the articles and texts listed in the Bibliography.

1. General Considerations

The final *specific activity* of a radiopharmaceutical (as opposed to the radionuclide) is determined by losses in specific activity that occur during the chemical synthesis of the radiopharmaceutical. This is particularly an issue for isotopes of elements that have high natural abundances. For example, the theoretical maximum specific activity for ^{11}C is 3.5×10^8 MBq/μmol (CFSA from Equation 4-22), whereas the specific activity of ^{11}C-labeled radiopharmaceuticals actually obtained in practice is approximately 10^5 MBq/μmol. This is largely because of the presence of stable carbon in the air (as CO_2) and in the materials of the reaction vessels and tubing used in the chemical synthesis procedure.

Radiochemical purity is the fraction of the radioactivity in the sample that is present in the desired chemical form. Radiochemical impurities usually stem from competing chemical reactions in the radiolabeling process or from decomposition (chemical or radiation induced) of the sample. Radiochemical impurities are problematic in that their distribution in the body is generally different, thus adding a background to the image of the desired compound. The typical radiochemical purity for radiopharmaceuticals is higher than 95%. *Chemical purity* (the fraction of the sample that is present in the desired chemical

form) is also important, with desirable values of greater than 99%.

The dynamic time course of the radiopharmaceutical in the body must be considered. Some radiopharmaceuticals have rapid uptake and clearance, whereas others circulate in blood with only slow uptake into tissues of interest. The rate of clearance of the radiopharmaceutical from the body is called the *biologic half-life*. Together with the physical half-life of the radionuclide, this determines the number of radioactive decays that will be observed from a particular region of tissue as a function of time. These two factors also are important factors in determining the radiation dose to the subject (see Chapter 22, Section B). It is important that radiopharmaceuticals be labeled with radionuclides with half-lives that are long enough to encompass the temporal characteristics of the biologic process being studied. For example, labeled antibodies generally require hours to days before significant uptake in a target tissue is reached and blood levels have dropped sufficiently for the target to be visualized. Short-lived radionuclides with half-lives of minutes or less would not be useful in this situation.

The radiopharmaceutical must not be toxic at the mass levels administered. This requirement usually is straightforward in nuclear medicine studies because of the relatively high specific activity of most radiopharmaceuticals, resulting in typical injections of microgram to nanogram quantities of material. Generally, milligram levels of materials are required for pharmacologic effects. Safety concerns also require that all radiopharmaceuticals be sterile and pyrogen-free prior to injection. Organisms can be removed by filtration through a sterile filter with a pore size of 0.22 μm or better. Use of pharmaceutical-grade chemicals, sterile water, and sterilized equipment can minimize the risk of pyrogens. Finally, the pH of the injected solution should be appropriate.

2. Labeling Strategies

There are two distinct strategies for labeling of *small molecules* with radionuclides. In *direct substitution,* a stable atom in the molecule is replaced with a radioactive atom of the same element. The compound has exactly the same biologic properties as the unlabeled compound. This allows many compounds of biologic relevance to be labeled and studied in vivo using radioactive isotopes of elements that are widely found in nature (e.g., hydrogen, carbon, nitrogen, and oxygen). An example is replacing a ^{12}C atom in glucose with a ^{11}C atom to create ^{11}C-glucose. This radiopharmaceutical will undergo the same distribution and metabolism in the body as unlabeled glucose.

The second approach is to *create analogs*. This involves modifying the original compound. Analogs allow the use of radioactive isotopes of elements that are not so widely found in nature but that otherwise have beneficial imaging properties (e.g., fluorine and iodine). Analogs also allow chemists to beneficially change the biologic properties of the molecule by changing the rates of uptake, clearance, or metabolism. For example, replacing the hydroxyl (OH) group on the second carbon in glucose with ^{18}F results in FDG, an analog of glucose. This has the advantage of putting a longer-lived radioactive tag onto glucose compared with ^{11}C; and even more important, FDG undergoes only the first step in the metabolic pathway for glucose, thus making data analysis much more straightforward (see Chapter 21, Section E.5). FDG is now a widely used radiopharmaceutical for measuring metabolic rates for glucose. The downside to analogs are that they behave differently from the native compound, and these differences need to be carefully understood if the analog is used to provide a measure of the biologic function related to the native molecule.

An alternative approach to labeling materials that is possible only for *larger biomolecules* is to keep the radioactive label away from the biologically active site of the molecule. Thus large molecules (e.g., antibodies, peptides, and proteins) may be labeled with many different radionuclides, with minimal effect on their biologic properties.

3. Technetium-99m-Labeled Radiopharmaceuticals

The ^{99}Mo-^{99m}Tc generator produces technetium in the form of ^{99m}TcO$_4^-$. A number of "cold kits" are available that allow different ^{99m}Tc complexes to be produced by simply mixing the ^{99m}TcO$_4^-$ and the contents of the cold kit together. The cold kit generally contains a reducing agent, usually stannous chloride, which reduces the ^{99m}Tc to lower oxidation states, allowing it to bind to a complexing agent (also known as the *ligand*) to form the radiopharmaceutical. Using these kits, a range of ^{99m}Tc-labeled radiopharmaceuticals that are targeted to different organ systems and different biologic processes can be prepared quickly and conveniently in the hospital setting. Table 5-5 lists a few examples of ^{99m}Tc radiopharmaceuticals that are prepared from kits.

4. Radiopharmaceuticals Labeled with Positron Emitters

Positron emitters such as ^{11}C, ^{13}N, and ^{15}O can be substituted for stable atoms of the same elements in compounds of biologic importance. This results in radiolabeled compounds with exactly the same biochemical properties as the original compound. Alternatively, ^{18}F, another positron-emitting radionuclide, can be substituted for hydrogen to produce labeled analogs. Several hundreds of compounds have been synthesized with ^{11}C, ^{13}N, ^{15}O, or ^{18}F labels for imaging with positron emission tomography. The short half-life of ^{11}C, ^{13}N, and ^{15}O requires in-house radionuclide production in a biomedical cyclotron and rapid synthesis techniques to incorporate them into radiopharmaceuticals. On the other hand, the

TABLE 5-5
SOME ^{99m}Tc-LABELED RADIOPHARMACEUTICALS PREPARED FROM KITS

Compound	Abbreviation Stands for	Applications
^{99m}Tc-MDP	Methylene diphosphonate	Bone scans
^{99m}Tc-DMSA	2,3-Dimercaptosuccinic acid	Renal imaging
^{99m}Tc-DTPA	Diethylenetriaminepenta acetic acid	Renal function
^{99m}Tc-sestamibi	2-Methoxy-2-methylpropyl isonitrile	Myocardial perfusion, breast cancer
^{99m}Tc-HMPAO	Hexamethylpropylene-amine oxime	Cerebral perfusion
^{99m}Tc-HIDA	*N*-(2,6-dimethylphenol-carbamoylmethyl)-iminodiacetic acid	Hepatic function
^{99m}Tc-ECD	N,N'-1,2-ethylenediyl-*bis*-L-cysteine diethylester	Cerebral perfusion

relatively longer half-life of ^{18}F permits its distribution within a radius of a few hundred miles from the site of production, thus obviating the need of a cyclotron in the nuclear medicine imaging facility.

The most widely used positron-labeled radiopharmaceutical is the glucose analog FDG. Glucose is used by cells to produce adenosine triphosphate, the energy "currency" of the body, and accumulation of FDG in cells is proportional to the metabolic rate for glucose. Because the energy demands of cells are altered in many disease states, FDG has been shown to be a sensitive marker for a range of clinically important conditions, including neurodegenerative diseases, epilepsy, coronary artery disease, and most cancers and their metastases.

5. Radiopharmaceuticals for Therapy Applications

Other radiopharmaceuticals are designed for therapy applications. These are normally labeled with a β^- emitter, and the radiopharmaceutical is targeted against abnormal cells, commonly cancer cells. The β^- emitter deposits radiation only within a small radius (typically 0.1 to 1 mm) and selectively kills cells in this region through radiation damage. If the radiopharmaceutical is more readily accumulated by cancer cells than normal cells, a therapeutic effect can be obtained.

6. Radiopharmaceuticals in Clinical Nuclear Medicine

Many different radiopharmaceuticals have been approved for use in clinical nuclear medicine studies. Each of these radiopharmaceuticals is targeted to measuring a specific biologic process, and therefore what is measured depends directly on which radiopharmaceutical is administered to the patient. Some of the more common radiopharmaceuticals are listed in Table 1-1 and Table 5-5.

Most radiopharmaceuticals are used in conjunction with imaging systems that can determine the location of the radiopharmaceutical within the body. Often, the rate of change of radiopharmaceutical localization within a specific tissue (the rate of uptake or clearance) is also important and is measured by acquiring multiple images as a function of time. The imaging systems used in nuclear medicine studies are discussed in Chapters 13, 14, and 17-19.

REFERENCES

1. Browne E, Firestone RB: *Table of Radioactive Isotopes,* New York, 1986, John Wiley.
2. Wang Y: *Handbook of Radioactive Nuclides,* Cleveland, 1969, Chemical Rubber Company.
3. Schwartz SW, Gaeble GG, Welch MJ: Accelerators and positron emission tomography radiopharmaceuticals. In Sandler MP, Coleman RE, Patton JA, et al, editors: *Diagnostic Nuclear Medicine,* ed 4, Philadelphia, 2003, Lippincott, Williams & Wilkins, pp 117-132.
4. Helus F, Colombetti LG: *Radionuclides Production,* Vols I, II, Boca Raton, 1983, CRC Press.
5. Holland ME, Deutsch E, Heineman WR: Studies on commercially available ^{99}Mo/^{99m}Tc radionuclide generators—II. Operating characteristics and behavior of ^{99}Mo/^{99m}Tc generators. *Appl Radiat Isot* 37:173-180, 1986.
6. Boyd RE: Molybdenum-99: Technetium-99m generator. *Radiochim Acta* 30:123-145, 1982.
7. *Medical Isotope Production Without Highly Enriched Uranium.* Washington, D.C., 2009, National Academies Press.
8. Murray RL: *Nuclear Energy,* ed 5, Boston, 2001, Butterworth Heinemann.

BIBLIOGRAPHY

Further information on radionuclide production and radiopharmaceutical preparation can be found in the following:

Knapp FF, Mirzadeh S: The continuing role of radionuclide generator systems for nuclear medicine. *Eur J Nucl Med* 21:1151-1165, 1994.

Lieser KH: *Nuclear and Radiochemistry,* ed 2, Weinheim, Germany, 2001, Wiley VCH.

Sampson CB: *Textbook of Radiopharmacy: Theory and Practice,* ed 2, New York, 1994, Gordon & Breach.

Tewson TJ, Krohn KA: PET radiopharmaceuticals: State-of-the-art and future prospects. *Semin Nucl Med* 28:221-234, 1998.

Welch MJ, Redvanly CS: *Handbook of Radiopharmaceuticals: Radiochemistry and Applications.* Chichester, England, 2003, John Wiley & Sons.

Interaction of Radiation with Matter

The two most important general types of radiation emitted during radioactive decay are *charged particles*, such as α particles and β particles, and *electromagnetic radiation* (photons), such as γ rays and x rays. These radiations transfer their energy to matter as they pass through it. The principle mechanisms for energy transfer are ionization and excitation of atoms and molecules. Most of this energy ultimately is degraded into heat (atomic and molecular vibrations); however, the ionization effect has other important consequences. For this reason, the radiations emitted during radioactive decay often are called *ionizing radiations*. The processes by which ionizing radiations transfer their energy to matter are fundamental to the detection of radiation, discussed in Chapter 7. As well, they are important for radiation dosimetry, discussed in Chapter 22. In this chapter, we discuss those processes in some detail. Because the mechanisms differ, they are discussed separately for particulate versus electromagnetic radiation.

A. INTERACTIONS OF CHARGED PARTICLES WITH MATTER

1. Charged-Particle Interaction Mechanisms

High-energy charged particles, such as α particles or β particles, lose energy and slow down as they pass through matter, as a result of collisions with atoms and molecules. High-energy electrons, which also are charged particles, are a byproduct of these collisions. In addition, high-energy electrons are generated when γ rays and x rays interact with matter, and they are emitted in internal conversion (see Chapter 3, Section E) and in the Auger effect (see Chapter 2, Section C.3).

For these reasons, this section emphasizes the interactions of electrons with matter. Except for differences in sign, the forces experienced by positive and negative electrons (e.g., β+ and β− particles) are identical. There are minor differences between the ionizing interactions of these two types of particles, but they are not of importance to nuclear medicine and are not discussed here. In this chapter, the term *electrons* is meant to include both the positive and negative types. The annihilation effect, which occurs when a positive electron (positron) has lost all of its kinetic energy and stopped, is discussed in Chapter 3, Section G.

The "collisions" that occur between a charged particle and atoms or molecules involve electrical forces of attraction or repulsion rather than actual mechanical contact. For example, a charged particle passing near an atom exerts electrical forces on the orbital electrons of that atom. In a close encounter, the strength of the forces may be sufficient to cause an orbital electron to be separated from the atom, thus causing ionization (Fig. 6-1A). An ionization interaction looks like a collision between the charged particle and an orbital electron. The charged particle loses energy in the collision. Part of this energy is used to overcome the binding energy of the electron to the atom, and the remainder is given to the ejected *secondary electron* as kinetic energy. Ionization involving an inner-shell electron eventually leads to the emission of characteristic x rays or Auger electrons; however, these effects generally are very small, because most ionization interactions involve outer-shell electrons. The ejected electron may be sufficiently energetic to cause *secondary ionizations* on its own. Such an electron is called a *delta (δ) ray*.

A less-close encounter between a charged particle and an atom may result in an orbital electron being raised to an *excited state,* thus

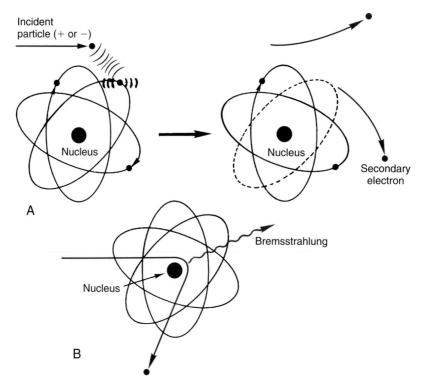

FIGURE 6-1 Interactions of charged particles with atoms. *A,* Interaction with an orbital electron resulting in ionization. Less-close encounters may result in atomic excitation without ionization. *B,* Interaction with a nucleus, resulting in bremsstrahlung production. [*Repulsion by orbital electron (A) and attraction toward nucleus (B) indicates incident particles are negatively charged in the examples shown.*]

causing atomic or molecular *excitation.* These interactions generally result in smaller energy losses than occur in ionization events. The energy transferred to an atom in an excitation interaction is dissipated in molecular vibrations, atomic emission of infrared, visible, ultraviolet radiation, and so forth.

A third type of interaction occurs when the charged particle actually penetrates the orbital electron cloud of an atom and interacts with its nucleus. For a heavy charged particle of sufficiently high energy, such as an α particle or a proton, this may result in nuclear reactions of the types used for the production of radionuclides (see Chapter 5); however, for both heavy charged particles and electrons, a more likely result is that that particle will simply be deflected by the strong electrical forces exerted on it by the nucleus (Fig. 6-1B). The particle is rapidly decelerated and loses energy in the "collision." The energy appears as a photon of electromagnetic radiation, called *bremsstrahlung* (German for "braking radiation").

The energy of bremsstrahlung photons can range anywhere from nearly zero (events in which the particle is only slightly deflected) up to a maximum equal to the full energy of the incident particle (events in which the particle is virtually stopped in the collision). Figure 6-2 shows the energy spectrum for bremsstrahlung photons generated in aluminum by particles from a ^{90}Sr-^{90}Y source mixture (E_β^{max} = 2.27 MeV) and illustrates that most of the photons are in the lower energy range.

2. Collisional Versus Radiation Losses

Energy losses incurred by a charged particle in ionization and excitation events are called *collisional losses,* whereas those incurred in nuclear encounters, resulting in bremsstrahlung production, are called *radiation losses.* In the nuclear medicine energy range, collisional losses are by far the dominating factor (See Fig. 6-5). Radiation losses increase with increasing particle energy and with increasing atomic number of the absorbing medium. An approximation for percentage radiation losses for β particles having maximum energy E_β^{max} (MeV) is

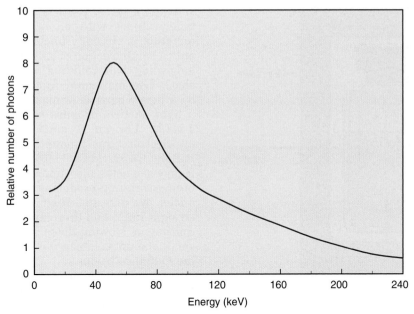

FIGURE 6-2   Bremsstrahlung spectrum for β particles emitted by ^{90}Sr + ^{90}Y (E_β^{max} = 2.27 MeV) mixture in aluminum. (*Adapted from Mladjenovic M:* Radioisotope and Radiation Physics. *New York, 1973, Academic Press, p 121.*)

percentage radiation losses $\approx (ZE_\beta^{max}/3000)$

$$\times 100\% \tag{6-1}$$

where Z is the atomic number of the absorber. This approximation is accurate to within approximately 30%. For a mixture of elements, an "effective" atomic number for bremsstrahlung production should be used

$$Z_{eff} = \sum_i f_i Z_i^2 \Big/ \sum f_i Z_i \tag{6-2}$$

where $f_1, f_2, \ldots$ are the fractions by weight of the elements $Z_1, Z_2, \ldots$ in the mixture.

EXAMPLE 6-1

Calculate the percentage of radiation losses for ^{32}P β particles in water and in lead.

Answer
E_β^{max} = 1.7 MeV for ^{32}P. Water comprises $^2\!/_{18}$ parts hydrogen (Z = 1, AW ≈ 1) and $^{16}\!/_{18}$ parts oxygen (Z = 8, AW ≈ 16); thus its effective atomic number for bremsstrahlung production is (Equation 6-2)

$$Z_{eff} = \frac{[(1/9)(1)^2(8/9)(8)^2]}{[(1/9)(1)+(8/9)(8)]} = 7.9$$

The percentage of radiation losses in water are therefore (Equation 6-1)

$$(7.9 \times 1.7/3000) \times 100\% \approx 0.4\%$$

and in lead (Z = 82) they are

$$(82 \times 1.7/3000) \times 100\% \approx 4.6\%$$

The remaining 99.6% and 95.4%, respectively, are dissipated as collisional losses.

Example 6-1 demonstrates that high-energy electrons in the nuclear medicine energy range dissipate most of their energy in collisional losses. Bremsstrahlung production accounts for only a small fraction of their energy. Nevertheless, bremsstrahlung can be important in some situations, such as the shielding of relatively large quantities of an energetic β-particle emitter (e.g., hundreds of MBq of ^{32}P). The β particles themselves are easily stopped by only a few millimeters of plastic, glass, or lead (see Section B.2); however, the bremsstrahlung photons they generate are much more penetrating and may require additional shielding around the primary β-particle shielding. It is helpful in such situations to use a low-Z material, such as plastic, for the primary β-particle shielding, and then to surround this with a higher-Z material, such as lead, for bremsstrahlung shielding (Fig. 6-3). This arrangement minimizes bremsstrahlung production by the β particles in the shielding material.

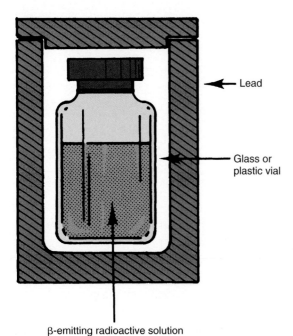

Lead →

← Glass or
plastic vial

β-emitting radioactive solution

FIGURE 6-3 Preferred arrangement for shielding energetic β-emitting radioactive solution. Glass or plastic walls of a vial stop the β particles with minimum bremsstrahlung production, and a lead container absorbs the few bremsstrahlung photons produced.

Bremsstrahlung production and radiation losses for α particles and other heavy charged particles are very small because the amount of bremsstrahlung production is inversely proportional to the mass of the incident charged particle. Alpha particles and protons are thousands of times heavier than electrons and therefore dissipate only a few hundredths of 1% or less of their energy as radiation losses. These particles, even at energies up to 100 MeV, dissipate nearly all of their energy as collisional losses.

3. Charged-Particle Tracks

A charged particle passing through matter leaves a track of secondary electrons and ionized atoms in its path. In soft tissue and materials of similar density, the tracks are typically approximately 100 μm wide, with occasionally longer side tracks generated by energetic δ rays. The tracks are studied in nuclear physics using film emulsions, cloud chambers,* and other devices.

When a heavy particle, such as an α particle, collides with an orbital electron, its direction is virtually unchanged and it loses only a small fraction of its energy (rather like a bowling ball colliding with a small lead shot). The maximum fractional energy loss by a heavy particle of mass M colliding with a light particle of mass m is approximately $4\,m/M$. For an α particle colliding with an electron, this amounts to only approximately 0.05% [$4 \times (1/1840)/4 \approx (1/2000)$]. Heavy particles also undergo relatively few bremsstrahlung-producing collisions with nuclei. As a result, their tracks tend to be straight lines, and they experience an almost continuous slowing down in which they lose small amounts of energy in a large number of individual collisions.

By contrast, electrons can undergo large-angle deflections in collisions with orbital electrons and can lose a large fraction of their energy in these collisions. These events are more like collisions between billiard balls of equal mass. Electrons also undergo occasional collisions with nuclei in which they are deflected through large angles and bremsstrahlung photons are emitted. For these reasons, electron tracks are tortuous, and their exact shape and length are unpredictable.

An additional difference between electrons and heavy particles is that for a given amount of kinetic energy, an electron travels at a much faster speed. For example, a 4-MeV α particle travels at approximately 10% of the speed of light, whereas a 1-MeV electron travels at 90% of the speed of light. As a result, an electron spends a much briefer time in the vicinity of an atom than does an α particle of similar energy and is therefore less likely to interact with the atom. Also, an electron carries only one unit of electrical charge, versus two for α particles, and thus exerts weaker forces on orbital electrons. For these reasons, electrons experience less frequent interactions and lose their energy more slowly than α particles; they are much less densely ionizing, and they travel farther before they are stopped than α particles or other heavy charged particles of similar energy.

To illustrate these differences, Figure 6-4 shows (in greatly enlarged detail) some

---

*A cloud chamber consists of a cylinder with a piston at one end and viewing windows at the other end and around the sides. The cylinder contains a water-alcohol vapor mixture under pressure. When the piston is rapidly withdrawn to suddenly decrease the pressure and temperature of the vapor, droplets of condensed liquid are formed around ionized nuclei. Ionization tracks existing in the chamber at the time thus can be observed and photographed through the viewing windows.

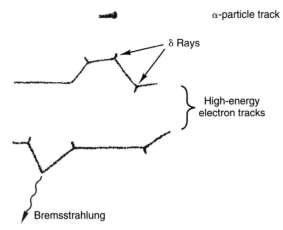

FIGURE 6-4   Representation of α particle and electron tracks in an absorber. Alpha particles leave short, straight, densely ionized tracks, whereas electron paths are tortuous and much longer; δ rays are energetic secondary electrons.

possible tracks for β particles and for α particles in water. The actual track lengths are on the order of microns for α particles and fractions of a centimeter for β particles. This is discussed further in Section B.

4. Deposition of Energy Along a Charged-Particle Track

The rate at which a charged particle loses energy determines the distance it will travel and the density of ionization along its track.

Energy loss rates and ionization densities depend on the type of particle and its energy and on the composition and density of the absorbing medium. Density affects energy loss rates because it determines the density of atoms along the particle path. In the nuclear medicine energy range ($\lesssim 10$ MeV), energy loss rates for charged particles increase linearly with the density of the absorbing medium. (At higher energies, density effects are more complicated, as discussed in the sources cited in the references and bibliography at the end of this chapter.)

Figure 6-5 shows collisional and radiation energy loss rates for electrons in the energy range of 0.01-10 MeV in water and in lead. Energy loss rates $\Delta E / \Delta x$ are expressed in MeV/g $\cdot$ cm^{-2} to normalize for density effects

$$\Delta E/\Delta x \ (\text{MeV/g} \cdot \text{cm}^{-2}) = \frac{\Delta E/\Delta x \ (\text{MeV/cm})}{\rho \ (\text{g/cm}^3)}$$

$$(6\text{-}3)$$

Thus for a given density ρ the energy loss rate in MeV/cm is given by

$$\Delta E/\Delta x \ (\text{MeV/cm}) = [\Delta E/\Delta x \ (\text{MeV/g} \cdot \text{cm}^{-2})]$$
$$\times \rho \ (\text{g/cm}^3)$$

$$(6\text{-}4)$$

Collisional loss rates $\Delta E/\Delta x_{\text{coll}}$ decrease with increasing electron energy, reflecting the velocity effect mentioned in Section A.3. Also,

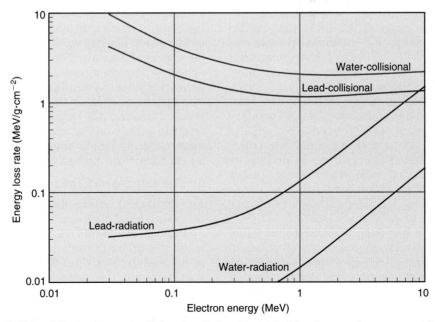

FIGURE 6-5   Collisional (ionization, excitation) and radiation (bremsstrahlung) energy losses versus electron energy in lead and in water. (*Adapted from Johns HE, Cunningham JR: The Physics of Radiology, 3rd ed. Springfield, IL, 1971, Charles C Thomas, p 47.*)

$\Delta E/\Delta x_{\text{coll}}$ decreases with increasing atomic number of the absorbing medium because in atoms of higher atomic number, inner-shell electrons are "screened" from the incident electron by layers of outer-shell electrons, making interactions with inner-shell electrons less likely in these atoms. Gram for gram, lighter elements are better absorbers of electron energy than are heavier elements.

Radiation loss rates $\Delta E/\Delta x_{\text{rad}}$ increase with increasing electron energy and increasing atomic number of the absorber. This is discussed in Section A.2.

The total energy loss rate of a charged particle, $\Delta E/\Delta x_{\text{total}}$, expressed in MeV/cm, is also called the *linear stopping power* (S_l). A closely related parameter is the *linear energy transfer* (LET), L, which refers to energy lost that is deposited "locally" along the track. L differs from S_l in that it does not include radiation losses. These result in the production of bremsstrahlung photons, which may deposit their energy at some distance from the particle track. For both electrons and α particles in the nuclear medicine energy range, however, radiation losses are small, and the two quantities S_l and L are practically identical.

The average value of the linear energy transfer measured along a charged-particle track, $\overline{L}$, is an important parameter in health physics (see Chapter 23, Section A.1.). $\overline{L}$ usually is expressed in units of keV/μm. For electrons in the energy range of 10 keV to 10 MeV traveling through soft tissue, $\overline{L}$ has values in the range of 0.2-2 keV/μm. Lower-energy electrons, for example, β particles emitted by ^{3}H ($\overline{E}_\beta = 5.6$ keV), have somewhat higher values of $\overline{L}$. Alpha particles have values of $\overline{L} \approx 100$ keV/μm.

Specific ionization (SI) refers to the total number of ion pairs produced by both primary and secondary ionization events per unit of track length along a charged particle track. The ratio of linear energy transfer divided by specific ionization is W, the *average energy expended per ionization event*:

$$W = L / \text{SI} \qquad (6\text{-}5)$$

This quantity has been measured and found to have a relatively narrow range of values in a variety of gases (25-45 eV per ion pair or, equivalently, per ionization) independent of the type or energy of the incident particle. The value for air is 33.7 eV per ion pair. W is not the same as the *ionization potential* (I), which is the average energy *required* to cause an ionization in a material (averaged over all

the electron shells). Ionization potentials for gases are in the range 10-15 eV. The difference between W and I is energy dissipated by a charged particle in nonionizing excitation events. Apparently, more than half of the energy of a charged particle is expended in this way. Similar ratios between W and I are found in semiconductor solids, except that in these materials the values of W and I are both approximately a factor of 10 smaller than for gases (see Table 7-1).

Because W does not change appreciably with particle type or energy, specific ionization is proportional to linear energy transfer L along a charged particle track. Figure 6-6 shows specific ionization in air for electrons as a function of their energy. The curve indicates that specific ionization increases with decreasing energy down to an energy of approximately 100 eV. This behavior reflects the fact that energy loss rates and L increase as the electron slows down. Below approximately 100 eV, the electron energy is inadequate to cause ionizations efficiently, and specific ionization decreases rapidly to zero.

Specific ionization values for α particles are typically 100 times greater than for electrons of the same energy because of their greater charge and much lower velocities. This leads to greater rates of energy loss, as discussed in previously in Section A.3.

The fact that specific ionization increases as a particle slows down leads to a marked increase in ionization density near the end of its track. This effect is especially pronounced for heavy particles. Figure 6-7 shows a graph of ionization density versus distance traveled for α particles in air. The peak near the end of the α-particle range is called the *Bragg ionization peak.** A similar increase in ionization density is seen at the end of an electron track; however, the peak occurs when the electron energy has been reduced to less than approximately 1 keV, and it accounts for only a small fraction of its total energy.

5. The Cerenkov Effect

An additional charged-particle interaction deserving brief mention is the *Cerenkov* (pronounced cher-en'-kof) *effect*. This effect occurs when a charged particle travels in a medium at a speed greater than the speed of light in that medium. The restriction that a particle

---

*Advantage is taken of this peak in radiation therapy using high-energy protons. The energy of protons is adjusted so that the Bragg peak occurs within the tumor or other treated tissue.

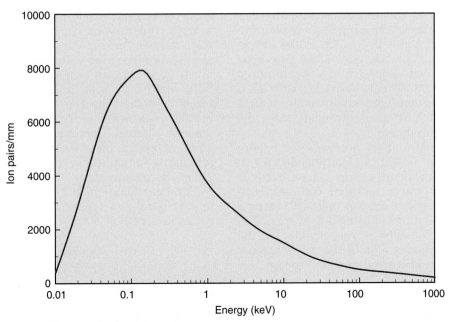

FIGURE 6-6   Specific ionization for electrons versus energy in water. (*Adapted from Mladjenovic M:* Radioisotope and Radiation Physics. *New York, 1973, Academic Press, p 145.*)

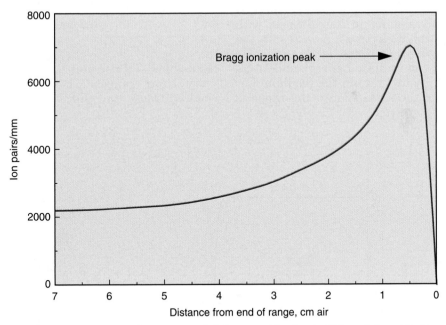

FIGURE 6-7   Specific ionization versus distance traveled for α particles in air. (*Adapted from Mladjenovic M:* Radioisotope and Radiation Physics. *New York, 1973, Academic Press, p 111.*)

cannot travel faster than the speed of light applies to the speed of light in a vacuum ($c \approx 3 \times 10^8$ m/sec); however, a 1-MeV β particle emitted in water travels with a velocity of $v \approx 0.8\,c$, whereas the speed of light in water (refractive index $n = 1.33$) is $c' = c/n \approx 0.75\,c$. Under these conditions, the particle creates an electromagnetic "shock wave" in much the same way that an airplane traveling faster than the speed of sound creates an acoustic shock wave. The electromagnetic shock wave appears as a burst of visible radiation, typically bluish in color, called *Cerenkov radiation.* The Cerenkov effect can occur for electrons with energies of a few hundred keV; however, for heavy particles such as α particles and protons, energies of several thousands of MeV are required to meet the velocity requirements.

The Cerenkov effect accounts for a very small fraction (<1%) of electron energies in the nuclear medicine energy range, but it is detectable in water solutions containing an energetic β-particle emitter (e.g., ^{32}P) using a liquid scintillation-counting apparatus. The Cerenkov effect also is responsible for the bluish glow that is seen in the water around the core of an operating nuclear reactor.

B. CHARGED-PARTICLE RANGES

1. Alpha Particles

An α particle loses energy in a more or less continuous slowing-down process as it travels through matter. The particle is deflected only slightly in its collisions with atoms and orbital electrons. As a result, the distance traveled, or *range*, of an α particle depends only on its initial energy and on its average energy loss rate in the medium. For α particles of the same energy, the range is quite consistent from one particle to the next. A transmission curve, showing percent transmission for α particles versus thickness of absorber, remains essentially flat at 100% until the maximum range is reached; then it falls rapidly to zero (Fig. 6-8). The *mean range* is defined as the thickness resulting in 50% transmission. There is only a small amount of range fluctuation, or *range straggling,* about the mean value. Typically, range straggling amounts to only approximately 1% of the mean range.

For α particles emitted in radioactive decay (E = 4-8 MeV), an approximation for the mean range in air is

$$R(\text{cm}) \approx 0.325 \, E^{3/2}(\text{MeV}) \qquad (6\text{-}6)$$

EXAMPLE 6-2

Calculate the mean range in air of α particles emitted by ^{241}Am (E_α = 5.49 MeV).

Answer

$$R(\text{cm}) = 0.325 \times (5.49)^{3/2}$$
$$= 0.325 \times (\sqrt{5.49})^3$$
$$\approx 4.2 \text{ cm}$$

Example 6-2 illustrates that α particles have very short ranges. They produce densely ionized tracks over this short range.

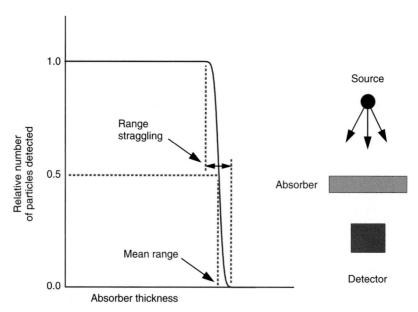

FIGURE 6-8 Relative number of particles detected versus absorber thickness in a transmission experiment with α particles. Range straggling is exaggerated for purposes of illustration.

EXAMPLE 6-3

Estimate the average value of specific ionization in air for α particles emitted by ^{241}Am.

Answer
$W = 33.7$ eV/ion pair in air. Therefore the number N of ionizations caused by an α particle of energy 5.49 MeV is

$$N = 5.49 \times 10^6 \text{ eV}/33.7 \text{ eV/ion pair}$$

$$\approx 1.63 \times 10^5 \text{ ion pairs}$$

Over a distance of travel of 4.2 cm, the average specific ionization is therefore

$$\overline{\text{SI}} \approx 1.63 \times 10^5 \text{ ion pairs}/4.2 \text{ cm}$$

$$\approx 3.9 \times 10^4 \text{ ion pairs/cm}$$

$$\approx 3.9 \times 10^3 \text{ ion pairs/mm}$$

Compare the result in Example 6-3 with the values shown in Figures 6-6 and 6-7. Only near the very end of their ranges (E & 1 keV) do electrons have specific ionizations comparable to the *average* values for α particles.

Alpha particle ranges in materials other than air can be estimated using the equation

$$R_x = R_{air}\left(\frac{\rho_{air}\sqrt{A_x}}{\rho_x\sqrt{A_{air}}}\right) \tag{6-7}$$

where R_{air} is the range of the α particle in air, ρ_{air} (=0.001293 g/cm^3) and A_{air} (≈14) are the density and (average) mass number of air, and ρ_x and A_x are the same quantities for the material of interest. This estimation is accurate to within approximately 15%.

EXAMPLE 6-4

What is the approximate mean range of α particles emitted by ^{241}Am in soft tissue? Assume that $\rho_{tissue} = 1$ g/cm^3.

Answer
The elemental compositions of air and soft tissue are similar; thus $A_{air} \approx A_x$ may be assumed. From Example 6-2, $R_{air} \approx 4.2$ cm. Therefore the approximate range in soft tissue is

$$R_{tissue} \approx 4.2 \text{ cm} \times (0.001293 \text{ g/cm}^3)/(1 \text{ g/cm}^3)$$

$$\approx 0.0054 \text{ cm}$$

$$\approx 54 \text{ μm}$$

Examples 6-2 and 6-4 illustrate that α particles have very short ranges in air as well as in soft tissue and other solid materials. The very short ranges of α particles mean that they constitute an almost negligible hazard as an external radiation source. Only a few centimeters of air, a sheet of paper, or a rubber glove provides adequate shielding protection. Even those particles that do reach the skin deliver a radiation dose only to the most superficial layers of skin. Alpha particle emitters become a radiation hazard only when ingested; then, because of their densely ionizing nature, they become very potent radiation hazards. (See Chapter 22, Section A and Chapter 23, Section A.).

2. Beta Particles and Electrons

Alpha particles travel in straight lines. Thus their path lengths (total length of path traveled) and ranges (thickness of material required to stop them) are essentially equal. This does not apply to electrons, which can undergo sharp deflections along their path or be stopped completely in a single interaction. Electron ranges are quite variable from one electron to the next, even for electrons of exactly the same energy in the same absorbing material. Path lengths are an important parameter for calculating linear energy transfer. Ranges are important for radiation dosimetry (Chapter 22) and radiation protection (Chapter 23), and for determining the limiting spatial resolution of positron imaging devices (Chapter 18). For this reason, the following discussions focus on electron ranges.

A transmission experiment with β particles results in a curve of the type illustrated in Figure 6-9. Transmission begins to decrease immediately when absorber is added because even thin absorbers can remove a few electrons by the processes mentioned earlier. When the transmission curve is plotted on a semilogarithmic scale, it follows at first a more or less straight-line decline until it gradually merges with a long, relatively flat tail. The tail of the curve does not reflect β-particle transmission but rather represents the detection of relatively penetrating bremsstrahlung photons generated by the β particles in the absorber and possibly in the source and source holder. Extraneous instrument and radiation background also may contribute to the tail of the curve.

The thickness of absorber corresponding to the intersection between the extrapolation of the linearly descending portion and the tail of

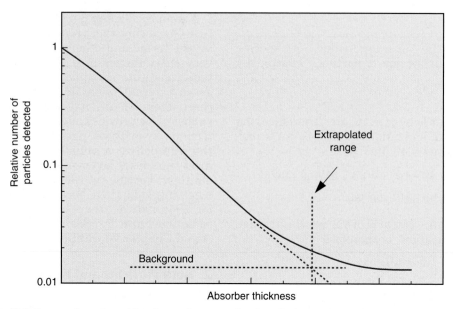

FIGURE 6-9   Relative number of particles detected versus absorber thickness in an electron absorption experiment. Compare with Figure 6-8.

the curve is called the *extrapolated range* R_e of the electrons. This is slightly less (perhaps by a few percent) than the maximum range R_m, which is the actual maximum thickness of absorber penetrated by the maximum energy β particles (Fig. 6-9); however, because the difference is small and because R_m is very difficult to measure, R_e usually is specified as the maximum β-particle range.

The extrapolated range for a monoenergetic beam of electrons of energy E is the same as that for a beam of β particles of maximum energy $E_\beta^{max} = E$. In both cases, range is determined by the maximum energy of electrons in the beam. The shapes of the transmission curves for monoenergetic electrons and for β particles are somewhat different, however. Specifically, the curve for β particles declines more rapidly for very thin absorbers because of rapid elimination of low-energy electrons in the β-particle energy spectrum (See Fig. 3-2).

Extrapolated ranges are found to be inversely proportional to the density ρ of the absorbing material. To normalize for density effects, electron ranges usually are expressed in g/cm² of absorber. This is the weight of a 1-cm² section cut from a thickness of an absorber equal to the range of electrons in it. The range in cm and in g/cm² are related according to

$$R_e(\text{g/cm}^2) = R_e(\text{cm}) \times \rho(\text{g/cm}^3)$$
$$R_e(\text{cm}) = R_e(\text{g/cm}^2)/\rho(\text{g/cm}^3)$$

(6-8)

It also is found that extrapolated ranges in different elements, when expressed in g/cm², are practically identical. There are small differences in electron energy loss rates in different elements, as discussed in Section A.4, but they have only a small effect on total ranges.

Figure 6-10 shows a curve for the extrapolated range of electrons in water, in centimeters, versus electron energy (or maximum β-particle energy, E_β^{max}). Because the density of water is 1, this curve is numerically equal to the extrapolated range in g/cm² of water, which has the same value for all absorbers. It can be used to determine extrapolated ranges for other absorbers by dividing by the absorber density, as indicated in Equation 6-8.

EXAMPLE 6-5

Using Figure 6-10, determine the range of 1-MeV electrons in air ($\rho = 0.001293$ g/cm³) and lead ($\rho = 11.3$ g/cm³).

Answer

From Figure 6-10, the range of a 1-MeV electron in water ($\rho = 1$ g/cm³) is 0.4 cm, or 0.4 g/cm². Thus

$$R_e(\text{air}) = (0.4 \text{ g/cm}^2)/0.001293 \text{ g/cm}^3$$
$$\approx 309 \text{ cm}$$
$$R_e(\text{lead}) = (0.4 \text{ g/cm}^2)/11.3 \text{ g/cm}^3$$
$$\approx 0.035 \text{ cm}$$

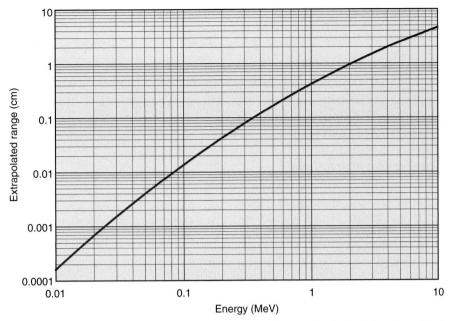

FIGURE 6-10 Extrapolated range in water versus electron energy. The curve is derived from Equation 3-3 in Chapter 21 of reference 1. Curve applies to other absorbers by dividing range in water by absorber density, ρ, in g/cm³.

Example 6-5 illustrates that extrapolated ranges can be several meters in air but that they are only a few millimeters or fractions of a millimeter in solid materials or liquids. Some ranges for β particles emitted by radionuclides of medical interest are summarized in Table 6-1.

The *average range* for electrons is the thickness required to stop 50% of an electron beam. From Figure 6-9, it is evident that this is much smaller than the extrapolated range.

It is found experimentally that the average range for a β-particle beam is given by[1]

$$\bar{D}_{1/2}(\text{cm}) \approx 0.108 \times [E_\beta^{max}]^{1.14}/\rho(\text{g/cm}^3) \qquad (6\text{-}9)$$

where E_β^{max} is the maximum energy of the β particles in MeV and ρ is the density of the absorbing material. The average range of positrons plays a significant role in imaging with positron-emitting radionuclides, where it places a fundamental limit on obtainable

TABLE 6-1
BETA-PARTICLE RANGES FOR SOME COMMONLY USED β^+ AND β^- EMITTERS*

Radionuclide	E_β^{max} (MeV)	Extrapolated Range (cm) in			Average Range (cm) in
		Air	Water	Aluminum	Water
³H	0.0186	4.5	0.00059	0.00022	—
¹¹C	0.961	302	0.39	0.145	0.103
¹⁴C†	0.156	21.9	0.028	0.011	0.013
¹³N	1.19	395	0.51	0.189	0.132
¹⁵O	1.723	617	0.80	0.295	0.201
¹⁸F	0.635	176	0.23	0.084	0.064
³²P	1.70	607	0.785	0.290	0.198
⁸²Rb	3.35	1280	1.65	0.612	0.429

*Extrapolated and average ranges calculated from Equations 3-3 and 3-7, respectively, in Chapter 21 of reference 1.
†Ranges for ³⁵S ($E_\beta^{max} = 0.167$ MeV) are nearly the same as those for ¹⁴C.

spatial resolution. This is discussed in Chapter 18, Section A.4. Average ranges in water (ρ = 1) also are listed in Table 6-1. Average ranges in soft tissue are essentially the same as for water.

C. PASSAGE OF HIGH-ENERGY PHOTONS THROUGH MATTER

1. Photon Interaction Mechanisms

High-energy photons (γ rays, x rays, annihilation radiation, and bremsstrahlung) transfer their energy to matter in complex interactions with atoms, nuclei, and electrons. For practical purposes, however, these interactions can be viewed as simple collisions between a photon and a target atom, nucleus, or electron. These interactions do not cause ionization directly, as do the charged-particle interactions; however, some of the photon interactions result in the ejection of orbital electrons from atoms or in the creation of positive-negative electron pairs. These electrons in turn cause ionization effects, which are the basis for mechanisms by which high-energy photons are detected and by which they cause radiobiologic effects. For these reasons, high-energy photons are classified as *secondary ionizing radiation*.

There are nine possible interactions between photons and matter, of which only four are of significance to nuclear medicine. These four interactions, and mathematical aspects of the passage of photon beams through matter, are discussed.

2. The Photoelectric Effect

The *photoelectric effect* is an atomic absorption process in which an atom absorbs totally the energy of an incident photon. The photon disappears and the energy absorbed is used to eject an orbital electron from the atom. The ejected electron is called a *photoelectron*. It receives kinetic energy E_{pe}, equal to the difference between the incident photon energy E_0 and the binding energy of the electron shell from which it was ejected. For example, if a K-shell electron is ejected, the kinetic energy of the photoelectron is

$$E_{pe} = E_0 - K_B \qquad (6\text{-}10)$$

where K_B is the K-shell binding energy for the atom from which it is ejected (see Chapter 2, Section C.2). The photoelectric effect looks like a "collision" between a photon and an orbital electron in which the electron is ejected

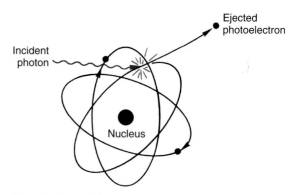

FIGURE 6-11 Schematic representation of the photoelectric effect. The incident photon transfers its energy to a photoelectron and disappears.

from the atom and the photon disappears (Fig. 6-11).

Photoelectrons cannot be ejected from an electron shell unless the incident photon energy exceeds the binding energy of that shell. (Values of K-shell binding energies for the elements are listed in Appendix B.) If sufficient photon energy is available, the photoelectron is most likely to be ejected from the innermost possible shell. For example, ejection of a K-shell electron is four to seven times more likely than ejection of an L-shell electron when the energy requirement of the K shell is met, depending on the absorber element.

The photoelectric effect creates a vacancy in an orbital electron shell, which in turn leads to the emission of characteristic x rays (or Auger electrons). In low-Z elements, binding energies and characteristic x-ray energies are only a few keV or less. Thus binding energy is a small factor in photoelectric interactions in body tissues. In heavier elements, however, such as iodine or lead, binding energies are in the 20- to 100-keV range, and they may account for a significant fraction of the absorbed photon energy.

The kinetic energy imparted to the photoelectron is deposited near the site of the photoelectric interaction by the ionization and excitation interactions of high-energy electrons described in Section A. Extrapolated ranges for photoelectrons of various energies in soft tissue can be determined from Figure 6-10.

3. Compton Scattering

Compton scattering is a "collision" between a photon and a loosely bound outer-shell orbital electron of an atom. In Compton scattering, because the incident photon energy greatly exceeds the binding energy of the electron to

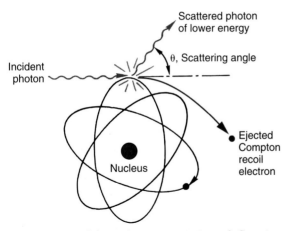

FIGURE 6-12 Schematic representation of Compton scattering. The incident photon transfers part of its energy to a Compton recoil electron and is scattered in another direction of travel (θ, scattering angle).

the atom, the interaction looks like a collision between the photon and a "free" electron (Fig. 6-12).

The photon does not disappear in Compton scattering. Instead, it is deflected through a scattering angle θ. Part of its energy is transferred to the *recoil electron*; thus the photon loses energy in the process. The energy of the scattered photon is related to the scattering angle θ by considerations of energy and momentum conservation according to[*]

$$E_{sc} = E_0/[1 + (E_0/0.511)(1 - \cos\theta)] \quad (6\text{-}11)$$

where E_0 and E_{sc} are the incident and scattered photon energies in MeV, respectively. The energy of the recoil electron, E_{re}, is thus

$$E_{re} = E_0 - E_{sc} \quad (6\text{-}12)$$

---

[*]Derivations of Compton energy-angle relationships can be found in Chapter 23 of reference 1.

The energy transferred does not depend on the density, atomic number, or any other property of the absorbing material. Compton scattering is strictly a photon-electron interaction.

The amount of energy transferred to the recoil electron in Compton scattering ranges from nearly zero for θ ≈ 0 degrees ("grazing" collisions) up to a maximum value E_{re}^{max} that occurs in 180-degree *backscattering events*. The minimum energy for scattered photons, E_{sc}^{min}, also occurs for 180-degree backscattering events. The minimum energy of Compton-scattered photons can be calculated from Equation 6-11 with θ = 180 degrees (cos 180° = −1):

$$E_{sc}^{min} = E_0/[1 + (2E_0/0.511)] \quad (6\text{-}13)$$

Thus

$$E_{re}^{max} = E_0 - E_{sc}^{min}$$
$$= E_0\left[1 - \frac{1}{[1 + (2E_0/0.511)]}\right] \quad (6\text{-}14)$$
$$= E_0^2/(E_0 + 0.2555)$$

The minimum energy of backscattered photons, E_{sc}^{min}, and the maximum energy transferred to the recoil electron, E_{re}^{max}, have characteristic values that depend on E_0, the energy of the incident photon. These energies are of interest in pulse-height spectrometry because they result in characteristic structures in pulse-height spectra (see Chapter 10, Section B.1).

Table 6-2 lists some values of E_{sc}^{min} and E_{re}^{max} for some γ-ray and x-ray emissions from radionuclides of interest in nuclear medicine. Note that for relatively low photon energies (e.g., ^{125}I, 27.5 keV), the recoil electron receives only a small fraction of the incident

TABLE 6-2
SCATTERED PHOTON AND RECOIL ELECTRON ENERGIES FOR 180-DEGREE COMPTON SCATTERING INTERACTIONS

Radionuclide	Photon Energy (keV)	E_{sc}^{min} (keV)	E_{re}^{max} (keV)
^{125}I	27.5	24.8	2.7
^{133}Xe	81	62	19
^{99m}Tc	140	91	49
^{131}I	364	150	214
β⁺ (annihilation)	511	170	341
^{60}Co	1330	214	1116
—	∞	255.5	—

photon energy, even in 180-degree scattering events. Thus photon energy changes very little in Compton scattering at low photon energies. The smallness of this energy change has important implications for the elimination of Compton-scattered photons by energy-discrimination techniques (See Fig. 10-10). At higher energies the energy distribution changes and E_{sc}^{min} approaches a maximum value of 255.5 keV. The remaining energy, which now accounts for most of the incident photon energy, is transferred to the recoil electron in 180-degree scattering events. Note also that the energy of Compton-scattered photons never is zero—that is, a photon cannot transfer all its energy to an electron in a Compton scattering event.

The angular distribution of Compton-scattered photons also depends on the incident photon energy. Figure 6-13 shows the relative probability of scattering at different angles per unit of solid angle. Solid angle is proportional to the area subtended on a sphere divided by the total area of the sphere (see also Fig. 11-1). Thus Figure 6-13 reflects the relative number of scattered photons that would be recorded by a detector of fixed area as it was moved about at a fixed distance from the scattering object at different angles relative to the incident beam (in the absence of attenuation, secondary scattering, etc). At relatively low energies (10-100 keV) the highest intensity of Compton-scattered

photons would be detected in either the forward or backward direction, with a minimum at right angles (90 degrees) to the direction of the incident photons. At higher energies ($\gg 0.5$ MeV), the highest intensity detected would be increasingly toward the forward direction (scattering angle ~0°).

4. Pair Production

Pair production occurs when a photon interacts with the electric field of a charged particle. Usually the interaction is with an atomic nucleus, but occasionally it is with an electron. In pair production, the photon disappears and its energy is used to create a positron-electron pair (Fig. 6-14). Because the positron and electron both have a rest mass equivalent to 0.511 MeV, a minimum photon energy of 2×0.511 MeV = 1.022 MeV must be available for pair production to occur. The difference between the incident photon energy E_0 and the 1.022 MeV of energy required to create the electron pair is imparted as kinetic energy to the positron (E_{e^+}) and the electron, (E_{e^-})

$$E_{e^+} + E_{e^-} = E_0 - 1.022 \text{ MeV} \qquad (6-15)$$

The energy sharing between the electron and positron is more or less random from one interaction to the next, usually within the 20% to 80% sharing range.

The electron and positron dissipate their kinetic energy primarily in ionization and excitation interactions. When the positron

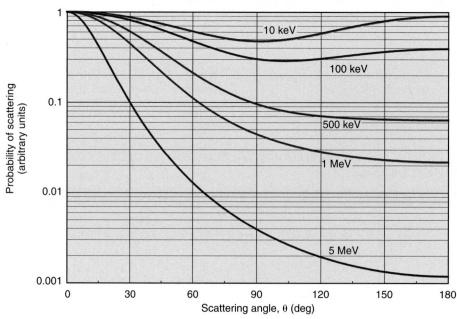

FIGURE 6-13 Relative probability of Compton scattering (arbitrary units) per unit of solid angle versus scattering angle θ for different incident photon energies.

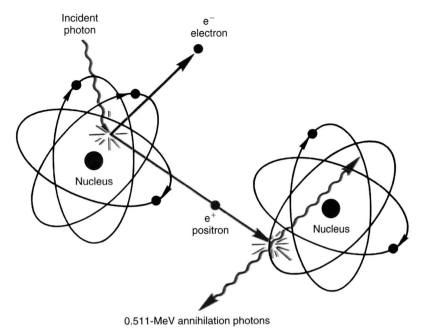

FIGURE 6-14 Schematic representation of pair production. Energy of incident photon is converted into an electron and a positron (total 1.022-MeV mass-energy equivalent) plus their kinetic energy. The positron eventually undergoes mutual annihilation with a different electron, producing two 0.511-MeV annihilation photons.

has lost its kinetic energy and stopped, it undergoes mutual annihilation with a nearby electron, and a pair of 0.511-MeV *annihilation photons* are emitted in opposite directions from the site of the annihilation event (see Chapter 3, Section G). Annihilation photons usually travel for some distance before interacting again. Thus usually only the kinetic energy of the electron and positron (Equation 6-15) is deposited at the site of the pair production event.

5. Coherent (Rayleigh) Scattering

Coherent or *Rayleigh scattering* is a type of scattering interaction that occurs between a photon and an atom as a whole. Because of the great mass of an atom (e.g., in comparison to the recoil electron in the Compton scattering process), very little recoil energy is absorbed by the atom. The photon is therefore deflected with essentially no loss of energy.

Coherent scattering is important only at relatively low energies ($\ll 50$ keV). It can be of significance in some precise photon transmission measurements—for example, in x-ray computed tomographic scanning—because it is a mechanism by which photons are removed from a photon beam. Coherent scattering also is an important interaction in x-ray crystallography; however, because it is not an effective mechanism for transferring photon

energy to matter, it is of little practical importance in nuclear medicine.

6. Deposition of Photon Energy in Matter

The most important interactions in the transfer of photon energy to matter are the photoelectric effect, Compton scattering, and pair production. The transfer of energy occurs typically in a series of these interactions in which energy is transferred to electrons, and, usually, secondary photons, of progressively less energy (Fig. 6-15). The products of each interaction are secondary photons and high-energy electrons (Table 6-3). The high-energy

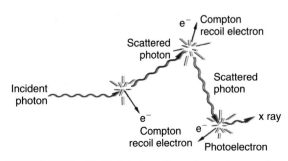

FIGURE 6-15 Multiple interactions of a photon passing through matter. Energy is transferred to electrons in a sequence of photon-energy degrading interactions.

electrons ultimately are responsible for the deposition of energy in matter. Ionization and excitation by these electrons are the mechanisms underlying all of the photon detectors described in Chapter 7. The electrons also are responsible for radiobiologic effects caused by γ-ray, x-ray, or bremsstrahlung radiation. Because of this, the average linear energy transfer of photons for radiobiologic purposes is the same as for electrons of similar energy, that is, 0.2-2 keV/μm (see Chapter 23).

TABLE 6-3
PRODUCTS OF THE THREE MAJOR PHOTON INTERACTION PROCESSES

Interaction	Secondary Photon(s)	High-Energy Secondary Electron(s)
Photoelectric	Characteristic x rays	Photoelectrons
		Auger electrons
Compton	Scattered photon	Recoil electron
Pair production	Annihilation photons	Positive-negative electron pair

D. ATTENUATION OF PHOTON BEAMS

1. Attenuation Coefficients

When a photon passes through a thickness of absorber material, the probability that it will experience an interaction depends on its energy and on the composition and thickness of the absorber. The dependence on thickness is relatively simple; the thicker the absorber, the greater the probability that an interaction will occur. The dependence on absorber composition and photon energy, however, is more complicated.

Consider the photon transmission measurement diagrammed in Figure 6-16. A beam of photons of intensity I (photons/cm$^2 \cdot$ sec) is directed onto an absorber of thickness Δx. Because of composition and photon energy effects, it will be assumed for the moment that the absorber is composed of a single element of atomic number Z and that the beam is monoenergetic with energy E. A photon detector records transmitted beam intensity. It is assumed that only those photons passing through the absorber without interaction are recorded. (The validity of this assumption is discussed further in Sections D.2 and D.3.)

For a "thin" absorber, such that beam intensity is reduced by only a small amount ($\lesssim 10\%$), it is found that the fractional decrease in beam intensity ($\Delta I/I$) is related to absorber thickness Δx according to

$$\Delta I/I \approx -\mu_l \times \Delta x \qquad (6\text{-}16)$$

the minus sign indicating beam intensity decreases with increasing absorber thickness. The quantity μ_l is called the *linear attenuation coefficient* of the absorber. It has dimensions (thickness)$^{-1}$ and usually is expressed in cm^{-1}. This quantity reflects the "absorptivity" of the absorbing material.

The quantity μ_l is found to increase linearly with absorber density ρ. Density effects are factored out by dividing μ_l by density ρ:

$$\mu_m = \mu_l/\rho \qquad (6\text{-}17)$$

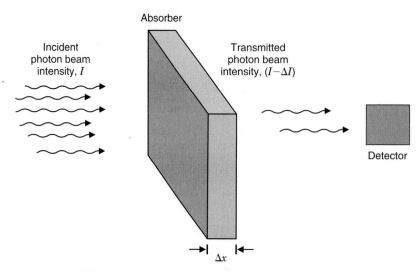

FIGURE 6-16 Photon-beam transmission measurement.

The quantity μ_m has dimensions of cm^2/g and is called the *mass attenuation coefficient* of the absorber. It depends on the absorber atomic number Z and photon energy E. This sometimes is emphasized by writing it as $\mu_m(Z, E)$.

It is possible to measure μ_m or μ_l in different absorber materials by transmission measurements with monoenergetic photon beams. Most tables, however, are based on theoretical calculations from atomic and nuclear physics. An extensive tabulation of values of μ_m versus photon energy for different absorber materials is found in reference 2. Some values of interest to nuclear medicine, taken from these tables, are presented in Appendix D. Usually, values of μ_m rather than μ_l are tabulated because μ_m does not depend on the physical state (density) of the absorber. Given a value of μ_m from the tables, μ_l for an absorber can be obtained from

$$\mu_l(cm^{-1}) = \mu_m(cm^2/g) \times \rho(g/cm^3) \quad (6\text{-}18)$$

The mass attenuation coefficient for a *mixture of elements* can be obtained from the values for its component elements according to

$$\mu_m(mix) = \mu_{m,1}f_1 + \mu_{m,2}f_2 + \cdots \quad (6\text{-}19)$$

where $\mu_{m,1}$, $\mu_{m,2}$ ... are the mass attenuation coefficients for elements 1, 2, ..., and $f_1, f_2, ...,$ are the fractions by weight of these elements in the mixture. For example, the mass attenuation coefficient for water (2/18 H, 16/18 O, by weight) is given by

$$\mu_m(water) = (2/18)\mu_m(H) + (16/18)\mu_m(O) \quad (6\text{-}20)$$

The mass attenuation coefficient μ_m can be broken down into components according to

$$\mu_m = \tau + \sigma + \kappa \quad (6\text{-}21)$$

where τ is that part of μ_m caused by the photoelectric effect, σ is the part caused by Compton scattering, and κ is the part caused by pair production. Thus, for example, τ would be the mass attenuation coefficient of an absorber in the absence of Compton scattering and pair production. Note that μ_m involves both absorption and scattering processes. Thus μ_m is properly called an *attenuation coefficient* rather than an *absorption coefficient*.

The relative magnitudes of τ, σ, and κ vary with atomic number Z and photon energy E. Figure 6-17 shows graphs of μ_m and its components, τ, σ, and κ versus photon energy from 0.01-10 MeV in water, NaI(Tl), and lead. The following points are illustrated by these graphs:

1. The photoelectric component τ decreases rapidly with increasing photon energy and increases rapidly with increasing atomic number of the absorber ($\tau \approx Z^3/E^3$). The photoelectric effect is thus the dominating effect in heavy elements at low photon energies. The photoelectric component also increases abruptly at energies corresponding to orbital electron binding energies of the absorber elements. At the K-shell binding energies of iodine ($K_B = 33.2$ keV) and lead ($K_B = 88.0$ keV), the increase is a factor of 5-6. These abrupt increases are called *K absorption edges*. They result from the fact that photoelectric absorption involving K-shell electrons cannot occur until the photon energy exceeds the K-shell binding energy. L absorption edges also are seen at $E \sim 13\text{-}16$ keV in the graph for lead. L absorption edges in water and iodine and the K absorption edge for water also exist, but they occur at energies less than those shown in the graphs.

2. The Compton-scatter component σ decreases slowly with increasing photon energy E and with increasing absorber atomic number Z. The changes are so small that for practical purposes σ usually is considered to be invariant with Z and E. Compton scattering is the dominating interaction for intermediate values of Z and E.

3. The pair-production component κ is zero for photon energies less than the threshold energy of 1.02 MeV for this interaction; then it increases logarithmically with increasing photon energy and with increasing atomic number of the absorber ($\kappa \approx Z \log E$). Pair production is the dominating effect at higher photon energies in absorbers of high atomic number.

Figure 6-18 shows the dominating (most probable) interaction versus photon energy E and absorber atomic number Z. Note that Compton scattering is the dominating interaction for $Z \lesssim 20$ (body tissues) over most of the nuclear medicine energy range.

2. Thick Absorbers, Narrow-Beam Geometry

The transmission of a photon beam through a "thick" absorber—that is, one in which the probability of photon interaction is not "small" ($\gtrsim 10\%$)—depends on the geometric

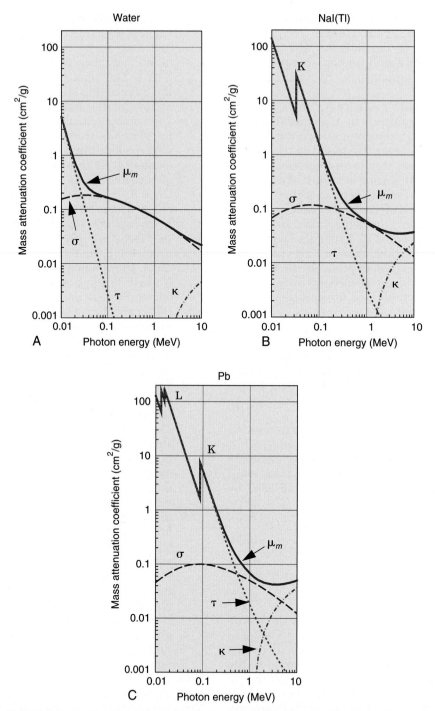

FIGURE 6-17 Photoelectric (τ), Compton (σ), pair-production (κ), and total (μ_m) mass attenuation coefficients (square centimeters per gram) for water (*A*), NaI(Tl) (*B*), and Pb (*C*) from 0.01 to 10 MeV. K and L are absorption edges. Data taken from reference 2. Curves for μ_l can be obtained by multiplying by the appropriate density values.

arrangement of the photon source, absorber, and detector. Specifically, transmission depends on whether scattered photons are recorded as part of the transmitted beam. An arrangement that is designed to minimize the recording of scattered photons is called *narrow-beam* *geometry.* Conversely, an arrangement in which many scattered photons are recorded is called *broad-beam geometry.* (They also are called *good geometry* and *poor geometry*, respectively.) Figure 6-19 shows examples of these geometries.

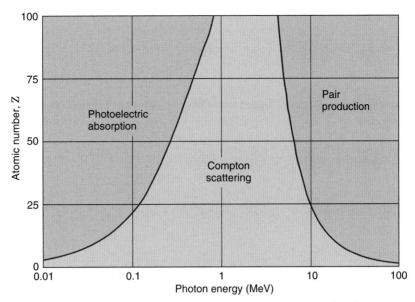

FIGURE 6-18 Predominating (most probable) interaction versus photon energy for absorbers of different atomic numbers. Curves were generated using values obtained from reference 2.

Conditions of narrow-beam geometry usually require that the beam be collimated with a narrow aperture at the source so that only a narrow beam of photons is directed onto the absorber. This minimizes the probability that photons will strike neighboring objects (e.g., the walls of the room or other parts of the measurement apparatus) and scatter toward the detector. Matching collimation on the detector helps prevent photons that are multiple-scattered in the absorber from being recorded. In addition, it is desirable to place the absorber approximately halfway between the source and the detector.

Under conditions of narrow-beam geometry, the transmission of a monoenergetic photon beam through an absorber is described by an exponential equation

$$I(x) = I(0) \, e^{-\mu_l x} \qquad (6\text{-}22)$$

where $I(x)$ is the beam intensity transmitted through a thickness x of absorber, $I(0)$ is the intensity recorded with no absorber present, and μ_l is the linear attenuation coefficient of the absorber at the photon energy of interest. In contrast to charged particles, photons do not have a definite maximum range. There is always some finite probability that a

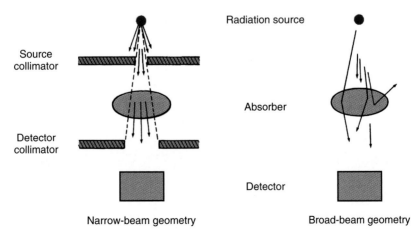

FIGURE 6-19 Narrow-beam and broad-beam geometries for photon-beam attenuation measurements. Narrow-beam geometry is designed to minimize the number of scattered photons recorded.

photon will penetrate even the thickest absorber [i.e., $I(x)$ in Equation 6-22 never reaches zero].

Equation 6-22 is exactly analogous to Equation 4-6 for the decay of radioactivity, with the attenuation coefficient μ_l replacing the decay constant λ and absorber thickness x replacing decay time t. Analogous to the concept of half-life in radioactive decay, the thickness of an absorber that decreases recorded beam intensity by one half is called the *half-value thickness* (HVT) or *half-value layer* (HVL). It is related to the linear attenuation coefficient according to

$$HVT = \ln 2/\mu_l$$
$$\mu_l = \ln 2/HVT \qquad (6\text{-}23)$$

where $\ln 2 \approx 0.693$. Compare these equations with Equations 4-8 and 4-9.

Some radiation-shielding problems require the use of relatively thick absorbers; for this purpose it is sometimes useful to know the *tenth-value thickness* (TVT)—that is, the thickness of absorber that decreases transmitted beam intensity by a factor of 10. This quantity is given by

$$TVT = \ln 10/\mu_l$$
$$\approx 3.32 \times HVT \qquad (6\text{-}24)$$

where $\ln 10 \approx 2.30$. Some HVTs for water and TVTs for lead are listed in Table 6-4.

The quantity

$$X_m = 1/\mu_l \qquad (6\text{-}25)$$

is called the *mean free path* for photons in an absorber. It is the *average distance* traveled by a photon in the absorber before experiencing an interaction. Mean free path is related to HVT according to

$$X_m = HVT/\ln 2$$
$$\approx 1.44 \times HVT \qquad (6\text{-}26)$$

Note the analogy to average lifetime, τ (Equation 4-12).

Table 6-5 compares mean free paths for photons in water against maximum ranges for electrons in water and α particles in air as

TABLE 6-4
HALF-VALUE THICKNESSES IN WATER AND TENTH-VALUE THICKNESSES IN LEAD (NARROW-BEAM CONDITIONS)

Radionuclide	Photon Energy (keV)	HVT in Water (cm)	TVT in Lead (mm)
^{125}I	27.5	1.7	0.06
^{133}Xe	81	4.3	1.0
^{99m}Tc	140	4.6	0.9
^{131}I	364	6.3	7.7
β^+ (annihilation)	511	7.1	13.5
^{60}Co	1330	11.2	36.2

HVT, half-value thickness; TVT, tenth-value thickness.

TABLE 6-5
COMPARISON OF PHOTON MEAN FREE PATHS AND MAXIMUM ELECTRON AND α-PARTICLE RANGES

Photon or Particle Energy (MeV)	Photon MFP (cm H_2O)	Electron Range (cm H_2O)	α-Particle Range (cm air)
0.01	0.20	0.00016	—
0.1	5.95	0.014	0.1
1	14.14	0.41	0.5
10	45.05	4.6	10.3

MFP, mean free path.

a function of their energy. Although the concepts of photon mean free path and charged particle ranges are different, the comparison gives an indication of relative penetration of photons versus particle radiation. Over the energy range 0.01-10 MeV, photons are much more penetrating than electrons or α particles. For this reason they sometimes are called *penetrating* radiation.

The quantity $e^{-\mu_l x}$ [or $I(x)]/I(0)$ in Equation 6-22], the fraction of beam intensity transmitted by an absorber, is called its *transmission factor*. The transmission factor can be determined using the methods described for determining decay factors in Chapter 4, Section C. For example, the graph shown in Figure 4-3 can be used with "decay factor" replaced by "transmission factor" and "number of half-lives" replaced by "number of HVTs."

EXAMPLE 6-6

Determine the transmission factor for 140-keV photons in 10 cm of soft tissue (water) by direct calculation.

Answer
From Table 6-4, HVT = 4.6 cm in water at 140 keV. Thus $\mu_l = 0.693/4.6 = 0.151$ cm^{-1}, and the transmission factor is

$$I(10)/I(0) = e^{-0.151 \times 10} = e^{-1.51}$$

Using a pocket calculator

$$e^{-1.51} = 0.221$$

Thus the transmission factor for 140-keV photons through 10 cm of water is 22.1%.

EXAMPLE 6-7

Estimate the transmission factor for 511-keV photons in 1 cm of lead using graphical methods (see Fig. 4-3).

Answer
From Table 6-4, the TVT of 511-keV photons in lead is 13.5 mm. From Equation 6-24, HVT ≈ TVT/3.32 so for 511-keV photons in lead, HVT = 1.35 cm/3.32 = 0.4 cm. Thus 1 cm = 2.5 HVTs. From Figure 4-3, the transmission (decay) factor for 2.5 HVTs ($T_{1/2}$) is approximately 0.18 (18% transmission).

It must be remembered that the answers obtained in Examples 6-6 and 6-7 apply only to narrow-beam conditions. Broad-beam conditions are discussed in the following section.

3. Thick Absorbers, Broad-Beam Geometry

Practical problems of photon-beam attenuation in nuclear medicine usually involve broad-beam conditions. Examples are the shielding of radioactive materials in lead containers and the penetration of body tissues by photons emitted from radioactive tracers localized in internal organs. In both of these examples, a considerable amount of scattering occurs in the absorber material surrounding or overlying the radiation source.

The factor by which transmission is increased in broad-beam conditions, relative to narrow-beam conditions, is called the *buildup factor B*. Thus the transmission factor T for broad-beam conditions is given by

$$T = Be^{-\mu_l x} \qquad (6\text{-}27)$$

where μ_l and x are the linear attenuation coefficient and thickness, respectively, of the absorber.

Buildup factors for various source-absorber-detector geometries have been calculated. Some values for water and lead for a source embedded in or surrounded by scattering and absorbing material are listed in Table 6-6. Note that B depends on photon energy and on the product $\mu_l x$ for the absorber.

EXAMPLE 6-8

In Example 6-7, the transmission factor for 511-keV photons in 1 cm of lead was found to be 18% for narrow-beam conditions. Estimate the actual transmission for broad-beam conditions (e.g., a vial of β⁺-emitting radioactive solution in a lead container of 1-cm wall thickness).

Answer
For 511-keV photons, HVT = 0.4 cm (Example 6-7). Thus $\mu_l = 0.693/(0.4$ cm$) \approx 1.73$ cm^{-1}, and, for $x = 1$ cm, $\mu_l x \approx 1.73$. Taking values for 0.5 MeV (≈ 511 keV) from Table 6-6 and using linear interpolation between values for $\mu_l x = 1$ ($B = 1.24$) and $\mu_l x = 2$ ($B = 1.39$), one obtains

$$B = 1.24 + (0.73)(1.39 - 1.24)$$

$$= 1.35$$

For $B = 1.35$, the transmission in broad-beam conditions is 35% greater than calculated for narrow-beam conditions. Thus the actual transmission factor is

$$T \approx 1.35 \times 0.18$$

$$\approx 0.24$$

or 24%.

TABLE 6-6
EXPOSURE BUILDUP FACTORS IN WATER AND IN LEAD*

Material	Photon Energy (MeV)	$\mu_r x$						
		1	2	4	7	10	15	20
Water	0.1	4.55	11.8	41.3	137	321	938	2170
	0.5	2.44	4.88	12.8	32.7	62.9	139	252
	1.0	2.08	3.62	7.68	15.8	26.1	47.7	74.0
	2.0	1.83	2.81	4.98	8.65	12.7	20.1	28.0
	4.0	1.63	2.24	3.46	5.30	7.16	10.3	13.4
	6.0	1.51	1.97	2.84	4.12	5.37	7.41	9.42
	10.0	1.37	1.68	2.25	3.07	3.86	5.19	6.38
Lead	0.5	1.24	1.39	1.62	1.88	2.10	2.39	2.64
	1.0	1.38	1.68	2.19	2.89	3.51	4.45	5.27
	2.0	1.40	1.76	2.52	3.74	5.07	7.44	9.08
	4.0	1.36	1.67	2.40	3.79	5.61	9.73	15.4
	6.0	1.42	1.73	2.49	4.13	6.61	13.7	26.6
	10.0	1.51	2.01	3.42	7.37	15.4	50.8	161

*Data taken from Schleien B (ed): *The Health Physics and Radiological Health Handbook*. Silver Spring, MD, 1992, Scinta.[3]

Example 6-8 illustrates that scatter effects can be significant in broad-beam conditions. The thickness of lead shielding required to achieve a given level of protection is greater than that calculated using narrow-beam equations.

EXAMPLE 6-9

Estimate the thickness of lead shielding required to achieve an actual transmission of 18% in the problem described in Example 6-8.

Answer
Because $B = 1.35$, it is necessary to further reduce transmission by approximately 1/1.35 ≈ 0.74 to correct for scattered radiation. According to Figure 4-3, this would require approximately 0.45 HVTs, or approximately 0.18 cm (1 HVT = 0.4 cm). This is only an estimate, because the HVT used applies to narrow-beam conditions. A more exact answer could be obtained by successive approximations.

Broad-beam conditions also arise in problems of internal radiation dosimetry—for example, when it is desired to calculate the radiation dose to an organ delivered by a radioactive concentration in another organ.

This issue is discussed further in Chapter 22, Section B.

4. Polyenergetic Sources

Many radionuclides emit photons of more than one energy. The photon transmission curve for such an emitter consists of a sum of exponentials, one component for each of the photon energies emitted. The transmission curve has an appearance similar to the decay curve for a mixed radionuclide sample shown in Figure 4-5. The transmission curve drops steeply at first as the lower-energy ("softer") components of the beam are removed. Then it gradually flattens out, reflecting greater penetration by the higher-energy ("harder") components of the beam. The average energy of photons remaining in the beam increases with increasing absorber thickness. This effect is called *beam hardening*.

It is possible to detect small amounts of a high-energy photon emitter in the presence of large amounts of a low-energy photon emitter by making use of the beam-hardening effect. For example, a 3-mm thickness of lead is several TVTs for the 140-keV γ rays of ^{99m}Tc, but it is only approximately 1 HVT for the 700- to 800-keV γ rays of ^{99}Mo. Thus a

3-mm-thick lead shield placed around a vial containing ^{99m}Tc solution permits detection of small amounts of ^{99}Mo contamination with minimal interference from the ^{99m}Tc γ rays. (See Chapter 5, Section C).

REFERENCES

1. Evans RD: *The Atomic Nucleus*, New York, 1972, McGraw-Hill, p 628. (Note: This reference contains useful discussions of many details of radiation interactions with matter.)
2. Berger MJ, Hubbell JH: XCOM: Photon Cross-Sections Database, NIST Standard Reference Database 8 (XGAM) Available at: http://www.nist.gov/pml/data/xcom/index.cfm. (Accessed August 17, 2011.).
3. Schleien B, editor: *The Health Physics and Radiological Health Handbook*, Silver Spring, MD, 1992, Scinta, pp 176-177. (Note: This reference also contains useful tabulations of charged-particle ranges and other absorption data.)

BIBLIOGRAPHY

Discussions of radiation interactions and their passage through matter are found in the following:

Johns HE, Cunningham JR: *The Physics of Radiology*, ed 4, Springfield, IL, 1983, Charles C Thomas, Chapter 6.

Lapp RE, Andrews HL: *Nuclear Radiation Physics*, ed 4, Englewood Cliffs, NJ, 1972, Prentice-Hall, pp 196-203, 233-247, 261-279.

A comprehensive tabulation of x-ray and γ-ray attenuation coefficients can be found in the following reference.

Hubbell JH, Seltzer SM: Tables of X-Ray Mass Attenuation Coefficients and Mass Energy-Absorption Coefficients 1 keV to 20 MeV for Elements Z = 1 to 92 and 48 Additional Substances of Dosimetric Interest. NISTIR 5632, Gaithersburg MD, 1995, US Department of Commerce. Available at: http://physics.nist.gov/PhysRefData/XrayMassCoef/cover.html (Accessed August 26th, 2011.)

Radiation Detectors

When radiations from a radioactive material pass through matter, they interact with atoms and molecules and transfer energy to them. The transfer of energy has two effects: *ionization* and *excitation.* Ionization occurs when the energy transferred is sufficient to cause an orbital electron to be stripped away from its parent atom or molecule, thus creating an *ion pair* (a negatively charged electron and a positively charged atom or molecule). Excitation occurs when electrons are perturbed from their normal arrangement in an atom or molecule, thus creating an atom or molecule in an *excited state.* Both of these processes are involved in the detection of radiation events; however, ionization is the primary event, and hence the term *ionizing radiation* is used frequently when referring to the emissions from radioactive material. Radiation interactions were discussed in detail in Chapter 6. In this chapter, we describe the basic principles of radiation detectors used in nuclear medicine.

A. GAS-FILLED DETECTORS

1. Basic Principles

Most gas-filled detectors belong to a class of detectors called *ionization detectors.* These detectors respond to radiation by means of ionization-induced electrical currents. The basic principles are illustrated in Figure 7-1. A volume of gas is contained between two electrodes having a voltage difference (and thus an electric field) between them. The negative electrode is called the *cathode,* the positive electrode the *anode.* The electrodes are shown as parallel plates, but they may be a pair of wires, concentric cylinders, and so forth. Under normal circumstances, the gas is an insulator and no electrical current flows between the electrodes. However, radiation passing through the gas causes ionization, both direct ionization from the incident

radiation and secondary ionization from δ rays (see Chapter 6, Section A.1). The electrons produced by ionization are attracted to the positive electrode and the ionized atoms to the negative electrode, causing a momentary flow of a small amount of electrical current.

Gas-filled detectors include *ionization chambers, proportional counters,* and *Geiger-Müller (GM) counters.* The use of these detectors in nuclear medicine is somewhat limited because their stopping power and detection efficiency for x rays and γ rays are quite low; however, they find some use for applications in which detection efficiency is not a major factor and for detection and measurement of nonpenetrating, particle-type radiations. Some of their applications are discussed in Chapters 12 and 23.

2. Ionization Chambers

In most ionization chambers, the gas between the electrodes is air. The chamber may or may not be sealed from the atmosphere. Many different designs have been used for the electrodes in an ionization chamber, but usually they consist of a wire inside of a cylinder or a pair of concentric cylinders.

For maximum efficiency of operation, the voltage between the electrodes must be sufficient to ensure complete collection of ions and electrons produced by radiation within the chamber. If the voltage is too low, some of the ions and electrons simply recombine with one another without contributing to electrical current flow. Figure 7-2 shows the effect of voltage difference between the electrodes on the electrical current recorded by an ionization chamber per ionizing radiation event detected. Recombination occurs at low voltages (*recombination region* of the curve). As the voltage increases there is less recombination and the response (electrical current) increases. When the voltage becomes sufficient to cause complete collection of all of the charges produced, the curve enters a plateau

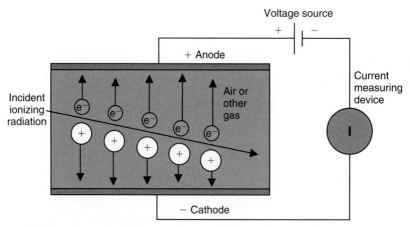

FIGURE 7-1 Basic principles of a gas-filled detector. Electrical charge liberated by ionizing radiation is collected by positive (anode) and negative (cathode) electrodes.

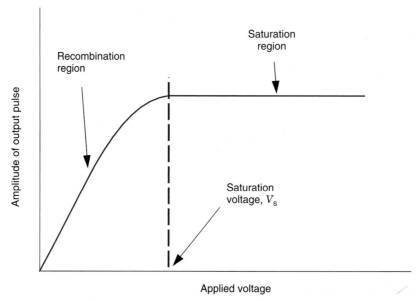

FIGURE 7-2 Voltage response curve (charge collected vs. voltage applied to the electrodes) for a typical ionization chamber. In usual operation, applied voltage exceeds saturation voltage V_s to ensure complete collection of liberated charge.

called the *saturation region.* The voltage at which the saturation region begins is called the *saturation voltage* (V_s). Typically, $V_s \approx$ 50-300 V, depending on the design of the chamber. Ionization chambers are operated at voltages in the saturation region. This ensures a maximum response to radiation and also that the response will be relatively insensitive to instabilities in the voltage applied to the electrodes.

The amount of electrical charge released in an ionization chamber by a single ionizing radiation event is very small. For example, the energy expended in producing a single

ionization event in air is approximately 34 eV.[*] Thus a 1-MeV β particle, for example, causes approximately $(10^6/34) \approx 3 \times 10^4$ ionizations in air and releases a total amount of electrical charge of only approximately 3×10^{-15} coulombs.

---

[*]The average energy expended in producing a single ionization event is symbolized by W. This is not the same as the average energy required to ionize an air molecule, but is the average energy expended per ionization by the ionizing particle, including both ionization and excitation effects. This is discussed in detail in Chapter 6, Section A.4. Values of W for some detector materials are listed in Table 7-1.

TABLE 7-1
SOME PROPERTIES OF DETECTOR MATERIALS USED AS IONIZATION DETECTORS

	Si(Li)	Ge(Li) or Ge	CdTe*	Air
ρ(g/cm³)	2.33	5.32	6.06	0.001297
Z	14	32	48 & 52	~7.6
W(eV)†	3.6	2.9	4.43	33.7

CdTe, cadmium telluride; Ge, germanium; Li, lithium; Si, silicon.

*Cadmium zinc telluride (CZT) is CdTe in which some of the Te atoms (typically 20%) are replaced by zinc atoms. CZT has properties similar to CdTe.

†Average energy expended per electron-hole pair created or per ionization.

FIGURE 7-3 A battery-powered radiation survey meter. An ionization chamber is contained in the base of the unit, with the entrance window on the bottom face of the device (not shown). The meter indicates radiation level. The rotary switch is used to select different scale factors. (*Courtesy Ludlum Measurements, Inc., Sweetwater, TX.*)

Because of the small amount of electrical charge or current involved, ionization chambers generally are not used to record or count individual radiation events. Instead, the total amount of current passing through the chamber caused by a beam of radiation is measured. Alternatively, the electrical charge released in the chamber by the radiation beam may be collected and measured.

Small amounts of electrical current are measured using sensitive current-measuring devices called *electrometers*. Two devices consisting of ionization chambers and electrometers in nuclear medicine are *survey meters* and *dose calibrators*. A typical ionization chamber survey meter is shown in Figure 7-3. The survey meter is battery operated and portable. The ionization chamber consists of an outer cylindrical electrode (metal or graphite-coated plastic) with a wire electrode running down its center. There is often a protective cap on the end of the chamber for most measurements; however, it is removed for measurement of nonpenetrating radiations such as α particles, β particles, and low-energy (≲10 keV) photons.

Survey meters are used to monitor radiation levels for radiation protection purposes (see Chapter 23, Section E). Ionization current is displayed on a front-panel meter. Many older units are calibrated to read traditional units of *exposure rate* in roentgens per hour (R/hr) or mR/hr. Newer units are calibrated to read Systeme International units of *air kerma* in grays per hour (Gy/hr), mGy/hr, and so forth, or have a switch-selectable option for choosing between the two systems of units. The definitions and relationships between these units are discussed in Chapter 23. A typical survey meter can measure exposure rates down to approximately 1 mR/hr or air kerma rates down to approximately 10 µGy/hr.

Dose calibrators are used to assay activity levels in syringes, vials, and so forth containing materials that are to be administered to patients. Unlike other types of ionization chambers discussed in this section, dose calibrators employ sealed and pressurized chambers filled with argon gas. This eliminates the effect of changing barometric pressure on output readings. Dose calibrators typically are calibrated to read directly in units of activity (becquerels or curies), with switches to set the display for different radionuclides. Dose calibrators are discussed in detail in Chapter 12, Section D.1.

A device that records total charge collected over time is the *pocket dosimeter*. The basic principles are illustrated in Figure 7-4. The ionization chamber electrodes are a central charging electrode and the outside case of the dosimeter. They are insulated electrically from one another and form an electrical capacitor. The capacitor is first charged to a reference voltage V by connecting the charging rod to a separate charging unit. If the capacitance between the charging electrode and the case is C, the charge stored on the capacitor is $Q = V \times C$. When the chamber is exposed to

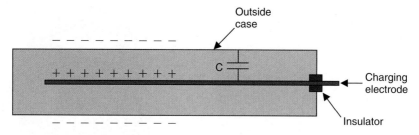

FIGURE 7-4 Schematic representation of a pocket dosimeter.

radiation, electrical charge ΔQ is collected by the electrodes, discharging the capacitor. The voltage change across the capacitor is measured and is related to the amount of electrical charge collected by the ionization chamber electrodes ($\Delta Q = \Delta V \times C$).

Pocket dosimeters are used in nuclear medicine to monitor radiation levels for radiation protection purposes. A typical system is shown in Figure 7-5. The ionization chamber is contained in a small metal or plastic cylinder (~1.5 cm diameter × 10 cm long) that can be clipped to a shirt pocket or collar. Electrodes recessed into one end of the chamber are used to connect the dosimeter to a separate charger unit to charge up the capacitor to the reference voltage. Voltage on the capacitor causes a fine wire within the chamber to be deflected. The position of the wire changes as the voltage on the capacitor changes. The wire is observed through a viewing window at one end of the chamber. Its position is read against a scale that has been calibrated in terms of the total radiation recorded by the chamber, usually in units of air kerma (gray) or exposure (roentgens) (see Chapter 23, Section E). Pocket

dosimeters are suitable for measuring radiation exposures down to approximately 10 mR (air kerma of 0.1 mGy) to an accuracy of approximately 20%.

A basic problem with ionization chambers is that they are quite inefficient as detectors for x rays and γ rays. Only a very small percentage (<1%) of x rays or γ rays passing through the chamber actually interact with and cause ionization of air molecules. Indeed, most of the electrical charge released in an ionization chamber by photon radiations comes from secondary electrons knocked loose from the walls of the chamber by the incident radiations rather than by direct ionization of air molecules. The relatively low detection efficiency of ionization chambers is not a serious limitation in the applications described earlier; however, it precludes their use for most other applications in nuclear medicine, such as imaging.

Two additional problems with ionization chambers should be noted. The first is that for x rays and γ rays, their response changes with photon energy because photon absorption in the gas volume and in the chamber walls (i.e., detection efficiency) and relative penetration

FIGURE 7-5 Pocket dosimeter with charging system. (*Courtesy Ludlum Measurements Inc., Sweetwater, Tx.*)

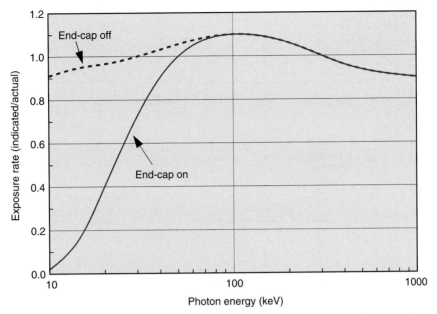

FIGURE 7-6 Energy response curve for a typical ionization chamber survey meter with and without a removable protective end cap.

of photons through the chamber walls are both energy-dependent processes. Figure 7-6 shows a typical energy-response curve for a survey meter. A second problem is that in *unsealed* chambers the density of the air in the chamber, and hence its absorption efficiency, changes with atmospheric pressure ($\rho \propto P$) and temperature ($\rho \propto 1/T$). Most chambers are calibrated to read accurately at sea-level pressure ($P_{ref} = 1.013$ N/m^2 = 760 mm Hg) and average room temperature ($T_{ref} = 22°C = 295$K). For other temperatures T and pressures P the chamber reading must be corrected (multiplied) by a temperature-pressure correction factor

$$C_{TP} = (P_{ref} \times T)/(P \times T_{ref}) \qquad (7\text{-}1)$$

Temperature must be expressed on the Kelvin scale in this equation (K = °C + 273). The correction is significant in some cases, for example, at higher elevations ($P \approx 0.85$ N/m^2 ≈ 640 mm Hg at 1600-meter elevation). Note that temperature-pressure corrections are *not* required with sealed chambers, such as in most dose calibrators. A defective seal on such an instrument obviously could lead to erroneous readings.

3. Proportional Counters

In an ionization chamber, the voltage between the electrodes is sufficient only to collect those charges liberated by direct action of the ionizing radiations. However, if the voltage is increased to a sufficiently high value, the electrons liberated by radiation gain such high velocities and energies when accelerated toward the positive electrode that they cause additional ionization in collisions with other atoms in the gas. These electrons in turn can cause further ionization and so on. This cascade process is called the *Townsend avalanche* or the *gas amplification* of charge. The factor by which ionization is increased is called the *gas amplification factor.* This factor increases rapidly with applied voltage, as shown in Figure 7-7. The gas amplification factor may be as high as 10^6, depending on the chamber design and the applied voltage.

Detectors that operate in the ascending portion of the curve shown in Figure 7-7 are called *proportional counters.* In this region, the ionization caused by an incident radiation event is multiplied (amplified) by the gas amplification factor. The total amount of charge produced is equal to the number of ionizations caused by the primary radiation event (at 34 eV/ionization in air) multiplied by the amplification factor. Thus the total charge produced is proportional to the total amount of energy deposited in the detector by the detected radiation event.

Actually, proportional counters are not simply ionization chambers operated at high voltages but are specially constructed chambers designed to optimize the gas amplification effect, both in terms of the amount of

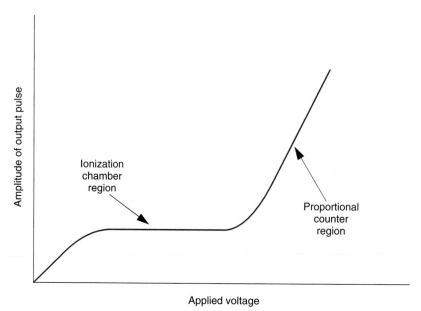

FIGURE 7-7 Voltage response curve for a proportional counter. With increasing applied voltage, the charge collected increases because of the gas amplification effect.

amplification and the uniformity of this amplification within the chamber. In particular, proportional counters are filled with gases that allow easy migration of free electrons, because this is critical for the amplification effect. Common fill gases are the noble gases, with argon and xenon being the most popular.

The major advantage of proportional counters versus ionization chambers is that the size of the electrical signal produced by an individual ionizing radiation event is much larger. They are, in fact, useful for detecting and *counting* individual radiation events. Furthermore, because the size of an individual current pulse is proportional to the amount of energy deposited by the radiation event in the detector, proportional counters can be used for energy-sensitive counting, such as to discriminate between radiation events of different energies on the basis of electrical pulse size (see Chapter 10). They are still inefficient detectors for higher energy x rays and γ rays. Consequently, they find very limited use in nuclear medicine. Proportional counters are used mostly in research applications for measuring nonpenetrating radiations such as α particles and β particles. A practical application is discussed in Chapter 12, Section D.2.

4. Geiger-Müller Counters

A *Geiger-Müller* (GM) *counter* is a gas-filled detector designed for maximum gas amplification effect. The principles of a GM counter

are shown in Figure 7-8. The center wire (anode) is maintained at a high positive voltage relative to the outer cylindrical electrode (cathode). The outer electrode may be a metal cylinder or a metallic film sprayed on the inside of a glass or plastic tube. Some GM counters have a thin radiation *entrance window* at one end of the tube. The cylinder of the tube is sealed and filled with a special gas mixture, typically argon plus a quenching gas (discussed later).

When ionization occurs in a GM counter, electrons are accelerated toward the center wire. Gas amplification occurs in the GM counter as in a proportional counter. In addition to ionizing gas molecules, the accelerating electrons also can cause excitation of gas molecules through collisions. These excited gas molecules quickly (~10^{-9} sec) return to the ground state through the emission of photons at visible or ultraviolet (UV) wavelengths. If a UV photon interacts in the gas, or at the cathode surface by photoelectric absorption (see Chapter 6, Section C.2), this releases another electron, which can trigger a further electron avalanche as it moves toward the anode (see Fig. 7-8). In this way, an *avalanche ionization* is propagated throughout the gas volume and along the entire length of the center wire.

As the avalanche progresses, the electrons, being relatively light, are quickly collected, but the heavy, slow-moving positive ions are not. Eventually, a "hose" of slow-moving

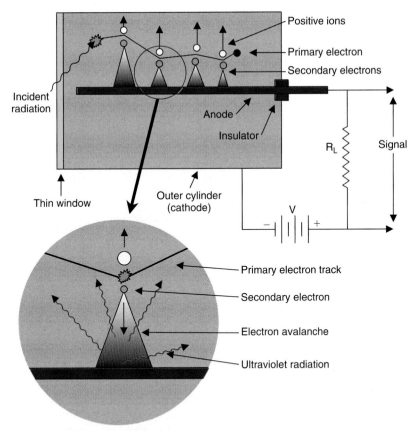

FIGURE 7-8 Operating principles of a Geiger-Müller counter. The incoming radiation produces ion pairs by direct ionization and through secondary fast electrons (δ rays) created in the ionization process. These ion pairs are then multiplied by an avalanche process that in turn triggers further avalanches through the emission of ultraviolet radiation. This process is terminated when a sufficient number of positive ions collect around the anode, effectively reducing the electric field experienced by the electrons owing to charge buildup at the anode.

positive charges is formed around the center wire. The avalanche then terminates because the positive ions reduce the effective electric field around the anode wire, eventually dropping it below the level required for gas multiplication.

The avalanche ionization in a GM tube releases a large and essentially constant quantity of electrical charge, regardless of voltage applied to the tube (Fig. 7-9) or the energy of the ionizing radiation event. The gas amplification factor may be as high as 10^{10}. The large electrical signal is easily detected with electronic circuits. Thus a GM counter, like a proportional counter, can be used to detect and count individual ionizing radiation events. However, because the size of the electrical signal output is constant, regardless of the energy of the radiation detected, a GM counter cannot be used to distinguish between radiation events of different energies.

Once the avalanche has terminated in a GM counter, an additional problem arises. The positive ion cloud moves toward the outer electrode. When the ion cloud is very close to the outer electrode, electrons are pulled out from it to neutralize the positive ions. Some of these electrons enter higher-energy orbits of the positive ions; when they eventually drop into the lower-energy orbits, UV radiation is emitted. This can cause the release of more electrons from the outer wall and set off another avalanche. Thus if no precautions are taken, a single ionizing radiation event can cause the GM counter to go into a pulsating series of discharges.

This problem is prevented by the introduction of a *quenching gas* into the GM counter gas mixture. Such GM counters are called *self-quenched*. Effective quenching gases have three properties: First, they tend to give up electrons easily. When the positive ion cloud is formed, molecules of the quenching gas

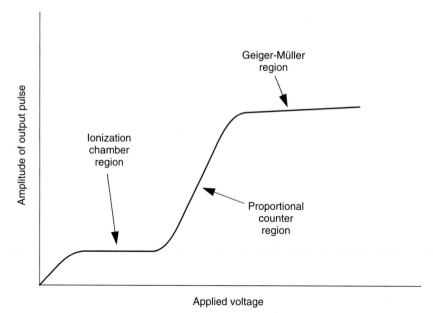

FIGURE 7-9   Voltage response curve (pulse amplitude vs. applied voltage) for a Geiger-Müller counter.

neutralize other ions by donating electrons to them. The ion cloud is thus converted into ionized molecules of quenching gas. Second, when the quenching gas molecules are neutralized by electrons entering higher energy orbits, they deenergize themselves by dissociating into molecular fragments rather than by emitting UV photons. Third, the quenching gas molecules are strong absorbers of UV radiation. Thus the few UV photons that are released during neutralization of the positive ion cloud are quickly absorbed before they can set off another avalanche.

Commonly used quenching gases include heavy organic vapors (e.g., alcohol) and halogen gases (e.g., Cl_2). The organic vapors are more effective quenching agents but have the disadvantage that their molecular fragments do not recombine after dissociation. Thus an organic quenching gas eventually is used up, typically after approximately 10^{10} radiations have been detected. Halogen gas molecules recombine after dissociation and thus have an essentially unlimited lifetime in a GM counter.

A certain minimum voltage is required between the electrodes of a GM counter to sustain an avalanche ionization and to raise the amplitude of the pulses to the threshold of the counting system. This voltage can be determined by exposing the GM counter to a constant source of radiation and observing the counting rate as a function of voltage

applied to the counter electrodes. Figure 7-10 shows the results of such an experiment. This curve is called the *counting curve* or *plateau curve* of the GM counter. As the high voltage is increased, the counting rate increases rapidly as more and more of the output pulses exceed the counter threshold. When the voltage is sufficient that essentially all pulses are above the threshold and are counted, a *plateau region* is reached. The point at which the plateau begins is called the knee of the curve. Further increases in voltage may still increase the amplitude of the output pulses; however, the counting rate remains constant as the radiation source is constant.[*]

When the voltage is increased to a very high value, the counting rate again begins to increase. This happens when the voltage is so high that spontaneous ionization begins to occur in the chamber. The curve then enters the *spontaneous discharge region*. GM counters should not be operated in the spontaneous discharge region because no useful information can be obtained there. Furthermore, if the counter contains an organic quenching gas, it is rapidly used up by the spontaneous discharges, thus shortening the

---

[*]Actually, for most GM counters the counting rate increases by 1% to 2% per 100 volts in the plateau region. This is of no practical consequence in nuclear medicine.

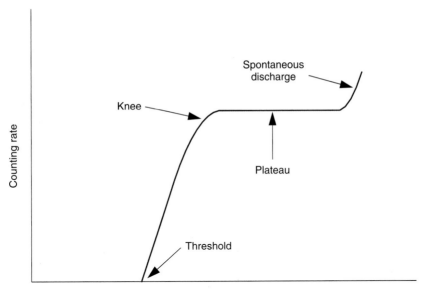

FIGURE 7-10 Counting curve (counting rate from a fixed radiation source vs. applied voltage) for a Geiger-Müller counter. As voltage increases, pulse amplitude increases above threshold of counting system electronics. When all events produce a signal above the threshold, a plateau is reached. At very high voltages, spontaneous discharge events occur within the chamber. These are not caused by radiation events but by electrical breakdown in the gas.

life of the counter. The proper operating voltage is the plateau region, about one third the distance from the knee to the spontaneous discharge region.

GM counters are simple, rugged, and relatively inexpensive radiation detectors. Much of the early (pre-1950s) work in nuclear medicine was done with GM counters; however, they have since been replaced for most applications by other types of detectors. The major disadvantages of GM counters are low detection efficiency (<1%) for γ rays and x rays and an inability to distinguish between radiation events of different energies on the basis of pulse size for energy-selective counting (because all pulses from a GM counter are the same size).

GM counters are used mostly in survey meters for radiation protection purposes. An example is shown in Figure 7-11. The detector in this survey meter is of the *pancake* type. The entrance window at the end of the counter tube is a thin layer of mica (0.01-mm thick) that is sufficiently thin to permit passage of particles and low-energy photons into the counter. The rather fragile window is protected by a wire screen. GM counters designed for counting only relatively penetrating radiations, such as γ rays and high-energy β particles, have thicker, sturdier windows,

FIGURE 7-11 Radiation survey meter with an external Geiger-Müller (GM) "pancake" counter radiation detector attachment. In addition to the external detector shown, some units have built-in GM counters. (*Courtesy Ludlum Measurements, Inc. Sweetwater, TX.*)

for example, 0.1-mm-thick aluminum or stainless steel. Many GM counters are provided with removable covers on the entrance window that can be used to distinguish between penetrating and nonpenetrating radiations by observing the difference between counting rates with and without the cover in place. GM survey meters are more sensitive than ionization chamber survey meters, typically by a factor of approximately 10.

B. SEMICONDUCTOR DETECTORS

Semiconductor detectors are essentially solid-state analogs of gas-filled ionization chambers. Because the solid detector materials used in semiconductor detectors are 2000 to 5000 times more dense than gases (see Table 7-1), they have much better stopping power and are much more efficient detectors for x rays and γ rays.

Semiconductor detectors normally are poor electrical conductors; however, when they are ionized by an ionizing radiation event, the electrical charge produced can be collected by an external applied voltage, as it is with gas-filled detectors. This principle could not be applied using a conducting material for the detector (e.g., a block of metal) because such a material would conduct a large amount of current even without ionizing events. Insulators (e.g., glass) are not suitable detector materials either, because they do not conduct even in the presence of ionizing radiation. Hence only semiconductor materials can function as "solid ionization chambers."

The most commonly used semiconductor detector materials are silicon (Si) and germanium (Ge). More recently, cadmium telluride (CdTe) or cadmium zinc telluride (CZT) have been used as the detector material in small nuclear medicine counting and imaging devices. Characteristics of these semiconductor materials are listed in Table 7-1. One ionization is produced per 3 to 5 eV of radiation energy absorbed. By comparison, this value for gases (air) is approximately 34 eV per ionization. Thus a semiconductor detector not only is a more efficient absorber of radiation but produces an electrical signal that is approximately 10 times larger (per unit of radiation energy absorbed) than a gas-filled detector. The signal is large enough to permit detection and counting of individual radiation events. Furthermore, the size of the electrical signal is proportional to the amount of radiation energy absorbed. Therefore semiconductor detectors can be used for energy-selective radiation counting. For reasons discussed in Chapter 10, Section C.1, they are in fact the preferred type of detector for this application.

In spite of their apparent advantages, semiconductor detectors have a number of problems that have limited their use in nuclear medicine. The first is that both Si and Ge (especially Ge) conduct a significant amount of thermally induced electrical current at room temperature. This creates a background "noise current" that interferes with detection of radiation-induced currents. Therefore Si detectors (usually) and Ge detectors (always) must be operated at temperatures well below room temperature.

A second problem is the presence of impurities even in relatively pure crystals of Si and Ge. Impurities (atoms of other elements) enter into and disturb the regular arrangement of Si and Ge atoms in the crystal matrix. These disturbances create "electron traps" and capture electrons released in ionization events. This results in a substantial reduction in the amount of electrical signal available and limits the thickness of a practical detector to approximately 1 cm. Because of the relatively low atomic numbers of Si and Ge, this restricts their efficiency for detection of γ rays.

Two approaches have been used to solve the impurity problem. One is to prepare very pure samples of the detector material. This has been accomplished only with Ge [*high-purity germanium* (HPGe)] and is, unfortunately, quite expensive. Also, the size of pure crystals is limited to approximately 5 cm in diameter by 1 cm thick. Detectors made of HPGe are sometimes called *intrinsic Ge* detectors. A second approach is to deliberately introduce into the crystal matrix "compensating" impurities that donate electrons to fill the electron traps created by other impurities. Lithium (Li) is commonly used in Si and Ge detectors for this purpose. Detectors made of "lithium-doped" materials are called *lithium-drifted* detectors, or Si(Li) or Ge(Li) detectors. Unfortunately, the process of preparing Si(Li) or Ge(Li) crystals is time consuming and expensive. Crystal sizes are limited to a few centimeters in diameter by approximately 1 cm thick for Si(Li) and approximately 5 cm diameter by 5 cm thick for Ge(Li).

An additional problem is that Li ions tend to "condense" within the crystal matrix at room temperature, especially in Ge. Therefore Si(Li) and Ge(Li) not only must be operated at low temperatures (to minimize thermally induced background currents) but Ge(Li) detectors must and Si(Li) detectors should also be *stored* at low temperatures.

Liquid nitrogen (T = 77K or −196°C) is used for detector cooling. Ge(Li) detectors can be ruined by only an hour or so at room temperature. Si(Li) detectors can tolerate elevated temperatures, but they provide optimum performance if they also are stored at liquid nitrogen temperatures.

Because of the difficulties inherent in Li-drifted detectors, HPGe has become the detector material of choice for γ-ray spectroscopy applications (see Chapter 10), and most manufacturers have now stopped producing Ge(Li) detectors. Si(Li) finds applications in low-energy x-ray and β-particle spectroscopy, in which its low atomic number is not a disadvantage.

Figure 7-12 shows schematically a typical semiconductor detector assembly. The detector consists of a thin, circular disc of the detector material [Si(Li), Ge(Li), or HPGe] with electrodes attached to its opposite faces for charge collection. One electrode is a thin metal foil fastened to the front surface ("entrance window"), whereas the other is a wire or set of wires embedded in the opposite surface of the crystal. Other detector shapes and electrode configurations also are used.

Figure 7-12 also shows in cross-section an apparatus used to cool the crystal with liquid nitrogen. A "coldfinger" extends from the liquid nitrogen container (a Dewar flask) to cool the detector. Some of the preamplifier electronic circuitry also is cooled to reduce electronic noise levels. Liquid nitrogen

evaporates and the container needs periodic refilling—typically every 2 to 3 days, depending on container size and insulation characteristics.

CdTe and CZT (which has properties very similar to CdTe) are more recently developed semiconductor materials that overcome two of the major disadvantages of Si and Ge: (1) they can be operated at room temperature without excessive electronic noise, and (2) their high atomic number means that even relatively thin detectors can have good stopping efficiency for detecting γ rays. Although CdTe and CZT are now being used in some nuclear medicine counting and imaging devices, their use has generally been restricted to small detectors, or detectors comprising multiple small elements, because of the difficulty and expense of growing large pieces of CdTe or CZT with the required purity. Additional discussion of their properties for pulse-height spectrometry is presented in Chapter 10, Section C.1.

C. SCINTILLATION DETECTORS

1. Basic Principles

As indicated earlier in this chapter, radiation from radioactive materials interacts with matter by causing ionization or excitation of atoms and molecules. When the ionized or excited products undergo recombination or deexcitation, energy is released. Most of the energy is dissipated as thermal energy, such as molecular vibrations in gases or liquids or lattice vibrations in a crystal; however, in some materials a portion of the energy is released as visible light.* These materials are called *scintillators,* and radiation detectors made from them are called *scintillation detectors.*

The scintillator materials used for detectors in nuclear medicine are of two general types: inorganic substances in the form of solid crystals and organic substances dissolved in liquid solution. The scintillation mechanisms are different for these two types and are described separately in later sections.

A characteristic common to all scintillators is that the amount of light produced following

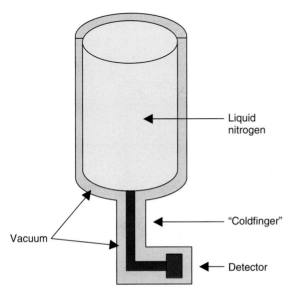

FIGURE 7-12 Schematic representation of a typical semiconductor detector assembly. "Coldfinger" is a thermal conductor for cooling the detector element.

---

*For simplicity, the term *visible light* is used to describe scintillation emission. In fact, the emissions from many scintillators extend into the UV portion of the spectrum as well.

the interaction of a single γ ray, β particle, or other ionizing radiation, is proportional to the energy deposited by the incident radiation in the scintillator. The amount of light produced also is very small, typically a few hundred to a few thousand photons for a single γ-ray interaction within the energy range of interest for nuclear medicine imaging (70-511 keV). In the early days of nuclear physics, it was common to study the characteristics of particles by observing and counting, in a darkened room, the scintillations produced by these particles on a zinc sulfide scintillation screen. The obvious limitations on counting speed and accuracy with this system have been eliminated in modern application with the introduction of ultrasensitive electronic light detectors called *photomultiplier (PM) tubes*.

2. Photomultiplier Tubes

PM tubes (also called *phototubes* and sometimes abbreviated *PMT*) are electronic tubes that produce a pulse of electrical current when stimulated by very weak light signals, such as the scintillation produced by a γ ray or β particle in a scintillation detector. Their basic principles are illustrated in Figure 7-13.

The inside front surface of the glass *entrance window* of the PM tube is coated with a photoemissive substance. A *photoemissive* substance is one that ejects electrons when struck by photons of visible light. Cesium antimony (CsSb) and other bialkali

compounds are commonly used for this material. The photoemissive surface is called the *photocathode*, and electrons ejected from it are called *photoelectrons*. The conversion efficiency for visible light to electrons, also known as the *quantum efficiency*, is typically 1 to 3 photoelectrons per 10 visible light photons striking the photocathode. The dependence of quantum efficiency on the wavelength of the light is shown for a conventional bialkali photocathode in Figure 7-14.

A short distance from the photocathode is a metal plate called a *dynode*. The dynode is maintained at a positive voltage (typically 200-400 V) relative to the photocathode and attracts the photoelectrons ejected from it. A *focusing grid* directs the photoelectrons toward the dynode. The dynode is coated with a material having relatively high secondary emission characteristics. CsSb also can be used for this material. A high-speed photoelectron striking the dynode surface ejects several *secondary electrons* from it. The electron multiplication factor depends on the energy of the photoelectron, which in turn is determined by the voltage difference between the dynode and the photocathode.

Secondary electrons ejected from the first dynode are attracted to a second dynode, which is maintained at a 50-150 V higher potential than the first dynode, and the electron multiplication process is repeated. This occurs through many additional dynode stages (typically 9 to 12 in all), until finally a

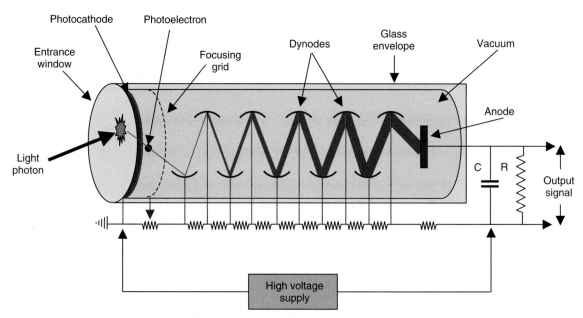

FIGURE 7-13 Basic principles of a photomultiplier tube.

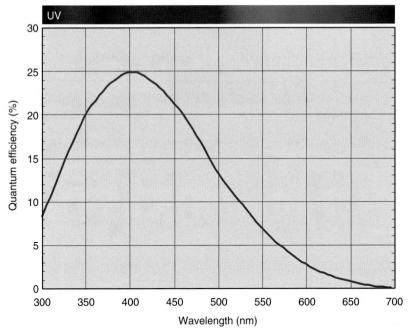

FIGURE 7-14 Quantum efficiency as a function of wavelength for a typical bialkali photocathode. The peak occurs at approximately 400 nm, which is well-matched to the emission wavelength of many scintillators. UV, ultraviolet light.

shower of electrons is collected at the *anode*. Typical electron multiplication factors are ×3 to ×6 per dynode. The total electron multiplication factor is very large—for example, 6^{10} ($\sim 6 \times 10^7$) for a 10-stage tube with an average multiplication factor of 6 at each dynode. Thus a relatively large pulse of current is produced when the tube is stimulated by even a relatively weak light signal. Note that the amount of current produced is proportional to the intensity of the light signal incident on the photocathode and thus also to the amount of energy deposited by the radiation event in the crystal.

PM tubes require a high-voltage supply. For example, as shown in Figure 7-13, if the tube has 10 dynodes, with the first at +300 V relative to the photocathode and the remaining 9 dynodes and the anode at additional +100 V increments, a voltage of +1300 V is needed. Furthermore, the voltage supply must be very stable because the electron multiplication factor is very sensitive to dynode voltage changes. Typically a 1% increase in high voltage applied to the tube increases the amount of current collected at the anode by approximately 10%. This is of considerable importance in applications where pulse size is being measured, such as in pulse-height spectrometry to determine γ-ray energies (see Chapter 10).

PM tubes are sealed in glass and evacuated. Electrical connections to the dynodes, the photocathode, and the anode are made through pins in the tube. The focusing of the electron beam from one dynode to the next can be affected by external magnetic fields. Therefore PM tubes often are wrapped in metal foil for magnetic shielding. "Mu-metal," an alloy composed of iron, nickel, and small amounts of copper and chromium, is commonly used for this purpose. PM tubes come in various shapes (round, square, and hexagonal) and sizes (Fig. 7-15). Most of those used in nuclear medicine have photocathodes in the range of 1- to 7.5-cm diameter. There are also position-sensitive and multichannel PM tubes available that have the ability to determine the location of incident light on the photocathode.

3. Photodiodes

In some applications, the PM tube may be replaced by a light-sensitive semiconductor detector, such as a Si photodiode. Note that in this case, the semiconductor is not being used to detect the γ rays directly but to detect the visible light emitted from a scintillator material in which a γ ray has interacted. The photons from the scintillator have sufficient energy to cause ionization within Si, and the total charge produced is proportional to the

FIGURE 7-15 Assortment of photomultiplier tubes illustrating their wide variety of shapes and sizes. (*Courtesy Hamamatsu Corp., Bridgewater, NJ.*)

number of scintillation light photons incident on the photodiode. These photodiode detectors have the advantage that they can be made very small in area and that they are typically only a few millimeters thick, including the packaging. They also have significantly higher quantum efficiency than PM tubes, with values ranging typically between 60% and 80%. However, conventional Si photodiodes have unity gain (compared with 10^6 to 10^7 for a PM tube), requiring very low-noise electronics for readout.

A related device, the Si avalanche photodiode (APD) uses a very high internal electric field such that each electron produced within the device gains enough energy between collisions to create further ionization. This is analogous to the proportional region for gas-filled detectors that was discussed previously in Section A.3. APD detectors can reach gains of 10^2 to 10^3 but still require low-noise electronics for successful operation. The gain of these devices also is a very strong function of bias voltage and temperature, and these parameters therefore must be very carefully controlled for stable operation. These types of solid-state light detectors are used presently only in specialized nuclear medicine systems, such as small animal scanners (see Chapter 17, Section A.4, and Chapter 18, Section B.5) and dual-modality positron emission tomography–magnetic resonance imaging (PET-MRI) systems (see Chapter 19, Section F).

APDs also can be operated at higher bias voltages in *geiger mode*. This is analogous to the gas-filled GM counter (Section A.4), and the output signal of the APD becomes very large and independent of the energy of the incident radiation. The gain of such devices can be as high as 10^7. Light-sensitive detectors consisting of a large number of tiny (20-50 μm) *geiger-mode APDs* that are incorporated into a single device are being developed for use in scintillation detectors.

4. Inorganic Scintillators

Inorganic scintillators are crystalline solids that scintillate because of characteristics of their crystal structure. Individual atoms and molecules of these substances do not scintillate. They are scintillators only in crystalline form. Table 7-2 summarizes the properties of a number of inorganic scintillators of interest for nuclear medicine applications.

Some inorganic crystals are scintillators in their pure state; for example, pure NaI crystals are scintillators at liquid nitrogen temperatures. Most are "impurity activated," however. These are crystals containing small amounts of "impurity" atoms of other elements. Impurity atoms in the crystal matrix cause disturbances in its normal structure. Because they are responsible for the scintillation effect, the impurity atoms in the crystal matrix are sometimes called *activator centers*. Some impurity-activated scintillators that have been used in radiation detectors include sodium iodide [NaI(Tl)] and cesium iodide [CsI(Tl)]. In each case, the element in parentheses is the impurity that is added to create activator centers in the crystal.

The most commonly used scintillator for detectors in nuclear medicine is NaI(Tl). Pure

TABLE 7-2
PROPERTIES OF SOME SCINTILLATOR MATERIALS USED IN NUCLEAR MEDICINE

Property	NaI(Tl)	BGO	LSO(Ce)	GSO(Ce)	CsI(Tl)	LuAP(Ce)	LaBr$_3$(Ce)	Plastic*
Density (g/cm^3)	3.67	7.13	7.40	6.71	4.51	8.34	5.3	1.03
Effective atomic number	50	73	66	59	54	65	46	12
Decay time (nsec)	230	300	40	60	1000	18	35	2
Photon yield (per keV)	38	8	20-30	12-15	52	12	61	10
Index of refraction	1.85	2.15	1.82	1.85	1.80	1.97	1.9	1.58
Hygroscopic	Yes	No	No	No	Slightly	No	Yes	No
Peak emission (nm)	415	480	420	430	540	365	358	Various

*Typical values—there are many different plastic scintillators available.
BGO, Bi$_3$Ge$_4$O$_{12}$; GSO(Ce), Gd$_2$SiO$_5$(Ce); LSO(Ce), Lu$_2$SiO$_5$(Ce); LuAP(Ce), LuAlO$_5$(Ce)

NaI crystals are scintillators only at liquid nitrogen temperatures. They become efficient scintillators at room temperatures with the addition of small amounts of thallium. Single crystals of NaI(Tl) for radiation detectors are "grown" from molten sodium iodide to which a small amount of thallium (0.1-0.4 mole percent) has been added. Crystals of relatively large size are grown in ovens under carefully controlled temperature conditions. For example, crystals for gamma cameras (see Chapter 13) use NaI(Tl) crystals that are typically 30-50 cm in diameter by 1-cm thick.

Figure 7-16 shows the construction of a typical scintillation detector consisting of a NaI(Tl) crystal and PM tube assembly. The crystal is sealed in an aluminum or stainless-steel jacket with a transparent glass or plastic optical *window* at one end to permit the exit of scintillation light from the crystal to the PM tube. A transparent optical "coupling grease" is used between the crystal and the PM tube to minimize internal reflections at this interface. The crystal and PM tube are hermetically sealed in a light-tight jacket to keep out moisture and extraneous light and for mechanical protection. The inside surface of the radiation entrance window and sides of the crystal are coated with a highly reflective diffuse material to maximize the light collected by the PM tube photocathode. With efficient optical coupling, good reflective surfaces, and a crystal free of cracks or other opacifying defects, approximately 30% of the light emitted by the crystal actually reaches the cathode of the PM tube. Some NaI(Tl) detectors have very

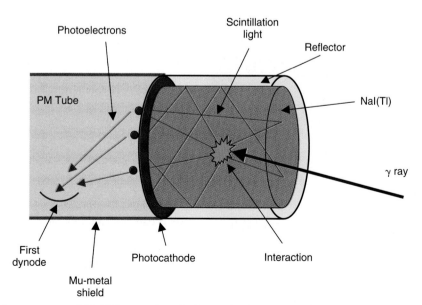

FIGURE 7-16 Arrangement of NaI(Tl) crystal and photomultiplier (PM) tube in a typical scintillation detector assembly.

thin aluminum or beryllium foil "entrance windows" to permit detection of radiations having relatively low penetrating power, such as low-energy x rays and γ rays (E ≲ 10 keV) and β particles; however, most NaI(Tl) detectors have thicker entrance windows of aluminum or stainless steel and are best suited for detecting higher-energy γ rays (E ≳ 50 keV). Figure 7-17 shows some typical integral NaI(Tl) crystal and PM tube assemblies.

Some reasons for the usefulness of NaI(Tl) scintillation detectors include the following:

1. It is relatively dense ($\rho = 3.67 \text{ g/cm}^3$) and contains an element of relatively high atomic number (iodine, $Z = 53$). Therefore it is a good absorber and a very efficient detector of penetrating radiations, such as x rays and γ rays in the 50- to 250-keV energy range. The predominant mode of interaction in this energy range is by photoelectric absorption.

2. It is a relatively efficient scintillator, yielding one visible light photon per approximately 30 eV of radiation energy absorbed.

3. It is transparent to its own scintillation emissions. Therefore there is little loss of scintillation light caused by self-absorption, even in NaI(Tl) crystals of relatively large size.

4. It can be grown relatively inexpensively in large plates, which is advantageous for imaging detectors.

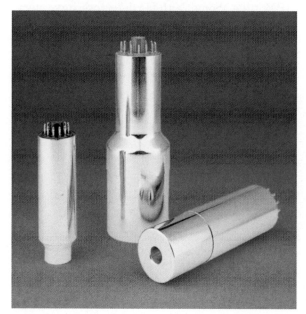

FIGURE 7-17 NaI(Tl) crystal and photomultiplier tube assemblies. (*Courtesy Crystals and Detectors Division, Saint-Gobain Ceramics and Plastics, Inc., Newbury, OH.*)

5. The scintillation light is well-matched in wavelength to the peak response of the PM tube photocathode (see Fig. 7-14). The emission spectrum of light from NaI(Tl) is shown in Figure 7-18.

Some disadvantages of NaI(Tl) detectors are the following:

1. The NaI(Tl) crystal is quite fragile and easily fractured by mechanical or thermal stresses (e.g., rapid temperature changes). Fractures in the crystal do not necessarily destroy its usefulness as a detector, but they create opacifications within the crystal that reduce the amount of scintillation light reaching the photocathode.

2. Sodium iodide is hygroscopic. Exposure to moisture or a humid atmosphere causes a yellowish surface discoloration that again impairs light transmission to the PM tube. Thus hermetic sealing is required.

3. At higher γ-ray energies (≳ 250 keV), the predominant mechanism of interaction is by Compton interaction, and larger volumes of NaI(Tl) are required for adequate detection efficiency.

Other types of detectors have advantages over NaI(Tl) detectors in certain areas. For example, gas-filled detectors are cheaper (but have much lower detector efficiency), and semiconductor detectors have better energy resolution (but are expensive to use in large-area imaging cameras). However, the overall advantages of NaI(Tl) have made it the detector of choice for almost all routine applications in nuclear medicine involving γ-ray detection in the 50-250-keV energy range.

At higher energies, particularly for detection of the 511-keV emissions from positron emitters, denser scintillators generally are preferred. Bismuth germanate ($Bi_4Ge_3O_{12}$; BGO) is a commonly used scintillator in PET imaging, because of its excellent detection efficiency at 511 keV. Lutetium oxyorthosilicate [$Lu_2SiO_5(Ce)$; LSO] is slightly less efficient at 511 keV than BGO, but is brighter and faster and may sometimes offer advantages over BGO when the counting rate on the detector is high, when fast timing information is needed, or when small scintillator elements are to be decoded in an imaging system (see Chapter 18, Section B). However, LSO is rather expensive to grow, because of its high melting point and its raw material costs. A related material, lutetium yttrium orthosilicate (LYSO—LSO in which a small fraction of the lutetium atoms are replaced

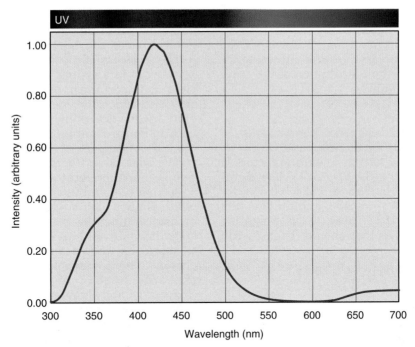

FIGURE 7-18 The emission spectrum of NaI(Tl) scintillator at room temperature. UV, ultraviolet light. (*Data courtesy Kanai Shah and Jarek Glodo, Radiation Monitoring Devices Inc., Watertown, MA.*)

with yttrium) has scintillation properties very similar to LSO. BaF_2 and CsF also have historically been used in positron cameras because of their very fast decay time (which is important for timing of γ-ray interactions in coincidence detection) (see Chapter 18, Section A.2). However, their low detection efficiency compared with BGO and LSO has prevented any widespread application.

New scintillator materials continue to be discovered and developed. Among the more promising recent candidates for nuclear medicine applications are LuAP [$LuAlO_3(Ce)$], lanthanum bromide [$LaBr_3(Ce)$], and lanthanum chloride [$LaCl_3(Ce)$].

5. Considerations in Choosing an Inorganic Scintillator

The ability of a scintillator to stop high-energy γ rays ($\gtrsim 100$ keV) is of particular importance in nuclear medicine, especially for imaging applications. As discussed in Chapter 6, Section D.1, the photoelectric component of the mass-attenuation coefficient increases strongly with the atomic number, Z, of a material, whereas the Compton component is essentially independent of atomic number. As illustrated in Fig. 6-18, although the Compton component is significant over much of the nuclear-medicine energy range, the

photoelectric component dominates in high-Z materials over much of that range. As well, high-Z materials tend to be denser than low-Z materials, thus increasing their linear attenuation coefficients and enhancing their stopping power per unit of detector thickness. Thus, high-Z detector materials generally are more efficient than low-Z materials for detecting high-energy γ rays.

Most detector materials consist of a mixture of elements. The effective atomic number, Z_{eff}, is a useful and convenient parameter for comparing the ability of such mixtures to stop high-energy γ rays. It is given by

$$Z_{eff} = \sqrt[x]{w_1 Z_1^x + w_2 Z_2^x + \cdots w_n Z_n^x} \qquad (7\text{-}2)$$

where w_i is a weighting factor proportional to the fractional number of electrons per gram for element i and can be calculated as

$$w_i = \frac{m_i Z_i}{\sum_{i=1}^{n} m_i Z_i} \qquad (7\text{-}3)$$

where m_i represents the number of atoms of element i present in the compound or mixture. The power x depends on the energy of the γ rays. For γ rays in the 100- to 600-keV range, x is typically taken to be between 3 and 3.5.[1]

EXAMPLE 7-1

Calculate the effective atomic number of BGO ($Bi_4Ge_3O_{12}$). Compare this with the value for NaI(Tl). (Ignore the contribution of the trace amounts of thallium.)

Answer

There are three elements contributing to BGO: Bi (Z = 83), Ge (Z = 32), and O (Z = 8). The denominator for the weighting factors is $(83 \times 4) + (32 \times 3) + (8 \times 12) = 524$. The weighting factor for Bi is $(83 \times 4) / 524 = 0.634$, for Ge is $(32 \times 3) / 524 = 0.183$, and for O is $(8 \times 12) / 524 = 0.183$. These three weighting factors add to 1, as they should. Then Z_{eff} can be calculated as

$$Z_{eff} = (0.634 \times 83^{3.5} + 0.183 \times 32^{3.5} + 0.183 \times 8^{3.5})^{1/3.5}$$
$$= 73.1$$

NaI(Tl) consists of Na (Z=11) and I (Z=53) atoms that are present in equal quantities. The weighting factor for Na is $11/(11 + 53) = 0.172$ and for I is $53/(11 + 53) = 0.828$. Therefore Z_{eff} for NaI(Tl) is

$$Z_{eff} = (0.172 \times 11^{3.5} + 0.828 \times 53^{3.5})^{1/3.5}$$
$$= 50.2$$

In addition to choosing a scintillator that has sufficient stopping power for the efficient detection of γ rays with a particular energy, the other properties listed in Table 7-2 also influence the choice of scintillator materials for a specific application. The *decay time* of the scintillator is important in two respects. Firstly, it determines the precision with which the time of γ-ray interaction in the scintillator can be determined. Faster light production within the scintillator (faster decay time) results in better timing precision. This is important in nuclear medicine applications in which timing is important, most notably in PET in which coincident 511 keV annihilation photons are detected (see Chapter 18, Section A). Secondly, the decay time of the scintillator is a limiting factor in how many γ-ray interactions a detector can process per unit time. To unambiguously detect two interactions, they should be separated by roughly 2-3 times the decay time; otherwise events "pile up" on top of each other, leading to dead time (see Chapter 11, Section C). As a rough rule of thumb, if the scintillator has a decay time of τ, the maximum event rate that a detector made using that scintillator can handle is approximately 1/2τ. Faster scintillators detectors therefore can handle higher event rates on the detector.

The *efficiency* of the scintillator in converting a γ ray into visible light photons (photon yield) is important in determining the precision with which the energy of the interacting γ ray can be determined. This becomes relevant in many counting and imaging systems in which it is important to distinguish between γ rays that have Compton scattered in the sample (and therefore lost energy; see Chapter 6, Section C.3) and those that remain unscattered. Higher photon yield also is important in determining the positioning accuracy in many imaging systems in which it is common to share the limited number of scintillation photons among multiple PM tubes to determine the location of an interaction (see Chapter 13 and Chapter 18, Section B).

The *index of refraction* of a scintillator plays a role in determining how efficiently scintillation light can be coupled from a scintillator crystal into a PM tube. The index of refraction of the glass entrance window on a PM tube is ~1.5. Therefore for best transmission of light from the scintillator into the PM tube with minimal internal reflection at the scintillator crystal and PM tube interface, the scintillator should have an index of refraction as close to 1.5 as possible. In practice, most scintillators of interest have indices of refraction significantly higher than 1.5 and this is one reason why only a fraction of the scintillation light produced actually reaches the PM tube.

Lastly, it is important that the *emission spectrum* of the light produced by the scintillator is a good match for the quantum efficiency of the photodetector that is used to convert the scintillation light into an electronic pulse. In the case of a PM tube with a standard bialkali photocathode (see Fig. 7-14), it is apparent that scintillators that have their peak light emission in the range of 350-475 nm are optimal. NaI(Tl) is an example of a scintillator that has an emission spectrum (see Fig. 7-18) that matches the quantum efficiency of PM tubes very well. The transmission of the scintillation light through the glass used in the PM tube entrance window also must be considered. Many glasses are efficient absorbers of UV light at wavelengths significantly shorter than ~350 nm.

6. Organic Scintillators

In contrast with inorganic scintillators, the scintillation process in organic scintillators is an inherent molecular property. The scintillation mechanism is one of molecular excitation (e.g., by absorbing energy from a γ ray or β particle) followed by a deexcitation process in which visible light is emitted. These substances are scintillators whether they are in solid, liquid, or gaseous forms.

Certain plastics (e.g., see Table 7-2) are organic scintillators and have been used for direct detection of β particles emitted from radionuclides, particularly in compact probes designed for surgical use (see Chapter 12, Section F.2). A more common application for organic scintillators, however, is to employ them in liquid form for *liquid scintillation (LS) counting.* In these systems, the scintillator is dissolved in a solvent material in a glass or plastic vial and the radioactive sample is added to this mixture. The vial is then placed in a light-tight enclosure between a pair of PM tubes to detect the scintillation events (Fig. 7-19).

LS solutions consist of four components:

1. An organic *solvent* comprises most of the solution. The solvent must dissolve not only the scintillator material but also the radioactive sample added to it. The solvent actually is responsible for most of the direct absorption of radiation energy from the sample. High-speed electrons generated by ionizing radiation events in the solvent transfer energy to the scintillator molecules, causing the scintillation effect. Commonly used solvents include di-iso-propylnapthalene (DIN) and phenylxylylethane (PXE), which are replacing traditional, more environmentally harsh solvents such as toluene and xylene.

2. The *primary solute* (or primary *fluor*) absorbs energy from the solvent and emits light. Some common primary scintillators include *p*-bis-(omethylstyryl) benzene (abbreviated as bis-MSB) and 2,5-diphenyloxazole (also known as PPO).

3. The emissions of the primary solute are not always well matched to the response characteristics of PM tubes. Therefore a *secondary solute,* or *waveshifter,* is sometimes added to the solution. The function of this material is to absorb emissions of the primary solute and reemit photons of longer wavelength, which are better matched to the PM tube response. 1,4-di-(2-5-phenyloxazole) benzene (also known as POPOP) is a commonly used secondary scintillator.

4. LS solutions frequently contain *additives* to improve some aspect of their performance, such as the efficiency of energy transfer from the solvent to the primary solute. Solubilizers (e.g., hyamine hydroxide) are sometimes added to improve the dissolution of added samples such as blood.

The precise "cocktail" of solvent, primary and secondary solutes, and additives depends on the sample type that is being measured. A wide variety of different LS cocktails optimized for different applications are available commercially.

Because of the intimate relationship between sample and detector, LS counting is the method of choice for efficient detection of particles, low-energy x rays and γ rays, and other nonpenetrating radiations. It is widely used for measurement of ^{3}H and ^{14}C. In medical applications, it is used primarily for sensitive assay of radioactivity in biologic specimens, such as blood and urine.

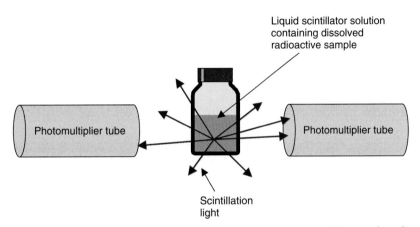

FIGURE 7-19 Arrangement of sample and detector for liquid scintillation counting. The sample is dissolved in a liquid scintillator solution in a glass or plastic vial.

Although well suited for counting non-penetrating radiations in biologic samples, LS counters have numerous drawbacks as general-purpose radiation detectors. They are inefficient detectors of penetrating radiations such as γ rays and x rays of moderate energy because the detector solution is composed primarily of low-density, low-Z materials. In addition, liquid scintillators generally have low light output, only about one third that of NaI(Tl). This problem is worsened by the relatively inefficient light coupling from the scintillator vial to the PM tubes as compared with NaI(Tl) integral detectors.

For sample counting, special sample preparation may be required to dissolve the sample. Problems in sample preparation are discussed in Chapter 12, Section C.6. Also, the sample itself is "destroyed" when it is added to the scintillator solution.

Finally, all LS counting suffers from the problem of *quenching*. Quenching in this context refers to any mechanism that reduces the amount of light output from the sample (not to be confused with the electrical quenching that occurs in GM counters; see Section A.4). There are basically three types of LS quenching:

1. *Chemical quenching* is caused by substances that compete with the primary fluor for absorption of energy from the solvent but that are themselves not scintillators. Dissolved oxygen is one of the most troublesome chemical quenchers.
2. *Color quenching* is caused by substances that absorb the emissions of the primary or secondary solute. Blood and other colored materials are examples. Fogged or dirty containers can also produce a type of color quenching.
3. *Dilution quenching* occurs when a relatively large volume of sample is added to the scintillator solution. The effect is to reduce the concentration of primary and secondary solutes in the final solution, thus reducing the scintillator output efficiency.

Quenching can be minimized in various ways. For example, dissolved oxygen may be purged by ultrasound, and hydrogen peroxide may be added for color bleaching. However, there are no convenient ways to eliminate all causes of quenching; therefore a certain amount must be accepted in all practical LS counting. Quenching becomes a serious problem when there are wide variations in its extent from one sample to the next. This causes unpredictable variations in light output, for the same amount of radiation energy absorbed, from one sample to the next. *Quench correction* methods are employed in LS counters to account for this effect (Chapter 12, Section C.5).

REFERENCE

1. Johns HE, Cunningham JR: *The Physics of Radiology*, ed 4, Springfield, IL, 1983, Charles C Thomas, pp 241-243.

BIBLIOGRAPHY

A comprehensive reference for many different radiation detectors is the following:

Knoll GF: *Radiation Detection and Measurement*, ed 4, New York, 2010, John Wiley.

A detailed reference for inorganic scintillator mechanisms, properties, growth, and applications is the following:

Lecoq P, Annenkov A, Getkin A, Korzhik M, Pedrini C: *Inorganic Scintillators for Detector Systems: Physical Principles and Crystal Engineering*. Berlin, 2006, Springer.

A detailed general reference for scintillation detectors is the following:

Birks JB: *The Theory and Practice of Scintillation Counting*. London, 1967, Pergamon Press.

Electronic Instrumentation for Radiation Detection Systems

Most of the radiation detectors used in nuclear medicine are operated in a "pulse mode"; that is, they generate pulses of electrical charge or current that are counted to determine the number of radiation events detected. In addition, by analyzing the amplitude of pulses from the detector, it is possible with energy-sensitive detectors, such as scintillation and semiconductor detectors and proportional counters, to determine the energy of each radiation event detected. Selection of a narrow energy range for counting permits discrimination against events other than those of the energy of interest, such as scattered radiation and background radiation or the multiple emissions from a mixture of radionuclides.

Figure 8-1 shows in schematic form the basic electronic components of a nuclear radiation-counting instrument. These components are present in systems ranging from the most simple sample counters to complex imaging instruments. The purpose of this chapter is to describe the basic principles of these components. The electronics for specific systems also are described in Chapters 10, 12 to 14, and 18. Basic principles of electricity and electronics are reviewed in the sources cited in the Bibliography at the end of this chapter.

A. PREAMPLIFIERS

Table 8-1 summarizes the pulse output characteristics of detectors used in nuclear medicine. Most of them produce pulse signals of

relatively small amplitude. In addition, most of the detectors listed have relatively high output impedance, that is, a high internal resistance to the flow of electrical current. In handling electronic signals, it is important that the impedance levels of successive components be matched to one another, or electronic interferences that distort the pulse signals may develop and system performance will be degraded.

The purposes of a *preamplifier* (or preamp) are threefold: (1) to amplify, if necessary, the relatively small signals produced by the radiation detector, (2) to match impedance levels between the detector and subsequent components in the system, and (3) to shape the signal pulse for optimal signal processing by the subsequent components.

There are two main types of preamplifier configurations used with radiation detectors: the *voltage-sensitive* preamplifier and the *charge-sensitive* preamplifier. Figure 8-2 shows a simplified diagram of these two configurations. The symbol —$\boxed{A}$— represents the signal (pulse)-amplifying component. The resistor (R) and capacitor (C) provide pulse shaping. The signal from the detector is typically a sharply rising pulse of electrical current of relatively short duration ($\lesssim 1$ μsec, except for Geiger-Müller (GM) counters; see Table 8-1). The voltage-sensitive preamp amplifies any voltage that appears at its input. Because radiation detectors are charge-producing devices, this input voltage, V_i, is given by the ratio of the charge, Q, and the

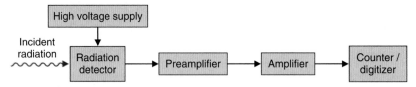

FIGURE 8-1   Schematic representation of the electronic components for a nuclear radiation counting system.

intrinsic capacitance of the detector and other components in the input circuit, C_i:

$$V_i = \frac{Q}{C_i} \qquad (8\text{-}1)$$

With energy-sensitive detectors, the amount of charge, Q, and thus the amplitude of the voltage V_i are proportional to the energy of the radiation event detected. The output voltage, V_o, in this configuration is approximately

$$V_o \approx -\frac{R_2}{R_1} V_i \qquad (8\text{-}2)$$

in which R_1 and R_2 are as shown in Figure 8-2. The minus sign indicates that the polarity of the pulse has been changed.

In semiconductor detectors, the input capacitance of the detector is sensitive to operating conditions, particularly temperature. Therefore the proportionality between charge and the voltage seen at the preamp input may not be stable. The charge-sensitive preamplifier overcomes this undesirable feature by using a feedback capacitor of capacitance C_f to integrate the charge from the radiation

detector. The resulting output voltage, given by

$$V_o \approx -\frac{Q}{C_f} \qquad (8\text{-}3)$$

is seen to be independent of the input capacitance, C_i.

The electrical charge leaks off the feedback capacitor through the resistor of resistance R_f, causing the voltage on the capacitor and at the outputs of the amplifier element to decrease exponentially with time t according to

$$V = V_o e^{-t/R_f C_f} \qquad (8\text{-}4)$$

The product $R_f \times C_f$ is called the *time constant* τ of the pulse-shaping circuit. The voltage decreases exponentially, dropping by 63% of its initial value during one time constant interval (see Fig. 8-2C). When R_f is given in ohms and C_f in farads, the time constant is given in seconds. Typical preamplifier time constants for nuclear medicine detectors (excepting those applications that require fast timing signals) are 20 to 200 μsec.

The amount of amplification provided by the amplifier element of the preamplifier

TABLE 8-1
TYPICAL SIGNAL OUTPUT AND PULSE DURATION OF VARIOUS RADIATION DETECTORS

Detector	Signal (V)	Pulse Duration (μsec)
Sodium iodide scintillator with photomultiplier tube	10^{-1}-1	0.23*
Lutetium oxyorthosilicate scintillator with photomultiplier tube	10^{-1}-1	0.04*
Liquid scintillator with photomultiplier tube	10^{-2}-10^{-1}	10^{-2}*
Lutetium oxyorthosilicate scintillator with avalanche photodiode	10^{-5}-10^{-4}	0.04*
Direct semiconductor detector	10^{-4}-10^{-3}	10^{-1}-1
Gas proportional counter	10^{-3}-10^{-2}	10^{-1}-1
Geiger-Müller counter	1-10	50-300

*Mean decay time.

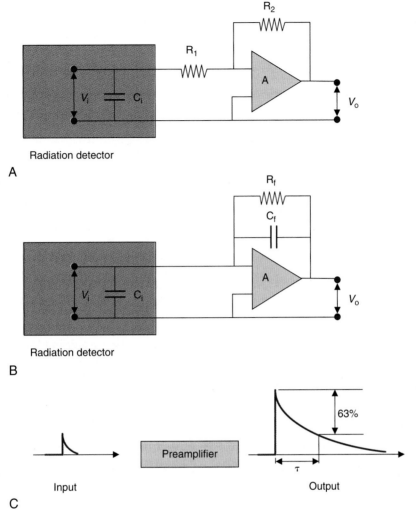

FIGURE 8-2 *A*, Simplified circuit diagram of a voltage-sensitive preamplifier. The output voltage is determined by the amount of charge from the radiation detector, the input capacitance C_i, and the resistances R_1 and R_2. *B*, Simplified circuit diagram of a charge-sensitive preamplifier. The output voltage is determined by the charge from the radiation detector and the value of the feedback capacitor C_f. The symbol ─|A⟩─ represents a voltage or current amplifying element. *C*, Input and output pulse signals for a charge-sensitive preamplifier. $\tau = (R_f \times C_f)$ is the time constant of the pulse-shaping circuit.

varies with the type of detector. With scintillation detectors that use photomultiplier (PM) tubes, the PM tube already provides a considerable degree of amplification (10^6-10^7); thus relatively little additional amplification may be needed. Typically, a preamplifier *gain factor* (ratio of output to input amplitudes) of 5-20 is used for these detectors; however, some NaI(Tl):PM tube systems employ no preamplifier gain (gain factor of 1).

Detectors producing smaller signals, such as semiconductor detectors, may require a relatively high level of preamplifier gain, perhaps in the range of 10^3-10^4. It is not a trivial problem to design an amplifier that

provides this amount of gain without introducing "noise signals" and temperature-related gain instabilities. Most of the modern high-gain preamplifiers employ *field-effect transistors*, which provide the desired low-noise and temperature-stability characteristics.

For energy-sensitive detectors, the preamplifier must operate in a *linear* fashion; that is, the amplitude of the signal out must be directly proportional to the amount of charge delivered to it by the detector. This preserves the relationship between pulse amplitude and energy of the radiation event detected, so that subsequent energy analysis may be applied to the pulse signals.

For the best results, the preamplifier component should be located as close as physically possible to the detector component. This maximizes the electronic signal-to-noise ratio (SNR) by amplifying the signal before additional noise or signal distortion can occur in the long cable runs that frequently separate the detector from the rest of the signal-processing components. This is particularly critical for detectors with small output signals (e.g., semiconductor detectors or scintillation detectors used for detecting low-energy radiations). It also is important for applications in which energy resolution is critical (see Chapter 10, Section B.7). Frequently, detectors and preamplifiers are packaged and sold as single units.

B. AMPLIFIERS

1. Amplification and Pulse-Shaping Functions

The amplifier component of a nuclear counting instrument has two major functions: (1) to amplify the still relatively small pulses from the preamp to sufficient amplitude (volts) to drive auxiliary equipment (pulse-height analyzers, scalers, etc.), and (2) to reshape the slow decaying pulse from the preamp into a narrow one to avoid the problem of pulse pile-up at high counting rates and to improve the electronic SNR.

The gain factor on an amplifier may range from ×1 to ×1000. Usually it is adjustable, first by a coarse adjustment (i.e., ×2, ×4, ×8) and then by a fine gain adjustment providing gain factors between the coarse steps. The coarse gain adjustment permits amplification of pulses over a wide range of amplitudes from different detectors and preamplifiers to the maximum output capability of the amplifier. The fine gain adjustment permits precise calibration of the relationship between amplifier output pulse amplitude (volts) and radiation energy absorbed (keV or MeV). For example, a convenient ratio might be 10 V of pulse amplitude per 1 MeV of radiation energy absorbed in the detector.

Pulse shaping—i.e., pulse shortening—is an essential function of the amplifier. The output of the preamp is a sharply rising pulse that decays with a time constant of about 50 μsec, returning to baseline after approximately 500 μsec. Thus if a second pulse occurs within 500 μsec, it rides on the tail of the previous pulse, providing incorrect amplitude information (Fig. 8-3). The system could not operate at counting rates exceeding a few hundred events per second without introducing this type of amplitude distortion.

The pulse-shaping circuits of the amplifier must provide an output of cleanly separated pulses, even though the output pulses from the preamp overlap. It must do this without distorting the information in the preamplifier signal, which is, mainly, (1) pulse amplitude

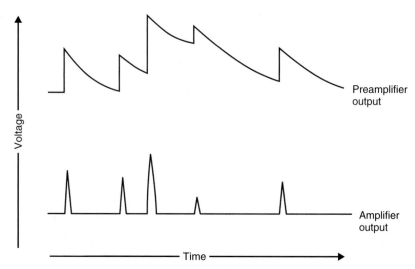

FIGURE 8-3 Sequence of pulse signals in a radiation counting system. *Top,* Relatively long preamplifier time constant results in overlapping of pulse signals. *Bottom,* Amplifier output pulses have been shortened but without significant loss of amplitude or timing information.

(proportional to the energy of radiation event for energy-sensitive detectors) and (2) rise time (time at which the radiation event was detected). An additional function of the pulse-shaping circuits is to discriminate against electronic noise signals, such as microphonic pickup and 50- to 120-Hz power line frequency.

The most common methods for amplifier pulse shaping are resistor-capacitor (RC), gaussian, and delay-line methods. The RC technique, commonly referred to as *RC shaping,* is described to illustrate the basic principles. More detailed circuit descriptions are found in the sources cited in the Bibliography at the end of this chapter.

2. Resistor-Capacitor Shaping

Basic RC pulse-shaping circuits are shown in Figure 8-4. When a sharply rising pulse of relatively long duration (e.g., preamplifier output pulse) is applied to a capacitor-resistor (CR), or *differentiation circuit* (see Fig. 8-4A), the output is a rapidly rising pulse that decays with a time constant τ_d determined by the RC product of the circuit components (Equation 8-4). The amplitude of the output pulse depends on the amplitude of the sharply rising portion of the input pulse and is insensitive to the "tail" of any preceding pulse. Note that a CR differentiation circuit also is used for pulse shaping in the preamplifier; however, the time constants used in the preamplifier circuits are much longer than those used in the amplifier. Figure 8-4A also

illustrates how the CR circuit discriminates against low-frequency noise signals.

Figure 8-4B shows an RC, or *integration* circuit. (Note that differentiation and integration differ only by the interchanging of the resistor R and the capacitor C.) When a sharply rising pulse is applied to this circuit, the output is a pulse with a shape described by

$$V_o(t) = V_i(1 - e^{-t/RC}) \qquad (8\text{-}5)$$

where V_i is the amplitude of the input pulse and RC = τ_i is the integration time constant of the circuit. This circuit discriminates effectively against high-frequency noise, as illustrated in Figure 8-4B.

Figure 8-5A shows a pulse-shaping circuit combining differentiation and integration stages. When the time constants of the two circuits are equal ($\tau = \tau_i = \tau_d$), the output is a pulse that rises to a maximum value in a time equal to 1.2τ and then decays to approximately zero in 7τ. The maximum amplitude of the output pulse is determined by the amplitude of the input pulse. For scintillation and semiconductor detectors, a time constant in the range $\tau \sim 0.25\text{-}5.0$ μsec usually is chosen. Thus the output pulse is shortened considerably relative to the pulse from the preamplifier (50-500 μsec) and is suitable for high counting rate applications. Except for a very small negative overshoot at the end of the pulse, the output pulse from this circuit has only one polarity (positive in Fig. 8-5A) and is called a *unipolar* output.

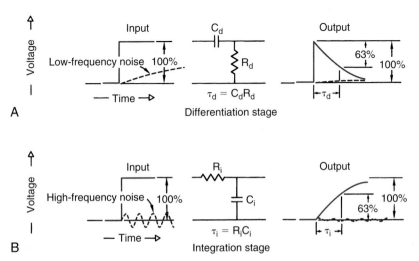

A

B

FIGURE 8-4 Basic resistor-capacitor pulse-shaping circuits. *A,* Differentiation provides a sharply rising output signal that decays with time constant τ_d and discriminates against low-frequency noise. *B,* Integration circuit provides an output pulse that rises with time constant τ_d and discriminates against high-frequency noise.

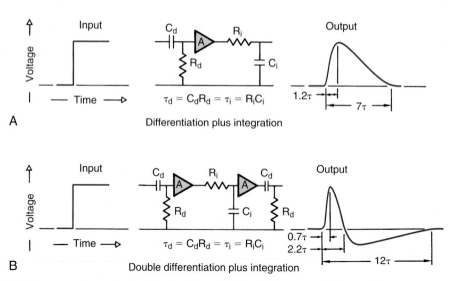

FIGURE 8-5 Resistor-capacitor pulse-shaping circuits combining differentiation and integration stages. *A*, Differentiation followed by integration. *B*, Differentiation-integration-differentiation circuit.

Figure 8-5B illustrates another type of shaping, called *double differential* shaping. The output pulse from this circuit has both positive and negative components and therefore is a *bipolar* pulse. For equal time-constant values, the bipolar output pulse has a shorter rise time and positive portion and a longer total duration than the unipolar output pulse. Unipolar pulses are preferred for signal-to-noise characteristics and are used where energy resolution is important. Bipolar pulses are preferred for high counting rate applications.

Research-grade amplifiers generally are provided with adjustable pulse-shaping time constants. A longer time constant provides better pulse amplitude information and is preferred in applications requiring optimal energy resolution, for example, with semiconductor detectors (Chapter 10, Section C.1). A shorter time constant is preferred in applications requiring more precise event timing and higher counting rate capabilities, such as scintillation cameras (Chapters 13 and 14) and coincidence detection of positron annihilation photons (Chapter 18).

3. Baseline Shift and Pulse Pile-Up

Baseline shift and *pulse pile-up* are two practical problems that occur in all amplifiers at high counting rates. Baseline shift is caused by the negative component that occurs at the end of the amplifier output pulse. A second pulse occurring during this component will be slightly depressed in amplitude (Fig. 8-6A).

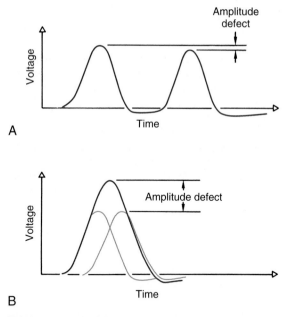

FIGURE 8-6 *A*, Schematic representation of baseline shift, caused by a pulse riding on the "tail" of a preceding pulse. *B*, Pulse pile-up effects for two pulses occurring very close together in time.

Inaccurate pulse amplitude and an apparent shift (decrease) in energy of the detected radiation event are the result (see Fig. 10-9).

Special circuitry has been developed to minimize baseline shift. This is called *pole zero cancellation*, or *baseline restoration*. This type of circuitry is employed in modern scintillation cameras to provide a high counting rate capability, particularly for cardiac studies. These circuits are described in the sources cited in the Bibliography at the end of this chapter.

At high counting rates, amplifier pulses can occur so close together that they fall on top of each other. This is referred to as *pulse pile-up* (Fig. 8-6B). When this happens, two pulses sum together and produce a single pulse with an amplitude that is not representative of either. Pulse pile-up distorts energy information and also contributes to counting losses (dead time) of the detection system, because two pulses are counted as one (Chapter 11, Section C).

Both baseline shift and pulse pile-up can be decreased by decreasing the width of the amplifier pulse (i.e., the time constant of the amplifier); however, shortening of the time constant usually produces poorer SNR and energy resolution. It is generally true that all the factors that provide high count rate capabilities in amplifiers also degrade energy resolution (Chapter 10, Section B.7).

Generally, amplifiers with double differentiation or double delay-line bipolar outputs are employed with NaI(Tl):PM tube detectors that must handle high counting rates. The bipolar output helps to avoid baseline shift problems, allowing good pulse-height determination at high counting rates. In addition, short time constants of 0.025-0.5 μsec are used. The relatively poor inherent energy resolution of NaI(Tl):PM tube detectors is not affected significantly by this type of amplifier, and a high counting rate capability is provided. Semiconductor detectors usually require much more sophisticated amplifiers, with unipolar pulse shaping, longer time constants (0.5-8 μsec), and circuits for stabilizing the baseline to maintain their exceptionally good energy resolution at high counting rates (Chapter 10, Section C.1).

C. PULSE-HEIGHT ANALYZERS

1. Basic Functions

When an energy-sensitive detector is used [e.g., NaI(Tl):PM tube or a semiconductor detector], the amplitude of the voltage pulse from the amplifier is proportional to the amount of energy deposited in the detector by the detected radiation event. By examining the amplitudes of amplifier output pulses, it is possible to determine the energies of detected radiation events. Selective counting of only those pulses within a certain *amplitude* range makes it possible to restrict counting to a selected *energy* range and to discriminate against background, scattered radiation, and so forth outside the desired energy range (see Fig. 10-6).

A device used for this purpose is called a *pulse-height analyzer* (PHA). A PHA is used to select for counting only those pulses from the amplifier falling within selected voltage amplitude intervals or "channels." If this is done for only one channel at a time, the device is called a *single-channel analyzer* (SCA). A device that is capable of analyzing simultaneously within many different intervals or channels is called a *multichannel analyzer* (MCA). Basic principles of these instruments are discussed in the following sections.

2. Single-Channel Analyzers

An SCA is used to select for counting only those pulses from the amplifier that fall within a selected voltage amplitude range. At this stage in the system voltage amplitude is proportional to radiation energy deposited in the detector, so it is equivalent to selecting an energy range for counting. Modern amplifiers produce output pulses with amplitudes in the range of 0-10 V. Therefore the voltage selection provided by most SCAs is also in the 0- to 10-V range.

An SCA has three basic circuit components (Fig. 8-7): a *lower-level discriminator* (LLD), an *upper-level discriminator* (ULD), and an *anticoincidence* circuit. The LLD sets a threshold voltage amplitude V (or energy E) for counting. The ULD sets an upper voltage limit $V + \Delta V$ (or $E + \Delta E$). The difference between these voltages (or energies), ΔV (or ΔE), is called the *window width*. Usually the LLD and ULD voltages are selected by means of potentiometer or other electronic controls that are adjusted to select some fraction of a 10-V reference voltage.

The LLD and ULD establish voltage levels in electronic circuits called *comparators*. As their name implies, these circuits compare the amplitude of an input pulse with the LLD and ULD voltages. They produce an output pulse only when these voltages are exceeded. Pulses from the comparator circuits are then sent to the anticoincidence circuit, which produces an output pulse when only one (LLD) but not both (ULD and LLD) pulses are present (see Fig. 8-7). Thus only those input pulses with amplitudes between V and $V + \Delta V$ (i.e., within the selected energy window) cause output pulses from the SCA.

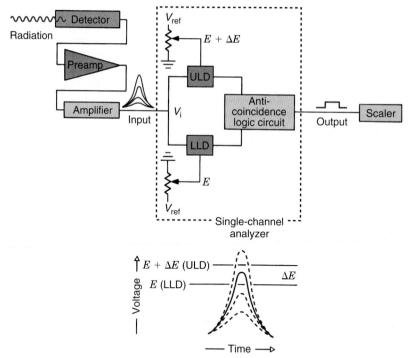

FIGURE 8-7 Principles of a single-channel pulse-height analyzer. *Top,* Electronic components that are used to generate an output pulse only when pulse amplitude falls between voltages established by lower-level discriminator (LLD) and upper-level discriminator (ULD) circuits. These voltages are an adjustable portion of a reference voltage V_{ref}. *Bottom,* LLD and ULD voltages in effect establish an energy range (E to $E + \Delta E$) for counting because pulse voltage amplitude V is proportional to radiation event energy E. Only pulse signals within the ΔE bracket (solid line) are counted.

The SCA output pulses are used to drive counters, rate meters, and other circuits. The output pulses from the SCA are all of the same amplitude and shape (typically 4-V amplitude, 1-μsec duration). Their amplitudes no longer contain information about radiation energy, because this information has already been extracted by the SCA.

Commercially made SCAs frequently have two front-panel controls: a lower-level (voltage V or energy E) control and a window (ΔV or ΔE) control. The LLD control is also called the *base level* on some instruments. The upper-level voltage is determined by electronic summation of lower-level and window voltages on these instruments.

Some instruments include "percent window" selections. With these instruments, the window width voltage is selected as a certain percentage of the window center voltage. (The window center voltage is the lower level voltage V plus one half of the window voltage, $\Delta V/2$.) For example, if one were to set the window center at 2 V with a 20% window, the window width would be 0.4 V (20% of 2 V), and the window would extend from 1.8 to 2.2 V.

On many nuclear medicine instruments, manufacturers have provided pushbuttons to select automatically the analyzer lower level and window voltages appropriate for commonly used radionuclides. In these systems, the pushbuttons insert calibrated resistance values into the SCA circuitry in place of the variable resistances shown in Figure 8-7.

Another possibility on some instruments is to remove the upper-level voltage limit entirely. Then all pulses with amplitudes exceeding the lower-level voltage result in output pulses. An analyzer operated in this mode is sometimes called a *discriminator.* Many auxiliary counting circuits (e.g., scalers and rate meters) have a built-in discriminator at their inputs to reject low-level electronic noise pulses.

3. Timing Methods

Accurate time placement of the radiation event is important in some nuclear medicine applications. For example, in the scintillation camera (Chapter 13), accurate timing is required to identify the multiple phototubes involved in detecting individual radiation events striking the NaI(Tl) crystal (i.e., for determining the location of each event with the position logic of the camera). Even more critical timing problems occur in coincidence counting of positron annihilation photons (Chapter 18) and in the liquid scintillation counter (Chapter 12, Section C).

Most SCAs used in nuclear medicine employ *leading-edge* timing. With this method, as shown in Figure 8-8A, the analyzer output pulse occurs at a fixed time T_D following the instant at which the rising portion of the input pulse triggers the LLD. This type of timing is adequate for many applications;

however, it suffers a certain amount of inaccuracy [5 to 50 nsec with NaI(Tl) coupled with a PM tube] because the timing of the output pulse depends on the amplitude of the input pulse. This timing variation Δt is called *timing walk*.

More precise timing is obtained with analyzers employing fast timing techniques. One such method is called *zero-crossover* timing (Fig. 8-8B). This method requires a bipolar input pulse to the SCA. The output pulse occurs at the time of crossover of the bipolar pulse from a positive to a negative voltage

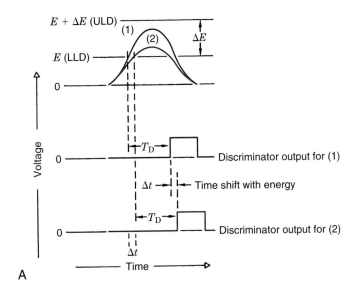

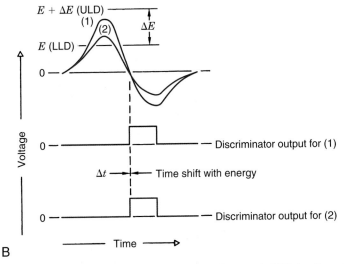

FIGURE 8-8 Examples of timing methods used in pulse-height analyzers. *A,* With leading-edge timing, the output pulse occurs at a fixed time T_D after the leading edge of the pulse passes through the lower-level discriminator (LLD) voltage. *B,* With zero-crossover timing, output pulse occurs when the bipolar input pulse passes through zero. The latter is preferred for precise timing because there is very little time shift with different pulse amplitudes (energy). ULD, upper-level discriminator.

value. The zero-crossover method is much less sensitive to pulse amplitude than the leading-edge method and can provide timing accuracy to within ± 4 nsec with NaI(Tl):PM tube detectors. Other fast-timing methods include *peak detection* and *constant fraction* techniques. They are discussed in the sources cited in the Bibliography at the end of this chapter.

4. Multichannel Analyzers

Some applications of pulse-height analysis require simultaneous recording of events in multiple voltage or energy windows. One approach is to use many SCAs, each with its own voltage window. For example, some imaging devices have two or three independent SCAs to record simultaneously the multiple γ-ray energies emitted by nuclides such as ^{67}Ga; however, this approach is unsatisfactory when tens or even thousands of different windows are required, as in some applications

of pulse-height spectroscopy (Chapter 10). Multiple SCAs would be expensive, and the adjusting and balancing of many different analyzer windows would be a very tedious project.

A practical solution is provided by an MCA. Figure 8-9 demonstrates the basic principles. The heart of the MCA is an *analog-to-digital converter* (ADC), which measures and sorts out the incoming pulses according to their amplitudes. The pulse amplitude range, usually 0-10 V, is divided by the ADC into a finite number of discrete intervals, or *channels,* which may range from 100 in small analyzers to as many as 65,536 (2^{16}) in larger systems. Thus, for example, the ADC in a 1000-channel analyzer would divide the 0- to 10-V amplitude range into 1000 channels, each 10 V/1000 = 0.01 V wide: 0-0.01 V corresponding to channel 1, 0.01-0.02 V to channel 2, and so forth. The ADC converts an *analog* signal (volts of pulse amplitude),

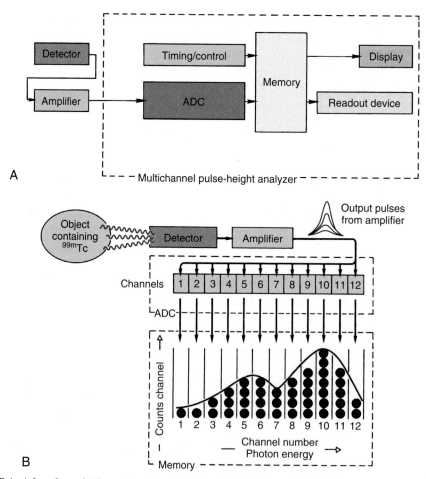

FIGURE 8-9 Principles of a multichannel analyzer (MCA). *A,* Basic components. *B,* Example of pulse sorting according to amplitude for radiation events detected from an object containing ^{99m}Tc. ADC, analog-to-digital converter.

which has an essentially infinite number of possible different values, into a *digital* one (channel number), which has only a finite number of integer values (see Fig. 8-9). In addition to their use in counting systems, ADCs are also used in the interface between nuclear medicine imaging detectors and computer systems.

For each analyzer channel, there is a corresponding storage location in the MCA *memory*. The MCA memory counts and stores the number of pulses recorded in each analyzer channel. The number of memory storage locations available determines the number of MCA channels. The sorting and storage of the energy information from radiation detectors with an MCA are used to record the *pulse-height spectrum* (counts per channel versus channel number, or energy), as shown in Figure 8-9B.

MCAs also are available as boards that plug into personal computers. The computer is used to program the settings on the MCA (i.e., number of channels to be used and voltage range to be selected) and also to control acquisition of data from the detector (acquisition start time and acquisition duration). The computer also is used to display the resulting data (number of counts per MCA channel for the measurement period) that are transferred from the MCA card onto the computer's hard disk. Many MCA boards are capable of receiving data from several inputs at once and can therefore be used to acquire and display data from several detector units simultaneously.

Two types of ADCs are commonly used in nuclear medicine for MCAs and for interfaces between scintillation cameras and computers. In the *Wilkinson,* or ramp, converter (Fig. 8-10), an input pulse from the radiation detector and amplifier causes an amount of charge to be deposited onto a capacitor at the ADC input. The amount of charge deposited depends on the pulse amplitude or energy. The capacitor discharges through a resistor, with a relatively long RC time constant. While the capacitor is discharging, a gate pulse activates a clock oscillator to produce a train of pulses that are counted in a counting circuit. When the capacitor has been discharged, the gate pulse is terminated and the clock oscillator is turned off. The number of clock pulses counted is determined by the capacitor discharge time, which in turn is determined by the initial amount of charge deposited on the capacitor and thus depends on the amplitude of the input pulse. The MCA control circuits

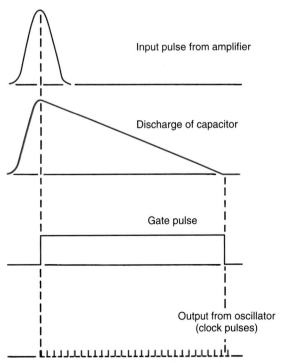

FIGURE 8-10 Principles of analog (pulse amplitude) to digital (channel number) conversion in the Wilkinson, or ramp, converter. Input pulse is used to charge a capacitor, and discharge time, which is proportional to pulse amplitude, is measured using a clock oscillator.

increment by one count the memory channel corresponding to the number of clock pulses counted, then clear the input circuitry and prepare the MCA to accept the next input pulse.

In the *successive approximation* (SA) converter, digitization occurs by comparing the pulse amplitude with a selected sequence of voltage levels. The first comparison level is equal to one half of the full-scale (maximum) value. If the pulse amplitude is greater than this level, the first digital "bit" is set to "1"; if not, it is set to "0." The comparison voltage level then is either increased or decreased by one half of its initial level, (i.e., to 25% or 75% of full scale) depending on whether the pulse amplitude did or did not exceed the initial level. The comparison is repeated and the second digital bit is recorded as "1" or "0," depending on whether the pulse amplitude is greater or smaller than the new comparison voltage level. The comparisons are repeated through several steps, each time decreasing the voltage increment by one half. The final set of bits provides a binary (base 2) representation for the amplitude of the input pulse.

For both the ramp and SA converters, the output is represented as a binary number between 0 and 2^n. The value of n determines the number of possible digital levels into which the input pulse amplitude can be converted. For example, an 8-bit converter, for which $n = 8$, divides the input range into 256 digital levels ($2^8 = 256$), a 10-bit converter into 1024 levels ($2^{10} = 1024$), and so forth. The larger the number of bits, the more precisely the ADC can determine the pulse amplitude. Thus an 8-bit converter can determine amplitude to a precision of one part in 256, a 10-bit converter to one part in 1024, and so forth. Generally, a larger number bit is favored for precision, but the digital conversion process then requires somewhat more time and the digitized values for pulse amplitude require greater amounts of computer storage space. Most nuclear medicine studies can be performed with 8-bit converters, but 10- and 12-bit converters also are used for situations in which precision is a prime concern (e.g., high-resolution energy spectroscopy with semiconductor detectors; see Chapter 10, Section C.1).

A finite amount of time is required for the digital conversion processes described earlier. For example, for a 10-bit (1024-channel) ramp converter with a 100-MHz (10^8 cycle/sec) clock, the capacitor discharge time required for an event in the 1000th channel (1000 clock pulses) is 1000 pulses ÷ 10^8 pulse/sec = 10^{-5} sec or 10 µsec. For an SA converter, time is needed for each of the voltage comparisons; for example, a 10-bit SA converter must perform a sequence of 10 voltage comparisons, each requiring a fraction of a microsecond to complete.

In addition to the conversion process, time is required to increment the memory location, reset the clock pulse counter on comparison voltage levels, and so on. The ADC can therefore be a "bottleneck" in MCAs as well as in the digital conversion process for signals from a scintillation camera. Modern ADCs, however, can digitize events at rates in excess of 1 million counts/sec; therefore ADC speed need not be a limiting factor for applications involving NaI(Tl) detectors, for which the primary time limitation is the decay time of the individual scintillation events.

Most MCAs have additional capabilities, such as offset or expansion of the analyzer voltage range and time histogram capabilities. These are discussed in detail in MCA operator manuals. Some scintillation cameras, well counters, and liquid scintillation counters contain MCAs that are used to examine and select energy windows of interest.

D. TIME-TO-AMPLITUDE CONVERTERS

In certain applications it is useful to be able to measure the distribution of time differences between incoming pulses from a detector, much in the same way that an MCA measures the distribution of energies deposited in the radiation detector. For example, we might wish to use two opposing scintillation detectors to view a positron-emitting radionuclide and to measure the time difference between the detection of two annihilation photons. If the difference is "small" ($\lesssim$ a few nanoseconds), they are highly likely to arise from a single positron annihilation event, whereas if the difference is not small, they probably reflect two independent events.

The *time-to-amplitude converter* (TAC) produces an output signal with a voltage proportional to the time difference between two *logic pulses* supplied to the input. The logic pulses typically come from the output of a discriminator or SCA (see Section C.2) attached to a radiation detector and have a standard box shape with a well-defined amplitude and duration. The concepts of a TAC are illustrated in Figure 8-11. The first logic pulse (known as the *START signal*) is used to start the charging of a capacitor by a constant current source. The second logic pulse (the *STOP signal*) is used to terminate the charging of the capacitor. Because the capacitor is charged from a constant current source, the voltage across the capacitor increases linearly with time and is therefore proportional to the time interval between the START and the STOP signals.

The voltage across the capacitor determines the amplitude of the output voltage pulse of the TAC and is therefore also proportional to the time interval between the two logic pulses. The output pulses from the TAC can be fed to a standard MCA to produce a histogram of the distribution of time differences between the two logic pulses. The MCA is calibrated in terms of time units by supplying the TAC with pulses with a known time interval between them. Alternatively, they can be used to set a timing threshold for accepting or rejecting two detected events as being coincident (e.g., originating from the same nuclear decay). Following the STOP

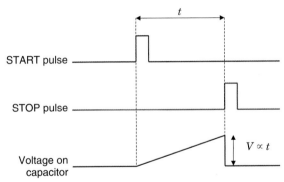

FIGURE 8-11 Principles of a time-to-amplitude converter (TAC). A START pulse is used to start the charging of a capacitor by a constant current source, which is terminated by the STOP pulse. The voltage developed across the capacitor is proportional to the time interval between the START and STOP pulses.

signal, the TAC is reset by discharging the capacitor, so it is ready for the next START signal.

E. DIGITAL COUNTERS AND RATE METERS

1. Scalers, Timers, and Counters

Digital counters are used to count output signals from radiation detectors after pulse-height analysis of the signals. A device that counts only pulses is called a *scaler*. An auxiliary device that controls the scaler counting time is called a *timer*. An instrument that incorporates both functions in a single unit is called a *scaler-timer*. These devices are often referred to under the generic name of *counters*. The number of counts recorded and the elapsed counting time may be displayed on a visual readout, or, more commonly, the output of the scaler-timer may be interfaced to a personal computer automated data processing. Computer-driven counters, which reside on a board that is placed inside of the computer, are also common.

Figure 8-12 shows schematically the basic elements of a scaler-timer. The input pulse must pass through an electronic "gate" that is opened or closed by front-panel switches or pushbutton controls that select the mode of operation. When the gate is open, the pulses pass through to decimal counter assemblies (DCAs). Each DCA records from zero to nine events. The tenth pulse resets the counter assembly to zero and sends a pulse to the next DCA in the series. The number of counter

assemblies determines the number of decades of scaler capacity. Thus a six-decade scaler has six DCAs and a counting capacity from 0 to 999,999 counts. (Usually the "1-millionth count" resets the scaler to "0" and turns on an overflow light). Data from each DCA are transferred to the display for continuous visual readout of the number of counts recorded during the counting interval.

As shown in Figure 8-12, the scaler gate can be controlled in a number of different ways. In *preset-time* mode, the gate is controlled by a timer circuit (usually an oscillator-driven clock circuit) that opens the gate for a counting time selected by front-panel switches, or by a computer. The counting interval begins when a "start" button is depressed and is terminated automatically when the selected counting time has elapsed. In *preset-count* (PSC) mode, the counting interval ends when a preselected number of counts has been recorded. PSC mode is used when one wants to achieve the same degree of statistical reliability for all measurements in a series of counting measurements (see Chapter 9). When the PSC mode is used, a method must be available to determine the elapsed time for each counting measurement (e.g., a visual display or printout of elapsed counting time) so that counting rates for each measurement can be determined (preset counts/elapsed time).

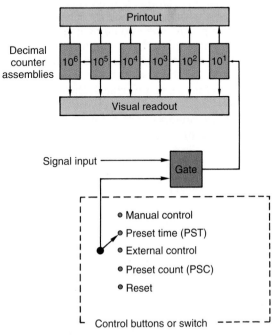

FIGURE 8-12 Schematic representation of components and controls for a scaler-timer.

External control of the scaler gate may be provided by an external timer or a sample-changer assembly. *Manual control* permits the operator to start and stop the counting interval by depressing front-panel "start" and "stop" buttons. In computer-controlled counters, all these parameters are controlled by keyboard entry and appropriate interface software.

The maximum counting rate capability depends on the minimum time separation required between two pulses for the scaler to record them as separate events. A 20-MHz scaler (2×10^7 counts/sec) can separate pulses that are spaced by 50 nsec, or 5×10^{-8} sec apart (2×10^7 counts/sec is equivalent to 1 count/5×10^{-8} sec). Most modern scalers are capable of 20- to 50-MHz counting rates, which means they can count at rates of several hundred thousand counts per second with losses of 1% or less caused by pulse overlap (see Chapter 11, Section C). Because pulse resolving times of most radiation detectors and their associated preamplifiers and amplifiers are on the order of 1 µsec, the counting rate limits of modern scalers are rarely of practical concern.

2. Analog Rate Meters

An *analog rate meter* is used to determine the average number of events (e.g., SCA output pulses) occurring per unit of time. The average is determined continuously, rather than during discrete counting intervals, as would be the case with a scaler-timer. The output of a rate meter is a continuously varying voltage level proportional to the average rate at which pulses are received at the rate meter input. The output voltage can be displayed on a front-panel meter or interfaced through a continuously sampling ADC to a personal computer. Rate meters are commonly used in radiation monitors (see Figs. 7-3 and 7-11).

Figure 8-13 shows the basic components of an analog rate meter. Input pulses pass through a pulse shaper, which shapes them to a constant amplitude and width. Each shaped pulse then causes a fixed amount of charge, Q, to be deposited on the capacitor C. The rate at which the charge discharges through the resistor R is determined by the product R × C, which is called the *rate meter time constant* τ.

Suppose that input pulses arrive at an average rate $\bar{n}$ pulses per second. The capacitor discharge then produces an average current I through the resistor R given by

$$I = \bar{n}Q \qquad (8\text{-}6)$$

By Ohm's law, this causes an average voltage

$$V = \bar{n}Q\text{R} \qquad (8\text{-}7)$$

to appear at the input to amplifier A. If the amplification factor of this amplifier is k, the average output voltage V_o is given by

$$V_o = k\bar{n}Q\text{R} \qquad (8\text{-}8)$$

Thus if k, Q, and R are constant factors for a given measurement, average output voltage V_o is proportional to average input counting rate $\bar{n}$.

The output voltage V_o can be used to drive a meter to read the average counting rate. The calibration usually is performed by adjusting the amplifier gain factor k. This factor is adjusted to select different full-scale ranges for the readout device, for example, 0-1000 cpm, 0-10,000 cpm, and so on.

A rate meter that follows the relationship described by Equation 8-8 is called a *linear* rate meter. For some applications it is desirable to have a logarithmic relationship:

$$V_o = k \log(\bar{n}Q\text{R}) \qquad (8\text{-}9)$$

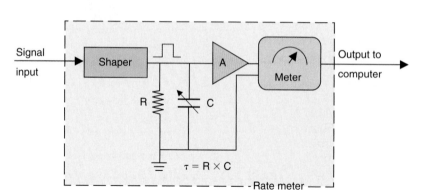

FIGURE 8-13 Schematic representation of an analog rate meter. Adjustable capacitor C provides variable *rate meter time constant*, τ.

The logarithmic conversion usually is performed by a logarithmic amplifier. *Logarithmic rate meters* have the advantage of a very wide range of counting rate measurement, typically 4 or 5 decades, without the need to change range settings as with a linear rate meter; however, it is more difficult to discern small changes in counting rate with a logarithmic rate meter.

The voltage relationships described by Equations 8-8 and 8-9 apply to *average* values only. When the input pulse rate changes, the rate meter output voltage does not respond instantaneously but responds during a period determined by the rate meter time constant τ. Figure 8-14 illustrates the response characteristic of a linear rate meter. The relationship between indicated counting rate R_i and the new average counting $\bar{R}_a$, following a change occurring at time $t = 0$ from a previous average value, $\bar{R}_0$, is given by

$$R_i = \bar{R}_a - (\bar{R}_a - \bar{R}_0)e^{-t/\tau} \qquad (8\text{-}10)$$

The rate meter reading (or output voltage) approaches its new average value exponentially with time t. Typically, three to five time constants are needed to reach a new stable value.

The rate meter time constant is selected by a front-panel switch (usually by adjusting the capacitor value C) and may range from 100ths of a second to 10s of seconds. Figure 8-14 shows that a rate meter actually provides a distorted representation of counting rate versus time (rounded edges and delayed response). This distortion can be minimized by choosing a very short time constant (Fig. 8-14 A). A long time constant has the advantage of smoothing out statistical fluctuations in counting rate, but it produces a more distorted representation of changes in counting rates (Fig. 8-14B).

F. COINCIDENCE UNITS

Coincidence units are logic units that produce a pulse only if two or more input pulses occur within a particular coincidence time window. One method for doing this is to sum the input pulses and pass them through a discriminator that is set just below the amplitude that would be seen if two or more pulses occurred simultaneously. As shown in Figure 8-15, the unit supplies an output pulse only when two or more pulses overlap in time and the discriminator threshold is exceeded. The *coincidence timing window* is the maximum time interval between two pulses for them to be counted as being in coincidence. In this illustration, this is twice the width of the input pulses (2τ).

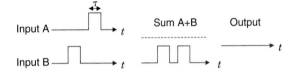

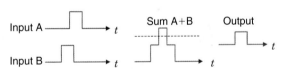

FIGURE 8-15   Principles of a coincidence unit. The signals from the inputs are combined and passed through a discriminator set just below the threshold required for simultaneous pulses on the two inputs. In this example, the coincidence window is approximately twice the width of the input pulses (2τ).

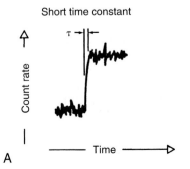

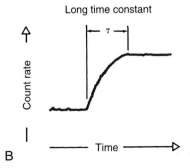

FIGURE 8-14   Rate meter response to a sudden change in counting rate for different rate meter time constants. A short time constant (*A*) reflects the change more accurately, but a long time constant (*B*) provides better averaging of statistical noise fluctuations.

Coincidence units often have up to four inputs and permit selection of two-way, three-way, and four-way coincidences between the input pulses. Of most interest in nuclear medicine is the use of coincidence units to identify the two-way coincidence events resulting from the detection of the annihilation photons from positron-emitting radionuclides (Chapter 18, Section A.1). In practice, most positron imaging systems record and compare the time for each detected event using digital electronics, rather than using the analog coincidence units described previously. In this case, the coincidence timing window is just the maximum time difference allowed between two events for them still to be considered in coincidence. Another use of coincidence units is to minimize background in liquid scintillation counting (Chapter 12, Section C).

G. HIGH-VOLTAGE POWER SUPPLIES

The high-voltage (HV) power supply provides the charge collection voltage for semiconductor, gas proportional, and GM detectors and the accelerating voltage for electron multiplication in the PM tubes used with scintillation detectors such as NaI(Tl) and liquid scintillators. The HV power supply converts the alternating current voltage provided by the line source into a constant or direct current (DC) voltage.

Whereas variation of the HV has little effect on the output pulse amplitude with semiconductor and GM detectors, changes in the HV with gas proportional or scintillation detectors strongly affect their output pulse amplitude. For example, a 1% change in the HV on a scintillation detector PM tube can change the output pulse amplitude by 10% or more because the HV on the PM tube (and on gas proportional counters) determines the multiplication factor for the number of electrons caused by an ionization event in those detectors (Chapter 7, Section C.2).

Instabilities in HV power supplies can arise from a number of factors, such as temperature changes, variations in line voltage, and the amount of current drawn by the detector (commonly referred to as the *output load*). The output can also drift over time. In a well-regulated HV power supply suitable for scintillation detectors, drifting of the output with time and temperature are more important than the effects of line voltage and current loads (unless maximum current ratings are exceeded); however, the former

problems are still relatively small, because modern HV supplies are very stable for long periods and over wide temperature ranges.

The output current rating of the HV power supply must be sufficient for the particular detector system. Most scintillation detectors draw about 1 mA of current, for which the 0- to 10-mA rating of most commercial HV supplies is adequate. If the current load is inadvertently increased above this limit, it will affect the stability and may even damage the HV supply. Thus the current requirements of the detector or detectors should be within the specified limits for the HV supply. The current requirements need to be specified at the intended operating voltage of the detector, because the current load drawn by the detector will increase with the applied voltage. Most commercial HV supplies have an overload protection circuit that will shut off the unit if the recommended current load is exceeded.

Superimposed on the DC output of the HV supply is a time-varying component, usually of relatively small amplitude, referred to as "ripple." The amplitude of ripple ranges from 10 to 100 mV in most commercial units. Ripple in the HV supply can be a serious problem with high-resolution semiconductor detectors, because it produces noise in the detector output and reduces the energy resolution of the detector. HV supplies used in conjunction with high-resolution semiconductors usually have a ripple of less than 10 mV.

H. NUCLEAR INSTRUMENT MODULES

Most of the counting and imaging instruments used in nuclear medicine are dedicated to specific and well-defined tasks. Usually, they are designed as self-contained "hard-wired" units, with no capability for interchanging components, such as amplifiers, SCAs, or scalers, between different instruments. Although these integrated circuits generally result in an efficiently designed and attractively packaged instrument, there are some applications, especially in research, for which interchangeability of components is highly desirable. For example, most scalers, timers, and rate meters can be used with any detector system, but different detectors may require different amplifiers, and different types of PHAs may be desired for different pulse-timing requirements.

Flexibility and interchangeability of components are provided by the *nuclear instrument module* (NIM). Individual NIM components (such as scalers and amplifiers) slide into slots in a master "bin" from which they draw their operating power. They have standard input and output signals and are interconnectable with standard cables and connectors.

A NIM system generally is more expensive than a dedicated system with the same capabilities; however, it has the advantage that it can be upgraded or applied to different radiation detectors and counting problems by replacement of individual components rather than replacing the entire unit. A wide variety of component types and performance specifications are available in the NIM standard.

I. OSCILLOSCOPES

The oscilloscope is an instrument that displays as a function of time the amplitude (voltage) and frequency of signals. It is used for examining the pulses from the pulse-processing units described in the previous sections of this chapter and for testing, calibrating, and repairing electronic equipment in nuclear medicine.

1. Cathode Ray Tube

Analog oscilloscopes, as well as older nuclear medicine systems (gamma cameras, liquid scintillation counters, well counters, and MCAs) typically use a cathode ray tube (CRT) display. The CRT is an evacuated tube containing the basic components shown in Figure 8-16. The *electron gun* provides a focused source of electrons. Most CRTs use a hot, or thermionic emission, cathode. Electrons are boiled off the cathode by heating it with an electric current. The *control grid* is a cap that

fits over the cathode. The electrons pass through a small hole in its center. A negative potential on the grid can be varied to control the number of electrons that are allowed to pass. The *first anode,* or *accelerating anode,* is similar in shape to the grid except that its orientation is reversed. The flat end contains a small hole through which the electrons pass. It has a high positive potential that attracts the electrons and accelerates them to high velocities. Most of the electrons actually strike the front face of the first anode, but a small percentage pass through the opening and are accelerated down the CRT tube as a narrow beam.

The *second anode,* or *focusing anode,* further shapes the electron beam by focusing it to a sharp point where it strikes the phosphor-coated screen. A negative potential on the second anode is used to both compress and focus the beam of electrons. The diameter of the electron beam striking the phosphor screen is usually around 0.1 mm.

Deflection plates are used to move the electron beam across the screen. Electrostatic deflection employs two sets of plates mounted at right angles to each other. Voltages are applied to one pair to exert a force on the electron beam in the vertical direction and on the other pair for the horizontal direction on the display screen. The amount of deflection is proportional to the voltage applied to the deflection plates.

The *display screen* is a glass screen having an inside surface coated with a phosphorescent material. The high-velocity electrons striking the phosphor cause it to give off phosphorescent light. The brightness of the phosphorescent light depends on the intensity and energy of the electron beam. The lifetime of the light emission from the phosphor is

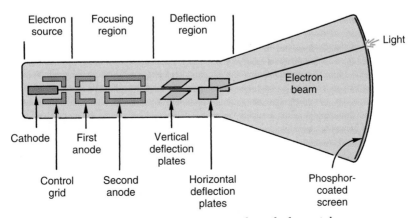

FIGURE 8-16　Basic components of a cathode ray tube.

referred to as the *persistence time* and is typically 0.5 msec on an oscilloscope display.

2. Analog Oscilloscope

A typical analog oscilloscope consists of a CRT, a signal amplifier for the vertical deflection plate of the CRT, and a time-sweep generator. An amplifier is provided so that small voltage inputs can be amplified and applied to the vertical deflection plate to display the amplitude of the input signals. The time-sweep generator is connected to the horizontal deflection plates of the CRT to sweep the electron beam across the screen at a constant speed and repetition rate. The horizontal sweep rate usually can be varied from nanoseconds (10^{-9} sec) to seconds per centimeter by a calibrated selector switch on the front panel of the oscilloscope. Thus the oscilloscope provides a visual display of time-varying electrical signals.

3. Digital Oscilloscope

Most modern oscilloscopes are digital, employing fast ADCs that digitize the amplified waveforms prior to display, and some form of microprocessor that allows pulses to be analyzed and manipulated. The CRT screen is typically replaced with a flat-panel liquid crystal display. Digital oscilloscopes have the advantage that pulses can be stored in computer memory for further analysis and are ideal for studying repetitive, regular pulses. The disadvantage of using a digital oscilloscope to look at the pulses from γ-ray detectors is that individual pulses generally are of a different amplitude (reflecting differing energies deposited in the detector), and that they arrive randomly in time. A digital oscilloscope shows only one pulse at a time. Some digital oscilloscopes now have a "persistence" function (essentially software or hardware that mimics the response of a phosphorescent screen), which allows many pulses to be viewed simultaneously, with appropriate intensity where pulses overlap. This allows the range of pulse amplitudes and shapes to be appreciated easily in a single glance and gives the digital oscilloscope the feel of an older analog oscilloscope with a fairly long (10^{-1} to 1 sec) persistence phosphor.

BIBLIOGRAPHY

Basic nuclear electronics are discussed in the following:

Knoll GF: *Radiation Detection and Measurement*, ed 4, New York, 2010, John Wiley.

Leo WR: *Techniques for Nuclear and Particle Physics Experiments*, ed 2, New York, 1994, Springer-Verlag.

A comprehensive reference for electronics is:

Horowitz P, Hill W: *The Art of Electronics*, ed 2, Cambridge, 1989, Cambridge University Press.

Nuclear Counting Statistics

All measurements are subject to measurement error. This includes physical measurements, such as radiation counting measurements used in nuclear medicine procedures, as well as in biologic and clinical studies, such as evaluation of the effectiveness of an imaging technique. In this chapter, we discuss the type of errors that occur, how they are analyzed, and how, in some cases, they can be minimized.

A. TYPES OF MEASUREMENT ERROR

Measurement errors are of three general types: blunders, systematic errors, and random errors.

Blunders are errors that are adequately described by their name. Usually they produce grossly inaccurate results and their occurrence is easily detected. Examples in radiation measurements include the use of incorrect instrument settings, incorrect labeling of sample containers, and injecting the wrong radiopharmaceutical into the patient. When a single value in the data seems to be grossly out of line with others in an experiment, statistical tests are available to determine whether the suspect value may be discarded (see Section E.3). Apart from this there is no way to "analyze" errors of this type, only to avoid them by careful work.

Systematic errors produce results that differ consistently from the correct result by some fixed amount. The same result may be obtained in repeated measurements, but it is the wrong result. For example, length measurements with a warped ruler, or activity measurements with a radiation detector that was miscalibrated or had some other persistent malfunction, could contain systematic errors. Observer bias in the subjective

interpretation of data (e.g., scan reading) is another example of systematic error, as is the use for a clinical study of two population groups having underlying differences in some important characteristic, such as different average ages. Measurement results having systematic errors are said to be *inaccurate.*

It is not always easy to detect the presence of systematic error. Measurement results affected by systematic error may be very repeatable and not too different from the expected results, which may lead to a mistaken sense of confidence. One way to detect systematic error in physical measurements is by the use of measurement *standards,* which are known from previous measurements with a properly operating system to give a certain measurement result. For example, radionuclide standards, containing a known quantity of radioactivity, are used in various quality assurance procedures to test for systematic error in radiation counting systems. Some of these procedures are described in Chapter 11, Section D.

Random errors are variations in results from one measurement to the next, arising from physical limitations of the measurement system or from actual random variations of the measured quantity itself. For example, length measurements with an ordinary ruler are subject to random error because of inexact repositioning of the ruler and limitations of the human eye. In clinical or animal studies, random error may arise from differences between individual subjects, for example, in uptake of a radiopharmaceutical. Random error *always* is present in radiation counting measurements because the quantity that is being measured—namely, the rate of emission from the radiation source—is itself a randomly varying quantity.

Random error affects measurement *repro-ducibility* and thus the ability to detect real differences in measured data. Measurements that are very reproducible—in that nearly the same result is obtained in repeated measurements—are said to be *precise.* It is possible to minimize random error by using careful measurement technique, refined instrumentation, and so forth; however, it is impossible to eliminate it completely. There is always some limit to the precision of a measurement or measurement system. The amount of random error present sometimes is called the *uncertainty* in the measurement.

It also is possible for a measurement to be precise (small random error) but inaccurate (large systematic error), or vice versa. For example, length measurements with a warped ruler may be very reproducible (precise); nevertheless, they still are inaccurate. On the other hand, radiation counting measurements may be imprecise (because of inevitable variations in radiation emission rates) but still they can be accurate, at least in an average sense.

Because random errors always are present in radiation counting and other measured data, it is necessary to be able to analyze them and to obtain estimates of their magnitude. This is done using methods of statistical analysis. (For this reason, they are also sometimes called *statistical errors.*) The remainder of this chapter describes these methods of analysis. The discussion focuses on applications involving nuclear radiation-counting measurements; however, some of the methods to be described also are applicable to a wider class of experimental data as discussed in Section E.

B. NUCLEAR COUNTING STATISTICS

1. The Poisson Distribution

Suppose that a long-lived radioactive sample is counted repeatedly under supposedly identical conditions with a properly operating counting system. Because the disintegration rate of the radioactive sample undergoes random variations from one moment to the next, the numbers of counts recorded in successive measurements (N_1, N_2, N_3, etc.) are not the same. Given that different results are obtained from one measurement to the next, one might question if a "true value" for the measurement actually exists. One possible solution is to make a large number of

measurements and use the average $\bar{N}$ as an estimate for the "true value."

$$\text{True Value} \approx \bar{N} \tag{9-1}$$

$$\begin{aligned} \bar{N} &= (N_1 + N_2 + \cdots + N_n)/n \\ &= \frac{1}{n}\sum_{i=1}^{n} N_i \end{aligned} \tag{9-2}$$

where n is the number of measurements taken. The notation Σ indicates a sum that is taken over the indicated values of the parameter with the subscript i.

Unfortunately, multiple measurements are impractical in routine practice, and one often must be satisfied with only one measurement. The question then is, how good is the result of a single measurement as an estimate of the true value; that is, what is the uncertainty in this result? The answer to this depends on the *frequency distribution* of the measurement results. Figure 9-1 shows a typical frequency distribution curve for radiation counting measurements. The solid dots show the different possible results (i.e., number of counts recorded) versus the probability of getting each result. The probability is peaked at a *mean value, m,* which is the true value for the measurement. Thus if a large number of measurements were made and their results averaged, one would obtain

$$\bar{N} \approx m \tag{9-3}$$

The solid dots in Figure 9-1 are described mathematically by the *Poisson distribution.* For this distribution, the probability of getting a certain result N when the true value is m is given by

$$P(N; m) = e^{-m} m^N / N! \tag{9-4}$$

where e (= 2.718 ...) is the base of natural logarithms and $N!$ (*N factorial*) is the product of all integers up to N (i.e., $1 \times 2 \times 3 \times \cdots \times N$) (Note that, by definition, $0! = 1$). From Figure 9-1 it is apparent that the probability of getting the exact result $N = m$ is rather small; however, one could hope that the result would at least be "close to" m. Note that the Poisson distribution is defined only for non-negative integer values of N (0, 1, 2, ...).

The probability that a measurement result will be "close to" m depends on the relative width, or dispersion, of the frequency distribution curve. This is related to a parameter called the *variance,* σ^2, of the distribution.

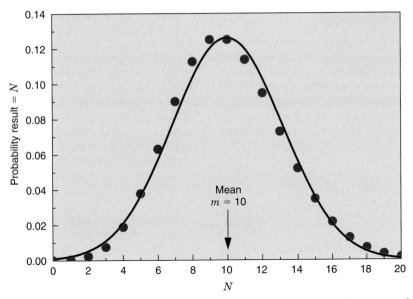

FIGURE 9-1   Poisson (●) and Gaussian (—) distributions for mean, m, and variance, $\sigma^2 = 10$.

The variance is a number such that 68.3% (~2/3) of the measurement results fall within $\pm\sigma$ (i.e., square root of the variance) of the true value m. For the Poisson distribution, the variance is given by

$$\sigma^2 = m \qquad (9\text{-}5)$$

Thus one expects to find approximately 2/3 of the counting measurement results within the range $\pm\sqrt{m}$ of the true value m.

Given only the result of a single measurement, N, one does not know the exact value of m or of σ; however, one can reasonably assume that $N \approx m$, and thus that $\sigma \approx \sqrt{N}$. One can therefore say that if the result of the measurement is N, there is a 68.3% chance that the true value of the measurement is within the range $N \pm \sqrt{N}$. This is called the "68.3% confidence interval" for m; that is, one is 68.3% confident that m is somewhere within the range $N \pm \sqrt{N}$.

The range $\pm\sqrt{N}$ is the uncertainty in N. The *percentage uncertainty* in N is

$$V = (\sqrt{N}/N) \times 100\%$$
$$\quad = 100\%/\sqrt{N} \qquad (9\text{-}6)$$

EXAMPLE 9-1

Compare the percentage uncertainties in the measurements $N_1 = 100$ counts and $N_2 = 10{,}000$ counts.

Answer

For $N_1 = 100$ counts, $V_1 = 100\% / \sqrt{100} = 10\%$ (Equation 9-6). For $N_2 = 10{,}000$ counts, $V_2 = 100\% / \sqrt{10{,}000} = 1\%$. Thus the percentage uncertainty in 10,000 counts is only 1/10 the percentage uncertainty in 100 counts.

Equation 9-6 and Example 9-1 indicate that *large numbers of counts have smaller percentage uncertainties and are statistically more reliable than small numbers of counts.*

Other confidence intervals can be defined in terms of σ or $\sqrt{N}$. They are summarized in Table 9-1. The 50% confidence interval (0.675σ) is sometimes called the *probable error* in N.

TABLE 9-1
CONFIDENCE LEVELS IN RADIATION COUNTING MEASUREMENTS

Range	Confidence Level for m (True Value) (%)
$N \pm 0.675\sigma$	50
$N \pm \sigma$	68.3
$N \pm 1.64\sigma$	90
$N \pm 2\sigma$	95
$N \pm 3\sigma$	99.7

2. The Standard Deviation

The variance σ^2 is related to a statistical index called the *standard deviation (SD)*. The standard deviation is a number that is calculated for a series of measurements. If n counting measurements are made, with results $N_1, N_2, N_3, \ldots, N_n$, and a mean value $\bar{N}$ for those results is found, the standard deviation is

$$SD = \sqrt{\frac{\sum_{i=1}^{n}(N_i - \bar{N})^2}{(n-1)}} \qquad (9\text{-}7)$$

The standard deviation is a measure of the dispersion of measurement results about the mean and is in fact an estimate of σ, the square root of the variance. For radiation counting measurements, one therefore should obtain

$$SD \approx \sqrt{N} \qquad (9\text{-}8)$$

This can be used to test whether the random error observed in a series of counting measurements is consistent with that predicted from random variations in source decay rate, or if there are additional random errors present, such as from faulty instrument performance. This is discussed further in Section E.

3. The Gaussian Distribution

When the mean value m is "large," the Poisson distribution can be approximated by the *Gaussian distribution* (also called the *normal distribution*). The equation describing the Gaussian distribution is

$$P(x; m, \sigma) = (1/\sqrt{2\pi\sigma^2})e^{-(x-m)^2/2\sigma^2} \qquad (9\text{-}9)$$

where m and σ^2 are again the mean and variance. Equation 9-9 describes a symmetrical "bell-shaped" curve. As shown by Figure 9-1, the Gaussian distribution for $m = 10$ and $\sigma = \sqrt{m}$ is very similar to the Poisson distribution for $m = 10$. For $m \gtrsim 20$, the distributions are virtually indistinguishable. Two important differences are that the Poisson distribution is defined only for nonnegative integers, whereas the Gaussian distribution is defined for any value of x, and that for the Poisson distribution, the variance σ^2 is equal to the mean, m, whereas for the Gaussian distribution, it can have any value.

The Gaussian distribution with $\sigma^2 = m$ is a useful approximation for radiation counting measurements when the only random error present is that caused by random variations in source decay rate. When additional sources of random error are present (e.g., a random error or uncertainty of ΔN counts caused by variations in sample preparation technique, counting system variations, and so forth), the results are described by the Gaussian distribution with variance given by

$$\sigma^2 \approx m + (\Delta N)^2 \qquad (9\text{-}10)$$

The resulting Gaussian distribution curve would be wider than a Poisson curve with $\sigma^2 = m$. The confidence intervals given in Table 9-1 may be used for the Gaussian distribution with this modified value for the variance. For example, the 68.3% confidence interval for a measurement result N would be $\pm \sqrt{N + \Delta N^2}$ (assuming $N \approx m$).

EXAMPLE 9-2

A 1-mL radioactive sample is pipetted into a test tube for counting. The precision of the pipette is specified as "$\pm 2\%$," and 5000 counts are recorded from the sample. What is the uncertainty in sample counts per mL?

Answer
The uncertainty in counts arising from pipetting precision is 2% × 5000 counts = 100 counts. Therefore,

$$\sigma^2 = 5000 + (100)^2 \approx 15,000,$$

and the uncertainty is $\sqrt{15,000} \approx 122$ counts. Compare this with the uncertainty of $\sqrt{5000} \approx 71$ counts that would be obtained without the pipetting uncertainty.

C. PROPAGATION OF ERRORS

The preceding section described methods for estimating the random error or uncertainty in a single counting measurement; however, most nuclear medicine procedures involve multiple counting measurements, from which ratios, differences, and so on are used to compute the final result. In the following four sections we describe equations and methods that apply when a result is obtained from a set of counting measurements, $N_1, N_2, N_3 \ldots$ In some cases, we present first the general equation applicable for measurements of any type, M_1, M_2, M_3, ..., having individual variances, $\sigma (M_1)^2$, $\sigma (M_2)^2$, $\sigma (M_3)^2$, ... The general equations can be used to compute the uncertainty in the result for whatever M might represent (e.g., a series of readings from a scale or a thermometer). We then apply these general equations to nuclear counting measurements. Note that in the following subsections it is assumed that random fluctuations in counting measurements arise *only* from random fluctuations in sample decay rate and that the individual measurements are *statistically independent* from one another. The latter condition would be violated if N_1 was in some way correlated with N_2, for example, if N_1 was the result for the first half of the counting period for the measurement of N_2.

1. Sums and Differences

For either sums or differences of a series of individual measurements, M_1, M_2, M_3, ..., with individual variances, $\sigma (M_1)^2$, $\sigma (M_2)^2$, $\sigma (M_3)^2$, ..., the general equation for the square root of the variance of the result is given by

$$\text{SD}(M_1 \pm M_2 \pm M_3 \pm \cdots)$$
$$\approx \sqrt{\text{SD}(M_1)^2 + \text{SD}(M_2)^2 + \text{SD}(M_3)^2 + \cdots}$$

$$(9\text{-}11)$$

Thus, for a series of counting measurements with individual results N_1, N_2, N_3, ... an estimate for the standard deviation of their sums or differences can be obtained using the standard deviations for individual results (Eq. 9-8) as follows:

$$\text{SD}(N_1 \pm N_2 \pm N_3 \pm \cdots) \approx \sqrt{N_1 + N_2 + N_3 + \cdots}$$
$$(9\text{-}12)$$

and the percentage uncertainty is

$$V(N_1 \pm N_2 \pm N_3 \pm \cdots)$$
$$= \frac{\sqrt{N_1 + N_2 + N_3 + \cdots}}{N_1 \pm N_2 \pm N_3 \cdots} \times 100\% \quad (9\text{-}13)$$

Note that these equations apply to mixed combinations of sums and differences.

2. Constant Multipliers

If a measurement M having variance $\sigma_M{}^2$ is multiplied by a constant k, the general equation for the variance of the product is

$$\text{SD}(kM) \approx k\,\text{SD}_M \quad (9\text{-}14)$$

Substituting the appropriate quantities for counting measurements, with $M = N$

$$\text{SD}(kN) \approx k\sqrt{N} \quad (9\text{-}15)$$

The percentage uncertainty V in the product kN is

$$V(kN) = [\sigma (kN)/kN] \times 100\%$$
$$= 100\%/\sqrt{N} \quad (9\text{-}16)$$

which is the same result as Equation 9-6. Thus there is no statistical advantage gained or lost in multiplying the number of counts recorded by a constant. The percentage uncertainty still depends on the actual number of counts recorded.

3. Products and Ratios

The uncertainty in the product or ratio of a series of measurements M_1, M_2, M_3, ... is most conveniently expressed in terms of the *percentage* uncertainties in the individual results, V_1, V_2, V_3, ... The general equation is given by

$$V(M_1 \overset{\times}{\div} M_2 \overset{\times}{\div} M_3 \overset{\times}{\div} \cdots) = \sqrt{V_1^2 + V_2^2 + V_3^2 + \cdots}$$
$$(9\text{-}17)$$

For counting measurements, this becomes

$$V(N_1 \overset{\times}{\div} N_2 \overset{\times}{\div} N_3 \overset{\times}{\div} \cdots)$$
$$= \sqrt{\frac{1}{N_1} + \frac{1}{N_2} + \frac{1}{N_3} + \cdots} \times 100\% \quad (9\text{-}18)$$

Again, this expression applies to mixed combinations of products and ratios.

4. More Complicated Combinations

Many nuclear medicine procedures, such as thyroid uptakes and blood volume determinations, use equations of the following general form.

$$Y = \frac{k(N_1 - N_2)}{(N_3 - N_4)} \quad (9\text{-}19)$$

The uncertainty in Y is expressed most conveniently in terms of its percentage uncertainty. Using the rules given previously, one can show that

$$V_Y = \sqrt{\frac{(N_1 + N_2)}{(N_1 - N_2)^2} + \frac{(N_3 + N_4)}{(N_3 - N_4)^2}} \times 100\% \quad (9\text{-}20)$$

EXAMPLE 9-3

A patient is injected with a radionuclide. At some later time a blood sample is withdrawn for counting in a well counter and N_p = 1200 counts are recorded. A blood sample withdrawn prior to injection gives a blood background of N_{pb} = 400 counts. A standard prepared from the injection preparation records N_s = 2000 counts, and a "blank" sample records an instrument background of N_b = 200 counts. Calculate the ratio of net patient sample counts to net standard counts, and the uncertainty in this ratio.

Answer
The ratio is

$$Y = (N_p - N_{pb})/(N_s - N_b)$$
$$= (1200 - 400)/(2000 - 200)$$
$$= 800/1800 = 0.44$$

The percentage uncertainty in the ratio is (Equation 9-20)

$$V_Y = \sqrt{\frac{(1200+400)}{(1200-400)^2} + \frac{(2000+200)}{(2000-200)^2}} \times 100\%$$
$$= 5.6\%$$

The uncertainty in Y is 5.6% × 0.44 ≈ 0.02; thus the ratio and its uncertainty are Y = 0.44 ± 0.02.

D. APPLICATIONS OF STATISTICAL ANALYSIS

1. Effects of Averaging

If n counting measurements are used to compute an average result, the average $\bar{N}$ is a more reliable estimate of the true value than any one of the individual measurements. The uncertainty in $\bar{N}$, $\sigma_{\bar{N}}$, can be obtained by combining the rules for sums (Equation 9-11) and constant multipliers (Equation 9-14).

$$\sigma_{\bar{N}} = \sqrt{\bar{N}/n} \qquad (9\text{-}21)$$

The uncertainty in $\bar{N}$ as an estimator of m therefore is smaller than the uncertainty in a single measurement by a factor $1/\sqrt{n}$.

2. Counting Rates

If N counts are recorded during a measuring time t, the average counting rate during that interval is $R = N/t$. Using Equation 9-15, the uncertainty in the counting rate R is

$$\sigma_R = (1/t)\sqrt{N}$$
$$= \sqrt{N/t^2} \qquad (9\text{-}22)$$
$$= \sqrt{R/t}$$

The percentage uncertainty in R is

$$V_R = (\sigma_R/R) \times 100\%$$
$$= 100\%/\sqrt{Rt} \qquad (9\text{-}23)$$

EXAMPLE 9-4

In a 2-min counting measurement, 4900 counts are recorded. What is the average counting rate R(cpm) and its uncertainty?

Answer
$$R = 4900/2 = 2450 \text{ cpm}$$

From Equation 9-22

$$\sigma_R = \sqrt{2450/2} = 35 \text{ cpm}$$

and from Equation 9-23

$$V_R = 100\%/\sqrt{2450 \times 2} \approx 1.4\%$$

Note from Equations 9-22 and 9-23 that *longer counting times produce smaller uncertainties in estimated counting rates.*

3. Significance of Differences Between Counting Measurements

Suppose two samples are counted and that counts N_1 and N_2 are recorded. The difference $(N_1 - N_2)$ may be due to an actual difference between sample activities or may be simply the result of random variations in counting rates. There is no way to state with absolute certainty that a given difference is or is not caused by random error; however, one can assess the "statistical significance" of the difference by comparing it with the expected random error. In general, differences of less than 2 σ [i.e., $(N_1 - N_2) < 2\sqrt{N_1 + N_2}$] are considered to be of marginal or no statistical significance because there is at least a 5% chance that such a difference is simply caused by random error (see Table 9-1). Differences greater than 3σ are considered significant (<1% chance caused by random error), whereas differences between 2σ and 3σ are in the questionable category, perhaps deserving repeat measurement or longer measuring times to determine their significance.

If two counting rates R_1 and R_2 are determined from measurements using counting times t_1 and t_2, respectively, the uncertainty in their difference $R_1 - R_2$, can be obtained by applying Equations 9-11 and 9-22.

$$\sigma(R_1 - R_2) = \sqrt{R_1/t_1 + R_2/t_2} \quad (9\text{-}24)$$

Comparison of the observed difference to the expected random error difference can again be used to assess statistical significance, as described in Section B.

4. Effects of Background

All nuclear counting instruments have background counting rates, caused by electronic noise, detection of cosmic rays, natural radioactivity in the detector itself (e.g., ^{40}K), and so forth. If the background counting rate, measured with no sample present, is R_b and the gross counting rate with the sample is R_g, then the net sample counting rate is

$$R_s = R_g - R_b \quad (9\text{-}25)$$

The uncertainty in R_s is (from Equation 9-24)

$$\sigma_{R_s} = \sqrt{R_g/t_g + R_b/t_b} \quad (9\text{-}26)$$

The percentage uncertainty in R_s is

$$V_{R_s} = \left[\sqrt{R_g/t_g + R_b/t_b} / (R_g - R_b) \right] \times 100\% \quad (9\text{-}27)$$

If the same counting time t is used for both sample and background counting,

$$\begin{aligned} \sigma_{R_s} &= \sqrt{R_g + R_b} / \sqrt{t} \\ &= \sqrt{R_s + 2R_b} / \sqrt{t} \end{aligned} \quad (9\text{-}28)$$

EXAMPLE 9-5

In 4-min counting measurements, gross sample counts are 6000 counts and background counts are 4000 counts. What are the net sample counting rate and its uncertainty?

Answer
$$R_g = 6000/4 = 1500 \text{ cpm}$$
$$R_b = 4000/4 = 1000 \text{ cpm}$$
$$R_s = 1500 - 1000 = 500 \text{ cpm}$$

From Equation 9-28
$$\begin{aligned} \sigma_{R_s} &= \sqrt{500 + (2 \times 1000)} / \sqrt{4} \\ &= \sqrt{2500} / \sqrt{4} = 50/2 \\ &= 25 \text{ cpm} \end{aligned}$$

Therefore $R_s = 500 \pm 25$ cpm ($\pm 5\%$). Compare this with the uncertainty in the gross counting rate R_g (from Equation 9-22)

$$\sigma_{R_g} = \sqrt{1500/4} \approx 19 \text{ cpm} \ (\sim 1\%)$$

and to the uncertainty in R_s that would be obtained if there were negligible background ($R_b \approx 0$),

$$\sigma_{R_s} = \sqrt{500/4} \approx 11 \text{ cpm} \ (\sim 2\%)$$

Example 9-5 illustrates two important points:
1. *High background counting rates are undesirable because they increase uncertainties in net sample counting rates.*
2. *Small differences between relatively high counting rates can have relatively large uncertainties.*

5. Minimum Detectable Activity

The minimum detectable activity (MDA) of a radionuclide for a particular counting system and counting time t is that activity that increases the counts recorded by an amount that is "statistically significant" in comparison with random variations in background counts that would be recorded during the same measuring time. In this instance, statistically significant means a counting rate increase of 3σ. Therefore, from Equation 9-22, the counting rate for the MDA is

$$\text{MDA} = 3\sqrt{R_b/t} \quad (9\text{-}29)$$

EXAMPLE 9-6

A standard NaI(Tl) well counter has a background counting rate (full spectrum) of approximately 200 cpm. The sensitivity of the well counter for ^{131}I is approximately 29 cpm/Bq (see Table 12-2). What is the MDA for ^{131}I, using 4-min counting measurements?

Answer
The MDA is that amount of ^{131}I giving $3 \times \sqrt{200 \text{ cpm}/4} \approx 3 \times 7 \text{ cpm} = 21$ cpm. Thus

$$\begin{aligned} \text{MDA} &= 21 \text{ cpm}/(29 \text{ cpm/Bq}) \\ &\approx 0.7 \text{ Bq (i.e., } <1 \text{ dps)} \end{aligned}$$

In traditional units (1 µCi = 37 kBq), the MDA is ~ 0.00002 µCi.

6. Comparing Counting Systems

In Section B.1 it was noted that larger numbers of counts have smaller percentage uncertainties. Thus in general it is desirable from a statistical point of view to use a counting system with maximum sensitivity (i.e., large detector, wide pulse-height analyzer window) so that a maximum number of counts is obtained in a given measuring time; however, such systems are also more sensitive to background radiation and give higher background counting rates as well, which, as shown by Example 9-5, tends to increase statistical uncertainties. The tradeoff between sensitivity and background may be analyzed as follows:

Suppose a counting system provides gross sample counts G_1, background counts B_1, and net sample counts $S_1 = G_1 - B_1$ and that a second system provides gross, background, and net counts G_2, B_2, and S_2 in the same counting time. One can compare the uncertainties in S_1 and S_2 to determine which system is statistically more reliable. The percentage uncertainty in S_1 is given by

$$V_1 = \frac{\sqrt{G_1 + B_1}}{S_1} \times 100\%$$
$$= \frac{\sqrt{S_1 + 2B_1}}{S_1} \times 100\% \tag{9-30}$$

Corresponding equations apply to the second system. The ratio of the percentage uncertainties for the net sample counts obtained with two systems is therefore

$$\frac{V_1}{V_2} = \frac{S_2}{S_1} \times \frac{\sqrt{S_1 + 2B_1}}{\sqrt{S_2 + 2B_2}} \tag{9-31}$$

If $V_1/V_2 < 1$, then $V_1 < V_2$, in which case system 1 is the statistically preferred system. Conversely, if $V_1/V_2 > 1$, system 2 is preferred.

If background counts are relatively small ($B_1 \ll S_1$, $B_2 \ll S_2$), Equation 9-31 can be approximated by

$$\frac{V_1}{V_2} \approx \frac{S_2 \sqrt{S_1}}{S_1 \sqrt{S_2}}$$
$$\approx \sqrt{\frac{S_2}{S_1}} \tag{9-32}$$

Thus when background levels are "small," only relative sensitivities are important. The system with the higher sensitivity gives the smaller uncertainty. Conversely, if background counts are large ($B_1 \gg S_1$, $B_2 \gg S_2$), Equation 9-31 is approximated by

$$\frac{V_1}{V_2} \approx \frac{S_2}{S_1} \sqrt{\frac{B_1}{B_2}} \tag{9-33}$$

Both sensitivity and background are important in this case. Note that Equations 9-31 through 9-33 also can be used with counting *rates* (cpm, cps) substituted for counts when equal counting times are used for all measurements.

EXAMPLE 9-7

A sample is counted in a well counter using a "narrow" pulse-height analyzer window and net sample and background counts are S_N = 500 counts and B_N = 200 counts, respectively. The sample is counted with the same system but using a "wide" window and the net sample and background counts are S_W = 800 counts and B_W = 400 counts, respectively. Which window setting offers the statistical advantage?

Answer

Background counts are neither "very small" nor "very large" in comparison with net sample counts; thus Equation 9-31 must be used:

$$\frac{V_N}{V_W} = \frac{800}{500} \times \frac{\sqrt{500 + (2 \times 200)}}{\sqrt{800 + (2 \times 400)}}$$
$$= (8/5) \times \sqrt{9/16} = (8/5) \times (3/4)$$
$$= 1.2$$

Thus $V_N/V_W > 1$ and the statistical advantage belongs to the wider window setting, in spite of its higher background counting rate.

7. Estimating Required Counting Times

Suppose it is desired to determine net sample counting rate R_s to within a certain percentage uncertainty V. Suppose further that the approximate net sample and background counting rates are known to be R'_s and R'_b, respectively (e.g., from quick preliminary measurements). If a counting time t is to be used for both the sample and background counting measurements, then the time required to achieve the desired level of statistical reliability is given by

$$t = [(R'_s + 2R'_b)/R'^2_s](100\%/V)^2 \tag{9-34}$$

EXAMPLE 9-8

Preliminary measurements in a sample counting procedure indicate gross and background counting rates of $R_g = 900$ cpm and $R_b = 100$ cpm, respectively. What counting time is required to determine net sample counting rate to within 5%?

Answer

$$R'_s = 900 - 100 = 800 \text{ cpm}$$

$$t = \{[800 + (2 \times 100)]/800^2\} \times (100/5)^2$$
$$= (1000/800^2) \times (100/5)^2$$
$$= 0.625 \text{ min}$$

This time is used for both sample and background counting. Therefore the total counting time required is 1.25 min.

8. Optimal Division of Counting Times

In the preceding section it was assumed that equal counting times were used for the sample and background measurements. This is not necessary; in fact, statistically advantageous results may be obtainable by using unequal times. The difference between two counting rates R_1 and R_2 is determined with the smallest statistical error if the total counting time $t = t_1 + t_2$ is divided according to

$$t_1/t_2 = \sqrt{R'_1/R'_2} \qquad (9\text{-}35)$$

where R'_1 and R'_2 are counting rates estimated from preliminary measurements. Applying this to gross sample and background counting rate estimates, one obtains

$$t_g/t_b = \sqrt{R'_g/R'_b} \qquad (9\text{-}36)$$

If $(R'_g \approx R'_b)$, approximately equal counting times are preferred; however, if the background counting rate is small $(R'_b \ll R'_g)$, it is better to devote most of the available time to counting the sample.

EXAMPLE 9-9

In Example 9-8, what is the optimal division of a 1.25-min total counting time and the resulting uncertainty in the net sample counting rate?

Answer

Applying Equation 9-36, with $R'_g = 900$ cpm and $R'_b = 100$ cpm,

$$t_g/t_b = \sqrt{900/100} = 3$$

$$t_g = 3t_b$$

$$t_g + t_b = 3t_b + t_b = 1.25 \text{ min}$$

$$t_b = 1.25 \text{ min}/4 \approx 0.3 \text{ min}$$

$$t_g \approx 1.25 - 0.3 = 0.95 \text{ min}$$

The percentage uncertainty in R_s given by Equation 9-27 is

$$V_{R_s} = \left[\sqrt{R_g/t_g + (R_b/t_b)} \times 100\%\right]/(R_g - R_b)$$
$$= \left[\sqrt{(900/0.95) + (100/0.3)} \times 100\%\right]/800$$
$$\approx 4.5\%$$

Thus a small statistical advantage (4.5% vs. 5%) is gained by using an optimal division rather than equal counting times in this example.

E. STATISTICAL TESTS

In Section D.3, an example was given of a method for testing the statistical significance of the difference between two counting measurements. The test was based on the assumption of underlying Poisson distributions for the two individual measurements, with variances $\sigma^2 \approx N$. In this section we consider a few other tests for evaluating statistical parameters of sets of counting measurements. The discussion focuses on applications of these tests to nuclear counting data; however, as noted in the discussion, the tests also are applicable to other experimental data for which the underlying random variability is described by a Poisson or Gaussian distribution. More detailed discussions of statistical tests are found in the references and suggested readings at the end of this chapter.

1. The χ^2 Test

The χ^2 (*chi-square*) test is a means for testing whether random variations in a set of measurements are consistent with what would be expected for a Poisson distribution. This is a particularly useful test when a set of counting measurements is suspected to contain sources of random variation in addition to Poisson counting statistics, such as those resulting from faulty instrumentation or other random variability between samples, animals, patients, and measurement techniques. The test is performed as follows:

1. Obtain a series of counting measurements (at least 20 measurements is desirable).
2. Compute the mean

$$\bar{N} = \sum_{i=1}^{n} N_i / n \qquad (9\text{-}37)$$

and the quantity

$$\chi^2 = \sum_{i=1}^{n} (N_i - \bar{N})^2 / \bar{N}$$
$$= (n-1)SD^2/\bar{N} \qquad (9\text{-}38)$$

where SD = standard deviation (Equation 9-7). Many pocket calculators have programs for calculating standard deviations; thus the second form in Equation 9-38 may be more convenient to use.

3. Refer to a χ^2 table or graph (Fig. 9-2). Locate the value corresponding to the number of measurements, n, on the horizontal axis.
4. Compare the computed value of χ^2 to the most closely corresponding P-value curve.

P is the probability that random variations observed in a series of n measurements from a Poisson distribution would *equal or exceed* the calculated χ^2 value. Conversely, $1 - P$ is the probability that smaller variations would be observed. A P value of 0.5 (50%) would be "perfect." It indicates that the observed χ^2 value is in the middle of the range expected for a Poisson distribution. (Note that this corresponds to $\chi^2 \approx n - 1$.)

A low P value (<0.01) indicates that there is only a small probability that a Poisson distribution would give the χ^2 value as large as the value observed and suggests that additional sources of random error are present. A high P value (>0.99) indicates that random variations are much smaller than expected and also is cause for concern. For example, it could indicate that periodic noise (e.g., 60-Hz line frequency) is being counted. Such signals are not subject to the same degree of random variation as are radiation counting measurements and therefore have very small χ^2 values. In general, a range $0.05 < P < 0.95$ is considered an acceptable result. If P falls outside the range (0.01-0.99), one usually can conclude that something is wrong with the measurement system. If $0.01 < P < 0.05$ or $0.95 < P < 0.99$, the results are suspicious but the experiment is considered inconclusive and should be repeated.

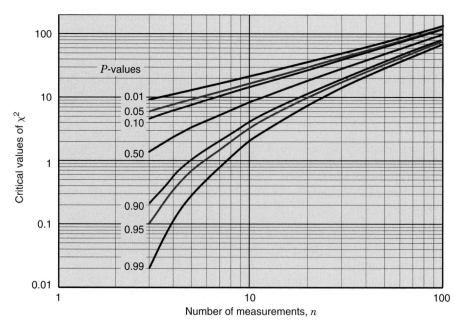

FIGURE 9-2 Critical values of χ^2 versus number of measurements, n. For a properly operating system, one should obtain $\chi^2 \approx (n-1)$. P values indicate probability of obtaining χ^2 larger than associated curve value. —, $P = 0.01$-0.99; —, $P = 0.05$-0.95; —, $P = 0.10$-0.90; —, $P = 0.5$.

EXAMPLE 9-10

Use the χ^2 test to determine the likelihood that the following set of 20 counting measurements were obtained from a Poisson distribution.

3875	3575
3949	4023
3621	3314
3817	3612
3790	3705
3902	3412
3851	3520
3798	3743
3833	3622
3864	3514

Answer

Using a pocket calculator or by direct calculation, it can be shown that the mean and standard deviation of the counting measurements are

$$\bar{N} = 3717$$

$$SD = 187.4$$

Thus, from Equation 9-38,

$$\chi^2 = 19 \times (187.4)^2 / 3717$$

$$\approx 179.5$$

Using Figure 9-2, the calculated value for χ^2 far exceeds the largest critical value shown for $n = 20$; (critical value ≈ 35 for $P = 0.01$). Hence, we conclude that the probability is very small that the observed set of counting measurements were obtained from a Poisson distribution ($P \ll 0.01$). The observed standard deviation, $SD = 187.4$, also far exceeds what would be expected for a Poisson distribution, $\sqrt{N} = 61$. These results suggest the presence of additional sources of random variation beyond simple counting statistics in the data.

Tables of χ^2 values are provided in most statistics textbooks. It is possible to determine more precise P values from these tables than can be read from Figure 9-2, especially for large values of n. However, it should be noted that χ^2 is itself a statistically variable quantity, having a standard deviation ranging from approximately 25% for $n \sim 30$ to approximately 15% for $n \sim 100$. Thus it is unwise to place too much confidence in χ^2 values that are within approximately 10% of a critical value, which is about the accuracy to which values can be read from Figure 9-2. When χ^2 values are close to critical values, it is recommended that the experiment be repeated. A useful discussion of the χ^2 statistic and applications to a variety of tests of nuclear counting systems can be found in reference 1.

2. The *t*-Test

The *t*-test (also sometimes called the *Student t-test*) is used to determine the significance of the difference between the means of two sets of data. In essence, the test compares the difference in means relative to the observed random variations in each set. Strictly speaking, the test is applicable only to Gaussian-distributed data; however, it is reasonably reliable for Poisson-distributed data as well (see Fig. 9-1).

Two different tests are used, depending on whether the two sets represent independent or paired data. *Independent* data are obtained from two *different* sample groups, for example, two different groups of radioactive samples, two different groups of patients or animals, and so forth. *Paired* data are obtained from the *same* sample group but at different times or under different measurement conditions, such as the *same* samples counted on two different instruments or a group of patients or animals imaged "before" and "after" a procedure. The test for paired data assumes that there is some degree of correlation between the two measurements of a pair. For example, in an experiment comparing two different radiopharmaceuticals that supposedly have an uptake proportional to blood flow, a subject with a "high" uptake for one radiopharmaceutical may have a "high" uptake for the other as well.

To test whether the difference between the means of two sets of *independent measurements* is significantly different from zero, the following quantity is calculated

$$t = \frac{\left| \bar{X}_1 - \bar{X}_2 \right|}{\sqrt{[(n_1 - 1)SD_1^2 + (n_2 - 1)SD_2^2]/(n_1 + n_2 - 2)}} \times \frac{1}{\sqrt{(1/n_1) + (1/n_2)}}$$

$$(9\text{-}39)$$

where $\bar{X}_1$ and $\bar{X}_2$ are the means of the two data sets, SD_1 and SD_2 are their standard deviations (calculated as in Equation 9-7), and n_1 = number of data values in set 1 and n_2 = number of data values in set 2. The vertical lines bracketing the difference of the means indicates that the absolute value should be used. For $n_1 \approx n_2$ and a reasonably large number of samples in each group ($\gtrsim 10$ in each), Equation 9-39 reduces to

$$t = \frac{|\bar{X}_1 - \bar{X}_2|}{\sqrt{2(SD_1^2 + SD_2^2)/(n_1 + n_2 - 2)}} \quad (9\text{-}40)$$

In either case, the calculated value of t then is compared with critical values of the t-distribution for the appropriate number of degrees of freedom, df $= n_1 + n_2 - 2$.

Figure 9-3 shows values of t that would be exceeded at various probability levels if the two sets of data actually were obtained from the *same* distribution. For example, for df $= 10$, a value of $t \gtrsim 2.2$ would be obtained by chance with a probability of only 5% ($P = 0.05$) if the underlying distributions actually were the same. This probability is sufficiently small that the difference between the means usually would be considered to be "statistically significant," that is, that the underlying distributions very likely have different means.

The t values given by Equations 9-39 and 9-40, and the derivation of associated P values from Figure 9-3 as described earlier, correspond to a *two-sided test*. The P values so obtained express the probability that the observed difference in means of measured data, whether positive or negative, would be

obtained if the true means of the underlying distributions actually were the same, that is, $m_1 = m_2$. If a "statistically significant difference" (i.e., a very low P value) is obtained, one concludes that $m_1 \neq m_2$, which could imply either that $m_1 > m_2$, or that $m_2 > m_1$. For example, a two-sided test would be appropriate if one were concerned only whether the uptakes of two radiopharmaceuticals were different.

A *one-sided test* is used when one is concerned only whether one mean is greater than the other (e.g., whether the uptake of one radiopharmaceutical is greater than that of the other). For example, if the experimental result is $\bar{X}_1 > \bar{X}_2$, one might ask whether this is consistent with $m_1 > m_2$. In this case, for a given t value, the P values in Figure 9-3 are reduced by a factor of 2. The P value then is interpreted as the probability that the observed difference in means of the data would be obtained if $m_1 \leq m_2$, that is, if $m_1 < m_2$ or $m_1 = m_2$. Statisticians generally do not recommend the use of one-sided tests. For example, a nonsignificant one-sided test for $m_1 > m_2$ may overlook the possibility that $m_1 < m_2$, which could be an equally important conclusion.

Note further that, as with the χ^2 statistic, t values have their own statistical variations from one experiment to the next. Thus t values that are within a few percent of a critical value should be interpreted with caution.

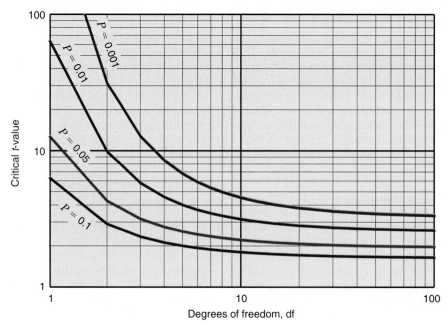

FIGURE 9-3   Critical values of t versus degrees of freedom (df) for different P values. Curves shown are for a two-sided test of significance.

For most practical situations, P values can be read with sufficient accuracy from Figure 9-3. More precise t and P values are provided in tables in statistics textbooks or by many pocket calculators.

EXAMPLE 9-11

Suppose the two columns of data in Example 9-10 represent counts measured on two different groups of animals, for the uptake of two different radiopharmaceuticals. Use the t-test to determine whether the means of the two sets of counts are significantly different (two-sided test).

Answer

Using a pocket calculator or by direct calculation, the means and standard deviations of the two sets of data are found to be (1 = left column, 2 = right column)

$$\bar{X}_1 = 3830$$

$$SD_1 = 87.8$$

$$\bar{X}_2 = 3604$$

$$SD_2 = 195.1$$

Thus, from Equation 9-40,

$$t = \frac{|3830 - 3604|}{\sqrt{2 \times (87.8^2 + 195.1^2)/18}}$$
$$\approx 3.17$$

From Figure 9-3, this comfortably exceeds the critical value of t for df = (10 + 10 − 2) = 18 and $P = 0.05$ (~2.1) and exceeds as well the value for $P = 0.01$ (~2.9). Thus we can conclude that it is very unlikely that the means of the two sets of data are the same ($P < 0.01$), and that they are in fact significantly different.

For *paired comparisons,* the same table of critical values is used but a different method is used for calculating t. In this case, the differences between pairs of measurements are determined, and t is calculated from

$$t = \frac{\left|1/n \sum_{i=1}^{n} (X_{1,i} - X_{2,i})\right|}{(SD_\Delta / \sqrt{n})} \quad (9\text{-}41)$$

The numerator is formed by computing the average of the paired differences and taking its absolute value. SD_Δ is the standard deviation of the paired differences (calculated as in Equation 9-7, with the N's replaced by Δ's) and n is the number of pairs of measurements. The sign of the difference between individual data pairs is significant and should be used in calculating the mean of the differences. The calculated value of t is compared with critical values in the t-distribution table using df = $(n − 1)$. Probability values are interpreted in the same manner as for independent data.

EXAMPLE 9-12

Suppose that the two columns of data in Example 9-10 represent counts measured on the same group of animals for the uptake of two different radiopharmaceuticals; that is, opposing values in the two columns represent measurements on the same animal. Use the t-test to determine whether there is a significant difference in average uptake of the two radiopharmaceuticals in these animals.

Answer

The first step is to calculate the difference in counts for each pair of measurements. Subtracting the data value in the right-hand column from that in the left for each pair, one obtains for the differences

$$3875 - 3575 = +300$$

$$3949 - 4023 = -74$$

$$\text{etc.}$$

The absolute value of the mean difference, $|\bar{\Delta}|$, and standard deviation of the differences are found to be

$$|\bar{\Delta}| = 240.8$$

$$SD_\Delta = 141.0$$

Using Equation 9-41

$$t = \frac{240.8}{(141.0 / \sqrt{10})} \approx 5.4$$

From Figure 9-3, the critical value of t for df = $n − 1 = 9$ and $P = 0.01$ is $t \approx 3.3$; thus, as in Example 9-11, we can conclude that the means of the two sets of data are significantly different.

This discussion of paired data applies for two-sided tests. One-sided tests may be performed using the methods outlined in the discussion of unpaired data.

3. Treatment of "Outliers"

Occasionally, a set of data will contain what appears to be a spurious, or "outlier," result, reflecting possible experimental or measurement error. Although generally it is inadvisable to discard data, statistical tests can be used to determine whether it is reasonable, from a *statistical* point of view, to do so. These tests involve calculating the standard deviation of the observed data set and comparing this with the difference between the sample mean $\bar{X}$ and the suspected outlier, X. The quantity calculated is

$$T = (X - \bar{X})/SD \qquad (9\text{-}42)$$

which then is compared with a table of critical values (Table 9-2). The interpretation of the result is the same as for the *t*-test; that is, the critical value is that value of T (also sometimes called the *Thompson criterion*) that would be exceeded by chance at a specified probability level if all the data values were obtained from the same Gaussian distribution. Rejection of data must be done with caution; for example, in a series of 20 measurements, it is likely that at least one of the data values will exceed the critical value at the 5% confidence level.

EXAMPLE 9-13

In the right-hand column of data in Example 9-10, the value 4023 appears to be an outlier, differing by several standard deviations from the mean of that column (see Example 9-11). Use the Thompson criterion to determine whether this data value may be discarded from the right-hand column of data.

Answer
From Example 9-11, the mean and standard deviation of the right-hand column of data are $\bar{X}_2 = 3604$, $SD_2 = 195.1$. Using Equation 9-42

$$T = (4023 - 3604)/195.1$$
$$= 419/195.1$$
$$= 2.15$$

According to Table 9-2, for 10 observations and $P = 0.05$, the critical value of T is 2.29. Because the observed value is smaller, we must conclude that there is a relatively high probability ($P > 0.05$) that the value could have been obtained by chance from the observed distribution, and therefore that it should not be discarded.

TABLE 9-2
CRITICAL VALUES OF THE THOMPSON CRITERION FOR REJECTION OF A SINGLE OUTLIER

Number of Observations, n	Level of Significance, P		
	.1	.05	.01
3	1.15	1.15	1.15
4	1.46	1.48	1.49
5	1.67	1.71	1.75
6	1.82	1.89	1.94
7	1.94	2.02	2.10
8	2.03	2.13	2.22
9	2.11	2.21	2.32
10	2.18	2.29	2.41
11	2.23	2.36	2.48
12	2.29	2.41	2.55
13	2.33	2.46	2.61
14	2.37	2.51	2.66
15	2.41	2.55	2.71
16	2.44	2.59	2.75
17	2.47	2.62	2.79
18	2.50	2.65	2.82
19	2.53	2.68	2.85
20	2.56	2.71	2.88
21	2.58	2.73	2.91
22	2.60	2.76	2.94
23	2.62	2.78	2.96
24	2.64	2.80	2.99
25	2.66	2.82	3.01
30	2.75	2.91	
35	2.82	2.98	
40	2.87	3.04	
45	2.92	3.09	
50	2.96	3.13	
60	3.03	3.20	
70	3.09	3.26	
80	3.14	3.31	
90	3.18	3.35	
100	3.21	3.38	

Adapted from Levin S: Statistical Methods. In Harbert J, Rocha AFG (eds): *Textbook of Nuclear Medicine*, Vol 1, ed 2. Philadelphia, 1984, Lea and Febiger, Chapter 4.

4. Linear Regression

Frequently, it is desired to know whether there exists a correlation between a measured quantity and some other parameter (e.g., counts versus time, radionuclide uptake versus organ weight, etc.). The simplest such relationship is described by an equation of the form

$$Y = a + bX \qquad (9\text{-}43)$$

Here, Y is the measured quantity and X is the parameter with which it is suspected to be correlated. The graph of Y versus X is a straight line, with Y-axis intercept a and slope b (Fig. 9-4).

To estimate values for a and b from a set of data, the following quantities are calculated.*

$$b = \frac{\left[n\sum X_i Y_i - \sum X_i \sum Y_i\right]}{\left[n\sum X_i^2 - \left(\sum X_i\right)^2\right]} \qquad (9\text{-}44)$$

$$a = \bar{Y} - b\bar{X} \qquad (9\text{-}45)$$

Here n is the number of pairs of data values; X_i and Y_i are individual values of these pairs

---

*The equations for regression parameters are interrelated and are expressed in a variety of ways in different textbooks. See recommended additional texts at the end of this chapter.

and $\bar{X}$ and $\bar{Y}$ are their means. The summations $\sum$ in Equation 9-44 extend over all values of i (1, 2, ... n).

The quantity $SD_{Y \cdot X}$ is "the standard deviation of Y given X," that is, the standard deviation of data values Y about the regression line. It is computed from

$$SD_{Y \cdot X}^2 = \frac{n-1}{n-2} \times (SD_Y^2 - b^2 SD_X^2) \quad (9\text{-}46)$$

where SD_X and SD_Y are the standard deviations of X and Y calculated by the usual methods. The estimated uncertainties (standard deviations) in b and a are given by

$$SD_b = SD_{Y \cdot X} / \left[SD_X \sqrt{n-1} \right]$$

$$SD_a = SD_{Y \cdot X} \sqrt{\frac{1}{n} + \frac{\bar{X}^2}{(n-1)SD_X^2}} \qquad (9\text{-}47)$$

Finally, the *correlation coefficient*, r, is computed from

$$r = b \, (SD_X / SD_Y) \qquad (9\text{-}48)$$

The correlation coefficient has a value between ± 1, depending on whether the slope b is positive or negative. A value near zero suggests no correlation between X and Y, (i.e., $b \approx 0$)

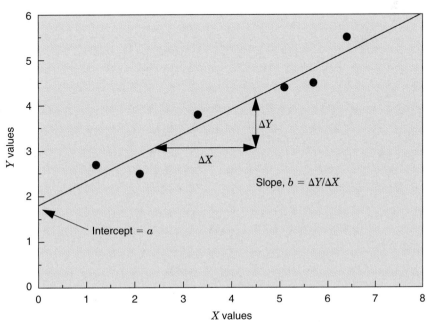

FIGURE 9-4 Hypothetical example of data and linear regression curve. ● = data values; ——— = calculated regression curve; $Y = a + bX$, a = Y-axis intercept; b = slope, $\Delta Y/\Delta X$.

and a value near ± 1 suggests a strong correlation.*

An alternative method for evaluating the strength of the correlation and its statistical significance is to determine whether b is significantly different from zero. This can be done by calculating

$$t = b/SD_b \qquad (9\text{-}49)$$

and comparing this to critical values of the t-distribution (see Fig. 9-3). The number of degrees of freedom is df $= (n - 2)$ in which n is the number of (X,Y) data pairs. If the calculated value of t exceeds the tabulated critical value at a selected significance level, one can conclude that the data support the hypothesis that Y is correlated with X. A similar analysis can be performed (using SD_a) to determine whether the intercept, a, is significantly different from zero.

REFERENCE

1. Tries MA, Skrable KW, French CS, Chabot GE: Basic applications of the chi-square statistic using counting data. *Health Phys* 77:441-454, 1999.

BIBLIOGRAPHY

Many useful general statistics texts are available. Some that were used for this chapter include the following:

Bevington PR: *Data Reduction and Error Analysis for the Physical Sciences*, New York, 1992, McGraw-Hill. [includes computer disk]
Crow EL, Davis FA, Maxfield MW: *Statistics Manual*, New York, 1960, Dover Publications.
Snedecor GW, Cochran WG: *Statistical Methods*, ed 8, Ames, IA, 1989, Iowa State University Press.

Additional discussion of nuclear counting statistics may be found in the following:

Evans RD: *The Atomic Nucleus*, New York, 1972, McGraw-Hill, Chapters 26 and 27.
Knoll GF: *Radiation Detection and Measurement*, ed 4, New York, 2010, John Wiley, Chapter 3.
Leo WR: *Techniques for Nuclear and Particle Physics Experiments*, ed 2, New York, 1994, Springer-Verlag, Chapter 3.

---

*An intuitively attractive interpretation of the correlation coefficient is that r^2 is the fraction of the observed variance of the data set Y that actually is attributable to variations in X and the dependence of Y on X. Thus, $r^2 = 0.64$ $(r = 0.8)$ implies that 64% of the observed variance SD_Y^2 actually is caused by the underlying variations in X, with the remaining 36% attributable to "other factors" (including random statistical variations).

Pulse-Height Spectrometry

Most of the radiation measurement systems used in nuclear medicine use pulse-height analysis (Chapter 8, Section C) to sort out the different radiation energies striking the detector. This is called *pulse-height* or *energy spectrometry*. It is used to discriminate against background radiation, scattered radiation, and so on, and to identify the emission energies of unknown radionuclides. In this chapter we discuss the basic principles of pulse-height spectrometry and some of its characteristics as applied to different types of detectors.

A. BASIC PRINCIPLES

Pulse-height spectrometry is used to examine the amplitudes of signals (electrical current or light) from a radiation detector to determine the energies of radiations striking the detector, or to select for counting only those energies within a desired energy range. This can be accomplished only with those detectors that provide output signals with amplitudes proportional to radiation energy detected, such as proportional counters, scintillation detectors, and semiconductor detectors (Chapter 7). A pulse-height, or energy, *spectrometer* consists of such a radiation detector and its high-voltage supply, preamplifier, amplifier, and pulse-height analyzer (Chapter 8, Section C). A pulse-height *spectrum* is a display showing the number of events detected ("counts") versus the amplitude of those events. This is provided most conveniently by a multichannel analyzer (Chapter 8, Section C.4 and Fig. 8-9).

The spectrum recorded from a radiation source depends not only on the energy of the emissions from the source but also on the type of radiation detector used. It also depends on the mechanisms by which the radiation energy is deposited in the detector. It is important to remember that the amplitude of the signal from a proportional, scintillation, or semiconductor detector depends on the amount of radiation energy *deposited in the detector,* which may be *less* than the full energy of the incident particle or photon.

In the case of particulate radiation (e.g., β particles or α particles), energy is transferred to the detector by collisions with atomic electrons in *primary ionization events*. These electrons may be given sufficient energy to cause *secondary ionizations* in collisions with other atomic electrons (Fig. 10-1A). Approximately 80% of the total ionization from particle-type radiation is the result of secondary ionization. The total amount of ionization produced (primary plus secondary) determines the amplitude of signal out of the detector (electrical current or light). Whether the full energy of the incident particle is deposited in the detector depends primarily on the range of the particle in the detector material. Particle ranges are very short in solids and liquids; thus the energy transfer is complete in most solid and liquid detectors—for example, sodium iodide [NaI(Tl)] and liquid scintillation detectors—and the amplitude of signal from the detector is thus proportional to particle energy. In gas-filled detectors (e.g., proportional counters), however, or in very thin solid detectors (e.g., some semiconductor detectors) that do not have sufficient thickness to stop the particle, the energy transfer may be incomplete. In this case, the amplitude of the signal from the detector will not reflect the total energy of the incident particle.

In the case of photons (γ rays, x rays, bremsstrahlung), energy is transferred to the detector primarily in photoelectric, Compton, or pair-production interactions. A portion of the incident photon energy is transferred as kinetic energy to photoelectrons, Compton electrons, or positive-negative electron pairs,

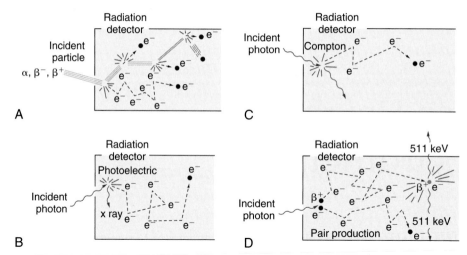

FIGURE 10-1 Deposition of radiation energy in a radiation detector. *A,* Energy transfer from an incident-charged particle to electrons in multiple ionization events. *Filled red circles* indicate electrons generated in primary ionization events and *dashed lines* are their trajectories; other electrons shown are released in secondary ionization events. *B-D,* Energy transfer from incident photon to electrons in photoelectric (*B*), Compton (*C*), and pair-production (*D*) interactions.

respectively, which in turn transfer their kinetic energy to the detector in secondary ionization events (Fig. 10-1B-D). Whether the amplitude of the signal out of the detector reflects the full energy of the incident photon depends on the fate of the remaining energy, which is converted into one or more *secondary photons* (characteristic x ray, Compton-scattered photon, or annihilation photons). A secondary photon may deposit its energy in the detector by additional interactions*; however, if it *escapes* from the detector, then the energy deposited in the detector and the amplitude of the signal from the detector do not reflect the full energy of the incident photon. The amplitude of the signal from the detector reflects only *the amount of energy deposited in it* by the radiation event.

B. SPECTROMETRY WITH NaI(Tl)

Because of its favorable performance-to-cost ratio, a NaI(Tl) scintillator [coupled to a photomultiplier (PM) tube, or in some cases to a photodiode] is the most commonly used detector in nuclear medicine (Chapter 7, Section C). The basic principles of pulse-height spectrometry are illustrated for this detector. Because NaI(Tl) is used almost

*Note that multiple interactions arising from a single incident photon occur so rapidly in the detector that they appear to be a single event.

exclusively for detecting photons (γ rays or x rays, primarily), only photon spectrometry is considered here.

1. The Ideal Pulse-Height Spectrum

Suppose that a monoenergetic γ-ray source is placed in front of a radiation detector. Assume, further, that the energy of the γ rays, E_γ, is less than 1.022 MeV, so that pair-production interactions do not occur. The principle γ-ray interactions with the detector will be by photoelectric absorption and Compton scattering. Most of the photoelectric interactions result in full deposition of the γ-ray energy in the detector (the characteristic x ray usually is also absorbed in the detector). Pulse amplitudes from these events are proportional to E_γ (Fig. 10-2A). With an *ideal* radiation detector, this would produce a single narrow line in the pulse-height spectrum, called the *photopeak*, at a location corresponding to the γ-ray energy E_γ (Fig. 10-2B). In Compton scattering, only a part of the γ-ray energy is transferred to the detector, via the Compton recoil electron. If the scattered γ ray also is absorbed in the detector, the event produces a pulse in the photopeak, whereas if the scattered γ ray escapes, the energy deposited in the detector is less than E_γ. According to Equation 6-14, the energy deposited in the detector in a single Compton scattering event ranges from near zero (small-angle scattering event), up to a maximum value E_{ce}, corresponding to the energy of the recoil electron for 180-degree Compton scattering events

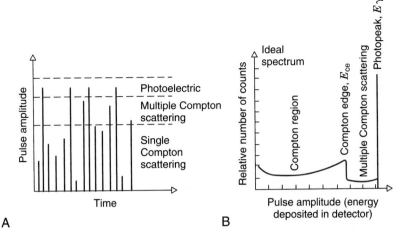

FIGURE 10-2 Elements of an ideal γ-ray pulse-height spectrum. *A,* Pulses from the detector representing different types of γ-ray interactions in the detector. *B,* Distribution (relative number) of pulses versus amplitude (or energy deposited in the detector). Only the photopeak represents deposition of the full energy of the γ ray in the detector.

$$E_{ce} = E_\gamma^2 /(E_\gamma + 0.2555) \qquad (10\text{-}1)$$

where E_γ and E_{ce} are in MeV. The ideal spectrum therefore includes a distribution of pulse amplitudes ranging from nearly zero amplitude up to some maximum amplitude corresponding to the energy given by Equation 10-1. As shown in Figure 10-2B, this part of the spectrum is called the *Compton region.* The sharp edge in the spectrum at E_{ce} is called the *Compton edge.*

Another possibility is that a Compton-scattered γ ray may experience additional Compton-scattering interactions in the detector. Multiple Compton scattering events produce the distribution of pulses with amplitudes in the "valley" between the Compton edge and the photopeak.

2. The Actual Spectrum

In practice, the actual spectrum obtained with a NaI(Tl) spectrometer is quite different from the ideal one shown in Figure 10-2B. For example, Figure 10-3 shows a spectrum obtained from a ^{137}Cs radiation source, which emits 662-keV γ rays and ~30-keV barium x

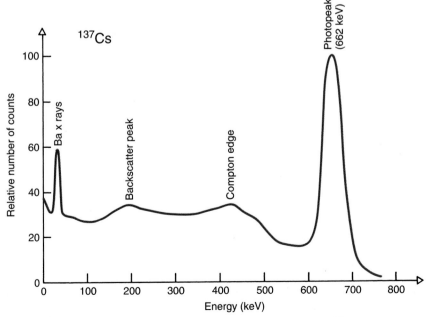

FIGURE 10-3 Actual pulse-height spectrum recorded with a NaI(Tl) detector and ^{137}Cs (662-keV γ rays, ~30 keV Ba x rays). Compare with Figure 10-2B.

rays. The spectrum was recorded with a multichannel analyzer, 0.01 V per channel, with the amplifier gain adjusted so that 662 keV of energy corresponds to 6.62 V of pulse amplitude. Thus the horizontal axis has been translated from pulse amplitude (~0-8 V) into energy (~0-800 keV).

The first feature noted is that the spectrum is "smeared out." The photopeak is not a sharp line, as shown in Figure 10-2B, but a somewhat broadened peak, and the Compton edge is rounded. This is caused by the imperfect energy resolution of the NaI(Tl) detector, discussed in Section B.7.

Another structure that may appear in the spectrum is a *backscatter peak*. This is caused by detection of γ rays that have been scattered toward the detector after undergoing a 180-degree scattering outside the detector. Certain detector configurations enhance the intensity of the backscatter peak. For example, in the well counter (Chapter 12, Section A), a γ ray may pass through the detector without interaction, then scatter back into the detector from the shielding material surrounding it and be detected.

Note that the energy of the backscatter peak, E_b, is the energy of the *scattered* γ ray after 180-degree scattering, whereas the energy of the Compton edge, E_{ce}, is the energy given to the *recoil electron* in a 180-degree scattering event. Therefore

$$E_b + E_{ce} = E_\gamma \qquad (10\text{-}2)$$

Equation 10-2 is helpful for identifying backscatter peaks.

Another structure that may appear is an *iodine escape peak*. This results from photoelectric absorption interactions with iodine atoms in the NaI(Tl) crystal, followed by escape from the detector of the characteristic iodine K-x ray, which has energy of approximately 30 keV. The iodine escape peak occurs at an energy approximately $E_\gamma - 30$ keV; that is, about 30 keV below the photopeak. Iodine escape peaks may be prominent with low-energy γ-ray emitters, for example, ^{197}Hg (Fig. 10-4). Low-energy γ rays are detected by absorption primarily in a thin layer close to the entrance surface of the NaI(Tl) crystal where there is a reasonable probability that the iodine x ray will escape from the detector. With increasing γ-ray energy, the interactions tend to occur deeper within the detector, and there is less likelihood that the x ray will escape. Also, the relative difference between the photopeak and escape peak energies

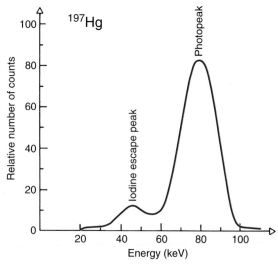

FIGURE 10-4 Pulse-height spectrum for ^{197}Hg ($E_\gamma = 77.3$ keV) recorded with NaI(Tl). Iodine escape peak (45-50 keV) is due to escape of characteristic iodine x ray (~30 keV) following a photoelectric absorption event in detector.

becomes smaller, and it becomes more difficult to distinguish between them.

Lead x-ray peaks sometimes are seen in spectra acquired with systems employing lead shielding and collimation. These peaks are caused by photoelectric interactions of the γ rays in the lead. These interactions are followed by emission characteristic 80- to 90-keV lead x rays, which may be recorded by the detector.

If the γ-ray energy exceeds 1.022 MeV, pair production interactions can occur. The kinetic energy given to the positive-negative electron pair is $E_\gamma - 1.022$ MeV (see Chapter 6, Section C.4). In most cases, the entire kinetic energies of both particles are deposited in the detector. When the positron comes to rest, it combines with an electron to create a pair of 511-keV annihilation photons. If both of these photons are absorbed in the detector, the event is recorded in the photopeak. If only one is absorbed, the event is recorded in the *single escape peak*, at energy $E_\gamma - 511$ keV (Fig. 10-5). If both escape, the event is recorded in the double escape peak, at $E_\gamma - 1.022$ MeV.

Scattering within or around the radiation source, or *object scatter,* changes the distribution of radiation energies striking the detector. This is especially important in counting measurements in vivo and in radionuclide imaging because substantial scattering of radiation occurs within the patient. Figure 10-6 shows spectra for ^{131}I with and without scattering material around the source. The

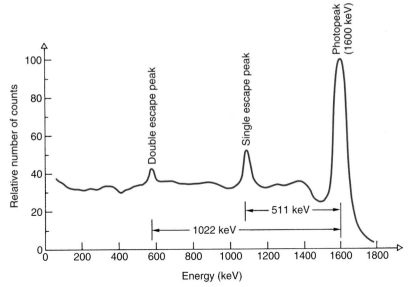

FIGURE 10-5 Pulse-height spectrum for a hypothetical 1.6-MeV (1600-keV) γ-ray emitter. Because γ-ray energy exceeds 1.022 MeV (1022 keV), pair-production interactions can occur in the detector. Escape peaks are due to escape of one or both annihilation photons from the detector following a pair-production interaction.

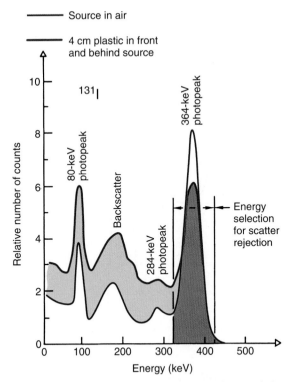

FIGURE 10-6 Effect of scattering material around the source on the pulse-height spectrum for ^{131}I. The *red curve* shows the spectrum with the source in air and the *blue curve* shows the spectrum after placing the source between 4-cm layers of plastic. For the blue curve, the *darker shaded area* represents counts within the photopeak and the *lighter shaded area* represents counts due to γ rays scattered in the plastic.

general effect of object scatter is to add events in the lower-energy region of the spectrum. It is possible to discriminate against scattered radiation by using a pulse-height analyzer to count only events in the photopeak, as shown in Figure 10-6.

Coincidence summing can occur when a radionuclide emits two or more γ rays per nuclear disintegration. Figure 10-7 shows spectra recorded with a NaI(Tl) well counter for ^{111}In, which emits a 173-keV and a 247-keV γ ray simultaneously. The peak at 420 keV seen when the source is inside the well counter results from simultaneous detection of these two γ rays. Summing between x rays and γ rays also can occur. With positron emitters, coincidence summing between the two 511-keV annihilation photons also may be observed. Coincidence summing is especially prominent with detector systems having a high geometric efficiency (see Chapter 11, Section A.2), that is, systems in which there is a high probability that both γ rays will be captured by the detector [e.g., well counters (Chapter 12, Section A)].

3. Effects of Detector Size

The larger the detector crystal size, the more likely it is that secondary photons (i.e., Compton-scattered γ rays and annihilation photons) will be absorbed in the crystal. Thus with increasing crystal size, the number of events in the photopeak versus Compton regions increases. Figure 10-8 shows this

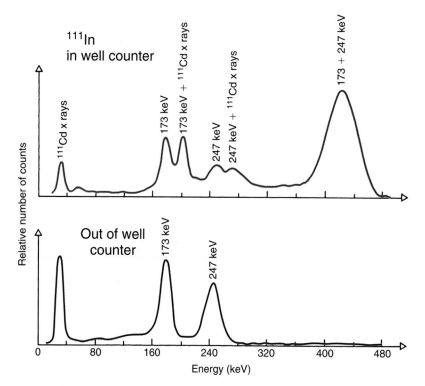

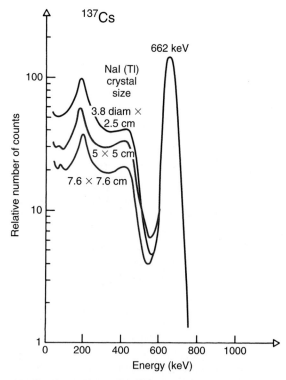

FIGURE 10-7 Pulse-height spectra recorded for ^{111}In with a NaI(Tl) well counter detector. *Top,* Coincidence summing between the x-ray and γ-ray emissions results in additional peaks in the spectrum when the source is inside the well. *Bottom,* When the source is outside the well, the probability of coincidence detection decreases and the coincidence peaks disappear.

effect on the spectrum for [137]Cs. Figure 10-8 also shows that the "valley" between the Compton edge and the photopeak at first increases with increasing detector size, due to greater likelihood of an incident photon undergoing multiple Compton interactions within the detector. However, the number of counts in this region eventually decreases due to greater likelihood of complete absorption of the incident photon's energy within the detector, thereby producing an event in the photopeak rather than in the valley. For γ-ray energies greater than 1.022 MeV, the size of annihilation escape peaks also decreases with increasing crystal size.

4. Effects of Counting Rate

Distortions of the spectrum occur at high counting rates as a result of overlap of detector output pulses. Pulse pile-up between two events can produce a single pulse with an amplitude equal to their sum (see Chapter 8, Section B.3). Pile-up between photopeak events and lower-energy events causes a general broadening of the photopeak (Fig. 10-9). This also is one of the causes of dead time losses (see Chapter 11, Section C). There also may be a shift of the photopeak toward lower energies because of baseline shift in the amplifier at high counting rates. Thus if a single-channel analyzer (SCA) is set up at low counting rates on the photopeak and the

FIGURE 10-8 Effect of NaI(Tl) crystal size on the pulse-height spectrum for [137]Cs. The spectra have been normalized to equal photopeak heights. In practice, the photopeak height also increases with increasing detector size because of increasing detection efficiency (Chapter 11, Section A).

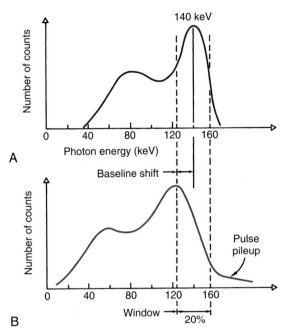

A

B

FIGURE 10-9 *A,* ^{99m}Tc spectrum at low counting rate. *B,* Spectral broadening and shift in apparent photopeak energy caused by pulse pileup and baseline shift in the spectrometer amplifier at high counting rate.

detector is used at very high counting rates, the photopeak can shift out of the SCA window and an incorrect reading may be recorded.

5. General Effects of γ-Ray Energy

Figure 10-10 shows pulse-height spectra for a number of radionuclides emitting γ rays of different energies. The solid lines are the spectra for unscattered γ rays, and the dashed lines are the spectra for object-scattered γ rays. In general, the relative number of events in the Compton region versus the photopeak region becomes larger with increasing γ-ray energy because the probability of Compton versus photoelectric interactions in the detector becomes larger. Also, as γ-ray energy increases, it becomes easier to separate object scatter from the photopeak. This is because the change in γ-ray energy with Compton scattering increases with γ-ray energy (see Chapter 6, Section C.3). For example, at 100 keV and at 500 keV, Compton scattering through 90 degrees produces scattered photon energies of 84 keV and 253 keV, respectively. In addition, as discussed in Section B.7 below, the energy resolution of NaI(Tl) detectors improves with increasing γ-ray energy, which provides further improvement in their ability to discriminate between scattered versus unscattered photons.

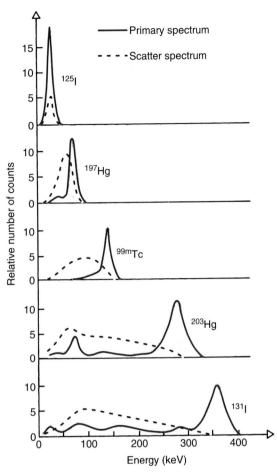

FIGURE 10-10 Pulse-height spectra recorded with a NaI(Tl) detector for different γ-ray energies. *Primary spectrum* refers to γ rays striking the detector without scattering from objects outside the detector. *Scatter spectrum* refers to γ rays that have been scattered by objects outside the detector, such as from tissues or other materials surrounding the source distribution. (*Adapted from Eichling JO, Ter Pogossian MM, Rhoten ALJ: Analysis of scattered radiation encountered in lower energy diagnostic scanning. In Gottschalk A, Beck RN, editors:* Fundamentals of Scanning. *Springfield, IL, 1968, Charles C Thomas.*)

6. Energy Linearity

Energy linearity refers to the proportionality between output pulse amplitude and energy absorbed in the detector. Figure 10-11, taken from early work on the basic properties of NaI(Tl) detectors, shows a typical relationship between apparent energy (pulse height) and actual γ-ray energy for a system calibrated with ^{137}Cs (662 keV). Most NaI(Tl) systems are quite linear for energies between 0.2 and 2 MeV, and a single-source energy calibration usually is acceptable in this range; however, one can run into problems by calibrating a spectrometer with a high-energy source (e.g., ^{137}Cs) and then attempting to use

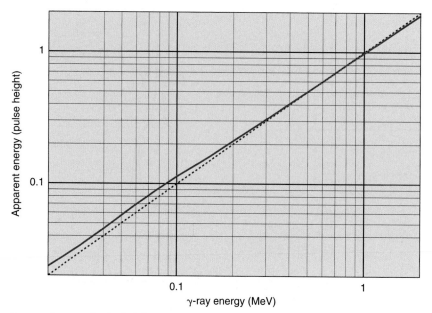

FIGURE 10-11 Apparent energy (pulse height) versus actual γ-ray energy for a NaI(Tl) scintillation detector calibrated for one unit of pulse height per MeV at 0.662 MeV (*solid line*). *Dashed line* is line of identity. With this calibration, detector nonlinearities can lead to 10% to 15% errors in apparent energy for $E_\gamma \lesssim 0.2$ MeV. (*Curve redrawn from Knoll GF:* Radiation Detection and Measurement, *ed 3. New York, 2000, John Wiley, p 339.*)

it for much lower-energy sources (e.g., ^{125}I or ^{99m}Tc) or vice versa. Modern spectrometers and gamma cameras frequently have pre-calibrated push buttons that are set for specific radionuclides and that take into account any energy nonlinearities. For systems that are not precalibrated, individual low- and high-energy sources should be used to calibrate a spectrometer for measurements that span a wide range of energies.

Energy linearity also is an important factor in energy resolution. This is discussed in the following section.

7. Energy Resolution

Sharp lines and sharp edges in the ideal spectrum (Fig. 10-2B) become broadened lines and rounded edges in actual spectra (Fig. 10-3). With NaI(Tl) detectors, this spectral blurring (or line broadening) is caused primarily by random *statistical variations* in the events leading to the formation of the output signal. For NaI(Tl) coupled to a PM tube, these include the following:

1. Statistical variations in the number of scintillation light protons produced per keV of radiation energy deposited in the crystal
2. Statistical variations in the number of photoelectrons released from the photocathode
3. Statistical variations in the electron multiplication factor of the dynodes in the PM tube

Causes of spectral blurring relating to fabrication of a NaI(Tl) detector assembly include the following:

4. Nonuniform sensitivity to scintillation light over the area of the PM tube cathode
5. Nonuniform light collection efficiency for light emitted from interactions at different locations within the detector crystal

An important but subtle cause of spectral blurring with scintillation detectors is the following:

6. Nonlinear energy response of the scintillator, such that the amount of light produced by the lower-energy Compton electrons in multiple Compton interactions generate a different total amount of light than is produced by a higher-energy photoelectron in a single high-energy photoelectric event, even when the total energy deposited in the crystal is the same (see Section B.6)

Electronic noise contributes to spectral blurring with all types of detectors. With scintillation detectors read out by a PM tube, the principal sources include the following:

7. Fluctuations in the high voltage applied to the PM tube
8. Electrical noise in the PM tube

Because of these factors, there are differences in the amplitude of the signal from a scintillation detector for events in which precisely the same amount of radiation energy is deposited in the detector. Instead of a narrow "line," the photopeak approximates a gaussian-shaped curve, as illustrated in Figure 10-3. The width of the photopeak, ΔE, measured across its points of half-maximum amplitude is the *energy resolution*. This is referred to as the *full width at half maximum* (FWHM). Usually the FWHM is expressed as a percentage of the photopeak energy E_γ:

$$\text{FWHM}(\%) = (\Delta E/E_\gamma) \times 100\% \quad (10\text{-}3)$$

Figure 10-12 illustrates this computation. Although FWHM can be computed for any γ-ray energy, it is customary to specify the value for the γ rays of a commonly used radionuclide when characterizing the performance of a particular detector. Examples are the 662-keV γ rays of ^{137}Cs, the 511-keV annihilation photons of positron emitters, or the 140-keV γ rays of ^{99m}Tc. For a gaussian-shaped curve, the FWHM is related to the standard deviation, *SD,* according to

$$\text{FWHM} \approx 2.35 \times SD \quad (10\text{-}4)$$

For NaI(Tl)-PM tube detectors, a major source of statistical variation in output pulse amplitude is in the number of photoelectrons released from the photocathode of the PM tube. On average, approximately 40 visible light photons are produced per keV of γ-ray energy absorbed in the crystal (see Table 7-2). With good-quality PM tubes and efficient optical coupling, approximately 25% of the light photons yield photoelectrons from the photocathode. Thus the average number of photoelectrons is approximately 10 per keV of radiation energy absorbed in the NaI(Tl) crystal. Complete absorption of a 662-keV γ ray from ^{137}Cs results in the release *on average* of approximately 6600 photoelectrons from the photocathode; however, the actual number varies from one γ ray to the next according to Poisson statistics, with a standard deviation of $\pm\sqrt{6600} \approx 81$ photoelectrons. This amounts to a variation of approximately $\pm 1.2\%$ in pulse amplitude (see Chapter 9, Section B), which translates into an FWHM of approximately 3% (Equation 10-4).

If this were the only source of variation in output pulse amplitude, the energy resolution of NaI(Tl) would be proportional to $1/\sqrt{E}$, because the number of photoelectrons is proportional to the energy deposited in the crystal. However, in practice, the effects of energy are smaller owing to the presence of other sources of pulse amplitude variation. This is evident from a simple comparison of FWHM achievable with a good-quality scintillation detector at 662 keV (about 6%) and the value predicted from simple photoelectron statistics (approximately 3%). The difference is due to other sources of amplitude variations listed earlier. Figure 10-13, showing the observed energy resolution for a NaI(Tl) detector versus a simple $1/\sqrt{E}$ relationship, illustrates this point.

Analyses suggest that photoelectron statistics, PM-tube noise (including electron multiplication), and nonlinear energy response of

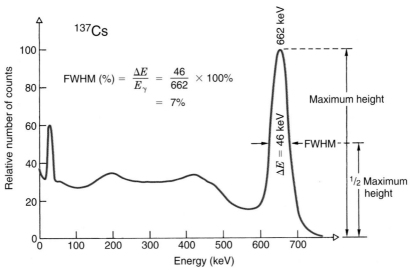

FIGURE 10-12 Calculation of full width at half maximum (FWHM) energy resolution of a NaI(Tl) detector for ^{137}Cs 662-keV γ rays.

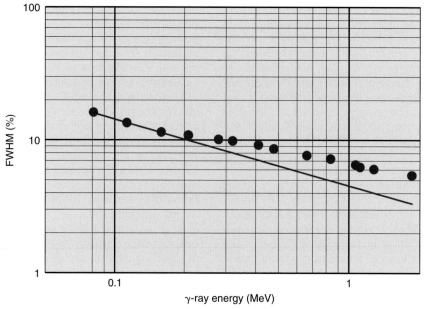

FIGURE 10-13   Energy resolution versus γ-ray energy for a 7.5-cm-diameter × 7.5-cm-thick NaI(Tl) scintillation detector. *Solid line* indicates theoretical $1/\sqrt{E}$ behavior, fitted to low-energy data points. *Experimental data points (●) from Birks JB: The Theory and Practice of Scintillation Counting. Oxford, England, 1964, Pergammon Press, p 159.*

the scintillator contribute about equally to overall energy resolution at 662 keV.[1,2] Significant improvements in PM tubes and optical coupling technology have yielded steady improvements in energy resolution during the past 3 to 4 decades. However, the nonlinear energy response of the scintillator may prove to be the limiting factor in achievable energy resolution for NaI(Tl), regardless of further technological improvements. Additional discussions of this complicated issue can be found in reference 1 and other recommended readings at the end of this chapter.

With good-quality PM tubes, energy resolution of 6% at 662 keV is achievable with NaI(Tl). These detectors have energy resolutions of approximately 10% for the 140-keV γ rays of [99m]Tc. With large-area crystals having multiple PM tubes [e.g., the gamma camera, (see Chapter 13)], the resolution for ^{99m}Tc can be degraded because of slightly different responses between PM tubes. However, modern gamma cameras employ electronic and software correction schemes to account for these variations and commonly achieve 10% energy resolution for ^{99m}Tc as well (Chapter 14, Section A.3).

Another factor that affects energy resolution is the integration time used to collect signal from the detected event. For routine imaging or spectrometry applications with NaI(Tl), the integration time typically is approximately 1 μsec, in which case the energy resolutions mentioned earlier may be achieved. However, for positron coincidence detection, the integration time may be shortened to only a few hundred nanoseconds to minimize the number of random coincidences between annihilation photons that do not actually arise from the same positron annihilation event (Chapter 18, Section A.9). With shorter integration times, the number of photoelectrons contributing to the detected signal is smaller; hence, energy resolution is degraded. Typically, the energy resolution at 511 keV (the energy of the annihilation photons) may be degraded from a value of 6% to 7% with "full integration" of the detected signal, to a value of approximately 10% with the shortened integration time used in positron coincidence mode.

Other factors that can degrade energy resolution include poor light coupling between the NaI(Tl) crystal and the PM tubes, which can cause a reduction in the number of photoelectrons released per keV. Energy resolution also may be degraded by other conditions that interfere with the efficient collection of light from the crystal by the PM tube. For example, a cracked detector crystal causes internal reflections and trapping of light in the detector crystal. A sudden degradation of energy resolution and loss of output pulse amplitude often are the first symptoms of a cracked crystal. Deterioration of the optical coupling grease between the detector crystal and PM tube has similar effects. Poor light collection also can occur with detectors having an unusual shape, such as a high aspect ratio (long and narrow).

Good energy resolution is a desirable characteristic for any spectrometer system because it permits precise identification and separation of γ rays with very similar energies, for example, for radionuclide identification or scatter rejection. The best energy resolution is obtained with semiconductor detectors, as discussed in the following section.

C. SPECTROMETRY WITH OTHER DETECTORS

1. Semiconductor Detector Spectrometers

The major advantage of [Si(Li)] and [Ge(Li)] semiconductor detectors (Chapter 7, Section B) is their superb energy resolution. It is typically 6-9 times better than proportional counters and 20-80 times better than NaI(Tl):PM tube detectors. The output signal from a semiconductor detector is a pulse of electrical current, the amplitude of which is proportional to the radiation energy deposited in the detector. The energy resolution of Si(Li) and Ge(Li) detectors is determined by statistical variations in the number of charges in this pulse. The average number is approximately 1 charge (electron) per 3 eV of radiation energy absorbed (see Table 7-1), as compared with only 10 photoelectrons per *keV* in a NaI(Tl):PM tube detector system. The much larger number of charges produced in these semiconductor detectors results in much smaller percentage statistical variations in signal amplitude and hence much better energy resolution than NaI(Tl). Figure 10-14 shows comparative NaI(Tl): PM Table and Ge(Li) spectra for ^{99m}Tc. The superior energy resolution of Ge(Li) permits almost complete elimination of scattered radiation by pulse-height analysis and clean separation of multiple photon emissions from single or multiple sources.

Despite their superior performance in terms of energy resolution, Si(Li) and Ge(Li) detectors have not found widespread usage in nuclear medicine. As explained in Chapter 7, they are available only in relatively small sizes. As well, Ge(Li) must be operated at liquid nitrogen temperatures, which poses practical inconveniences, and Si(Li) detectors are relatively inefficient for the γ-ray energies commonly used in nuclear medicine.

More recently developed "room temperature" semiconductor detectors such as cadmium telluride and cadmium zinc telluride (CZT) (Chapter 7, Section B) may provide more practical options for nuclear medicine. Although their energy resolution is not equal to that of Si(Li) or Ge(Li), owing to somewhat lower production of charge carriers, it is significantly better than NaI(Tl).

Figure 10-15 shows typical pulse-height spectra for ^{99m}Tc and ^{18}F (511-keV annihilation photons) obtained with a CZT detector. A number of interesting features are evident in these spectra. For ^{99m}Tc, the energy resolution is intermediate to that of Ge(Li) and NaI(Tl) (see Fig. 10-14). For both ^{99m}Tc and ^{18}F, there is evidence of a "tail" on the low-energy side of the photopeak. This is caused by "charge trapping" and incomplete charge collection within the CZT crystal. In addition to the main photopeak at 140 keV, a small photopeak is seen at approximately 20 keV. This corresponds to K-x rays of technetium emitted after internal conversion events (~7% emission frequency; see Appendix C). This peak is rarely, if ever, seen in NaI(Tl) spectra owing to attenuation of these x rays by the canning material housing the detector crystal.

The CZT spectrum for ^{18}F shows a well-defined Compton edge (E_{ce} = 341 keV) and backscatter peak (E_b – 170 keV). Also present are peaks at approximately 73 keV and 86 keV, which were caused by characteristic x rays of lead from shielding material placed around the source in this experiment.

Note finally that the energy resolution of the CZT spectra is essentially the same for ^{99m}Tc as for ^{18}F, in spite of a nearly fourfold difference in their γ-ray energies. With NaI(Tl), this would result in a significant difference in energy resolution, owing to a similar difference in the number of photoelectrons emitted by the photocathode of the PM tube. However, with CZT, the equivalent source of line broadening is in the number of charge carriers (electron-hole pairs) produced, which is a significantly larger number. The predominating causes of line broadening with CZT are leakage current through the detector itself and incomplete (and variable) collection of the charge carriers. These factors depend primarily on the operating

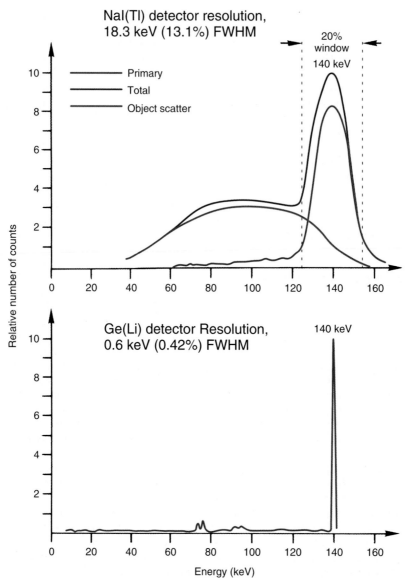

FIGURE 10-14 Comparative pulse-height spectra recorded from a ^{99m}Tc source with NaI(Tl) and Ge(Li) detectors. In the NaI(Tl) spectrum (*top*), the *blue* curve represents unscattered (primary) γ rays, the *orange* curve represents γ rays scattered by materials around the source, and the *red* curve represents the sum of the primary and scattered γ rays. For the Ge(Li) detector (*bottom*), only the spectrum for primary γ rays is shown. Separation of primary from scattered γ rays is much easier with the semiconductor detector.

voltage and on the specific detector configuration (such as electrode attachments). The next most important contributor is electronic noise. None of these factors depend directly on γ-ray energy. Thus the approximate $1/\sqrt{E}$ relationship seen with NaI(Tl) generally does not apply for room-temperature semiconductor detectors.

The performance of CZT detectors can be improved by operating them at low temperatures (thereby reducing background leakage current). This also would at least partially restore a $1/\sqrt{E}$ relationship in their energy resolution; however, this also would eliminate the practical benefits of room-temperature operation.

2. Liquid Scintillation Spectrometry

Although NaI(Tl) spectrometers are used in many different configurations and applications, both for in vivo and in vitro measurements, liquid scintillation spectrometers are used almost exclusively in a single configuration for in vitro sample counting (see Chapter

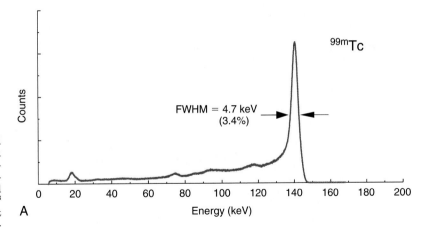

A

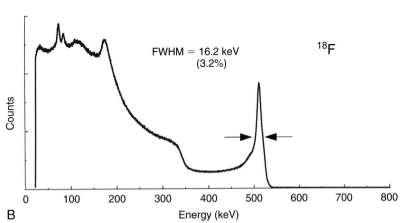

B

FIGURE 10-15 [99m]Tc (*A*) and [18]F (*B*) spectra obtained with a 5 × 5 × 5-mm cadmium zinc telluride (CZT) detector, with 0.6-mm-thick Al entrance window and CAPture electrode geometry.[3] The detector was operated at room temperature with an operating voltage of 1000 V for ^{99m}Tc and 1250 V for [18]F. FWHM, full width at half maximum. (*Data courtesy Paul Kinahan, University of Washington, Seattle, WA; eV Products, Saxonburg, PA; and James Wear of Lunar Corporation, Madison, WI.*)

12, Section C). Liquid scintillation detectors are used primarily for counting the low-energy β emissions from ^{3}H, ^{14}C, ^{35}S, [45]Ca, and [32]P.

Figure 10-16 shows pulse-height spectra recorded with a liquid scintillation system for a γ-ray emitter, ^{137}Cs, and for a β emitter, [14]C. Liquid scintillators provide poor energy resolution for γ rays because they produce relatively few scintillation light photons per keV of energy absorbed and hence produce relatively few photoelectrons at the PM tube photocathode in comparison with NaI(Tl). Another factor is the relatively inefficient transfer of light photons from the scintillator vial to the PM tubes. The spectrum for a β emitter has no sharp peak because the energy spectrum for β particles has a broad distribution from zero up to E_β^{max} for the radionuclide (compare Fig. 10-16 with Fig. 3-2).

3. Proportional Counter Spectrometers

Gas-filled proportional counters (Chapter 7, Section A.3) have found limited use for spectrometry in nuclear medicine. Their energy resolution is several times better than NaI(Tl).

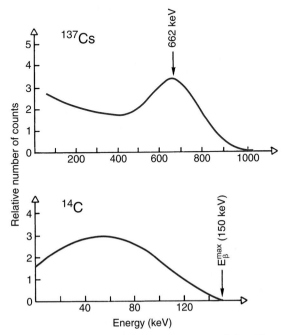

FIGURE 10-16 Pulse-height spectra recorded with a liquid scintillation detector, for a γ-ray emitter, [137]Cs (*top*), and a β emitter, [14]C (*bottom*).

Their major disadvantage is poor detection efficiency for γ rays (see Chapter 11, Section A.3). Some applications of proportional counter spectrometry are discussed in Chapter 12.

REFERENCES

1. Dorenbos P, de Haas JTM, van Eijk CWE: Non-proportionality of scintillation response and the energy resolution obtainable with scintillation crystals. *IEEE Trans Nucl Sci* 42:2190-2202, 1995.
2. Valentine JD, Rooney BD, Li J: The light yield non-proportionality component of scintillator energy resolution. *IEEE Trans Nucl Sci* 45:512-517, 1998.
3. Parnham K, Szeles C, Lynn KG, Tjossem R: Performance improvement of CdZnTe detectors using modified two-terminal electrode geometry. *SPIE Conference on Hard X-Ray, Gamma-Ray and Neutron Detector Physics*, Denver, CO, July 1999.

BIBLIOGRAPHY

Additional discussion of NaI(Tl) pulse-height spectrometry may be found in the following:

Birks JB: *The Theory and Practice of Scintillation Counting*, New York, 1964, MacMillan.

Hine GJ: Sodium iodide scintillators. In Hine GJ, editor: *Instrumentation in Nuclear Medicine*, Vol 1, New York, 1967, Academic Press, Chapter 6.

Spectrometry with Si(Li) and Ge(Li) semiconductor detectors is discussed in the following:

TerPogossian MM, Phelps ME: Semiconductor detector systems. *Semin Nucl Med* 3:343-365, 1973.

Spectrometry with room-temperature semiconductor detectors is discussed in the following:

Schlesinger TE, James RB, editors: Semiconductors for room temperature nuclear detector applications. In *Semiconductors and Semimetals*, Vol 43, San Diego, 1995, Academic Press. (Chapters 8, 9, and 14 are of particular interest.)

A useful general reference for pulse-height spectrometry is the following:

Knoll GF: *Radiation Detection and Measurement*, ed 4, New York, 2010, John Wiley.

Problems in Radiation Detection and Measurement

Nuclear medicine studies are performed with a variety of types of radiation measurement instruments, depending on the kind of radiation source that is being measured and the type of information sought. For example, some instruments are designed for in vitro measurements on blood samples, urine specimens, and so forth. Others are designed for in vivo measurements of radioactivity in patients (Chapter 12). Still others are used to obtain images of radioactive distributions in patients (Chapters 13, 14, and 17-19).

All these instruments have special design characteristics to optimize them for their specific tasks, as described in the chapters indicated above; however, some considerations of design characteristics and performance limitations are common to all of them. An important consideration for any radiation measurement instrument is its *detection efficiency*. Maximum detection efficiency is desirable because one thus obtains maximum information with a minimum amount of radioactivity.

Also important are the instrument's *counting rate limitations*. There are finite counting rate limits for all counting and imaging instruments used in nuclear medicine, above which inaccurate results are obtained because of data losses and other data distortions. Nonpenetrating radiations, such as β particles, have special detection and measurement problems. In this chapter, we discuss some of these general considerations in nuclear medicine instrumentation.

A. DETECTION EFFICIENCY

1. Components of Detection Efficiency

Detection efficiency refers to the efficiency with which a radiation-measuring instrument converts emissions from the radiation source into useful signals from the detector. Thus if a γ-ray-emitting source of activity A (Bq) emits η γ rays per disintegration, the emission rate ξ of that source is

$$\xi \,(\gamma \text{ rays/sec}) = A \,(\text{Bq}) \times 1 \,(\text{dps/Bq}) \\ \times \eta \,(\gamma \text{ rays/dis}) \quad (11\text{-}1)$$

If the counting rate recorded from this source is R [counts per second (cps)], then the detection efficiency D for the measuring system is

$$D = R/\xi \quad (11\text{-}2)$$

Alternatively, if the emission rate ξ and detection efficiency D are known, one can estimate the counting rate that will be recorded from the source from

$$R = D\xi \quad (11\text{-}3)$$

In general, it is desirable to have as large a detection efficiency as possible, so that a maximum counting rate can be obtained from a minimum amount of activity. Detection

155

efficiency is affected by several factors, including the following:

1. The *geometric efficiency*, which is the efficiency with which the detector intercepts radiation emitted from the source. This is determined mostly by detector size and the distance from the source to the detector.
2. The *intrinsic efficiency* of the detector, which refers to the efficiency with which the detector absorbs incident radiation events and converts them into potentially usable detector output signals. This is primarily a function of detector thickness and composition and of the type and energy of the radiation to be detected.
3. The fraction of output signals produced by the detector that are recorded by the counting system. This is an important factor in *energy-selective counting*, in which a pulse-height analyzer is used to select for counting only those detector output signals within a desired amplitude (energy) range.
4. *Absorption and scatter* of radiation within the source itself, or by material between the source and the radiation detector. This is especially important for in vivo studies, in which the source activity generally is at some depth within the patient.

In theory, one therefore can describe detection efficiency D as a product of individual factors,

$$D = g \times \varepsilon \times f \times F \qquad (11\text{-}4)$$

where g is the geometric efficiency of the detector, ε is its intrinsic efficiency, f is the fraction of output signals from the detector that falls within the pulse-height analyzer window, and F is a factor for absorption and scatter occurring within the source or between the source and detector. Each of these factors are considered in greater detail in this section. Most of the discussion is related to the detection of γ rays with NaI(Tl) detector systems. Basic equations are presented for somewhat idealized conditions. Complications that arise when the idealized conditions are not met also are discussed. An additional factor applicable for radionuclide imaging instruments is the collimator efficiency, that is, the efficiency with which the collimator transmits radiation to the detector. This is discussed in Chapter 13.

2. Geometric Efficiency

Radiation from a radioactive source is emitted *isotropically,* that is, with equal intensity in all directions. At a distance r from a point source of γ-ray-emitting radioactivity, the emitted radiation passes through the surface of an imaginary sphere having a surface area $4\pi r^2$. Thus the flux I of radiation passing through the sphere per unit of surface area, in units of γ rays/sec/cm^2, is

$$I = \xi / 4\pi\, r^2 \qquad (11\text{-}5)$$

where ξ is the emission rate of the source and r is given in centimeters. As distance r increases, the flux of radiation decreases as $1/r^2$ (Fig. 11-1). This behavior is known as the *inverse-square law*. It has important implications for detection efficiency as well as for radiation safety considerations (see Chapter 23). The inverse-square law applies to all types of radioactive emissions.

The inverse-square law can be used to obtain a first approximation for the geometric efficiency of a detector. As illustrated in Figure 11-1, a detector with surface area A placed at a distance r from a point source of radiation and facing toward the source will intercept a fraction $A/4\pi r^2$ of the emitted radiation. Thus its geometric efficiency g_p is

$$g_p \approx A/4\pi\, r^2 \qquad (11\text{-}6)$$

where the subscript p denotes a point source. The approximation sign indicates that the equation is valid only when the distance from the point source to the detector is large in comparison with detector size, as discussed in the following paragraphs.

EXAMPLE 11-1

Calculate the geometric efficiency for a detector of diameter $d = 7.5$ cm at a distance $r = 20$ cm from a point source.

Answer

The area, A, of the detector is

$$A = \pi\, d^2/4 = \pi[(7.5)^2/4]\ \text{cm}^2$$

Therefore, from Equation 11-6,

$$
\begin{aligned}
g_p &\approx A/4\pi\, r^2 \\
&\approx \pi\,(7.5)^2/[4 \times 4\pi\,(20)^2] \\
&\approx (7.5)^2/[16 \times (20)^2] \\
&\approx 0.0088
\end{aligned}
$$

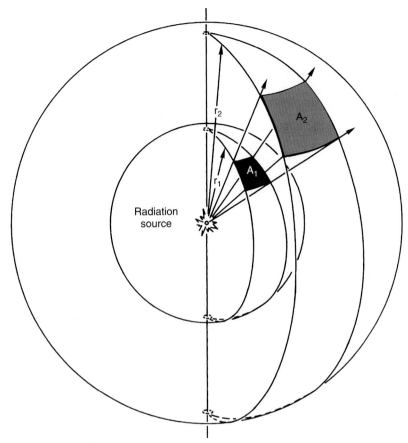

FIGURE 11-1 Illustration of the inverse-square law. As the distance from the radiation source increases from r_1 to r_2, the radiations passing through A_1 are spread out over a larger area A_2. Because $A \propto r^2$, the intensity of radiation *per unit area* decreases as $1/r^2$.

Thus the detector described in Example 11-1 intercepts less than 1% of the emitted radiation and has a rather small geometric efficiency, in spite of its relatively large diameter. At twice the distance (40 cm), the geometric efficiency is smaller by another factor of 4.

Equation 11-6 becomes inaccurate when the source is "close" to the detector. For example, for a source at $r = 0$, it predicts $g_p = \infty$. An equation that is more accurate at close distances for point sources located on the central axis of a circular detector is

$$g_p \approx (1/2)(1 - \cos\theta) \qquad (11\text{-}7)$$

where θ is the angle subtended between the center and edge of the detector from the source (Fig. 11-2). For example, when the radiation source is in contact with the surface of a circular detector, $\theta = 90$ degrees and $g_p = 1/2$ (Fig. 11-3A).

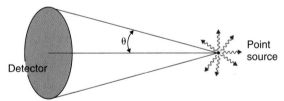

FIGURE 11-2 Point-source geometric efficiency for a circular large-area detector placed relatively close to the source depends on the angle subtended, θ (Equation 11-7).

Geometric efficiency can be increased by making θ even larger. For example, at the bottom of the well in a standard well counter (Chapter 12, Section A.2) the source is partially surrounded by the detector (Fig. 11-3B) so that $\theta \approx 150$ degrees and $g_p \approx 0.93$. In a liquid scintillation counter (see Chapter 12, Section C), the source is immersed in the detector material (scintillator fluid), so that $\theta = 180$ degrees and $g_p = 1$ (Fig. 11-3C).

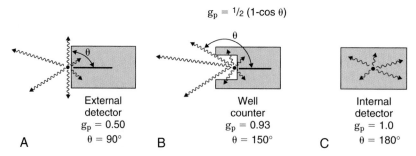

FIGURE 11-3 Examples of point-source geometric efficiencies computed from Equation 11-7 for different source-detector geometries.

Equation 11-7 avoids the obvious inaccuracies of Equation 11-6 for sources placed close to the detector; however, even Equation 11-7 has limitations when the attenuation by the detector is significantly less than 100%. This problem is discussed further in Section A.5.

The approximations given by Equations 11-6 and 11-7 apply to point sources of radiation located on the central axis of the detector. They also are valid for distributed sources having dimensions that are small in comparison to the source-to-detector distance; however, for larger sources (e.g., source diameter $\gtrsim 0.3r$) more complex forms are required.[1]

3. Intrinsic Efficiency

The fraction of radiation striking the detector that interacts with it is called the *intrinsic efficiency* ε of the detector:

$$\varepsilon = \frac{\text{no. of radiations interacting with detector}}{\text{no. of radiations striking detector}}$$
(11-8)

Intrinsic efficiency ranges between 0 and 1 and depends on the type and energy of the radiation and on the attenuation coefficient and thickness of the detector. For a point source located on the central axis of a γ-ray detector, it is given by

$$\varepsilon = 1 - e^{-\mu_l(E)\,x}$$
(11-9)

where $\mu_l(E)$ is the linear attenuation coefficient of the detector at the γ-ray energy of interest, E, and x is the detector thickness. In Equation 11-9 it is assumed that any interaction of the γ ray in the detector produces a potentially useful signal from the detector, although not necessarily all are recorded if energy-selective counting is used, as described in Section A.4.

The mass attenuation coefficient μ_m versus E for NaI(Tl) is shown in Figure 6-17. Numerical values are tabulated in Appendix D. Values of μ_l for Equation 11-9 may be obtained by multiplication of μ_m by 3.67 g/cm³, the density of NaI(Tl). Figure 11-4 shows intrinsic efficiency versus γ-ray energy for NaI(Tl) detectors of different thicknesses. For energies below approximately 100 keV, intrinsic efficiency is near unity for NaI(Tl) thicknesses greater than approximately 0.5 cm. For greater energies, crystal thickness effects become significant, but a 5-cm-thick crystal provides $\varepsilon > 0.8$ over most of the energy range of interest in nuclear medicine.

The intrinsic efficiency of semiconductor detectors also is energy dependent. Because of its low atomic number, silicon (Si, Z=14) is used primarily for low-energy γ rays and x rays ($\lesssim$100 keV), whereas germanium (Ge, Z=32) is preferred for higher energies. The effective atomic number of NaI(Tl) is approximately 50 (Table 7-2), which is greater than either Ge or Si; however, comparison with Ge is complicated by the fact that Ge has a greater density than NaI(Tl) ($\rho = 5.68$ g/cm³ vs. 3.67 g/cm³). The linear attenuation coefficient of NaI(Tl) is greater than that of Ge for $E \lesssim 250$ keV, but at greater energies the opposite is true; however, differences in cost and available physical sizes favor NaI(Tl) over Ge or Si detectors for most applications. The effective atomic numbers of cadmium telluride (CdTe) and cadmium zinc telluride (CZT) detectors are similar to that of NaI(Tl) (see Tables 7-1 and 7-2). They also have higher densities ($\rho \approx 6$ g/cm³). Thus for detectors of similar thickness, these detectors have somewhat greater intrinsic detection efficiencies than Na(Tl).

Gas-filled detectors generally have reasonably good intrinsic efficiencies ($\varepsilon \approx 1$) for particle radiations (β or α) but not for γ and x

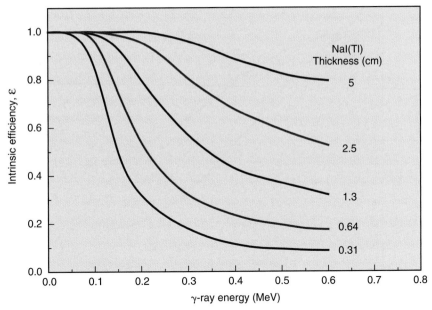

FIGURE 11-4   Intrinsic efficiency versus γ-ray energy for NaI(Tl) detectors of different thicknesses.

rays. Linear attenuation coefficients for most gases are quite small because of their low densities (e.g., $\rho \approx 0.0013$ g/cm^3 for air). In fact, most gas-filled detectors detect γ rays primarily by the electrons they knock loose from the walls of the detector into the gas volume rather than by direct interaction of γ and x rays with the gas. Intrinsic efficiencies for Geiger-Müller (GM) tubes, proportional counters, and ionization chambers for γ rays are typically 0.01 (1%) or less over most of the nuclear medicine energy range. Some special types of proportional counters, employing xenon gas at high pressures or lead or leaded glass γ-ray converters,* achieve greater efficiencies, but they still are generally most useful for γ- and x-ray energies below approximately 100 keV.

4. Energy-Selective Counting

The intrinsic efficiency computed from Equation 11-9 for a γ-ray detector assumes that all γ rays that interact with the detector produce an output signal; however, not all output signals are counted if a pulse-height analyzer is used for energy-selective counting. For example, if counting is restricted to the photopeak, most of the γ rays interacting with the detector by Compton scattering are not counted.

The fraction of detected γ rays that produce output signals within the pulse-height analyzer window is denoted by f. The fraction within the photopeak is called the *photofraction* f_p. The photofraction depends on the detector material and on the γ-ray energy, both of which affect the probability of photoelectric absorption by the detector. It depends also on crystal size (see Fig. 10-8) because with a larger-volume detector there is a greater probability of a second interaction to absorb the scattered γ ray following a Compton-scattering interaction in the detector (or of annihilation photons following pair production). Figure 11-5 shows the photofraction versus energy for NaI(Tl) detectors of different sizes.

If energy-selective counting is not used, then $f \approx 1$ is obtained. (Generally, some energy discrimination is used to reject very small amplitude noise pulses.) Full-spectrum counting provides the maximum possible counting rate and is used to advantage when a single radionuclide is counted, with little or no interference from scattered radiation. This applies, for example, to many in vitro measurements (see Chapter 12).

*A converter is a thin layer of material with relatively good γ-ray stopping power that is placed in front of or around the sensitive volume of a gas-filled detector. Recoil electrons ejected from γ-ray interactions in the converter are detected within the sensitive volume of the detector.

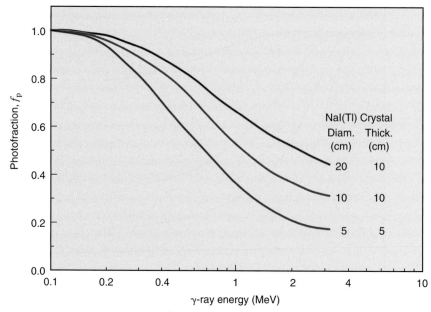

FIGURE 11-5 Photofraction versus γ-ray energy for cylindrical NaI(Tl) detectors of different sizes.

5. Some Complicating Factors

a. Nonuniform Detection Efficiency

Equations 11-6, 11-7, and 11-9 are somewhat idealized in that they assume that radiation is detected with uniform efficiency across the entire surface of the detector. In some cases, this assumption may be invalid. Figure 11-6 shows some examples for different trajectories from a point source of radiation. For trajectory A, the thickness of detector encountered by the radiation and employed for the calculation of intrinsic efficiency in Equation 11-9 conforms to what normally would be defined as the "detector thickness" in that equation. However, for trajectory B, a greater thickness is encountered and the intrinsic efficiency is larger. On the other hand, for trajectory C, near the edge of the detector, a smaller thickness of detector material is encountered and the intrinsic efficiency is smaller. Partial penetration of the beam for trajectory C sometimes is called an *edge effect*.

Thus, unless the attenuation by the detector is "very high" (essentially 100% within a thin layer near the surface), the intrinsic efficiency will vary across the surface of the detector. As well, the detector diameter (or area) used for the calculation of geometric efficiency in Equation 11-6 or 11-7 becomes ill-defined when edge effects are significant. When the complications illustrated in Figure 11-6 are significant, detection efficiency must be calculated by methods of integral calculus,

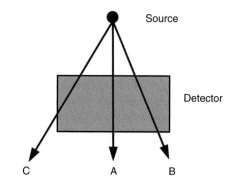

FIGURE 11-6 Three possible trajectories for radiations striking a detector from a point source, each having a different intrinsic detection efficiency.

rather than with the simplified equations described thus far. The calculations are complex and a complete analysis is beyond the scope of this text, but they have been analyzed in other books.[2] A few practical implications derived from more advanced calculations are presented here.

The nonuniform attenuation illustrated in Figure 11-6 affects both geometric efficiency (edge effects) and intrinsic efficiency. The parameter that accounts for both of these quantities is the *total detection efficiency*, ε_t. When idealized conditions apply, this can be obtained simply by multiplying the result of Equation 11-6 or 11-7 by the result from Equation 11-9

$$\varepsilon_t = g_p \times \varepsilon \qquad (11\text{-}10)$$

It is reasonable to use Equation 11-10 to compute total detection efficiency if the resulting discrepancy from a more exact calculation is "small," for example, less than 10%. If the discrepancy is larger, then one must consider using the more complex methods of integral calculus.

Figure 11-7 shows three detector profiles with different levels of effect for the trajectories shown in Figure 11-6. As compared with a "box" profile (i.e., one with equal thickness and width), a "wide" profile presents a greater range of potential detector thicknesses (trajectory B in Fig. 11-6), whereas a "narrow" profile has a greater fraction of its area affected by edge effects (trajectory C in Fig. 11-6). In addition to the profile of the detector, the extent of these effects depends on the attenuation properties of the detector (material and thickness) and on the source-to-detector distance. Thus one cannot provide a "one-size-fits-all"

rule of thumb for when it is necessary to use the more advanced equations instead of the simplified equations presented earlier. All of these parameters must be considered for making this determination.

Figure 11-8 presents a graph that can be helpful for this purpose. It applies to 1-cm-thick γ-ray detectors for two photon energies (140 keV and 511 keV) and two detector materials [NaI(Tl) and bismuth germanate (BGO)] that are used in nuclear medicine. In this graph, a "narrow" detector would lie toward the left end of the horizontal axis, and a "wide" one would lie toward the right. A "box" detector would have a diameter of 1 cm.

Also indicated on the graph are the intrinsic efficiencies, computed from Equation 11-9 for the central ray A in Figure 11-6, for different combinations of these photon energies and detectors. The curves indicate the minimum source-to-detector distance versus detector diameter at which the total detection

"Narrow"   "Box"   "Wide"

FIGURE 11-7 Examples of detector profiles with different complications for the computation of total detection efficiency.

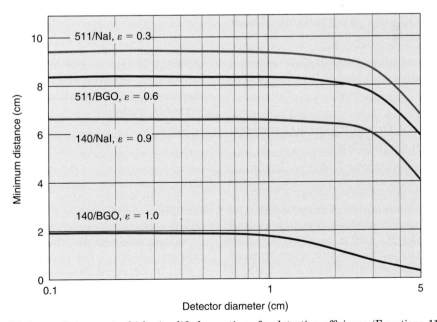

FIGURE 11-8 Minimum distance at which simplified equations for detection efficiency (Equations 11-7, 11-9, and 11-10) can be used with errors of less than 10% for 1-cm–thick detectors of different diameters and different combinations of photon energy-detector material. The graph assumes a point source of radiation is placed on the central axis of the detector. Photon energies are measured in keV. At closer distances, total detection efficiency must be computed from more-complicated mathematical models (see reference 2).

efficiency computed using Equations 11-7 and 11-9 in Equation 11-10 is accurate to within 10%. At distances closer than the minimum distance, the simplified calculations are inaccurate by *more* than 10%, and the more complicated methods of integral calculus should be used to compute the total detection efficiency, ε_t.[2]

Figure 11-8 shows that for detectors having a high value of intrinsic efficiency (e.g., BGO detector for 140-keV photons, $\varepsilon \approx 1$), the simplified equations can be used with less than 10% error at relatively close distances. Even for the "narrow" detector profile shown in Figure 11-7, they can be used within 2 cm of the detector. At the other extreme, for a detector having a low value of intrinsic efficiency [e.g., NaI(Tl) for 511-keV positron annihilation photons, $\varepsilon \approx 0.3$], the simplified equations fail within approximately 10 cm from a "narrow" detector and within about 5 cm from a "wide" one.

Figure 11-8 provides general guidance as to when the simplified equations can be used for estimating relative detection efficiencies on existing systems in a laboratory or for preliminary design work for a new detector system. It also can be used for guidance with other combinations of photon energy and detector material having dimensions and values of ε similar to those indicated on the graph. For more precise design work, it generally is preferable to go directly to the methods of integral calculus. Alternatively, Monte Carlo techniques, using a computer to simulate photon trajectories and interactions for a large number of individual photons originating from a radioactive source, can be used to estimate detection efficiency.

b. Detection of Simultaneously Emitted Radiations in Coincidence

Yet another complicating factor is that some radionuclides emit multiple γ rays in cascaded fashion from a single nuclear disintegration. In Figure 3-3, for example, β_1 may be followed by the emission of multiple γ rays (e.g., γ_5 and γ_2). In this example and in most other cases of cascaded γ emissions, the γ rays are emitted within a few nanoseconds of each other, which is well within the resolving time of most detectors (see Section C). If the two γ rays are detected simultaneously (*coincidence detection*), they are recorded as a single event having an apparent energy equal to the sum of the energies deposited in the detector by the individual γ rays. If energy-selective

counting is used, such as with the photopeak of one or the other γ ray, the pulse from the resulting event could be moved out of the selected analyzer window, thereby decreasing the counting rate for that γ ray (see Fig. 10-7). Note, however, that simultaneous detection does not occur when there is a significant delay before the emission of the second γ ray, such as in metastable states (see Chapter 3, Section E).

A full treatment of the problem of coincidence detection of cascaded emissions is beyond the scope of this text. However, the following discussion provides a first-level analysis and an indication of when it must be taken into consideration. Suppose that two γ rays, which we denote as γ_1 and γ_2, are emitted simultaneously, in cascaded fashion, with relative frequencies per disintegration η_1 and η_2. (Note that it is not necessary that $\eta_1 = \eta_2$; e.g., η_2 could be reduced by alternative decay pathways that result in nondetectable emissions, such as internal conversion.) Suppose further that the total (full-spectrum) detection efficiencies for the two γ rays are D_1 and D_2, respectively. The probability that a single nuclear disintegration will result in the detection of γ_1 is

$$p_1 = \eta_1 D_1 \qquad (11\text{-}11)$$

In the absence of coincidence detection, the counting rate recorded for γ_1 would be

$$R_1 = p_1 \times A \qquad (11\text{-}12)$$

where A is the source activity in Bq. Similarly, in the absence of coincidence detection, the counting rate resulting from detection of γ_2 events would be

$$R_2 = p_2 \times A \qquad (11\text{-}13)$$

Thus, if one did not account for the possibility of coincidence detection, the predicted full-spectrum counting rate for the source would be $(R_1 + R_2)$.

Taking into account the possibility of coincidence detection, the probability that γ_1 and γ_2 will be detected simultaneously is

$$p_{12} = \eta_1 D_1 \times \eta_2 D_2 \qquad (11\text{-}14)$$

and the counting rate for simultaneously detected events is

$$R_{12} = p_{12} \times A \qquad (11\text{-}15)$$

With full-spectrum counting, each coincidence event removes one event each from R_1 and R_2, replacing it with an event in R_{12}. Thus the recorded full-spectrum counting rate would be $(R_1 + R_2 - R_{12})$.

A similar analysis can be performed for photopeak counting. In this case, the total-spectrum detection efficiency is replaced by the photopeak detection efficiency for each γ ray. The counting rate in each photopeak is reduced by R_{12}; that is, they would be $(R_1 - R_{12})$ and $(R_2 - R_{12})$, and the summation photopeak counting rate would be R_{12}.

In actual practice, calculating the effect of detecting simultaneously emitted photons on recorded counting rates is somewhat more complicated than indicated by the previous equations. For example, for calculating the effect on photopeak counting rates, one should take into account the possibility of simultaneous detection of events in the Compton portion of the spectrum, which also could "move" events out of the photopeak, or possibly add up to create apparent photopeak events. A more detailed analysis also would include the possibility of angular correlations between the directions in which the two γ rays are emitted. In many cases, this would lead to different values of detection efficiencies for the two γ rays.

Nevertheless, the first-level analysis provided in the preceding discussion can give an indication of when the effects of coincidence detection can be significant. For example, from Equations 11-14 and 11-15, it can be seen that the effects depend on the values of η_1 and η_2 and on detection efficiencies D_1 and D_2. Thus if the primary emission of interest is γ_1, and, $\eta_1 \gg \eta_2$, the effect of γ_2 on R_1 (or $R_1 + R_2$) would be relatively small. As well, the effects are most severe in counting systems having high values of detection efficiency.

An example is the well counter, which generally has a high total detection efficiency. This is illustrated by Figure 10-7, showing how the coincidence sum peak for the two γ rays of ^{111}In increases dramatically when the source is moved from a location outside the well-counter detector where detection efficiency is low to a location inside it where it is high. When questions arise, the first-level analysis presented above can be used to estimate the relative magnitude of the effects of coincidence counting of cascaded γ rays. Experimental data can further help to resolve the issue, for example, by comparing the spectra with the source at different locations, as in Figure 10-7.

c. Attenuation and Scatter of Radiation Outside the Detector

A final complication that we consider is the possibility of absorption and scatter of radiation before it reaches the detector. The analysis to this point assumes that radiation passes unobstructed, without absorption or scattering, from the source to the detector (e.g., as in Fig. 11-1). However, when the radiation source is embedded at depth within an absorbing and scattering medium, as it is for most in vivo measurements, the calculation of detection efficiency is complicated by attenuation and scattering effects.

Absorption generally causes a decrease in the recorded counting rate, but scattered radiation may lead to a decrease or an increase, depending on whether there is more scattering away from or toward the detector. For example, the counting rate for a source at a shallow depth in a scattering medium actually may be greater than for the same source in air because the added contribution from backscattering may more than compensate for a small reduction in counting rate by absorption. (See also the discussion of the buildup factor in in Chapter 6, Section D.3). At greater depths absorption effects may predominate.

Corrections for attenuation and scattering for in vivo measurements are complicated because they depend on several factors, including the γ-ray energy, depth of the source in the absorbing and scattering medium, use of energy-selective counting, and so forth. Figure 11-9 shows the general effects versus γ-ray energy for a point source 7.5-cm deep in tissue-equivalent material and a NaI(Tl) counting system. The fraction of γ rays emitted from the source that are *neither scattered nor absorbed* on their way to the detector increases with γ-ray energy because absorption and scattering coefficients decrease with increasing energy. The fraction of γ rays *absorbed* in the tissue-equivalent material decreases with energy to a negligible fraction above approximately 100 keV.

Figure 11-9 also shows that the fraction of γ rays *scattered* at first increases with γ-ray energy because absorption effects decrease, leaving more γ rays to be scattered. This fraction reaches a maximum at approximately 100 keV, after which it also decreases with increasing energy. If energy-selective counting is used, the fraction of Compton-scattered γ rays recorded in the photopeak decreases with increasing γ-ray energy. This reflects the

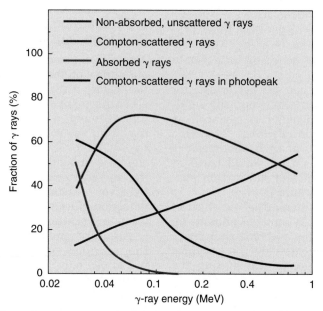

FIGURE 11-9   General effects of γ-ray energy on the fraction of γ rays scattered or absorbed from a point source 7.5 cm deep in tissue and on the fraction of unscattered γ rays and scattered γ rays having sufficient energy to be recorded with a photopeak window and NaI(Tl) detector. (*From Anger HO: Radioisotope cameras. In Hine GJ, editor:* Instrumentation in Nuclear Medicine, *Vol 1. New York, 1967, Academic Press, p 514.*)

increasing energy separation between scattered γ rays and the photopeak (see Fig. 10-10). With semiconductor detectors (Ge, Si, CdTe, or CZT), this fraction is much smaller because of their ability to clearly resolve scattered γ rays from the photopeak (see Figs. 10-14 and 10-15).

Another factor affecting detection efficiency is attenuation by the housing material of the detector. Most γ-ray detectors are fabricated with relatively thin entrance windows, such as thin layers of aluminum, so that their attenuation is negligible. Detectors designed for applications involving very-low-energy γ rays sometimes are constructed with ultra-thin (and fragile) entrance windows of alternative materials, such as beryllium. However, attenuation can become significant if the detector is used outside the range of its intended applications. Information provided by the manufacturer can be used to estimate this effect in questionable situations.

Attenuation by the detector housing can be severe in β-particle counting. This is discussed separately in Section B.

6. Calibration Sources

Detection efficiencies can be determined experimentally using *calibration sources*. A calibration source is one for which the activity or emission rate is known accurately. This determination usually is made by the commercial supplier of the source.

Detection efficiency can be determined by measuring the counting rate recorded from the calibration source and applying Equation 11-2. This method generally is satisfactory for systems in which a standard measuring configuration is used and for which the calibration source accurately simulates the shape and distribution of the sources usually measured with the system. For example, "rod standards" (Fig. 11-10) are used for determining detection efficiencies of well counters for test-tube samples.

Some γ-ray-emitting source materials that are available as calibration standards are listed in Table 11-1. Most are quite long-lived. Detection efficiencies for short-lived radionuclides can be estimated from measurements made on a calibration standard having similar emission characteristics. For example, ^{57}Co ($E_\gamma = 122$ keV and 136 keV) frequently is used to simulate ^{99m}Tc ($E_\gamma = 140$ keV). (Cobalt-57 is sometimes called "mock ^{99m}Tc.") For most detection systems, intrinsic efficiencies at these three energies are virtually identical. Therefore the detection efficiency per emitted γ ray as calculated from Equation 11-2 would be the same for ^{99m}Tc and ^{57}Co (assuming the same energy-selective counting conditions were used, e.g., photopeak counting for both).

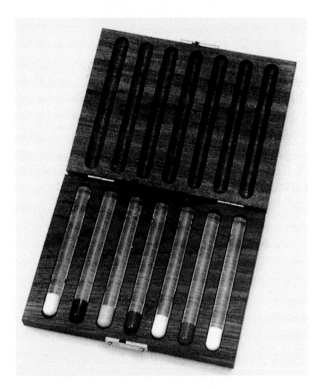

FIGURE 11-10 "Rod standards" containing accurately known quantities of different radionuclides used for determining the detection efficiencies of well counters. The sources are meant to simulate radioactivity in test tubes.

TABLE 11-1
PROPERTIES OF SOME γ-RAY SOURCES USED AS CALIBRATION STANDARDS

Radionuclide	Half-Life	γ-Ray or x-Ray Energy (keV)*	Emission Frequency (γ or x rays/dis)
^{22}Na	2.60 yr	511	1.798
		1274	0.999
^{54}Mn	312 d	834.8	1.000
^{57}Co	272 d	14.4	0.095
		122.1	0.856
		136.5	0.105
^{60}Co	5.27 yr	1173	0.999
		1333	1.000
^{68}Ge	271 d	511	1.780
^{85}Sr	64.9 d	514	0.980
^{109}Cd	463 d	22.0 (K$_\alpha$ x ray)	0.842
		24.9 (K$_\beta$ x ray)	0.178
		88.0	0.037
^{113}Sn	115 d	24.1 (K$_\alpha$ x ray)	0.794
		27.3 (K$_\beta$ x ray)	0.172
		391.7	0.649
^{129}I	15.7×10^6 yr	29.7 (K$_\alpha$ x ray)	0.571
		33.6 (K$_\beta$ x ray)	0.132
		39.6	0.075
^{137}Cs	30 yr	32.0 (K$_\alpha$ x ray)	0.057
		36.4 (K$_\beta$ x ray)	0.013
		661.7	0.851

Data adapted from NCRP Report No. 58: *A Handbook of Radioactivity Measurements Procedures,* ed 2, Bethesda, MD, 1985, National Council on Radiation protection and Measurements.

*Only predominant photon emissions are listed.

If the detection efficiency is determined on the basis of cps/Bq, one must take into account the differing emission frequencies of the two radionuclides. Cobalt-57 emits 0.96 γ/dis (γ rays per disintegration), whereas ^{99m}Tc emits 0.89 γ/dis (see Appendix C). Therefore the counting rate per Bq of ^{99m}Tc would be a factor of 0.89/0.96 = 0.93 smaller than that measured per Bq of ^{57}Co. This should be applied as a correction factor to the counting rate per Bq determined for ^{57}Co to obtain the counting rate per Bq for ^{99m}Tc.

Calibration sources also are used in phantoms simulating the human anatomy for estimating the detection efficiency for in vivo measurement systems; however, the result is only as accurate as the phantom and source distribution are accurate for simulating the human subject. For example, a 1-cm discrepancy between source depths in the phantom and in the human subject may result in a 10% to 20% difference in counting rate (see Chapter 12, Section F.1).

B. PROBLEMS IN THE DETECTION AND MEASUREMENT OF β PARTICLES

Because of their relatively short ranges in solid materials, β particles create special detection and measurement problems. These problems are especially severe with low-energy β-particle emitters, such as ^{3}H and ^{14}C. The preferred method for assay of these radionuclides is by liquid scintillation counting techniques (Chapters 7 and 12); however, these techniques are not applicable in all situations, such as when surveying a bench top with a survey meter to detect ^{14}C contamination (Chapter 23). A complete discussion of the problems arising in detection and assay of β-particle emitters is beyond the scope of this book; however, a few of the practical problems are described briefly.

A survey meter can be used to detect surface contamination by β-particle emitters provided it has an entrance window sufficiently thin to permit the β particles to enter the sensitive volume of the detector. Figure 11-11 shows relative counting rate versus entrance window thickness for two β-emitting radionuclides. Efficient detection of low-energy β emitters requires a very thin entrance window, preferably fabricated from a low-density material. A typical entrance window for a survey meter designed for ^{3}H and ^{14}C detection is 0.03-mm-thick Mylar (~1.3 mg/cm^2 thick).* Mica and beryllium also

*Thicknesses of detector windows often are specified in units of mass/area, for example, mg/cm^2. To obtain the window thickness, divide by the material density using the same units (e.g., mg/cm^3).

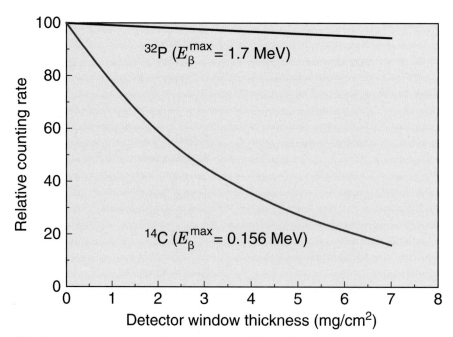

FIGURE 11-11 Relative counting rate versus detector window thickness for some β-emitting radionuclides. (*Adapted from Quimby EH, Feitelberg S, Gross W:* Radioactive Nuclides in Medicine and Biology. *Philadelphia, 1970, Lea & Febiger.*)

are used. Such thin windows are very fragile, and usually they are protected by an overlying wire screen. Beta particles that are more energetic (e.g., from ^{32}P) can be detected with much thicker and more rugged entrance windows; for example, 0.2-mm-thick aluminum ($\sim$50 mg/cm^2) provides approximately 50% detection efficiency for ^{32}P.

GM and proportional counters sometimes are used to assay the activities of β-emitting radionuclides in small trays ("planchets") or similar sample holders. Two serious problems arising in these measurements are *self-absorption* and *backscattering*, as illustrated in Figure 11-12. Self-absorption depends on the sample thickness and the β-particle energy. Figure 11-13 shows relative counting rate versus sample thickness for two β emitters. For ^{14}C and similar low-energy β emitters, self-absorption in a sample thickness of only a few mg/cm^2 is sufficient to cause a significant reduction of counting rate. (Note that for water, $\rho = 1$ g/cm^3; thus 1 mg/cm^2 is 0.001-cm thick.) Backscattering of β particles from the sample and sample holder tends to increase the sample counting rate and can amount to 20% to 30% of the total sample counting rate in some circumstances.

Accurate assay of β-emitting radioactive samples by external particle-counting techniques requires careful attention to sample preparation. If only relative counting rates

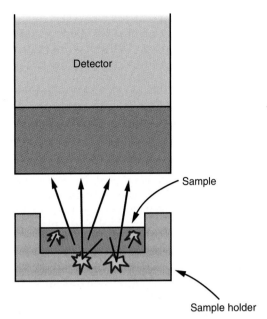

FIGURE 11-12　Self-absorption and backscattering in β-particle counting.

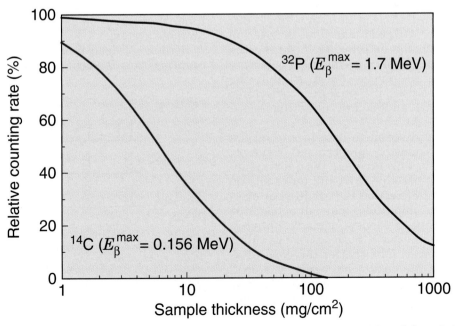

FIGURE 11-13　Effect of sample self-absorption on counting rate for two β emitters. (*Adapted from Quimby EH, Feitelberg S, Gross W: Radioactive Nuclides in Medicine and Biology. Philadelphia, 1970, Lea & Febiger.*)

are important, then it is necessary to have sample volumes and sample holders as nearly identical as possible. Other techniques for dealing with these difficult problems are discussed in reference 3.

Bremsstrahlung counting can be employed as an indirect method for detecting β particles using detectors that normally are sensitive only to more penetrating radiations such as x rays and γ rays, for example a NaI(Tl) well counter (Chapter 12). Bremsstrahlung counting also was employed in some early studies using ^{32}P for the detection of brain tumors and still is used occasionally to map the distribution of ^{32}P-labeled materials administered for therapeutic purposes. Bremsstrahlung counting is effective only for relatively energetic β particles (e.g., ^{32}P, $E_\beta^{max} = 1.7$ MeV, but not ^{14}C, $E_\beta^{max} = 0.156$ MeV) and requires perhaps 1000 times greater activity than a γ-ray emitter because of the very low efficiency of bremsstrahlung production.

C. DEAD TIME

1. Causes of Dead Time

Every radiation counting system exhibits a characteristic *dead time* or *pulse resolving time* τ that is related to the time required to process individual detected events. The pulses produced by a radiation detector have a finite time duration, such that if a second pulse occurs before the first has disappeared, the two pulses will overlap to form a single distorted pulse. With GM detectors, the overlap may occur in the detector itself, during the time that the "avalanche charge" is being collected from a previous pulse, so that the second pulse does not produce a detectable output signal and is lost (see Chapter 7, Section A.4).

With energy-sensitive detectors (scintillation, semiconductor, proportional counter), the overlap usually occurs in the pulse amplifier, causing baseline shift and pulse pile-up (see Chapter 8, Section B.3). Shifted or overlapped pulse amplitudes may fall outside the selected analyzer window, again resulting in a loss of valid events. Such losses are called *dead time losses*. The shorter the dead time, the smaller the dead time losses. The dead time for a GM tube is typically 50–200 μsec. Sodium iodide and semiconductor detector systems typically have dead times in the range of 0.5–5 μsec. Gas proportional counters and

liquid scintillation systems have dead times of 0.1–1 μsec.

Dead time losses also occur in pulse-height analyzers, scalers, computer interfaces, and other components that process pulse signals. Generally scalers and single-channel analyzers have dead times of much less than 1 μsec, whereas multichannel analyzer and computer interface dead times are on the order of a few microseconds. Usually the dead time is given for the counting system as whole; however, if one of the components has a dead time that is long in comparison to the other components, then system dead time is determined by that component.

2. Mathematical Models

Counting systems usually are classified as being of the paralyzable or nonparalyzable type. A *nonparalyzable system* is one for which, if an event occurs during the dead time τ of a preceding event, then the second event is simply ignored, with no effect on subsequently occurring events (Fig. 11-14). Digital counters, pulse-height analyzers, and computer interfaces frequently behave as nonparalyzable systems. A *paralyzable system* is one for which each event introduces a dead time τ whether or not that event actually is counted. Thus an event occurring during the dead time of a preceding event would not be counted but still would introduce its own dead time during which subsequent events could not be recorded. A paralyzable system may be thought of as one with an "extendable" dead time. Most radiation detectors behave as paralyzable systems.

Because of dead time losses, the *observed* counting rate R_o (cps) is less than the *true*

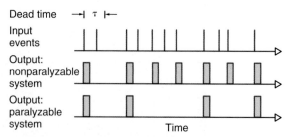

FIGURE 11-14 Difference in output signals between nonparalyzable and paralyzable systems. Both have dead time τ indicated in *top line*. *Second line* illustrates randomly occurring input events. With a nonparalyzable system (*third line*), events are lost if they occur within a time τ of a preceding *recorded* event, whereas with a paralyzable system (*fourth line*) events are lost if they occur within a time τ of *any* preceding event, regardless of whether that event has been recorded.

counting rate R_t (cps), where the latter is the counting rate that would be recorded if $\tau = 0$. The relationship among R_o, R_t, and τ depends on the type of dead time. For nonparalyzable systems,

$$R_o = R_t /(1 + R_t \tau) \qquad (11\text{-}16)$$

$$R_t = R_o /(1 - R_o \tau) \qquad (11\text{-}17)$$

where τ is given in seconds. If the system has a paralyzable dead time, then

$$R_o = R_t e^{-R_t \tau} \qquad (11\text{-}18)$$

There is no analytic equation for R_t as a function of R_o for the paralyzable case.

Figure 11-15 shows R_o versus R_t for the two types of systems. For a nonparalyzable system, the observed counting rate increases asymptotically toward a maximum value

$$R_o^{\max} = 1/\tau \qquad (11\text{-}19)$$

At very high true counting rates, greater than one count per dead time interval, the system simply records one event per dead time interval, ignoring all the others that occur during the dead time interval between counted events.

For a paralyzable system, the observed counting rate rises to a maximum value given by

$$R_o^{\max} = 1/e\tau \qquad (11\text{-}20)$$

where e (= 2.718 ...) is the base of natural logarithms. Then the observed counting rate actually *decreases* with a further increase in true counting rate. This is because additional events serve only to extend the already long dead time intervals without contributing to additional events in the observed counting rate. At very high true counting rates, the observed counting rate actually approaches zero. This is called *counter paralysis*.

Dead time losses are given by the difference between observed and true counting rates, $R_t - R_o$, and *percentage losses* are given by

$$\text{percentage losses} = [(R_t - R_o)/R_t] \times 100\% \qquad (11\text{-}21)$$

When the product $R_t \tau$ is "small" (≤ 0.1), the percentage losses are "small" (i.e., $\leq 10\%$), and they can be described by the same equation for both paralyzable and nonparalyzable systems

$$\text{percentage losses} \approx (R_t \tau) \times 100\% \qquad (11\text{-}22)$$

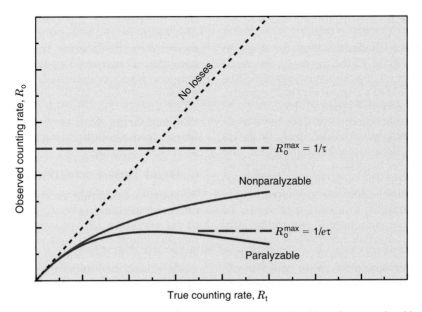

FIGURE 11-15 Observed (R_o) versus true (R_t) counting rate curves for paralyzable and nonparalyzable systems having the same dead time value, τ.

EXAMPLE 11-2

Calculate the percentage losses for a counting system having a dead time of 10 µsec at true counting rates of 10,000 and 100,000 cps.

Answer

At 10,000 cps, $R_t\tau = 10^4$ cps $\times 10^{-5}$ sec = 0.1. Because the losses are "small," Equation 11-22 can be used:

$$\text{percentage losses} \approx (0.1) \times 100\%$$
$$\approx 10\%$$

The observed counting rate would therefore be approximately 9000 cps, that is, 10% less than the true counting rate of 10,000 cps. At 100,000 cps, $R_t\tau = 10^5$ cps $\times 10^{-5}$ sec = 1.0; thus the losses are not "small." For a non-paralyzable system, the observed counting rate would be (Equation 11-16)

$$R_o = 100,000/(1 + 1.0) \text{ cps}$$
$$= 50,000 \text{ cps}$$

that is, the losses would be 50%. For a para-lyzable system (Equation 11-18)

$$R_o = 10^5 \times e^{-1.0} \text{ cps}$$
$$= 100,000 \times 0.368 \text{ cps}$$
$$= 36,800 \text{ cps}$$

The losses are therefore 100,000 − 36,800 cps = 63,200 cps, or 63.2% (of 100,000 cps).

Example 11-2 illustrates that for a given dead time and true counting rate, the dead time losses for a paralyzable system are greater than those of a nonparalyzable system. This is shown also by Figure 11-15.

Many nuclear medicine systems have multiple components in cascade, each with its own individual dead time. In some cases, one component of a cascaded system may be paralyzable (e.g., the scintillation detector) whereas the other may be nonparalyzable (e.g., a multichannel analyzer interface). In most cases, one component dominates the system and its behavior adequately describes the system behavior. However, if cascaded paralyzable and nonparalyzable components have similar dead times, both components contribute to dead time losses and the behavior is a hybrid of the two. The analysis of such systems is beyond the scope of this text; see references 4 and 5 for further details.

3. Window Fraction Effects

With NaI(Tl) and other detectors used for energy-selective counting, any detected event can cause pile-up with any other event in the pulse-height spectrum. Thus if a pulse-height analyzer is used, the number of events lost depends on the *total-spectrum counting rate,* not just on the counting rate within the selected analyzer window. With such systems, the apparent dead time may appear to change with pulse-height analyzer window setting. For example, if a certain fraction of detected events are lost with a given window setting, the same *fraction* will be lost when the analyzer window is narrowed, making it appear that the dead time *per event in the analyzer window* is longer when the narrower window is used.

An approximate equation for apparent dead time is[6]

$$\tau_a = \tau/w_f \qquad (11\text{-}23)$$

where τ is the actual dead time per detected event and w_f is the *window fraction,* that is, the fraction of detected events occurring within the selected analyzer window. For example, if a NaI(Tl) detector system has a dead time of 1 µsec (amplifier pulse duration) but a narrow window is used so that only 25% of detected events are within the window ($w_f = 0.25$), the apparent dead time will be (1/0.25) = 4 µsec. Window fractions also change with the amount of scattered radiation recorded by the detector because this also changes the energy spectrum of events recorded by the detector. In general, increased amounts of scattered radiation decrease the window fraction recorded with a photopeak window (see Fig. 10-6). The window fraction effect must be considered in specifying and comparing dead time values for systems using pulse-height analysis for energy-selective counting.

4. Dead Time Correction Methods

Measurements made on systems with a standardized measuring configuration, with little or no variation in window fraction from one measurement to the next, can be corrected for dead time losses using the mathematical models described in Section C.2. Some in vitro counting systems are in this category. Given an observed counting rate R_o and a dead time τ, the true counting rate can be determined from Equation 11-17 if the system is nonparalyzable or by graphical or approximation methods (Equation 11-18 and Fig. 11-15) if it is paralyzable.

Dead time τ can be determined using the *two-source method*. Two radioactive sources of similar activities, for which the dead time losses are expected to be 10% to 20%, are needed (Fig. 11-16). The counting rate for source 1 is determined, R_1 (cps). Without disturbing the position of source 1 (so as not to change the detection efficiency for source 1), source 2 is placed in position for counting and the counting rate for the two sources together is determined, R_{12} (cps). Then source 1 is removed (again, without disturbing source 2), and the counting rate for source 2 alone is determined, R_2 (cps). If the system is nonparalyzable, the dead time τ_n in seconds is given by

$$\tau_n \approx (R_1 + R_2 - R_{12})/(R_{12}^2 - R_1^2 - R_2^2) \qquad (11\text{-}24)$$

If the system is paralyzable, then

$$\tau_p \approx [2R_{12}/(R_1 + R_2)^2]\ln[(R_1 + R_2)/R_{12}] \qquad (11\text{-}25)$$

If a short-lived radionuclide is used, decay corrections can be avoided by making the three measurements R_1, R_{12}, and R_2, separated by equal time intervals.*

Additional measurements are required to determine whether Equation 11-24 or 11-25 is to be used. For example, a graph of observed counting rate versus activity might be constructed to determine which of the two curves in Figure 11-15 describes the system.

For measurements in which the window fraction is variable (e.g., most in vivo measurements), the equations given in Section C.2 can be used only if the window fraction is known. Another approach is to use a

---

*Some texts recommend also that a background measurement be made; however, background counting rates generally are negligibly small in comparison to the counting rates used in these tests.

fixed-rate pulser connected to the preamplifier of the radiation detector. The pulser injects pulses of fixed amplitude (usually larger than the photopeak pulses of interest) into the circuitry, and the counting rate for these events is monitored using a separate single-channel analyzer window (Fig. 11-17). The fractional loss of pulser events is equal to the fractional loss of radiation events because both are subject to the same loss mechanisms. The observed counting rate R_o from the γ-ray source is corrected by the ratio of true-to-observed pulser counting rates, P_t/P_o, to obtain the true γ-ray counting rate,

$$R_t = R_o\,(P_t/P_o) \qquad (11\text{-}26)$$

Dead time losses also affect counting statistics. For example, the standard deviation in observed counts, N_o, is not given by $\sqrt{N_o}$ if there are substantial dead time losses. Detailed discussions of counting statistics with dead time losses are presented in reference 7.

D. QUALITY ASSURANCE FOR RADIATION MEASUREMENT SYSTEMS

Radiation measurement systems are subject to various types of malfunctions that can lead to sudden or gradual changes in their

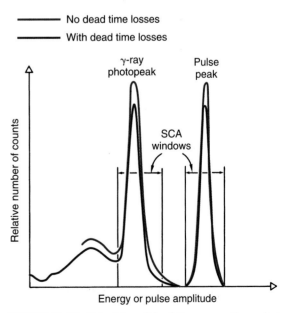

FIGURE 11-17 Principles of dead time correction using the fixed-rate pulser method. The fractional loss of events in the pulse peak (from the fixed-rate pulser) is assumed to equal the fractional losses of radiation events in the γ-ray photopeak window. *SCA*, Single-channel analyzer.

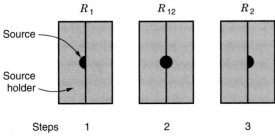

FIGURE 11-16 Illustration of the steps followed in determining dead time by the two-source method.

performance characteristics. For example, electronic components and detectors can fail or experience a progressive deterioration of function, leading to changes in detection efficiency, increased background, and so forth.

To ensure consistently accurate results, *quality assurance procedures* should be employed on a regular basis for all radiation measurement systems. These would include (1) daily measurement of the system's response to a standard radiation source (e.g., a calibration "rod standard" for a well counter or a "check source" for a survey meter); (2) daily measurement of background levels; and (3) for systems with pulse-height analysis capabilities, a periodic (e.g., monthly) measurement of system energy resolution.

Additional tests may be devised to evaluate other important characteristics on specific measuring systems. The results should be recorded in a log book for analysis when problems are suspected. In some cases, it is helpful to make a graph of the results (e.g., counting rate for a standard source or for background), with tolerance limits (e.g., ±2 standard deviations) to detect subtle, progressive changes in performance.

The statistical tests described in Chapter 9 can be used to assist in this analysis. For example, the χ^2 test, described in Chapter 9, Section E.1, is useful for detecting sporadic counting errors or other instabilities in system performance. Typically, a series of approximately 20 measurements are made and the χ^2 statistic is calculated. A result with $P < 0.01$ or $P > 0.99$ is taken as evidence of a system problem. A result with $0.05 < P < 0.95$ is considered acceptable. A result with a P value in the gaps between these ranges is considered equivocal, and the test should be repeated.

Quality assurance procedures also are used for imaging systems as described in Chapter 14, Section E, Chapter 17, Section C.4, and Chapter 18, Section E.

REFERENCES

1. Jaffey AH: Solid angle subtended by a circular aperture at point and spread sources: Formulas and some tables. *Rev Sci Instrum* 25:349-354, 1954.
2. NCRP Report No. 58: *A Handbook of Radioactivity Measurements Procedures*, ed 2. Bethesda, MD, 1985, National Council on Radiation Protection and Measurements, p 168.
3. Quimby EH, Feitelberg S, Gross W: *Radioactive Nuclides in Medicine and Biology*, Philadelphia, 1970, Lea & Febiger, Chapter 16.
4. Sorenson JA: Deadtime characteristics of Anger cameras. *J Nucl Med* 16:284-288, 1975.
5. Woldeselassie T: Modeling of scintillation camera systems. *Med Phys* 26:1375-1381, 1999.
6. Wicks R, Blau M: The effects of window fraction on the deadtime of Anger cameras. *J Nucl Med* 18:732-735, 1977.
7. Evans RD: *The Atomic Nucleus*, New York, 1955, McGraw-Hill, pp 785-793.

Counting Systems

Radiation counting systems are used for a variety of purposes in nuclear medicine. *In vitro* (from Latin, meaning "in glass") *counting systems* are employed to measure radioactivity in tissue, blood, and urine samples; for radioimmunoassay and competitive protein binding assay of drugs, hormones, and other biologically active compounds; and for radionuclide identification, quality control, and radioactivity assays in radiopharmacy and radiochemistry. In vitro counting systems range from relatively simple, manually operated, single-sample, single-detector instruments to automated systems capable of processing hundreds of samples in a batch with computer processing of the resulting data.

In vivo (from Latin, meaning "in the living subject") *counting systems* are employed for measuring radioactivity in human subjects or experimentally in animals. Different in vivo systems are designed for measuring localized concentrations in single organs (e.g., thyroid, kidney) and for measurements of whole-body content of radioactivity.

Most nuclear medicine counting systems consist of the following basic components: a detector and high voltage supply, preamplifier, amplifier, one or more single-channel analyzers (SCAs) or a multichannel analyzer (MCA) ("data analysis"), and a digital or analog scaler-timer, rate meter, or other data readout device. The majority of systems employ a computer or microprocessor for data analysis and readout.

At present, the most efficient and economical detector for counting γ-ray emissions[*] is a sodium iodide [NaI(Tl)] scintillation detector. The characteristics of various NaI(Tl) counting systems are discussed in Sections A and B in this chapter. Scintillation counters

for β particles and low-energy x rays or γ rays are presented in Section C later in this chapter. Counting systems based on gas detectors and semiconductor detectors are discussed in Sections D and E, respectively. Section F deals with counting systems for in vivo applications, including thyroid uptake, sentinel node detection, and intraoperative probes.

A. NaI(Tl) WELL COUNTER

1. Detector Characteristics

The detector for a NaI(Tl) well counter is a single crystal of NaI(Tl) with a hole in one end of the crystal for the insertion of the sample (Fig. 12-1A). Dimensions of some commonly used well detectors are given in Table 12-1. The 4.5-cm diameter × 5-cm long crystal with 1.6-cm diameter × 3.8-cm deep well, the *standard well-counter* detector, is the most frequently used in nuclear medicine. It is designed for counting of samples in standard-size test tubes. Very large well-counter detectors, up to 13-cm diameter × 25-cm length, have been employed for counting very small quantities of high-energy γ-ray emitters (e.g., ^{40}K and ^{137}Cs). Most well-counter systems employ 5 cm or greater thickness of lead around the detector to reduce background counting levels. A typical manually loaded well-counter system is shown in Figure 12-1B.

Light transfer between the NaI(Tl) crystal and the photomultiplier (PM) tube is less than optimal with well-type detectors because of reflection and scattering of light by the well surface inside the detector crystal. Energy resolution is therefore poorer [10% to 15% full width at half maximum (FWHM) for ^{137}Cs] than obtained with optimized NaI(Tl) detector designs (approximately 6% FWHM) (see Chapter 10, Section B.7).

---

[*]In this chapter the term *γ-ray emission* also includes other forms of ionizing electromagnetic radiation (e.g., x rays, bremsstrahlung, and annihilation radiation).

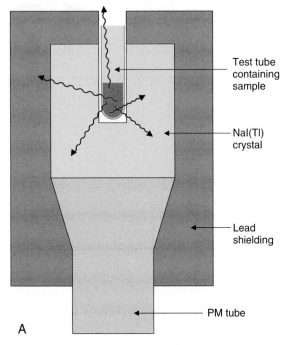

FIGURE 12-1   *A,* Cross-sectional view of a well-counter detector containing a radioactive sample. *B,* Photograph of a manually loaded well counter with a digital readout and printer output. (*Courtesy Capintec, Inc., Ramsey, NJ.*)

2. Detection Efficiency

The detection efficiency *D* (see Chapter 11, Section A) of the NaI(Tl) well counter for most γ-ray emitters is quite high, primarily because of their near 100% geometric efficiency *g*. The combination of high detection efficiency and low background counting levels makes the well counter highly suitable for counting samples containing very small quantities (Bq–kBq) of γ-ray-emitting activity. The geometric efficiency for small (≤1-mL) samples in the standard well counter is approximately 93% (see Fig. 11-3).

TABLE 12-1
DIMENSIONS OF TYPICAL NaI(Tl) WELL-COUNTER DETECTORS

Crystal Dimensions (cm)		Well Dimensions (cm)	
Diameter	**Length**	**Diameter**	**Depth**
4.5*	5.0*	1.6*	3.8*
5.0	5.0	1.6	3.8
7.6	7.6	1.7	5.2
10.0	10.0	3.8	7.0
12.7	12.7	3.8	7.0

*"Standard" well-counter detector.

The intrinsic efficiency ε (Equation 11-8) of well-counter detectors depends on the γ-ray energy and on the thickness of NaI(Tl) surrounding the sample; however, the calculation of intrinsic efficiency is complicated because different thicknesses of detector are traversed by γ rays at different angles around the source. Calculated intrinsic efficiencies (i.e., all pulses counted) versus γ-ray energy for 1-mL sample volumes and for different NaI(Tl) well-counter detectors are shown in Figure 12-2. Intrinsic efficiency is close to 100% for 1.3- to 4.5-cm wall thickness and E_γ ≤ 150 keV, but at 500 keV the intrinsic efficiencies range from 39% to 82%.

Intrinsic efficiency can be used to calculate the counting rate per kBq for a radionuclide if all pulses from the detector are counted; however, if only photopeak events are recorded, then the *photofraction* f_p also must be considered (see Chapter 11, Section A.4). The photofraction decreases with increasing γ-ray energy and increases with increasing well-detector size (Fig. 12-3). At 100 keV, f_p ≈ 100% for all detector sizes. At 500 keV, f_p ranges from 48% to 83% from the smallest to the largest common detector sizes (Table 12-1).

The *intrinsic photopeak efficiency* ε_p is the product of the intrinsic efficiency and photofraction

$$\varepsilon_p = \varepsilon \times f_p \qquad (12\text{-}1)$$

This may be used to estimate photopeak counting rates. Figure 12-4 shows ε_p versus γ-ray energy.

Table 12-2 lists some detection efficiencies, expressed as counts per minute (cpm) per becquerel, for full-spectrum counting of different

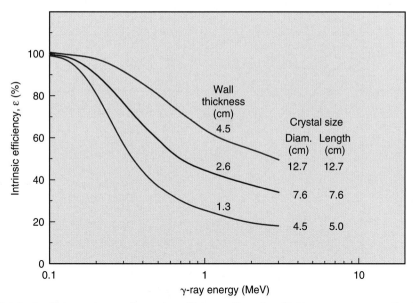

FIGURE 12-2 Intrinsic efficiency (γ-ray absorption efficiency, Equation 11-9) vs. γ-ray energy for different NaI(Tl) well-counter detectors.

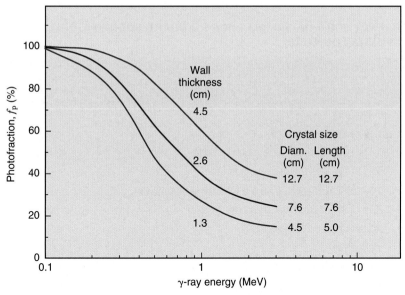

FIGURE 12-3 Photofraction versus γ-ray energy for different NaI(Tl) well-counter detectors.

radionuclides in the standard well counter. These values apply to 1-mL samples in standard test tubes.

3. Sample Volume Effects

The fraction of γ rays escaping through the hole at the end of the well depends on the position of the source in the well. The fraction is only about 7% near the bottom of the well but increases to 50% near the top and is even larger for sources outside the well. Thus the

geometric efficiency of a well counter depends on sample positioning. If a small volume of radioactive solution of *constant activity* in a test tube is diluted progressively by adding water to it, the counting rate recorded from the sample in a standard well detector progressively decreases, even though total activity in the sample remains constant (Fig. 12-5). In essence, the geometric efficiency for the sample decreases as portions of the activity are displaced to the top of the well.

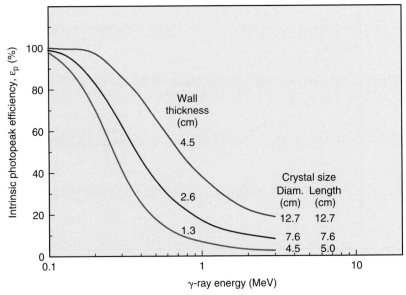

FIGURE 12-4   Intrinsic photopeak efficiency vs. γ-ray energy for different NaI(Tl) well-counter detectors.

TABLE 12-2
COUNTING EFFICIENCY FOR 1-mL SAMPLES IN A STANDARD SODIUM IODIDE WELL COUNTER (ASSUMING ALL PULSES COUNTED)

Radionuclide	γ-ray Energies (MeV) (% per Disintegration)	Counting Efficiency per Disintegration (%)	Counts per Minute per Becquerel
^{51}Cr	0.320 (8%)	4.3	2.6
^{60}Co	1.17 (100%) 1.33 (100%)	43	25.8
^{198}Au	0.411 (96.1%), 0.68 (1.1%), 1.09 (0.26%)	43.5	26.1
^{199}Au	0.051 (0.3%), 0.158 (41%), 0.209 (9%)	46	27.6
^{131}I	0.08 (2%), 0.28 (5%), 0.36 (80%), 0.64 (9%), 0.72 (3%)	48.3	28.9
^{59}Fe	0.19 (2.8%), 1.10(57%), 1.29 (43%)	27.3	16.4
^{203}Hg	0.073 (17%), 0.279 (83%)	67	40.3
^{42}K	1.53 (18%)	4.0	2.4
^{22}Na	0.511 (180%), 1.28 (100%)	81	48.6
^{24}Na	1.37 (100%), 2.75 (100%)	38	22.8

Adapted from Hine GJ: γ-ray sample counting. In Hine GJ (ed): *Instrumentation in Nuclear Medicine*. New York, 1967, Academic Press, p 282.

If the volume of a sample is increased by adding radioactive solution at a *constant* concentration, the counting rate first increases linearly with sample volume (or activity) but the proportionality is lost as the volume approaches and then exceeds the top of the well. Eventually there is little change with increasing sample volume, although the total activity is increasing (see Fig. 12-5). For example, an increase of sample volume in a standard test tube from 7 to 8 mL, a 14%

increase in volume, increases the counting rate by only about 1%.

Thus sample volume has significant effects on counting rate with well counters. Sample volumes should be the same when comparing two samples. One technique that is used when adequate sample volumes are available is to fill the test tubes to capacity because with full test tubes, small differences in total volume have only minor effects on counting rate (curve B in Fig. 12-5); however, this requires

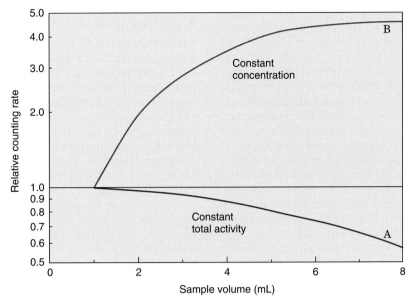

FIGURE 12-5 *A,* Change in counting rate in a standard NaI(Tl) well counter for a sample of constant *activity* but diluted to increasing sample volume in a test tube. *B,* Change in counting rate with volume for constant *concentration*.

that identical test tubes be used for all samples, so that the volume of activity inside the well itself does not differ between samples.

Absorption of γ rays within the sample volume or by the walls of the test tube is not a major factor except when low-energy sources, such as ^{125}I (27-35 keV) are counted. Identical test tubes and carefully prepared samples of equal volume should be used when comparing samples of these radionuclides.

4. Assay of Absolute Activity

A standard NaI(Tl) well counter can be used for assay of absolute activity (Bq or Bq/mL) in samples of unknown activity using the calibration data given in Table 12-2. Alternatively, one can compare the counting rate of the unknown sample to that of a calibration source (see Chapter 11, Section A.6). "Mock" sources containing long-lived radionuclides are used to simulate short-lived radionuclides, for example, a mixture of ^{133}Ba (356- and 384-keV γ rays) and ^{137}Cs (662-keV γ rays) for "mock ^{131}I." Frequently, such standards are calibrated in terms of "equivalent activity" of the radionuclide they are meant to simulate. Thus if the activity of a mock ^{131}I standard is given as "A(Bq) of ^{131}I," then the activity of a sample of ^{131}I of unknown activity X would be obtained from

$$X(\text{Bq}) = A(\text{Bq}) \times [R(^{131}\text{I})/R(\text{mock }^{131}\text{I})] \quad (12\text{-}2)$$

where $R(^{131}\text{I})$ and $R(\text{mock }^{131}\text{I})$ are the counting rates recorded in the well counter for the sample and the calibration standard, respectively.

Another commonly used mock standard is ^{57}Co (129 and 137 keV) for ^{99m}Tc (140 keV). If the ^{57}Co is calibrated in "equivalent Bq of ^{99m}Tc," then Equation 12-2 can be used for ^{99m}Tc calibrations also. If it is calibrated in becquerels of ^{57}Co, however, one must correct for the differing emission frequencies between ^{57}Co and ^{99m}Tc (0.962 γ rays/disintegration vs. 0.889 γ rays/disintegration, respectively). The activity X of a sample of ^{99m}Tc of unknown activity would then be given by

$$X(\text{Bq}) = A(\text{Bq}) \times [R(^{99m}\text{Tc})/R(^{57}\text{Co})]$$
$$\times (0.962/0.889)$$
$$(12\text{-}3)$$

where A is the calibrated activity of the ^{57}Co standard and $R(^{99m}\text{Tc})$ and $R(^{57}\text{Co})$ are the counting rates recorded from the ^{99m}Tc sample and the ^{57}Co standard, respectively.

5. Shielding and Background

It is desirable to keep counting rates from background radiation as low as possible with the well counter to minimize statistical uncertainties in counting measurements (see Chapter 9, Section D.4). Sources of background include cosmic rays, natural radioactivity in the detector (e.g., ^{40}K) and surrounding shielding materials (e.g., radionuclides of Rn, Th, and U in lead), and other radiation sources in the room. Additional sources of background in a hospital environment include patients who have been injected with radionuclides for nuclear medicine studies or for therapeutic purposes. These sources of radiation, although usually located some distance from the counter, can produce significant and variable sources of background. External sources of background radiation are minimized by surrounding the detector with lead. The thickness of the lead shielding is typically 2.5-7.5 cm; however, even with lead shielding it is still advisable to keep the counting area as free as possible of unnecessary radioactive samples.

In well counters with automated multiple-sample changers (Section A.9), it also is important to determine if high-activity samples are producing significant back-grounds levels in comparison with activity samples in the same counting rack. In many nuclear medicine procedures, background counting rates are measured between samples, but if the background counting rate becomes large (e.g., from a radioactive spill or contamination of the detector), it can produce significant statistical errors even when properly subtracted from the sample counting rate (see Chapter 9, Section D.4).

6. Energy Calibration

Energy selection in a well counter usually is accomplished by an SCA (Chapter 8, Section C.2). Commercial well-counter systems have push-button or computer selection of the appropriate SCA window settings for different radionuclides. In these systems compensation has been made by the manufacturer for the nonlinear energy response of the NaI(Tl) detector. However, because of the possibility of drifts in the electronics and the PM tube gain with time, the response of the well counter should be checked regularly with a long-lived standard source, such as ^{137}Cs, as a quality assurance measure. Some modern well counters incorporate MCAs, allowing the entire spectrum to be measured and analyzed.

7. Multiple Radionuclide Source Counting

When multiple radionuclides are counted simultaneously (e.g., from tracer studies with double labels), there is "crosstalk" interference because of overlap of the γ-ray spectra of the two sources, as shown in Figure 12-6 for ^{99m}Tc and ^{51}Cr. If SCA windows are positioned on the ^{99m}Tc (window 1) and ^{51}Cr (window 2)

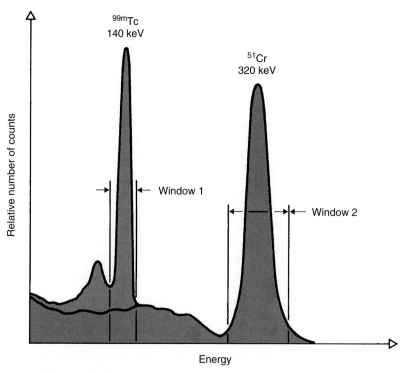

FIGURE 12-6 Window settings used for simultaneous measurement of ^{99m}Tc and ^{51}Cr in a mixed sample. Crosstalk from ^{51}Cr into the ^{99m}Tc window must be corrected for, using methods described in the text.

photopeaks, a correction for the interference can be applied as follows: A sample containing only ^{51}Cr is counted and the ratio R_{12} of counts in window 1 to counts in window 2 is determined. Similarly, a sample containing only ^{99m}Tc is counted and the ratio R_{21} of counts in window 2 to counts in window 1 is determined. Suppose then that a mixed sample containing unknown proportions of ^{99m}Tc and ^{51}Cr is counted and that N_1 counts are recorded in the ^{99m}Tc window (window 1) and that N_2 counts are recorded in the ^{51}Cr window (window 2). Suppose further that room and instrument background counts are negligible or have been subtracted from N_1 and N_2. Then the number of counts from ^{99m}Tc in window 1 [$N_1(^{99m}$Tc$)$] can be calculated from

$$N_1(^{99m}\text{Tc}) = (N_1 - R_{12}N_2)/(1 - R_{12}R_{21}) \quad (12\text{-}4)$$

and the number of counts from ^{51}Cr in window 2 [$N_2(^{51}$Cr$)$] from

$$N_2(^{51}\text{Cr}) = (N_2 - R_{21}N_1)/(1 - R_{12}R_{21}) \quad (12\text{-}5)$$

Equations 12-4 and 12-5 permit calculation of the number of counts that would be recorded in the photopeak window for each radionuclide in the absence of crosstalk interference from the other radionuclide. These equations can be used for other combinations of radionuclides and window settings with appropriate changes in symbols. For greatest precision, the ratios R_{12} and R_{21} should be determined to a high degree of statistical precision (e.g., $\pm 1\%$) so that they do not add significantly to the uncertainties in the calculated results. The technique is most accurate when crosstalk is small, that is, R_{12} and/or $R_{21} \ll 1$. Generally, the technique is *not* reliable for the in vivo measurements described in Section F, because of varying amounts of crosstalk caused by Compton scattering within body tissue.

EXAMPLE 12-1

A mixed sample containing ^{99m}Tc and ^{51}Cr provides 18,000 counts in the ^{99m}Tc window and 8000 counts in the ^{51}Cr window. A sample containing ^{51}Cr alone gives 25,000 counts in the ^{51}Cr window and 15,000 crosstalk counts in the ^{99m}Tc window, whereas a sample containing ^{99m}Tc alone gives 20,000 counts in the ^{99m}Tc window and 1000 crosstalk counts in the ^{51}Cr window. What are the counts due to each radionuclide in their respective photopeak windows? Assume that background counts are negligible.

Answer

The crosstalk interference factors are, for ^{51}Cr crosstalk in the ^{99m}Tc window

$$R_{12} = 15,000/25,000 = 0.6$$

and for ^{99m}Tc crosstalk in the ^{51}C window

$$R_{21} = 1000/20,000 = 0.05$$

Therefore the counts in the ^{99m}Tc window from ^{99m}Tc in the mixed sample are (Equation 12-4)

$$N_1(^{99m}\text{Tc}) = (18,000 - 0.6 \times 8000)/(1 - 0.6 \times 0.05)$$
$$= 13,200/0.97$$
$$\approx 13,608 \text{ counts}$$

and the counts in the ^{51}Cr window from ^{51}Cr are (Equation 12-5)

$$N_2(^{51}\text{Cr}) = (8000 - 0.05 \times 18,000)/(1 - 0.6 \times 0.05)$$
$$= 7100/0.97$$
$$\approx 7320 \text{ counts}$$

8. Dead Time

Because NaI(Tl) well counters have such high detection efficiency, only small amounts of activity can be counted (typically 10^2 to 10^4 Bq). If higher levels of activity are employed, serious dead time problems can be encountered (see Chapter 11, Section C). For example, if the dead time for the system (paralyzable) is 4 μsec, and 50 kBq of activity emitting one γ ray per disintegration is counted with 100% detection efficiency, then the true counting rate is 50,000 cps; however, the recorded counting rate would be approximately 41,000 cps because of 18% dead time losses (see Equation 11-18).

9. Automated Multiple-Sample Systems

Samples with high counting rates require short counting times and provide good statistical precision with little interference from normal background radiation. If only a few samples must be counted, they can be counted quickly and conveniently using manual techniques; however, with long counting times or large numbers of samples, the counting procedures become time consuming and cumbersome. Systems with automated sample changers have been developed to alleviate this problem (Fig. 12-7). Typically, these systems can accommodate 100 or more

FIGURE 12-7 A NaI(Tl) well counter with automated sample-changing capabilities. Hundreds of samples can be loaded and measured in a single run. This system also incorporates a multichannel analyzer for spectral analysis. (*Courtesy PerkinElmer, Inc. Waltham, MA.*)

samples, and each sample is loaded automatically into the counter in a sequential manner.

Most multiple sample systems use a variation of the well-counter detector known as the *"through-hole" detector*. As shown in Figure 12-8, the sample hole passes through the entire length of the NaI(Tl) crystal, and the PM tube is connected to the side of the scintillator. A key advantage of the through-hole detector is that samples can be automatically positioned at the center of the NaI(Tl) crystal, irrespective of sample volume. This results in the highest detection efficiency and minimizes efficiency changes with volume. Figure 12-9 shows the smaller changes in efficiency with volume for a through-hole versus a well-type counter for ^{59}Fe.

Systems with automated sample changers not only save time but also allow samples to be counted repeatedly to detect variations caused by malfunction of the detector or electronic equipment or changes in background counting rates. Background counting rates can be recorded automatically by alternating sample and blank counting vials. In these

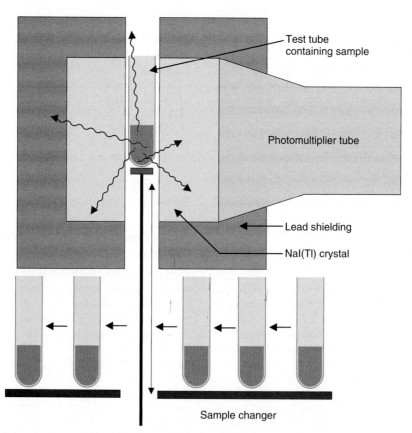

FIGURE 12-8 Schematic cross-sectional drawing of through-hole detector and sample-changing system. Placement of the sample can be automatically adjusted to center the sample volume in the detector.

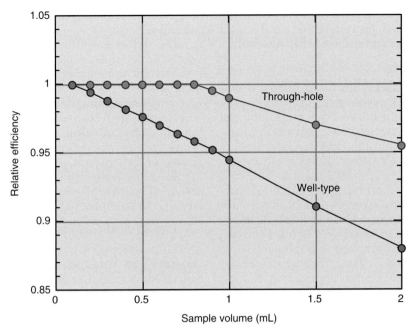

FIGURE 12-9 Efficiency of a well counter versus a through-hole counter for a constant total activity of ^{59}Fe. The efficiency of the through-hole detector shows less variation with sample volume because the sample can be centered in the detector. (*Adapted from Guide to Modern Gamma Counting. Packard Instrument Company, Meriden, CT, 1993.*)

systems, counting vials loaded into a tray or carriage are selected automatically and placed sequentially in the NaI(Tl) well counter. Measurements are taken for a preset time or a preset number of counts selected by the user. The well counter usually is shielded with 5 to 7.5 cm of lead, with a small hole in the lead shielding above and beneath the detector for insertion of the sample. One disadvantage of automated systems is that there is no lead shielding directly above or below the sample being counted. Therefore the system is not as well shielded as a manual well counter, which can cause an increase in background counting rates, particularly from other samples in the carriage. This can be a problem when low-activity samples are counted with high-activity samples in the carriage.

Commercial systems usually have MCAs or multiple SCAs to allow the selection of many different counting windows. The MCA also can be used to display the entire spectrum recorded by the NaI(Tl) detector on a computer. The displayed spectrum allows the user to inspect visually and select the positions of the single-channel windows for counting and to examine crosstalk interference when multiple radionuclides are counted simultaneously. It is also very useful for quickly and reliably checking to see if there are any significant photopeaks in the

spectrum from background sources, which could indicate a radioactive spill or contamination, or for checking the general condition of the NaI(Tl) detector.

Modern well-counter systems are interfaced to computers or have dedicated circuits that control sample changing, placement and counting time, and perform corrections for radionuclide decay and background. Programs for spectral analysis and correction of multiple isotope samples are also generally available. All interactions with the well-counter system generally are through the keyboard, where the user selects from a range of predefined protocols and provides information regarding the radionuclide, desired counting time, and sample volume.

For very high throughput, there are even multidetector systems that may contain as many as 10 NaI(Tl) scintillation detectors. This permits 10 samples to be counted simultaneously and many hundreds or even thousands of samples to be counted per hour. The individual detectors are carefully separated and shielded from each other by lead to prevent crosstalk; however, when counting high-energy γ rays ($\gtrsim$300 keV) some crosstalk may occur. This is in addition to the source of crosstalk described in Section A.7, which occurs from the samples waiting to be counted in the sample changer system. Background measurements in one detector while counting a sample in an adjacent detector can be used to estimate the magnitude of this crosstalk.

10. Applications

NaI(Tl) well counters are used almost exclusively to count x-ray or γ-ray-emitting radionuclides. Radionuclides with β emissions can be counted by detecting bremsstrahlung radiation, but the counting rate per becquerel is small because efficiency of bremsstrahlung production is very low (see Example 6-1). Well counters are used primarily for radioimmunoassays (e.g., measurement of thyroid hormones triiodothyronine and thyroxine), assay of radioactivity in blood and urine samples, radiochemical assays, and radiopharmaceutical quality control. They also are used for wipe tests (see Chapter 23, Section E.3) in radiation safety monitoring. Systems with multiple SCAs or MCAs allow multiple radionuclide sources to be counted simultaneously. These capabilities, combined with automated sample changing and automatic data processing, make the NaI(Tl) well counter an important tool for nuclear medicine in vitro assays.

B. COUNTING WITH CONVENTIONAL NaI(Tl) DETECTORS

1. Large Sample Volumes

The principal restriction on the use of most NaI(Tl) well counters is that they are useful only for small sample volumes (a few milliliters, typically) and small amounts of activity ($\leq$100 kBq). For activities greater than approximately 100 kBq of most radionuclides, the counting rate becomes so high that dead time losses may become excessive. Large sample volumes and larger amounts of activity can be counted using a conventional NaI(Tl) detector with the sample at some distance from the detector. Placing the sample at a distance from the detector decreases the geometric efficiency (see Example 11-1) and allows higher levels of activity to be counted than with the well detector. The sample-to-detector distance can be adjusted to accommodate the level of activity to be measured. Typically, shielding from background sources with these arrangements is not as good as with the well counter because the front of the detector is exposed; however, owing to the high counting-rate applications of these systems, background counting rates usually are not significant unless there are other

high-activity samples in the immediate vicinity.

The detection efficiency of a conventional detector depends on a number of factors, such as detector-to-sample distance, detector diameter, and sample size (see Chapter 11, Section A). If the sample-to-detector distance is large compared with the sample diameter, then usually the counting efficiency is relatively constant as the sample size is increased; however, this cannot always be assumed to be true, and sample size effects should be evaluated experimentally for specific counting conditions to be employed.

2. Liquid and Gas Flow Counting

NaI(Tl) detectors are used frequently as γ-ray monitors in conjunction with gas or liquid chromatographs. Chromatographs are used to separate and identify different chemical compounds by passing a gas or solution through columns containing beads that can selectively retain or control the rate of movement of different chemical species based on molecular size (gel filtration chromatography), net electric charge (ion exchange chromatography), or binding characteristics (affinity chromatography). By comparing the flow of radioactivity with the flow of chemical species, one can determine the radiochemical identity of different radioactive species (Fig. 12-10). The SCA typically is used to count only the photopeak to reduce background caused by scattered radiation from activity in the flow line outside the detector and in the chromatograph. MCAs or multiple SCAs also can be employed to detect multiple radionuclides simultaneously.

With simple systems the data output from the SCA is recorded with a ratemeter (digital or analog) and sent to some form of data recorder. With more sophisticated systems the data are collected with computers that have extended capability for data analysis.

C. LIQUID SCINTILLATION COUNTERS

1. General Characteristics

For *liquid scintillation* (LS) counting, the radioactive sample is dissolved in a scintillator solution contained in a counting vial and placed in a *liquid scintillation counter* (LSC) that consists of two PM tubes in a darkened counting chamber (Fig. 12-11). LSCs are used for counting β emitters, such as ^{3}H and ^{14}C, which would be strongly absorbed in the glass

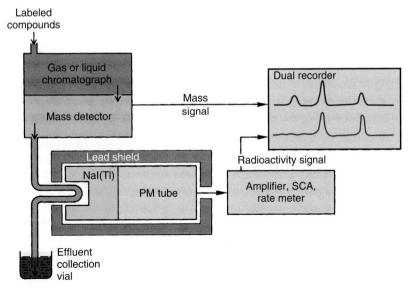

FIGURE 12-10 NaI(Tl) detector system used in conjunction with a gas or liquid chromatograph. The "mass detector" is used to detect chemical species, and the radiation detector is used to detect the radioactivity associated with these species for radiochemical identification.

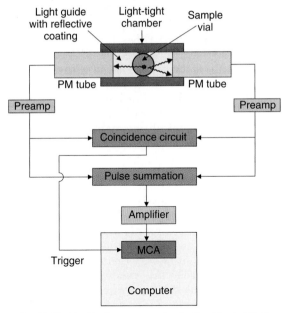

FIGURE 12-11 Basic components of a liquid scintillation counter.

or plastic of the test tube used to contain the sample in a standard well counter. They are also used for counting emitters of low-energy x rays and γ rays, which cannot be detected efficiently with NaI(Tl) detectors because of the thickness of canning material required around the detector.

The LS solution has a low atomic number (Z ~ 6 to 8) and density (ρ ~ 1) in comparison with other scintillators, such as NaI(Tl).

However, this is sufficient for high-efficiency detection of low-energy x rays, γ rays, and β particles. Because the radioactivity is in direct contact with the scintillator, LS counting is the preferred method for the detection of low-energy β-emitting radionuclides, such as ^{3}H and ^{14}C. Numerous other β-emitting radionuclides, including some (β,γ) emitters, also are counted with a LSC (Table 12-3). Positron (β^{+}) emitters, however, are generally

TABLE 12-3

RADIONUCLIDES COMMONLY COUNTED WITH LIQUID SCINTILLATION DETECTORS

Radionuclide	Half-Life	Maximum β Energy (MeV)
^{3}H	12.3 yr	0.019
^{14}C	5700 yr	0.156
^{35}S	87.5 d	0.167
^{45}Ca	163 d	0.257
^{65}Zn	243 d	0.325
^{59}Fe	45 d	0.467
^{22}Na	2.6 yr	0.546
^{131}I	8.06 d	0.606
^{36}Cl	3×10^{5} d	0.714
^{40}K	1.3×10^{9} yr	1.300
^{24}Na	15.0 hr	1.392
^{32}P	14.3 d	1.711

counted in a standard well counter, because of the penetrating 511-keV γ rays produced from the annihilation (see Chapter 3, Section G) of positrons with electrons in the sample, or in the walls of the tube containing the sample.

Because LSC systems are used primarily to count very low-energy particles, the system must have very low electronic noise levels. For example, with ^{3}H, the energy range of the β particle is 0-18 keV. Under optimal conditions, β particles from ^{3}H decay produce only 0 to 25 photoelectrons at the PM tube photocathode, with an average of only about eight ($\bar{E}_\beta \approx 1/3 E_\beta^{max}$). Background electronic noise is due mainly to spontaneous thermal emission of electrons from the photocathode of the PM tube. Background noise also is present from exposure to light of the scintillator solution during sample preparation. This exposure can produce light emission (phosphorescence), which persists for long periods (i.e., hours).

Several methods are employed in LS detectors to reduce this noise or background count rate. Thermal emission is reduced by refrigeration of the counting chamber to maintain the PM tubes at a constant low temperature (typically about −10° C). Constant PM tube temperature is important because the photocathode efficiency and electronic gain of the PM tube are temperature dependent, and variations in temperature produce variation in the amplitude of the output signal.

Pulse-height analysis also may be used to discriminate against noise because true radiation events usually produce larger signals than thermal emission noise; however, thermal emission noise still is superimposed on the radiation signals, which can cause deterioration of the energy resolution and linearity of the system.

The most effective reduction of noise is achieved by *coincidence detection* techniques (see Fig. 12-11). When a scintillation event occurs in the scintillator, light is emitted in all directions. Optical reflectors placed around the counting vial reflect the light into two opposing PM tubes to maximize light collection efficiency. Pulses from each of the PM tubes are routed to separate preamplifiers and a *coincidence circuit* (see Chapter 8, Section F). The coincidence circuit rejects any pulse that does not arrive simultaneously with a pulse from the other PM tube (i.e., within approximately 0.03 μsec). Noise pulses are distributed randomly in time; therefore the probability of two noise pulses occurring simultaneously in the two PM tubes is very

small. Random coincidence rates R_r (cps) can be determined from

$$R_r = (2\tau)R_n^2 \qquad (12\text{-}6)$$

where 2τ is the resolving time of the coincidence circuit and R_n is the noise pulse rate for each PM tube (assumed to be equal) caused by PM tube noise and phosphorescence in the sample. For $2\tau = 0.03$ μsec and $R_n = 1000$ cps, one obtains $R_r = 3 \times 10^{-8} \times (10^3)^2 = 0.03$ cps. Thus most of the noise pulses are rejected by the coincidence circuit.

The output signals from the two PM tubes and preamplifiers are fed into the coincidence circuit as described earlier and also into a summation circuit, which adds the two signals together to produce an output signal proportional to the total energy of the detected event (see Fig. 12-11). The output signals from the summing circuit are sent to an amplifier to boost the signal, which is then digitized in an MCA. The output from the coincidence circuit is fed to the MCA to enable data collection only when both PM tubes have registered a pulse, thus rejecting noise. The MCA provides a spectrum of the energies of the detected events, which can be further processed by a computer, including routines for separating the counts from two radionuclides that are being counted simultaneously and for performing quench corrections as discussed in Section C.5.

2. Pulse-Height Spectrometry

Pulse-height analysis is used with LS counting to further reduce the background counting rate by selecting only the energy region corresponding to the radiation of interest or to select different energy regions when simultaneous sources are being counted. An example of *two-energy window analysis* for a source containing ^{3}H and ^{14}C is shown in Figure 12-12. Because of the continuous energy distribution in β decay, pulse-height analysis cannot separate completely the two spectra, and there is crosstalk interference. Methods to correct for this situation are discussed in Section C.4.

3. Counting Vials

Counting vials containing the radioactivity and the liquid scintillator solution usually are made of polyethylene or low-potassium-content glass. The low-potassium-content glass is used to avoid the natural background of ^{40}K. When standard laboratory glass vials (lime

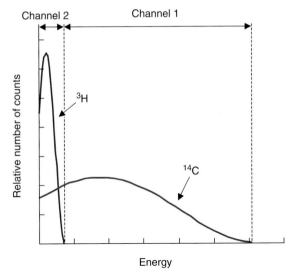

FIGURE 12-12   Example of pulse-height spectra obtained from ^{3}H and ^{14}C by liquid scintillation counting.

Modern LS counters have prestored calibrations that enable them to convert the detected cpm into disintegrations per minute for a wide range of radionuclides. These calibrations, however, depend on the material composition and thickness of the sample vial and on the effects of quenching, which are discussed in Section C.5).

Frequently, samples containing a mixture of two radionuclides (e.g., ^{3}H and ^{14}C) are counted. By selecting separate energy windows on each of the β spectra (see Fig. 12-12), the activities of each of the radionuclides can be determined. The optimal window for each radionuclide is determined individually by using separate ^{3}H and ^{14}C sources. If possible, the energy windows should be adjusted so that counts from the lower-energy emitter are not included in the window used for the higher-energy emitter. The method and equations used to correct for crosstalk interference—described in Section A.7 for well-counter applications—can be also used on the LS counter. There are also a number of increasingly sophisticated methods for dealing with samples containing radionuclides with very similar spectra. These methods are described in detail in reference 1.

5. Quench Corrections

Quenching refers to any process that reduces the amount of scintillation light produced by the sample or detected by the PM tubes. The causes of quenching in LS counting were described in Chapter 7, Section C.6. The principal effect of quenching is to cause an apparent shift of the energy spectrum to lower energies (Fig. 12-13). This results in a

glass) are used, the background for ^{3}H and ^{14}C is increased by 30-40 cpm because of ^{40}K in the glass. Polyethylene vials frequently are used to avoid this problem and also to increase light transmission from the liquid scintillator to the PM tubes. Polyethylene vials are excellent for dioxane solvents but should not be used with toluene as the scintillator solvent because toluene will cause the vials to distort and swell, which may jam the sample changer. Materials such as quartz, Vicor, and others also are used for counting vials.

Exposure of the vial and liquid scintillator solution to strong sunlight produces a background of phosphorescence that may take hours to decay; therefore samples frequently are stored temporarily in a darkened container before counting. This is referred to as *dark adaptation* of samples.

4. Energy and Efficiency Calibration

Beta emission results in a continuous spectrum of β-particle energies from zero to a maximum β-particle energy E_{β}^{max} that is characteristic of the nuclide, with a mean value at approximately $\bar{E}_{\beta} \approx 1/3 E_{\beta}^{max}$. Usually, most of the β-particle spectrum lies above the electronic noise, allowing almost the entire spectrum to be used and resulting in detection efficiencies of 80% or higher. An exception is ^{3}H. The low-energy β emission of ^{3}H ($E_{\beta}^{max} = 18$ keV) reduces the counting efficiency to approximately 40% to 60% because some of the events produce pulses below typical noise pulse amplitudes that are rejected by pulse-height analysis (see Section C.1).

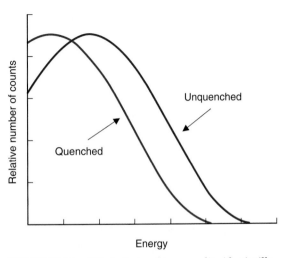

FIGURE 12-13   Effect of quenching on a liquid scintillation counter pulse-height spectrum.

loss of counts because events can either be shifted below the noise levels of the LS counter, or if pulse-height analysis is used, they may be shifted out of the energy window. Thus inaccurate counting rates are recorded. The error depends on the amount of quenching, which may vary from one sample to the next.

To obtain accurate results, it is necessary to correct the observed counting rate for quench-caused spectral shifts. Several methods have been developed.

With the *internal standardization method,* the sample-counting rate is determined; then a known quantity of the radionuclide of interest (from a calibrated standard solution) is added to the sample and it is recounted. The counting efficiency, ε_c, is calculated by

$$\varepsilon_c = \frac{cps(standard + sample) - cps(sample)}{standard(Bq)} \tag{12-7}$$

From the efficiency, the activity of the sample is obtained from

$$sample(Bq) = \frac{cps(sample)}{\varepsilon_c} \tag{12-8}$$

With internal standardization, the sample must be counted twice and the added activity of the standard must be distributed in the scintillator solution in the same manner as the sample. The method is not accurate if the sample and standard are not dissolved in the same way in the scintillator. Also, self-absorption of the emitted β particle by the labeled molecule might not be accounted for unless the standard is also in the form of the labeled molecule.

A second approach is called the *channel ratio method.* One channel is set to count an unquenched sample as efficiently as possible (i.e., channel 1 in Fig. 12-14), and a second channel is set to accumulate counts in the lower-energy region of the spectrum (channel 2 in Fig. 12-14). When the spectrum shifts to the left because of quenching, the lower channel gains counts, and the ratio of counts in the two channels changes. A series of standards of known activity are counted, each quenched deliberately a little more than the preceding one by adding a quenching agent, to obtain a quench curve relating counting efficiency (cps per becquerel) to the channel ratio. Then for subsequently measured samples, the channel ratio is used to determine the quench-corrected counting efficiency.

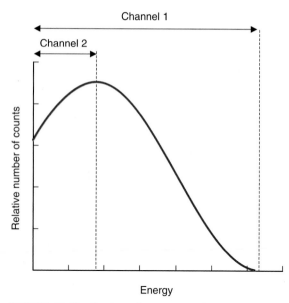

FIGURE 12-14   Setting of energy windows for quench corrections using the channel ratio method.

Once the correction curve has been obtained, only one (dual-channel) counting measurement per sample is required to determine counting efficiency. All causes of quenching are corrected by the channel ratio method. A disadvantage of the method is that at very low counting rates statistical errors in the value determined for the channel ratio can be large, which may result in significant errors in the estimated quench correction factor. Longer counting times may be employed to minimize this source of error.

A third approach is called the *automatic external standardization (AES) method.* This method incorporates features of both internal standardization and channel ratio. The sample is first counted and then recounted (usually for 1 min or less) with an external standard γ-ray source (usually ^{137}Cs) placed close to the sample (some counters count the sample plus standard first). Positioning of the standard is automatic. Compton recoil electrons produced by interactions of the γ rays with the scintillator solution are counted in two channels and a channel ratio determined, or

$$AES\,(ratio) = \frac{cpm\,(sample + STD) - cpm\,sample\,channel\,2}{cpm\,(sample + STD) - cpm\,sample\,channel\,1} \tag{12-9}$$

where STD refers to the standard γ-ray source and channels 1 and 2 are as indicated in Figure 12-14.

A series of quenched standards containing known amounts of the radionuclide of interest is prepared, and counting efficiency is related to the AES ratio. The AES ratio is then used to correct for quenching on subsequently measured samples.

The external standard method generally provides a high counting rate and thus small statistical errors in the determination of the quench correction factor while maintaining the sensitivity of the channel ratio method for detecting quenching effects. The disadvantage of the AES method is that only chemical and color quenching are corrected; β-particle self-absorption effects or losses caused by sample distribution effects are not. For example, the AES method might not be accurate with multiphase solutions in which the sample is not soluble in the counting solution.

Representative AES quench curves and crosstalk correction factors are shown in Figure 12-15 for ^{14}C and ^{3}H for double-label studies, that is, both radionuclides counted simultaneously. It is apparent from Figure 12-15 that as quenching increases (AES ratio decreases), the efficiency for counting both ^{14}C and ^{3}H decreases. Thus even though the true efficiency may be determined accurately from the quench correction curve, counting efficiency deteriorates with increased quenching, resulting in increased statistical errors.

6. Sample Preparation Techniques

Samples can be combined with scintillator solution in several different ways, depending on the composition, state (liquid or solid) and polarity of the sample compound or material. The medium into which the sample is placed is known as the *LS cocktail,* of which there are two main groups: *Emulsifying cocktails,* also known as *aqueous cocktails,* consist of an organic aromatic solvent, an emulsifier and the scintillator. *Organic cocktails,* also known as *nonaqueous* or *lipophilic cocktails,* consist of an organic aromatic solvent and the scintillator.

Liquid scintillators were discussed in Chapter 7, Section C.6. The most popular and widely used is a combination of 2,5-diphenyloxazole (also known as PPO) and *p*-bis-(*o*-methylstyryl)benzene (abbreviated as bis-MSB). Traditional aromatic solvents include toluene and xylene, and although these are still used, they are gradually being replaced with more environmentally friendly solvents such as di-isopropylnaphthalene (DIN) and phenylxylylethane (PXE).

Most radionuclides are present in aqueous form and therefore are not readily miscible with aromatic solvents. Detergents or emulsifiers are used to form a microemulsion, in which the aqueous solution is dispersed in tiny droplets through the solvent. Commonly used

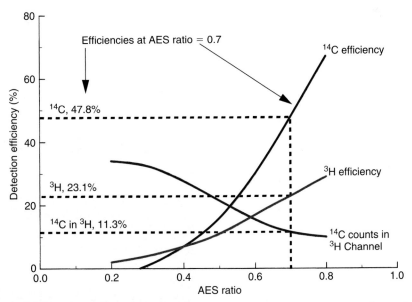

FIGURE 12-15 Representative quench correction curves based on the automatic external standardization (AES) method for counting mixed ^{14}C-^{3}H samples. For example, with an AES ratio of 0.7 (*vertical dashed line*) counts in the ^{3}H channel must first be corrected for ^{14}C crosstalk by 11.3% of the counts in the ^{14}C channel. The counting efficiency for the corrected ^{3}H counts is 23.1% and for the ^{14}C counts is 47.8% relative to an unquenched sample. Note that the AES ratio decreases with increasing quenching.

detergents include the alkyl phenol ethoxylates, alkyl and alkylaryl sulfonates, alcohol sulfates, and phosphate esters. Polar compounds also can be used by forming insoluble suspensions. For example, $^{14}CO_2$ can be precipitated as barium carbonate and then suspended in the scintillator solution with the addition of thixatropic jelling agents. Silica gels from thin-layer chromatography also can be counted in this manner. Samples deposited on filter paper such as from paper chromatography frequently are counted by placing the paper strip in the liquid scintillator. The scintillator solution also can be dissolved or suspended in the sample itself.

Another straightforward approach to counting complicated organic compounds such as proteins or sections of acrylamide gel columns with high efficiency is to combust the sample. The $^{14}CO_2$ and ^3H_2O released may be collected, dissolved in scintillator solution, and then counted.

Numerous other techniques have been developed for LS sample preparation. More discussion of these techniques is presented in reference 1. Careful sample preparation is critical for accurate application of the LS technique.

7. Cerenkov Counting

High-energy beta emitters may also be assayed in LSC systems without the use of a liquid scintillator solution by detecting optical Cerenkov radiation (see Chapter 6, Section A.5). Beta particles with an energy in excess of 263 keV will produce Cerenkov light in a water solution, which can be detected by the PM tubes in a LSC system. The calibration of the LSC system for measuring activity from the detected Cerenkov light must account for the directionality of the light cone produced and the spectral characteristics of the Cerenkov light, which is weighted toward the blue end of the visible spectrum. Because the production of Cerenkov light is a physical phenomenon, there is no chemical quenching of the signal. However, color quenching still must be accounted for. Cerenkov counting is used primarily to measure samples containing ^{32}P ($E_\beta^{max} = 1710$ keV) in which the counting efficiency can be in excess of 50%.

8. Liquid and Gas Flow Counting

In addition to counting individual samples, LSC systems also can be used for continuous monitoring of gas streams or flowing liquids. In these systems, the vial of LS solution is replaced with a cell filled with finely dispersed solid scintillator crystals through which the radioactively labeled gas or liquid is allowed to flow. The most common scintillator material for this purpose is the organic scintillator anthracene. This technique is used primarily for β-emitting radionuclides, typically ^{14}C and ^3H. The β particles interact with the anthracene crystals, and the resulting scintillation is detected in the same manner as from the LS vial.

These systems have been used for monitoring the effluent from amino acid analyzers, liquid chromatographs, and gas chromatographs. To monitor the effluent from gas chromatographs, the compounds usually are passed through a gas combustion furnace to convert them into $^{14}CO_2$ or ^3H_2O (vapor). Carrier gas from the gas chromatograph (e.g., He) is used to sweep the $^{14}CO_2$ or ^3H_2O through the counting cell.

Counting rates in these systems depend on the activity concentration and the flow rate. If fast flow rates and low-activity concentrations are required, the result may be data of poor statistical quality. Data from flow counting represent the time course of some process and usually are displayed as time-activity curves.

9. Automated Multiple-Sample LS Counters

LSC may be used for counting large numbers of samples or for counting low-level samples for long counting times. To expedite this and to remove the tedious job of manually counting multiple samples, automated multiple-sample LSC systems have been developed. These systems have automated sample changers that frequently can handle 100 or more counting vials (Fig. 12-16). A number of

FIGURE 12-16 A liquid scintillation counter with automated sample loading can efficiently count and analyze hundreds of samples. (*Courtesy Beckman Coulter, Inc., Brea, CA.*)

different sample-changing mechanisms have been developed, but the most common ones employ either trays or an endless belt for transport of the samples. Sample vials are selected automatically and loaded into the light-tight LS counting chamber. The samples are counted sequentially in serial fashion. Empty positions in the sample changer can be bypassed, and samples below a selectable low-level counting rate may be rejected automatically to avoid long counting times on samples that contain an insignificant amount of activity when preset counts are selected.

Modern automated multiple-sample LSC systems are provided with many different ways of handling and presenting the recorded data. Computer-based systems allow automatic implementation of quench corrections, efficiency corrections, background subtraction, statistical analysis, and calculations of parameters for radioimmunoassay or other assay analysis.

10. Applications

LSC systems are used in nuclear medicine for radioimmunoassays and protein-binding assays of drugs, hormones, and other biologically active compounds. LSC systems also are commonly used in studies of metabolic or physiologic processes with ^{3}H-or ^{14}C-labeled metabolic substrates or other physiologically important molecules. They are also used for wipe tests for radiation-monitoring purposes (see Chapter 23, Section E.3).

D. GAS-FILLED DETECTORS

1. Dose Calibrators

Although they are inefficient detectors for most γ-ray energies encountered in nuclear medicine, gas-filled detectors still find some specialized applications. A dose calibrator is essentially a well-type ionization chamber that is used for assaying relatively large quantities (i.e., MBq range) of γ-ray-emitting radioactivity (Fig. 12-17). Dose calibrators are used for measuring or verifying the activity of generator eluates, patient preparations, shipments of radioactivity received from suppliers, and similar quantities of activity too large for assay with NaI(Tl) detector systems.

The detector for a dose calibrator typically is an argon-filled chamber, sealed and pressurized to avoid variations in response with ambient barometric pressure (see Chapter 7, Section A.2). Ionization chamber

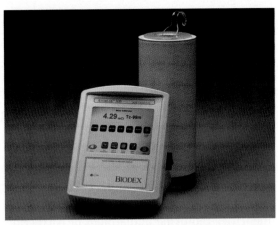

FIGURE 12-17 An ionization chamber dose calibrator. Samples are inserted into the well in the sealed ionization chamber. The current is measured and displayed on a digital readout. (*Courtesy Biodex Medical Systems, Shirley, NY.*)

dose calibrators assay the total amount of activity present by measuring the total amount of ionization produced by the sample. Plug-in resistor modules, pushbuttons, or other selector mechanisms are used to adjust the electrometer readout to display the activity of the selected radionuclide directly in MBq or kBq units. Because ionization chambers have no inherent ability for energy discrimination, they cannot be used to select different γ-ray energies for measurement, as is possible with detectors having pulse-height analysis capabilities. One approach that is used to distinguish low-energy versus high-energy γ-ray emitters (e.g., ^{99m}Tc vs. ^{99}Mo) is to measure the sample with and without a few millimeters of lead shielding around the source. Effectively, only the activity of the high-energy emitter is recorded with the shielding in place, whereas the total activity of both emitters is recorded with the shielding absent. This technique can be used to detect tens of kBq quantities of ^{99}Mo in the presence of tens or even hundreds of MBq of ^{99m}Tc.

As with the NaI(Tl) well counter, dose calibrators are subject to sample volume effects (see Section A.3). These effects should be investigated experimentally when a new dose calibrator is acquired, so that correction factors can be applied in its use, if necessary. For example, a quantity of activity can be measured in a very small volume (e.g., 0.1 mL in a 1-mL syringe), and that activity can be diluted progressively to larger volumes in larger syringes and then in beakers, and so forth to determine the amount by which the instrument reading changes with sample volume.

Another parameter worth evaluating is linearity of response versus sample activity. This may be determined conveniently by recording the reading for a ^{99m}Tc source of moderately high activity (e.g., 1 GBq, or whatever the approximate maximum amount of activity the dose calibrator will be used to assay), then recording the readings during a 24- to 48-hour period (4-8 half-lives) to determine whether they follow the expected decay curve for ^{99m}Tc. Deviations from the expected decay curve may indicate instrument electronic non-linearities requiring adjustment or correction of readings. In applying this technique, it is necessary to correct for ^{99}Mo contamination using the shielding technique described earlier, especially after several ^{99m}Tc half-lives have elapsed.

2. Gas Flow Counters

Gas-filled detectors also are used in gas flow counters, primarily for measurement of β-emitting activity. The detector in these systems usually can be operated in either proportional counter or Geiger-Müller mode. The most frequent application for these systems in nuclear medicine is for monitoring the effluent from gas chromatographs. Gases labeled with ^{3}H or ^{14}C in helium carrier gas from the chromatograph are passed through a combustion furnace to convert them to ^{3}H$_2$O or ^{14}CO$_2$, which then is allowed to flow through the counter gas volume itself with the counting gas (usually 90% He plus 10% methane). This permits a time-course analysis of the outflow from the chromatograph. These systems have good geometric and intrinsic detection efficiencies for low-energy β emitters, such as ^{3}H and ^{14}C; however, their intrinsic efficiency for γ-ray detection is only approximately 1%. Gases labeled with β emitters are therefore analyzed using NaI(Tl) detectors.

E. SEMICONDUCTOR DETECTOR SYSTEMS

1. System Components

Semiconductor detectors [germanium (Ge) and silicon (Si)] (see Chapter 7, Section B) created revolutionary advances in nuclear physics, nuclear chemistry, radiation chemistry, nondestructive materials analysis (e.g., x-ray fluorescence and neutron activation),

and other fields. To date, however, they have had limited effect on nuclear medicine. Their disadvantages of small size and high cost outweigh their advantage of superior energy resolution in comparison with other detection systems [e.g., NaI(Tl)] for general-purpose applications; however, the energy resolution of semiconductor detectors allows the separation of γ rays differing in energy by only a few keV as opposed to 20-80 keV with NaI(Tl) (Fig. 12-18; see also Fig. 10-14). Therefore in applications in which energy resolution is the critical factor and the relatively small size of the semiconductor detector is not completely restrictive, Ge or Si detectors are the system of choice.

Semiconductor detectors are used extensively as charged-particle and γ-ray spectrometers in physics. Their principal application in nuclear medicine is for assessment of radionuclide purity. Si has a lower atomic number and density than Ge and therefore a lower intrinsic detection efficiency for γ rays with energies $\gtrsim 40$ keV (see Chapter 11, Section A.3). Thus Si detectors are used primarily for detection of low-energy x rays and Ge, cadmium telluride (CdTe), and cadmium zinc telluride (CZT) are used for γ rays.

The basic configuration of a semiconductor system for in vitro analysis is shown in Figure 7-12. Except for a special low-noise high-voltage supply, preamplifier, and amplifier, the system components are the same as those of NaI(Tl) counting systems. Usually an MCA is employed rather than an SCA with semiconductor detectors because the detectors most commonly are used to resolve complex spectra of multiple emissions and multiple radionuclides (see Fig. 12-18).

The superior energy resolution of semiconductor detectors may result in a significant advantage in sensitivity [i.e., minimum detectable activity (MDA)] (see Chapter 9, Section D.5) in comparison with NaI(Tl) detectors for some applications. MDA depends on the ratio $S/\sqrt{B}$, in which S is the net sample counting rate and B is the background counting rate. Because the energy resolution of a semiconductor detector is 20 to 80 times better than NaI(Tl), a photopeak window 20 to 80 times narrower can be used, resulting in typically 20 to 80 times smaller background counting rate. Considering background alone, then, the MDA for a semiconductor detector could be a factor $\sqrt{20}$ to $\sqrt{80}$ smaller than a NaI(Tl) detector of comparable size. This advantage is partially offset by the larger available detector sizes with NaI(Tl) and,

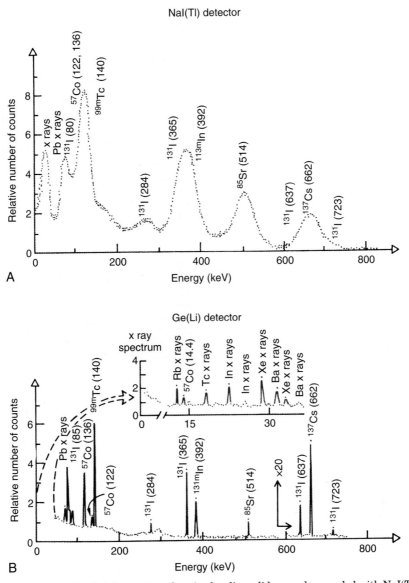

FIGURE 12-18 Comparative pulse-height spectra of a mixed radionuclide sample recorded with NaI(Tl) (*A*) and Ge(Li) (*B*) detectors. Because of its superior energy resolution, the Ge(Li) detector clearly resolves multiple γ rays and x rays of similar energies that appear as single peaks with NaI(Tl).

above approximately 200 keV, by the greater intrinsic photopeak efficiency of NaI(Tl) for comparable detector thicknesses (see Chapter 11, Section A.3); however, for lower-energy γ rays, measured in a configuration having a high geometric efficiency (e.g., sample placed directly against the detector), there is usually an advantage in MDA favoring the semiconductor detector. For higher-energy γ rays, CdTe or CZT semiconductors provide the advantage of both excellent energy resolution and good photopeak efficiency, although the cost per unit detector volume is much higher

than NaI(Tl), limiting them to situations in which small detector sizes are acceptable.

2. Applications

The major in vitro applications of semiconductor detectors in nuclear medicine have been for tracer studies employing many radionuclides simultaneously and for the assay of radionuclidic purity of radiopharmaceuticals. In both of these applications the superior energy resolution of semiconductor detectors, illustrated by Figure 12-18, offers a distinct advantage. The energy resolution of the Ge detector allows unequivocal identification of radionuclides, whereas the NaI(Tl) spectrum is ambiguous. Another application of semiconductor detectors is for analysis of samples in neutron activation analysis.

F. IN VIVO COUNTING SYSTEMS

In vivo counting systems are used to measure radioactive concentrations in patients and, occasionally, in experimental animals. Systems designed to monitor radioactivity in single organs or in localized parts of the body are called *probe systems*. For example, *single-probe* systems, employing only one detector, are used for measuring thyroidal uptake of radioactive iodine and for sentinel node detection in breast cancer. *Multiprobe* systems, although less common, have been used for renal function studies, for lung clearance studies, for obtaining washout curves from the brain, and so forth. Probe systems provide some degree of measurement localization but without the detail of imaging techniques discussed in Chapters 13-19. Because the radiation must in general pass through several centimeters of soft tissue to reach the detector, most in vivo counting systems are designed to detect γ rays.

1. NaI(Tl) Probe Systems

The simplest probe system consists of a collimated NaI(Tl) detector mounted on a stationary or mobile stand that can be oriented and positioned over an area of interest on the patient (Fig. 12-19). Such detectors are commonly used in diagnostic tests for thyroid disease. The detector is connected to the usual NaI(Tl) electronics, including an SCA for energy selection and a digital counter or computer that records the number of counts per second. A typical probe system employs a 5-cm diameter × 5-cm thick NaI(Tl) crystal, with a cylindrical or conically shaped collimator, 15-25 cm long, in front of the detector.

When calibrating a probe system for in vivo measurements, it is important to account for the effects of attenuation and scatter on the recorded counting rate (see Chapter 11, Section A.5). Usually, the depth of the source distribution within the patient is not known accurately. Because the linear attenuation coefficient for soft tissue is in the range $\mu_l = 0.1$ to 0.2 cm^{-1} for most γ-ray energies in nuclear medicine, a 1- to 2-cm difference in source depth can result in a 10% to 40% difference in recorded counting rate. The

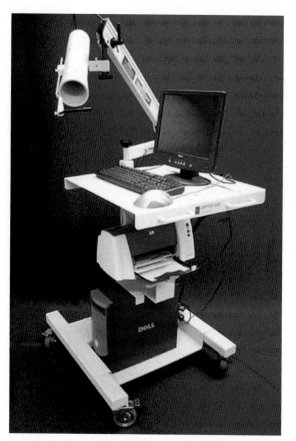

FIGURE 12-19   Typical NaI(Tl) probe system for measuring thyroid uptake of radioactive iodine. (*Courtesy Capintec, Inc., Ramsey, NJ.*)

intensity of scattered radiation is another important variable. For example, a source lying outside the direct field-of-view of the collimator can contribute to the recorded counting rate by Compton scattering in the tissues surrounding the source distribution. To minimize the contribution from scattered radiation, measurements usually are made with the SCA window set on the photopeak of the γ-ray emission to be counted. Even this is not completely effective for eliminating all the variable effects of scattered radiation on the measurement, however, especially when low-energy γ rays are counted (see Figs. 10-6 and 10-10).

2. Miniature γ-Ray and β Probes for Surgical Use

Miniature, compact γ-ray probes are designed for use in conjunction with surgical procedures, primarily in cancer applications. The most important application is the detection of the *sentinel lymph node* in patients with breast cancer and melanoma. The sentinel node is the most likely initial site for metastatic spread of the cancer; thus biopsy of the sentinel node is important for patient management. The sentinel node is identified by direct injection of a [99mTc]-labeled *colloid* (a suspension of fine particles labeled with [99mTc]) into the tumor. This colloid is trapped in the first lymph node draining the tumor. During surgery, the γ-ray probe is used to identify the sentinel node, from which a biopsy sample is taken and sent to a pathology laboratory for analysis.

The second broad class of applications is in radioguided surgery. Here, tumor-seeking radiopharmaceuticals are injected into the patient. The radiopharmaceutical agent is designed to target and bind to cancer cells with high selectivity. After waiting for an appropriate length of time for selective uptake of the radiopharmaceutical agent into the tumor, the patient goes to surgery and the surgeon uses the γ-ray probe to assist in locating and removing the cancerous tissue, while sparing as much healthy tissue as possible. This procedure has been applied in parathyroid surgery and colorectal cancer, and in detecting lymph node involvement for a range of other cancers.

The requirements for γ-ray probes for intraoperative use are that they have high efficiency (so that radiolabeled tissue can be found quickly in the surgical environment), that they be lightweight and easy to use, and that they pose no hazard to the patient. The probe of choice for high-energy γ emitters, such as ^{111}In, ^{131}I, and ^{18}F, is a scintillation detector. A typical probe consists of a 5-mm diameter × 10-mm high cesium iodide [CsI(Tl)] scintillator crystal, coupled to a Si photodiode. The Si photodiode is a light-sensing semiconductor detector that replaces the PM tube found in conventional scintillation detectors and converts the scintillation light into an electrical signal (see Chapter 7, Section C.3). It is preferred in this application because of its compact size and low weight compared with a PM tube. CsI(Tl) is used in place of NaI(Tl) as the scintillator because its emission wavelengths are better matched to the spectral response of the Si photodiode. For lower-energy γ emitters, such as [99mTc], a semiconductor detector made from CZT or CdTe (see Chapter 7, Section B) that directly converts the γ rays to electric charge is typically used. This is an ideal application for these semiconductor detectors, because the required detector area is small. CZT and CdTe are better for this application than Si or Ge because they have higher stopping power for γ rays and can be operated at room temperature. A small collimator is used in front of the probe to provide directionality.

Figure 12-20 shows the components of a typical γ-ray probe system. The output signals from the probes are amplified and sent to an MCA. Discriminator levels are set automatically for each different radionuclide. The counting rate is presented on a digital display. Many systems also have an audible output proportional to the counting rate. The whole unit is battery powered and can run for many hours on a single charge, eliminating the need for power cords. Wireless probes also are

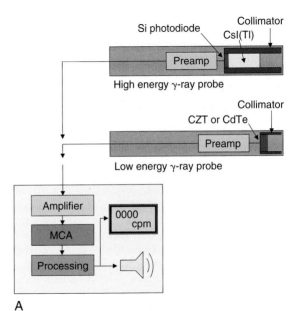

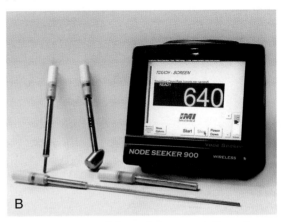

FIGURE 12-20 *A,* Schematic representation of γ-ray probes for intraoperative use. *B,* Four different wireless gamma probes shown with control unit. The geometry of the probes are tailored to suit specific clinical applications. (*Figure B courtesy IntraMedical Imaging, Los Angeles, CA.*)

available, further facilitating their use in the surgical environment. References 2 and 3 provide a detailed review of counting probe systems for intraoperative use.

Probes for β-particle detection also have been developed. These are typically used in conjunction with tumor-seeking, positron-emitting radiopharmaceuticals to aid in locating tumors during surgery or to map tumor margins during surgical resection, helping to ensure that the tumor is completely removed while sparing normal tissue. They differ from the γ-ray probes described previously in that these probes directly detect β+ particles (positrons) rather than the 511-keV annihilation photons. Because of the short range of positrons in tissue (see Chapter 6, Section B.2), they can only detect radioactivity that is very superficial (1-2 mm) at the surgical site, but have the advantage over γ-ray probes of being very insensitive to radioactivity that may be contained in adjacent tissues and organs and that could interfere with the local measurement.

3. Whole-Body Counters

Another class of in vivo measurement systems are *whole-body counters,* which are designed to measure the total amount of radioactivity in the body, with no attempt at localization of the activity distribution. Many (but not all) of these systems employ NaI(Tl) detectors. They are used for studying retention, turnover, and clearance rates with nuclides such as ^{60}Co and ^{57}Co (labeled vitamin B$_{12}$), ^{24}Na, ^{42}K, ^{47}Ca, and ^{59}Fe. Most of these radionuclides emit high-energy γ rays, and several have quite long half-lives. Thus it is important that a whole-body counter have good detection efficiency, so that very small amounts of activity (≤ 50 kBq) can be detected and measured accurately.

Another application for whole-body counting is the measurement of naturally occurring ^{40}K, which can be used to estimate total-body potassium content. This is another high-energy γ emitter present in very small quantities, requiring good detection efficiency for accurate measurement. Whole-body counters also are used for detecting and monitoring possible accidental ingestion of radioactive materials.

Most whole-body counters employ relatively large NaI(Tl) detectors, 15 to 30 cm in diameter × 5 to 10 cm thick, to obtain good geometric efficiency as well as good intrinsic efficiency for high-energy γ rays. Several such detectors may be employed. Also the "counting chamber" is well shielded with lead, concrete, steel, and other materials to obtain minimal background levels, thus ensuring minimum statistical error caused by background counting rates (see Chapter 9, Section D.4). Shielding materials are selected carefully for minimum contamination with background radioactivity.

REFERENCES

A detailed reference on in vitro counting systems is the following:

1. L'Annunciata MF: *Handbook of Radioactivity Analysis,* ed 2, San Diego, 2003, Academic Press.

The design and application of miniature γ probes for surgical use are reviewed in detail in the following:

2. Hoffman EJ, Tornai MP, Janacek M, et al: Intraoperative and imaging probes. *Eur J Nucl Med* 26:913-935, 1999.
3. Povoski SP, Neff RL, Mojzisik CM, et al: A comprehensive overview of radioguided surgery using gamma detection probe technology. *World J Surg Oncol* 7:11, 2009.

chapter

13

The Gamma Camera: Basic Principles

Radionuclide imaging is the most important application of radioactivity in nuclear medicine. Radionuclide imaging laboratories are found in almost every hospital, performing hundreds and even thousands of imaging procedures per month in larger institutions.

In this chapter, we discuss briefly some general aspects of radionuclide imaging, and we describe the basic principles of the most widely used imaging device, the *gamma camera,* also known as the *Anger scintillation camera,* named after its inventor, Hal Anger (see Chapter 1, Section C and Fig. 1-3). The performance characteristics of this instrument are discussed in Chapter 14. The use of the gamma camera for tomographic imaging is described in Chapter 17.

A. GENERAL CONCEPTS OF RADIONUCLIDE IMAGING

The purpose of radionuclide imaging is to obtain a picture of the distribution of a radioactively labeled substance within the body after it has been administered (e.g., by intravenous injection) to a patient. This is accomplished by recording the emissions from the radioactivity with external radiation detectors placed at different locations outside the patient. The preferred emissions for this application are γ rays in the approximate energy range of 80 to 500 keV (or annihilation photons, 511 keV). Gamma rays of these energies are sufficiently penetrating in body tissues to be detected from deep-lying organs, can be stopped efficiently by dense scintillators, and are shielded adequately with reasonable thicknesses of lead (see Fig. 6-17—soft tissue has attenuation properties similar to water). Alpha particles and electrons (β particles, Auger and conversion electrons) are of

little use for imaging because they cannot penetrate more than a few millimeters of tissue. Therefore they cannot escape from within the body and reach an external radiation detector, except from very superficial tissues. Bremsstrahlung (see Fig. 6-1) generated by electron emissions is more penetrating, but the intensity of this radiation generally is very weak.

Imaging system detectors must therefore have good detection efficiency for γ rays. It is also desirable that they have energy discrimination capability, so that γ rays that have lost positional information by Compton scattering within the body can be rejected based on their reduced energy (see Chapter 6, Section C.3). A sodium iodide [NaI(Tl)] scintillation detector (see Chapter 7, Section C) provides both of these features at a reasonable cost; for this reason it is currently the detector of choice for radionuclides with γ-ray emissions in the range of 80-300 keV.

The first attempts at radionuclide "imaging" occurred in the late 1940s. An array of radiation detectors was positioned on a matrix of measuring points around the head. Alternatively, a single detector was positioned manually for separate measurements at each point in the matrix. These devices were tedious to use and provided only very crude mappings of the distribution of radioactivity in the head (e.g., left-side versus right-side asymmetries).

A significant advance occurred in the early 1950s with the introduction of the rectilinear scanner by Benedict Cassen (see Fig. 1-2). With this instrument, the detector was scanned mechanically in a raster-like pattern over the area of interest. The image was a pattern of dots imprinted on a sheet of paper by a mechanical printer that followed the scanning motion of the detector, printing the dots as the γ rays were detected.

The principal disadvantage of the rectilinear scanner was its long imaging time (typically many minutes) because the image was formed by sequential measurements at many individual points within the imaged area. The first gamma-ray "camera" capable of recording at all points in the image at one time was described by Hal Anger in 1953. He used a pinhole aperture in a sheet of lead to project a γ-ray image of the radionuclide distribution onto a radiation detector composed of a NaI(Tl) screen and a sheet of x-ray film. The film was exposed by the scintillation light flashes generated by the γ rays in the NaI(Tl) screen. Unfortunately, this detection system (especially the film component) was so inefficient that hour-long exposures and therapeutic levels of administered radioactivity were needed to obtain satisfactory images.

In the late 1950s, Anger replaced the film-screen combination with a single, large-area, NaI(Tl) crystal and a photomultiplier (PM) tube assembly to greatly increase the detection efficiency of his "camera" concept. This instrument, the *Anger scintillation camera,*[1] or *gamma camera,* has been substantially refined and improved since that time. Although other ideas for nuclear-imaging instruments have come along since then, none, with the exception of modern positron emission tomography systems (see Chapter 18), has matched the gamma camera for a balance of image quality, detection efficiency, and ease of use in a hospital environment. The gamma camera has thus become the most widely used nuclear-imaging instrument for clinical applications.

B. BASIC PRINCIPLES OF THE GAMMA CAMERA

1. System Components

Figure 13-1 illustrates the basic principles of image formation with the gamma camera. The major components are a collimator, a large-area NaI(Tl) scintillation crystal, a light guide, and an array of PM tubes. Two features that differ from the conventional NaI(Tl) counting detectors described in Chapter 12 are crucial to image formation. The first is that an imaging *collimator* is used to define the direction of the detected γ rays. The collimator most commonly consists of a lead plate containing a large number of holes. By controlling which γ rays are accepted, the collimator forms a projected

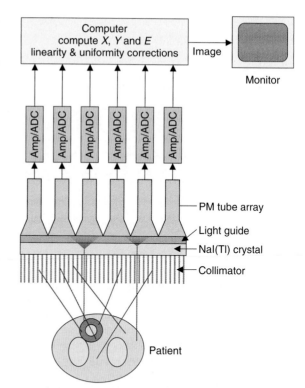

FIGURE 13-1 Basic principles and components of a modern gamma camera. The outputs of each photomultiplier (PM) tube are amplified and digitized using an analog-to-digital converter (ADC). The X-Y locations for each gamma ray that interacts in the NaI(Tl) crystal are computed from the digitized signals. The energy deposited by the gamma ray, *E*, which is proportional to the total measured pulse amplitude, also is computed by summing the individual PM tube signals. If *E* falls within the selected energy window, the event is accepted and placed at the appropriate X-Y location in the image.

image of the γ-ray distribution on the surface of the NaI(Tl) crystal (see Section B.3). The second is that the NaI(Tl) crystal is viewed by an array of PM tubes, rather than a single PM tube. Signals from the PM tubes are fed to electronic or digital *position logic circuits,* which determine the X-Y location of each scintillation event, as it occurs, by using the weighted average of the PM tube signals (see Section B.2).

Individual events also are analyzed for energy, *E*, by summing the signals from all PM tubes. When the pulse amplitude of an event falls within the selected energy window, it is accepted and the *X* and *Y* values are binned into a discrete two-dimensional array of image elements, or *pixels.* An image is formed from a histogram of the number of events at each possible X-Y location. Large numbers of events are required to form an interpretable image because each pixel must have a sufficient number of counts to achieve an acceptable signal-to-noise level. Because images often are formed in 64- × 64-pixel or 128- × 128-pixel arrays, the counting requirements are some 10^3 to 10^4 times higher than for a simple counting detector.

Images are displayed on a computer monitor, where image brightness and contrast may be manipulated and different color tables may be employed. More sophisticated digital image processing is discussed in Chapter 20.

Most modern gamma cameras are completely digital, in the sense that the output of each PM tube is directly digitized by an analog-to-digital converter (ADC). The calculation of X-Y position and pulse-height are performed in software based on the digitized PM tube signals, and errors in energy and positioning caused by noise and pulse distortions caused by the analog positioning circuitry are eliminated. This approach also permits improved handling of pulse pile-up at high counting rates, as described in Section B.2.

The gamma camera can be used for *static* imaging studies, in which an image of an unchanging radionuclide distribution can be recorded over an extended imaging time (e.g., minutes). Single contiguous images of the whole body can be obtained by scanning the gamma camera across the entire length of the patient. This can be achieved by moving either the bed or the gamma camera while adjusting the event positioning computation to account for this movement. Clinically important whole-body studies include bone scans of the skeleton, and the localization of tumors or their metastases in the body.

The gamma camera also can be used for *dynamic* imaging studies, in which changes in the radionuclide distribution can be observed, as rapidly as several images per second. This allows physiologic information to be obtained, such as the rate of tracer uptake or clearance from an organ of interest. Images also can be synchronized to electrocardiogram signals, permitting images of the heart in different phases of the cardiac cycle to be formed. These *gated* images can provide important information on cardiac function.

2. Detector System and Electronics

The gamma camera employs a single, large-area, rectangular NaI(Tl) detector crystal, usually 6- to 12.5-mm thick with sizes of up to 60 × 40 cm. Round crystals of 25 to 50 cm in diameter were used in many older systems. The NaI(Tl) crystal is surrounded by a highly

reflective material such as TiO_2 to maximize light output and hermetically sealed inside a thin aluminum casing to protect it from moisture. An optical glass window on the back surface of the casing permits the scintillation light to reach the PM tubes. A cross section of a typical gamma camera crystal assembly is shown in Figure 13-2. The choice of thickness of the NaI(Tl) crystal is a trade-off between its detection efficiency (which increases with increasing thickness) and, as shown in Chapter 14, Section A.1, its intrinsic spatial resolution (which deteriorates with increasing thickness). Most general-purpose gamma cameras have crystal thicknesses of approximately 9.5 mm. For lower-energy γ emitters, such as ^{99m}Tc and ^{201}Tl, however, detection efficiency is adequate even with 6-mm-thick detector crystals.

An array of PM tubes is coupled optically to the back face of the crystal with a silicone-based adhesive or grease. Round PM tubes are arranged in a hexagonal pattern to maximize the area of the NaI(Tl) crystal that is covered. Some cameras use hexagonal (or rarely, square) cross-section PM tubes for better coverage of the NaI(Tl) crystal. Typical PM tube sizes are 5 cm in diameter. Most modern cameras employ between 30 and 100 PM tubes. Figure 13-3 shows a photograph of a 30-tube model. The PM tubes are encased in a thin magnetic shield (Chapter 7, Section C.2) to prevent changes in the gain caused by changes in the orientation of the gamma camera relative to the earth's magnetic field. The ultrasensitivity of PM tubes to magnetic fields also makes gamma cameras susceptible to the stray fields from magnetic resonance imaging systems.

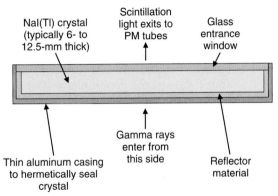

FIGURE 13-2 Schematic cross-section of a NaI(Tl) crystal assembly for a gamma camera.

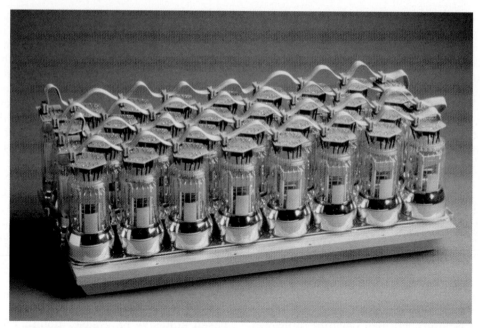

FIGURE 13-3 A rectangular gamma camera detector with the cover removed showing the photomultiplier (PM) tubes mounted on the NaI(Tl) crystal. In this example, the gamma camera detector measures 50×15 cm and is read out by 30 PM tubes 5 cm in diameter. This is a digital camera in which each of the PM tube outputs is individually digitized. (*Courtesy Dr. Joel Karp, University of Pennsylvania, Philadelphia, PA.*)

Many manufacturers employ plastic light guides between the detector crystal and PM tubes, whereas others couple the PM tubes directly to the crystal. The functions of the light guide are to increase the light collection efficiency, by channeling scintillation light away from the gaps between the PM tubes, and to improve the uniformity of light collection as a function of position. The latter effect is achieved by painting or etching a carefully designed pattern onto the entrance face of the light guide. The use of the PM tubes with hexagonal or square cross-sections that can be tiled without gaps on the NaI(Tl) crystal may in some cases allow elimination of the light guide, assuming there is sufficient spreading of the scintillation light in the glass entrance window of the PM tube for accurate positioning.

The detector crystal and PM tube array are enclosed in a light-tight, lead-lined protective housing. In most modern cameras, most of the electronics (such as preamplifiers, pulse-height analyzers, automatic gain control, pulse pile-up rejection circuits and ADCs) are mounted directly on the individual PM tube bases within the detector housing to minimize signal distortions that can occur in long cable runs between the detector head and control console.

The amount of light detected by a particular PM tube is inversely related to the lateral distance between the interaction site and the center of that PM tube. This is illustrated in one dimension in Figure 13-4. Ideally, the relationship between signal amplitude and location with respect to the center of a PM tube would be linear. This would enable the position of an event to be determined by taking a weighted average or centroid of the PM tube signals using the simple relationships shown in Figure 13-4. In practice, however, the response is more complex, with a plateau directly beneath the PM tube (because the PM tube is not a "point" detector) and long, flat tails caused by reflections of light from the back and side surfaces of the NaI(Tl) crystal. Therefore a calibration for spatial nonlinearity is required (see Chapter 14, Section B).

Figure 13-5A shows a schematic drawing for an eight-PM tube version of the gamma camera and is used to illustrate the principles of scintillation event localization in an *analog* detector. The position is determined by splitting the signal from each PM tube onto four output lines, whose signals are denoted X^+, X^-, Y^+, and Y^- (Fig. 13-5B). The fraction of the PM tube current that goes to each output line is determined by the value of the resistors

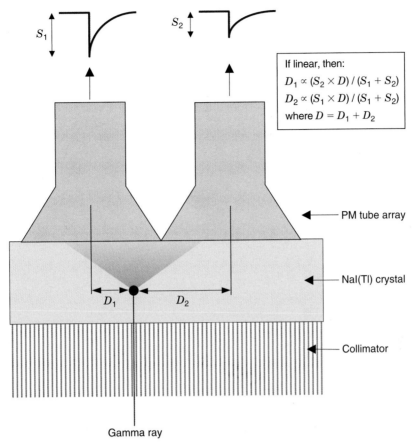

FIGURE 13-4 Illustration of light sharing between photomultiplier (PM) tubes. The PM-tube signal, *S*, is inversely related to the distance of the interaction site, *D*, from the center of the PM tube. Equations for a linear relationship are shown.

(R) that are used. By Ohm's law, this current is proportional to 1/R. A separate circuit sums the outputs of all the PM tubes to form the Z-signal. The Z-signal is proportional to the total amount of light produced by a scintillation event in the crystal, and therefore the total energy deposited by the gamma ray, and is used for pulse-height analysis.

The X^+, X^-, Y^+, and Y^- signals are combined to obtain *X-position* and *Y-position* signals. The X-position of the scintillation event is given by the difference in the X^+ and X^- signals, divided by the total X signal ($X^+ + X^-$)

$$X = (X^+ - X^-)/(X^+ + X^-) \quad (13\text{-}1)$$

Similarly, for the Y-position

$$Y = (Y^+ - Y^-)/(Y^+ + Y^-) \quad (13\text{-}2)$$

The X- and Y-position signals are normalized to the total X and Y signals, so that the calculated position of interaction does not depend on the pulse height. Note that the possible range of X and Y values is from −1 to +1. The resistor values shown in Figure 13-5C were chosen such that the calculated X- and Y-position signals vary linearly with distance in the X and Y directions. In a perfect gamma camera, measured (X, Y) values would change linearly from (−1, −1) in the bottom left-hand corner to (+1, +1) at the top right-hand corner of the camera face. The X and Y values can be scaled by the detector size to determine the absolute position of an event on the gamma camera face.

However, Equations 13-1 and 13-2 do not give a perfect mapping of source position because, as was discussed previously, the PM tubes signal does not actually vary linearly with interaction position. This gives rise to a "pincushion" artifact, which is illustrated in Figure 14-9. There are also effects caused by nonuniformities in the crystal, light reflections at the edge of the crystal, and nonuniform response across the face of the PM tubes that can cause further nonlinearities in

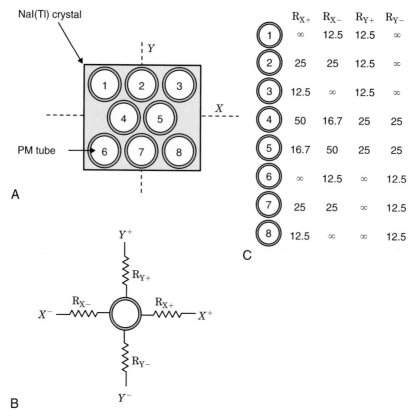

FIGURE 13-5 Illustration of analog positioning in a gamma camera. *A,* Schematic representation of an eight-photomultiplier (PM) tube camera. *B,* Signals from individual PM tubes are split using resistors onto four output lines, designated X^+, X^-, Y^+, and Y^-. *C,* Representative resistor values (in kΩ) for the eight PM tubes. Resistor values are chosen such that the X and Y positions computed from Equations 13-1 and 13-2 vary linearly with interaction position in the detector, ranging from a value of -1 in the bottom left hand corner to $+1$ in the top right corner.

position determination. These effects and correction techniques for them are discussed in Chapter 14, Section B.

In *digital* cameras, the output signal from each PM tube is digitized and the event position is calculated in software. Often, this is simply analogous to the resistor readout described earlier; the inverse of the resistor values are used as weighting factors for the individual PM tube signals, and Equations 13-1 and 13-2 are used to determine the X and Y values. However, digital cameras also can use more sophisticated algorithms that incorporate information regarding the nonlinearity of PM tube response with position into the weighting factors to provide better positioning accuracy.

A commonly used tactic that is employed in both digital and analog cameras to improve the positioning accuracy is to include in the position calculation only PM tubes with signals above a certain threshold. This has two important benefits. By using the signal only from those PM tubes that produce a significant pulse amplitude, the noise from the PM tubes that produce negligible signal amplitude (and that therefore contribute little to position information) is not included in the position calculation. Second, with signal thresholding, only a small number of PM tubes surrounding the interaction location are used for position determination. This allows a gamma camera to detect multiple events simultaneously when they occur in different portions of the gamma camera and their light cones (the projection of the scintillation light on the PM tube array) do not significantly overlap. This improves the counting rate performance of the gamma camera, reducing dead time losses.

Energy selection is important for imaging because it provides a means to discriminate against γ rays that been scattered within the body and therefore lost their positional information. By choosing a relatively narrow pulse-height analyzer window that is centered on the photopeak, only γ rays that undergo no scatter or small-angle scatter will be accepted.

Two different methods can be used to select the photopeak events. The first approach uses simple energy discrimination on the Z-signal. However, because of nonuniformities in the NaI(Tl) crystal (small variations in light production with position), in light collection efficiency and in PM tube gains, the position of the photopeak varies somewhat from position to position in the detector. If a single discriminator level is applied across the whole detector, the window must be widened to accommodate the fluctuations in photopeak position, thus accepting more scatter (Fig. 13-6, *top*).

In the second method, suitable only for digital cameras, the photopeak positions and appropriate discriminator level settings are computed and stored for many different locations across the detector face (Fig. 13-6, *bottom*). When an event is detected, the X, Y values are calculated based on Equations 13-1 and 13-2, and a look-up table is used to find the appropriate discriminator levels for that location. If the event amplitude Z falls within the pulse-height analyzer settings, the event is accepted.

A modern gamma camera has an energy resolution of 9% to 10% at 140 keV (^{99m}Tc). Typically, the *energy window* (the difference between upper-level and lower-level discriminators) is set to 14%, or 20 keV, centered around 140 keV. The gamma camera software adjusts the discriminator levels for radionuclides other than ^{99m}Tc based on the relationship (approximately linear over a small energy range) between the γ-ray energy deposited and the light output of NaI(Tl) (see Fig. 10-11).

3. Collimators

To obtain an image with a gamma camera, it is necessary to project γ rays from the source distribution onto the camera detector. Gamma rays cannot be focused; thus a "lens" principle similar to that used in photography cannot be applied. Therefore most practical γ-ray imaging systems employ the principle of *absorptive collimation* for image formation.* An absorptive collimator projects an image of the source distribution onto the detector by allowing only those γ rays traveling along certain directions to reach the detector.

---

*An important exception is imaging of the two 511-keV annihilation photons from positron-emitting radionuclides, in which electronic coincidence detection can be used to replace the collimator as described in Chapter 18, Section A.1.

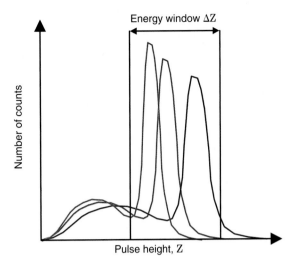

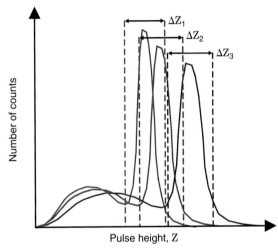

FIGURE 13-6 The pulse-height response is not uniform across the whole face of the gamma camera, leading to variation in the pulse height of photopeak events. Using a global energy window requires a wide window that leads to the inclusion of larger amounts of scatter and also results in nonuniform efficiency (*top*). Using local energy thresholds, in which the energy window is tailored to the event location, improves scatter rejection and uniformity (*bottom*).

Gamma rays not traveling in the proper direction are absorbed by the collimator before they reach the detector. This "projection by absorption" technique is an inherently inefficient method for using radiation because most of the potentially useful radiation traveling toward the detector actually is stopped by the absorptive walls between the collimator holes. This is one of the underlying reasons for the relatively poor quality of radionuclide images (e.g., as compared to radiographic images), as discussed in Chapter 15.

Four basic collimator types are used with the gamma camera: pinhole, parallel-hole, diverging, and converging. The different types of collimator are introduced subsequently. Their effects on the spatial resolution and sensitivity of the gamma camera are discussed in Chapter 14 , Sections C and D.

A *pinhole* collimator (Fig. 13-7A) consists of a small pinhole aperture in a piece of lead, tungsten, platinum, or other heavy metal absorber. The pinhole aperture is located at the end of a lead cone, typically 20 to 25 cm from the detector. The size of the pinhole can be varied by using removable inserts and is typically a few millimeters in diameter.

The imaging principle of a pinhole collimator is the same as that employed with inexpensive "box cameras." Gamma rays passing through the pinhole project an inverted image of the source distribution onto the detector crystal. The image is magnified when the distance b from the source to the pinhole is smaller than the collimator cone length f; it is minified when the source distribution is farther away. The image size I and object (source) size O are related according to

$$I/O = f/b \qquad (13\text{-}3)$$

The size of the imaged area also changes with distance from the pinhole collimator. If the detector diameter is D and the magnification (or minification) factor is I/O (Equation 13-3), the diameter of the image area projected onto the detector, D', is

$$D' = \frac{D}{I/O} \qquad (13\text{-}4)$$

Thus a large magnification factor, obtained at close source-to-collimator distances, results in a small imaged area.

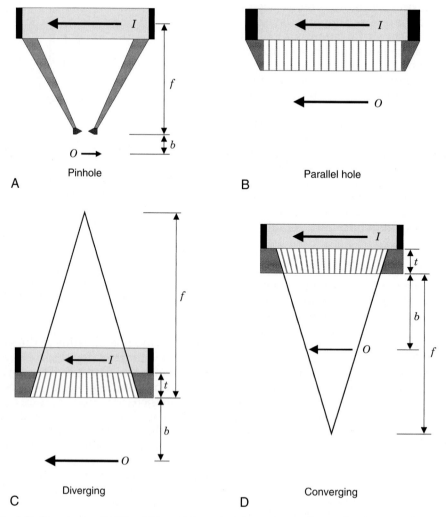

A   Pinhole   B   Parallel hole

C   Diverging   D   Converging

FIGURE 13-7 *A-D,* Four types of collimators used to project "γ-ray images" onto the detector of a gamma camera. *O,* Radioactive object; *I,* its projected image.

Image size changes with object-to-pinhole distance b. Therefore the pinhole collimator provides a somewhat distorted image of three-dimensional objects because source planes at different distances from the collimator are magnified by different amounts. Pinhole collimators are used primarily for magnification imaging of small organs (e.g., thyroid and heart) and for small-animal imaging.

Another type of pinhole collimator, the *multi-pinhole collimator,* has an array of multiple pinholes, typically seven, arranged in a hexagonal pattern. This collimator was employed in the past for tomographic imaging. This type of tomography now is seldom used clinically; however, multi-pinhole approaches are being widely employed for some small-animal imaging applications.

The *parallel-hole collimator* (Fig. 13-7B) is the "workhorse" collimator in most imaging laboratories. Parallel holes are drilled or cast in lead or are shaped from lead foils. The lead walls between the holes are called collimator *septa.* Septal thickness is chosen to prevent γ rays from crossing from one hole to the next (see Chapter 14, Section C.2). A magnified view of a parallel-hole collimator is shown in Figure 13-8. The parallel-hole collimator projects a γ-ray image of the same size as the source distribution onto the detector. A variation of the parallel-hole collimator is the slant-hole collimator, in which all of the holes are parallel to each other but angled, typically by approximately 25 degrees, from the perpendicular direction. This type of collimator has characteristics that are similar to those of the parallel-hole type. Because it views the source distribution from an angle rather than

directly "head-on," it can be positioned closer to the patient for better image detail in some imaging studies (e.g., left anterior oblique cardiac views).

A *diverging collimator* (Fig. 13-7C) has holes that diverge from the detector face. The holes diverge from a point typically 40-50 cm behind the collimator, projecting a *minified, noninverted* image of the source distribution onto the detector. The degree of minification depends on the distance f from the front of the collimator to the convergence point, the distance b from the front of the collimator to the object (source), and the collimator thickness t

$$I/O = (f - t)/(f + b) \qquad (13\text{-}5)$$

where I and O are image and object size, respectively. The useful image area becomes larger as the image becomes more minified (Equation 13-4).

EXAMPLE 13-1

What is the minification factor for a diverging collimator 5-cm thick, with $f = 45$ cm, and a source distribution 15 cm from the collimator? If the detector diameter is 30 cm, what is the imaged area at this distance?

Answer

From Equation 13-5,

$$I/O(\text{minification factor})$$
$$= (45 - 5)/(45 + 15) = 0.67$$

From Equation 13-4,

$$\text{Diameter of imaged area}$$
$$= 30 \text{ cm}/0.67 = 44.8 \text{ cm}$$

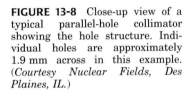

FIGURE 13-8 Close-up view of a typical parallel-hole collimator showing the hole structure. Individual holes are approximately 1.9 mm across in this example. (*Courtesy Nuclear Fields, Des Plaines, IL.*)

As shown by Example 13-1, a typical diverging collimator decreases the size of the image on the detector and increases the diameter of the imaged area, by approximately one-third as compared with a parallel-hole collimator. As with the pinhole collimator, image size changes with distance; thus there is a certain amount of image distortion. Diverging collimators are used primarily on cameras with smaller detectors to permit imaging of large organs such as the liver or lungs on a single view.

A *converging collimator* (Fig. 13-7D) has holes that converge to a point 40-50 cm in front of the collimator. For objects between the collimator face and the convergence point, the converging collimator projects a *magnified,* noninverted image of the source distribution. Image size I and object size O are related according to

$$I/O = (f + t)/(f - b) \qquad (13-6)$$

where f is the distance from the collimator face to the convergence point, b is the distance from the collimator face to the object, and t is collimator thickness.

Some manufacturers provide a single, invertible collimator insert that can be used in either converging or diverging mode.

EXAMPLE 13-2

Suppose the collimator described in Example 13-1 is inverted and used as a converging collimator to image a source distribution 15 cm in front of the collimator, also with a 30-cm diameter detector. What are the image magnification factor and the size of the imaged area?

Answer
When the collimator is inverted, the back face becomes the front face, and the convergence distance f becomes (45 – 5 cm) = 40 cm. Thus from Equation 13-6

I/O (magnification factor)

$= (40 + 5)/(40 - 15) = 1.8$

From Equation 13-4,

Diameter of imaged area

$= 30 \text{ cm}/1.8 = 16.7 \text{ cm}$

Again, because magnification depends on distance, there is some image distortion with the converging collimator (Fig. 13-9). Converging collimators are used primarily with

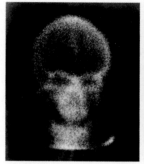 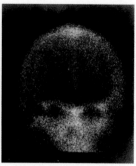

Parallel-hole collimator   Converging collimator

FIGURE 13-9 Examples of geometric image distortions created by the converging collimator.

cameras having large-area detectors to permit full use of the available detector area for imaging of small organs.

Converging collimators project an inverted *magnified* image when the object is located between the convergence point and twice the convergence length of the collimator, and an inverted *minified* image beyond that distance; however, they are used rarely at distances beyond the convergence point.

One consequence of the magnification or minification effects of these collimators is that the contribution of the intrinsic detector resolution (see Chapter 14, Section A) to the resulting image resolution may be reduced (magnification > 1) or increased (magnification < 1). Thus magnifying collimators can be useful in situations in which high spatial resolution is required, for instance in imaging of small organs such as the thyroid and in small-animal imaging applications.

4. Event Detection in a Gamma Camera

There are four types of events that may be detected by a gamma camera, as illustrated in Figure 13-10. Of these, only one provides correct positional information. The four events types (labeled to correspond with Fig. 13-10) are the following:

A: *valid event*—a γ ray is emitted parallel to the collimator holes, passes through a hole and interacts photoelectrically in the NaI(Tl) crystal, depositing all of its energy at a single location.

B: *detector scatter event*—a γ ray is emitted parallel to the collimator holes, passes through a hole and interacts by Compton scattering in the NaI(Tl) crystal. The scattered γ ray can either interact a

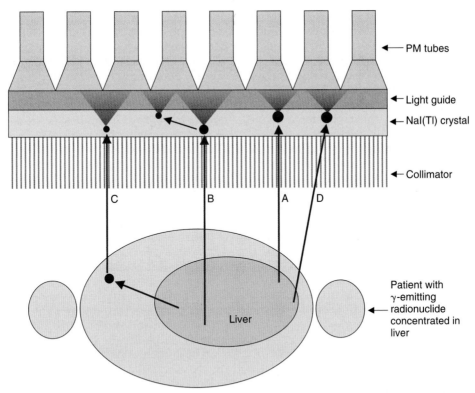

FIGURE 13-10 Illustration of different types of events that may be detected by a gamma camera. *Red circles* indicate locations of γ ray interactions. A, Valid event. B, Detector scatter event. C, Object scatter event. D, Septal penetration.

second time in the detector (as illustrated in Fig. 13-10), in which case the full energy of the γ ray is deposited, or it may escape the detector, in which case only part of the γ-ray energy is deposited. In the former case, energy discrimination cannot be used to reject the event, and the event will be mispositioned between the two interaction locations. In the latter case, it is likely that the event will be rejected because it does not satisfy the event energy criteria established by the upper- and lower-level discriminators. As discussed in Chapter 14, Section A.1, these events are relatively rare.

C: *object scatter event*—the γ ray is not emitted toward the collimator holes but is scattered within the body, then passes through a collimator hole and subsequently is detected. The γ ray loses energy during scattering and will therefore produce a smaller signal in the detector. Some of these events will be rejected by energy discrimination, but if the angle of scatter is small (≤45 degrees), the energy loss is small and the event may be accepted. In this case

the event is mispositioned, often many centimeters from the original site of emission. These events lead to a low-spatial-frequency background in the images that results in a loss of contrast. (See Chapter 15, Section C.). In clinical imaging situations, a large fraction of the detected events can be due to object scatter, and good energy resolution in the gamma camera is extremely important (see Chapter 14, Section A.3). The collimator itself can also be a cause of scatter leading to similar effects.

D: *septal penetration*—in this case a γ ray is emitted toward the collimator, but not parallel to it. Because of incomplete attenuation by the thin collimator walls (*septal penetration*), there is a finite chance that the γ ray will reach the NaI(Tl) crystal and interact with it. This again leads to blurring of the image, because all events are considered to have come from a direction perpendicular to the collimator face (for parallel-hole collimators). This effect becomes increasingly important when using high-energy γ emitters or high-resolution collimators with thin septa.

Considerable effort is expended in the design of gamma cameras to reduce or eliminate the detection of the events B, C, and D just described, each of which is a cause of blurring and a loss of contrast in the image. Collimators also are carefully designed for specific energies to minimize septal penetration while maximizing sensitivity for a given γ-ray energy (Chapter 14, Section C).

In addition to the simple cases illustrated in Figure 13-10, a combination of these event types can occur (e.g., scatter in the body and septal penetration, or septal penetration followed by Compton interaction in the detector). Finally, further complications arise when pulse pile-up occurs—that is, two or more events occur almost simultaneously in the gamma camera. This can also lead to event mispositioning and is discussed in detail in Chapter 14, Section A.4. Pile-up events can arise from of any combination of the event types described earlier.

C. TYPES OF GAMMA CAMERAS AND THEIR CLINICAL USES

The most common type of gamma camera is the *single-headed* system (Fig. 13-11). It consists of a gamma camera detector mounted on a gantry that allows the camera head to be positioned in a flexible way over different regions of the patient's body. Often, a moving bed is incorporated to permit imaging studies of the whole body. The gamma camera head often is mounted on a rotating gantry, allowing it to take multiple views around the patient. This feature also is necessary for producing *tomographic* images, or cross-sectional images through the body, as discussed in Chapters 16 and 17.

Dual-headed gamma cameras are becoming increasingly popular. In these systems, two gamma camera heads are mounted onto the gantry as shown in Figure 13-12. Usually, the two heads can be positioned at a variety of locations on the circular gantry. An obvious advantage of a dual-headed camera is that two different views of the patient can be acquired at the same time. For example, in whole-body imaging, the two detector heads can be placed at 180 degrees to each other to provide anterior and posterior views simultaneously. Triple-headed systems also exist, primarily for tomographic studies, as described in Chapters 16 and 17.

An example of a planar image acquired with a gamma camera system is presented in Figure 13-13. Dynamic processes can also be measured by taking multiple planar images

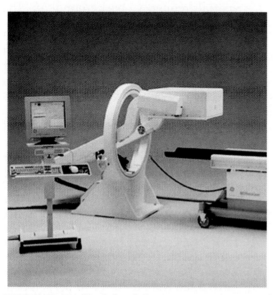

FIGURE 13-11 Single-headed gamma camera mounted on a rotating gantry. The camera is operated from the computer (*left*). The flexible positioning of the camera head and the bed (*right*) allows the system to obtain images of many different parts of the body. (*Courtesy GE Medical Systems, Milwaukee, WI.*)

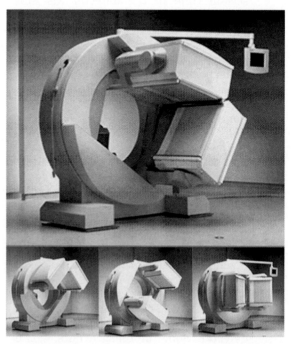

FIGURE 13-12 A dual-headed gamma camera system (*top*). Note that the camera heads can be placed in different orientations to provide two simultaneous views of an organ or the body (*bottom*). (*Courtesy Siemens Medical Systems, Inc., Hoffman Estates, IL.*)

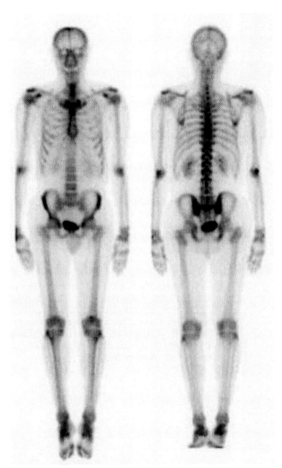

FIGURE 13-13 Whole-body bone scan obtained using ^{99m}Tc-MDP. These planar images were obtained with a dual-headed gamma camera on which both anterior (*left*) and posterior (*right*) views can be acquired simultaneously. The entire body was imaged by translating the patient bed through the gamma camera system. (*Courtesy Siemens Medical Systems, Inc., Hoffman Estates, IL.*)

over time. An example of a dynamic study is shown in Figure 13-14.

Single- and dual-headed gamma cameras are the workhorses of clinical nuclear medicine laboratories. However, a range of specialty gamma cameras have been or are being developed for specific imaging tasks. Examples are systems designed specifically for small-organ imaging (e.g., heart, breast, and thyroid) and mobile systems for use on patients who are too sick to be moved to the nuclear medicine department (e.g., from intensive care). These systems typically have smaller detector heads and may not have a built-in bed. An example of a compact gamma camera for breast imaging and representative images from it are shown in Figure 13-15. The detector typically ranges from 10×10 cm^2 to 20×20 cm^2. A number of different detector technologies are being exploited for these small-detector cameras, including traditional NaI(Tl)/PM tube systems, cameras based on pixellated NaI(Tl) or cesium iodide [CsI(Tl)] scintillator arrays (see Fig. 13-15B), and CsI(Tl) scintillator arrays with read-out by silicon photodiode arrays. There are also systems being developed that employ arrays of cadmium zinc telluride elements (see Chapter 7, Section B) for direct detection of γ rays, eliminating the need for a scintillator-photodetector combination.

High-resolution gamma cameras also have been developed for small-animal imaging. The goal is to provide a tool that biologists can use to monitor radiotracers in vivo, particularly in rats and mice. Most approaches involve the use of very small pinhole

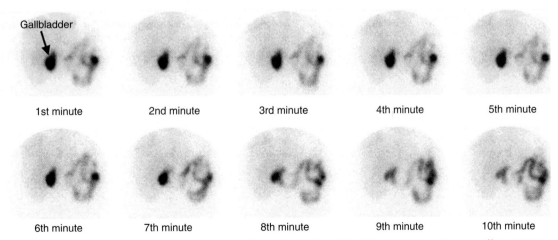

FIGURE 13-14 Planar gamma camera images over the region of the gallbladder following injection of ^{99m}Tc-HIDA. At approximately 7 minutes, cholecystokinin was given to the patient to stimulate emptying of the gallbladder. The rate and extent of emptying can be measured from this dynamic sequence of planar images. (*Courtesy GE Medical Systems, Milwaukee, WI.*)

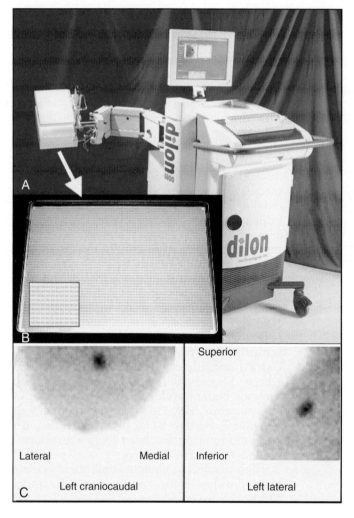

FIGURE 13-15   Example of a compact, mobile gamma camera system. *A,* The gamma camera head is attached to a cantilevered arm for easy and flexible positioning. *B,* Pixellated NaI(Tl) scintillator array that is coupled to small position-sensitive photomultiplier tubes to form the detector head. Each element in the array is approximately 2 × 2 mm and is separated from its neighbor by a reflective material. The *inset* shows a magnified view of the pixel elements. *C,* Clinical images of a breast cancer patient acquired with this camera following the injection of ^{99m}Tc-sestamibi. A tumor is seen as a "hot spot" against the low background uptake of the normal breast. (*Photographs and images courtesy Dilon Technologies, Inc., Annapolis, MD.*)

collimators to provide high spatial resolution. Tomographic small animal imaging systems based on this approach are discussed in Chapter 17, Section A.3.

REFERENCE

1. Anger HO: Scintillation camera. *Rev Sci Instr* 29:27-33, 1958.

BIBLIOGRAPHY

The principles of the gamma camera are discussed in greater detail in the following:

Simmons GH: *The Scintillation Camera*, New York, 1988, Society of Nuclear Medicine.

The Gamma Camera: Performance Characteristics

The performance of a gamma camera system is defined by the sharpness and detail of the images it produces, the efficiency with which it detects incident radiation, its ability to measure the energy of the incident γ rays (to minimize scatter), and the counting rate it can handle without significant dead time losses. A gamma camera is not capable of producing "perfect" images of the radionuclide distribution. Certain inherent imperfections arise from the performance characteristics of the detector, its associated electronic circuitry, and the collimator. Image artifacts also can be caused by malfunctions of various camera components. In this chapter, we describe the major factors that determine gamma camera performance and examine the limitations that can lead to artifacts in gamma camera images and their correction. Standard tests of gamma camera performance also are summarized.

A. BASIC PERFORMANCE CHARACTERISTICS

1. Intrinsic Spatial Resolution

Spatial resolution is a measure of the sharpness and detail of a gamma camera image. Sharp edges or small, pointed objects produce blurred rather than sharply defined images. Part of the blurring arises from collimator characteristics discussed in Sections C and D and part arises in the sodium iodide [NaI(Tl)] detector and positioning electronics. The limit of spatial resolution achievable by the detector and the electronics, ignoring additional blurring caused by the collimator, is called the *intrinsic spatial resolution* of the camera.

Intrinsic resolution is limited primarily by two factors. The first is *multiple scattering* of γ-ray photons within the detector. If a photon undergoes Compton scattering within the detector crystal and the residual scattered photon also is detected, but at some distance away, the two events are recorded as a single event occurring at a location along the line joining the two interaction sites. This is not a serious cause of degraded resolution for photon energies ≤300 keV in which multiple scatter Compton interactions in NaI(Tl) are almost negligible. Even at 662 keV, Anger calculated that for a detector thickness of 6.4 mm, less than 10% of photons are misplaced by more than 2.5 mm as a result of multiple scattering events.[1]

The second, and primary, cause of limited intrinsic resolution is statistical fluctuation in the distribution of light photons among photomultiplier (PM) tubes from one scintillation event to the next. The problem is exactly analogous to the statistical fluctuations observed in radioactive decay, discussed in Chapter 9. If a certain PM tube records, on average, N light photons from scintillation events occurring at a certain location in the detector crystal, the actual number recorded from one event to the next varies with a standard deviation given by $\sqrt{N}$. Thus if a very narrow beam of γ rays is directed at a point on the detector, the position of each event as determined by the positioning circuitry or computer algorithm is not precisely the same. Rather, they are distributed over a certain

209

area, the size of which depends on the magnitude of these statistical fluctuations.

A detailed method for measuring and characterizing intrinsic spatial resolution is discussed in Section E.1. Typically, a lead mask containing a number of narrow (~1 mm) slits is placed on the face of the gamma camera (without the collimator) and the camera is irradiated using a ^{99m}Tc (140-keV) point source. The resulting image is a series of lines corresponding to the locations of the slits (e.g., see Fig. 14-10A). The resolution is calculated as the full width at half maximum (FWHM) of a profile drawn perpendicular to the image of the lines at various locations in the field of view. The intrinsic spatial resolution of modern large field-of-view gamma cameras measured with ^{99m}Tc in this manner is in the range of 2.9- to 4.5-mm FWHM. Because the resolution is considerably worse than the width of the slits, the contribution of the slits themselves to the measured resolution is very small ($\leq$10% for measured resolution $\gtrsim$2.5 mm).

Intrinsic resolution becomes worse with decreasing γ-ray energy because lower-energy γ rays produce fewer light photons per scintillation event, and smaller numbers of light photons result in larger relative statistical fluctuations in their distribution (Chapter 9, Section B.1). As a rule of thumb, intrinsic resolution is proportional to $1/\sqrt{E}$, in which E is the γ-ray energy. This follows because the number of scintillation light photons produced, N, is roughly proportional to E and the relative statistical fluctuations in their distribution are therefore proportional to $1/\sqrt{N}$. This causes noticeably greater blurring at lower γ-ray energies. An example of the change of intrinsic spatial resolution as a function of γ-ray energy is shown in Figure 14-1.

Intrinsic resolution also depends on detector crystal thickness. Thicker detectors result in greater spreading of scintillation light before it reaches the PM tubes. Furthermore, there is a greater likelihood of detecting multiple Compton-scattered events in thicker detectors, particularly with higher-energy radionuclides. These are the primary reasons why gamma cameras use relatively thin detectors in comparison with NaI(Tl) systems that are used for counting applications. Figure 14-2 shows an example of the intrinsic spatial resolution versus crystal thickness for 140-keV γ rays.

Intrinsic resolution improves with increased efficiency of collection of scintillation photons. Modern cameras are substantially improved over earlier versions in this regard because of the use of more efficient PM tubes and of better techniques for optical coupling between the detector crystal and the PM tubes. The use of greater numbers of smaller PM tubes (5-cm-diameter tubes have become the standard, and some gamma cameras have as many as 110 PM tubes per head) and improved electronics also have contributed to this improvement. Accurate corrections for nonlinearity (see Section B.1) and nonuniformity (see Section B.2) have also resulted directly in improvements in intrinsic resolution, as

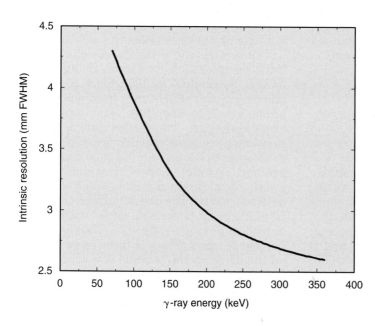

FIGURE 14-1 Intrinsic spatial resolution of a gamma camera as a function of γ-ray energy for a 6.3-mm-thick NaI(Tl) crystal. (*Compiled with data from Sano RM, Tinkel JB, LaVallee CA, Freedman GS: Consequences of crystal thickness reduction on gamma camera resolution and sensitivity. J Nucl Med 19:712-713, 1978; and Muehllehner G: Effect of crystal thickness on scintillation camera performance. J Nucl Med 20:992-993, 1979.*)

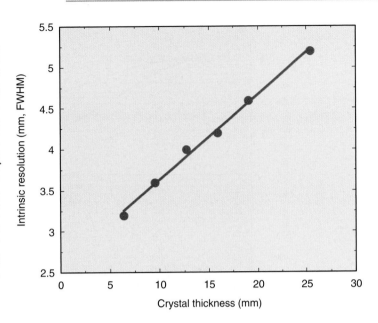

FIGURE 14-2 Intrinsic spatial resolution of a gamma camera at 140 keV as a function of crystal thickness. (*Compiled with data from Sano RM, Tinkel JB, LaVallee CA, Freedman GS: Consequences of crystal thickness reduction on gamma camera resolution and sensitivity.* J Nucl Med 19:712-713, 1978; Muehllehner G: Effect of crystal thickness on scintillation camera performance. J Nucl Med 20:992-993, 1979; Royal HD, Brown PH, Claunch BC: Effects of reduction in crystal thickness on Anger camera performance. J Nucl Med 20:977-980, 1979; Chapman D, Newcomer K, Berman D, et al: Half-inch versus quarter-inch Anger camera technology: Resolution and sensitivity differences at low photopeak energies. J Nucl Med 20:610-611, 1979; and unpublished data from Dr. Joel Karp, University of Pennsylvania, Philadelphia, PA.)*

discussed in the following sections. The best reported intrinsic resolution for a large field-of-view gamma camera is just below 3 mm FWHM at 140 keV (^{99m}Tc). Significant improvements beyond approximately 2 mm FWHM will be difficult to achieve, owing to the ultimate limitation of the light photon yield of NaI(Tl). In most practical situations, however, the intrinsic spatial resolution makes a negligible contribution to the overall system resolution of the gamma camera, which is largely determined by the resolution of the collimator (see Sections C and D).

2. Detection Efficiency

The gamma camera employs a sodium iodide crystal that is relatively thin in comparison with most other sodium iodide detectors used in nuclear medicine: 6.4 to 12.7 mm versus 2 to 5 cm for probe counting systems, scanners, and so on. The trade-off in gamma cameras is between *detection efficiency* (which improves with thicker crystals) and intrinsic spatial resolution (which improves with thinner crystals—see Fig. 14-2). The gamma camera is designed to provide acceptable detection efficiency while maintaining high intrinsic spatial resolution in the energy range of 100-200 keV. As a result, the detection efficiency of the gamma camera detector is somewhat less than would be desirable at higher γ-ray energies.

Figure 14-3 shows photopeak detection efficiency versus γ-ray energy for the gamma camera detector for a range of NaI(Tl) crystal thicknesses. The gamma camera is nearly 100% efficient for energies up to approximately 100 keV for all crystal thicknesses, but then shows a rather marked decrease in efficiency at higher energies, depending on crystal thickness. At 140 keV (γ-ray energy of ^{99m}Tc), the difference in efficiency between 6.4-mm and 12.7-mm-thick crystals is approximately 20% and the photopeak detection efficiency is in the 70% to 90% range. At approximately 500 keV, the standard gamma camera (detectors 0.64-0.95-cm-thick) is less than 20% efficient at converting incident γ rays into photopeak pulses.

At high energies, the performance of gamma cameras with 0.64- to 1.27-cm-thick crystals is limited by decreasing detection efficiency (as well as increasing collimator septal penetration—see Section C.2). Deteriorating intrinsic spatial resolution becomes the limiting factor at lower energies. Because of these tradeoffs, the optimal γ-ray energy range is approximately 100 to 200 keV for most gamma cameras. Some gamma cameras are now fitted with thicker crystals (12.7-25.4 mm), enabling them to achieve improved efficiency for imaging positron-emitting radionuclides at 511 keV (Chapter 18, Section B.4). This comes at the expense of some loss of intrinsic spatial resolution (see Fig. 14-2) when these systems are used in the 100-200-keV energy range.

3. Energy Resolution

It is not unusual in a typical patient study for there to be more Compton-scattered than unscattered γ rays striking the detector (see

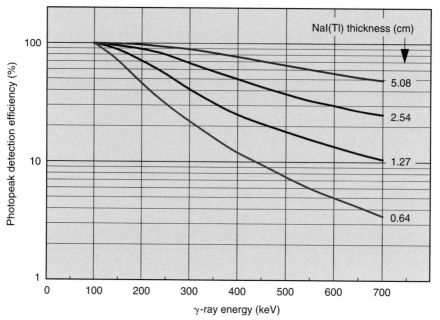

FIGURE 14-3 Photopeak detection efficiency versus γ-ray energy for NaI(Tl) detectors of different thicknesses. (*Adapted from Anger HO: Radioisotope cameras. In Hine GJ [ed]:* Instrumentation in Nuclear Medicine, *Vol 1. New York, 1967, Academic Press, p 506.*)

Fig. 11-9). Because the Compton-scattered photons have lower energy, it is possible to discriminate against them using pulse-height analysis. The energy resolution of the detector determines the efficiency with which this can be accomplished. Good energy resolution is perhaps the most important performance feature of the camera system for this purpose.

Energy resolution, like intrinsic spatial resolution, depends largely on statistical fluctuations in the number of light photons collected from a scintillation event (Chapter 10, Section B.7). Thus good light collection efficiency is a prerequisite for good energy resolution. As well, because the number of light photons released in a scintillation event increases almost linearly with γ-ray energy, E, (Fig. 10-11), energy resolution improves approximately in proportion to $1/\sqrt{E}$ (Fig. 10-13).

The energy resolution for gamma cameras is typically in the 9% to 11% range for ^{99m}Tc. Figure 14-4 shows a typical gamma camera spectrum for ^{99m}Tc with the pulse-height analyzer (PHA) window set to 130 to 150 keV. This corresponds to approximately a 15% energy window, which is a common setting for clinical studies. As illustrated by the figure, most of the events in the photopeak are accepted within this window. According to Equation 6-11, a low-energy threshold of

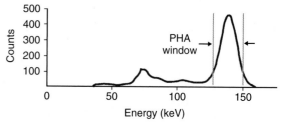

FIGURE 14-4 Energy spectrum from a gamma camera measured using a point source of ^{99m}Tc in air. The energy resolution at 140 keV in this example is 10.5%. A typical 15% energy window (approximately 130 to 150 keV) is shown superimposed on the spectrum. PHA, pulse-height analyzer. (*Data courtesy Dr. Magnus Dahlbom, UCLA School of Medicine, Los Angeles, CA.*)

130 keV should reject 140-keV γ rays that have been scattered through angles greater than approximately 45 degrees. However, because the spectrum for scattered γ rays is blurred in the same way as the spectrum for unscattered ones, the rejection efficiency for this scattering angle is only approximately 50%; half of the events produce pulses above the threshold, and half below it. This percentage would apply for 45-degree scattered 140-keV γ rays and a 130-keV lower energy level, regardless of the energy resolution of the detector. Gamma rays scattered through greater angles are rejected more efficiently,

and those scattered through smaller angles are rejected less efficiently.

Two advantages are obtained with improved energy resolution. First, the photopeak becomes narrower, resulting in more efficient detection of unscattered photons within the chosen energy window. This increases the number of valid events recorded and improves the statistical quality of the image. Second, γ rays scattered through large angles are rejected more efficiently, because their energy spread within the pulse-height spectrum is also smaller. Thus image contrast is improved. It also is true that γ rays scattered through smaller angles are detected somewhat more efficiently, because of the narrowing of their distribution as well. However, the increased efficiency for recording photopeak events more than offsets this effect, in terms of contrast-to-noise ratio (Chapter 15, Section D.2). Alternatively, one can take advantage of the improved energy resolution to use a narrower PHA window, trading back some of the increased efficiency for recording photopeak events for improved rejection of small-angle scatter. Either way, improved energy resolution results in better image quality.

4. Performance at High Counting Rates

At high counting rates, there is increased likelihood of recording two events at the same time. The most troublesome effect is known as *pulse pile-up* (Chapter 8, Section B.3). Pulse pile-up has two undesirable effects on gamma camera performance: *counting losses* and *image distortion.*

Counting losses cause inaccurate counting rates to be recorded at higher counting rates. The inaccuracies are described by conventional dead time models (Chapter 11, Section C) and may be significant in some high-count-rate quantitative studies, such as first-pass cardiac studies. Dead time corrections can be applied; however, these corrections generally become increasingly inaccurate as counting losses increase.

Because pulse pile-up can occur between any two events in the pulse-height spectrum, system counting losses are determined by total-spectrum counting rates. Most gamma cameras behave as paralyzable systems. The apparent dead time for a selected energy window depends on the *window fraction,* that is, the fraction of the total spectrum counting rate occurring within that window. The smaller the window fraction, the larger the apparent dead time. Thus the apparent dead time is longer when a photopeak window is used than when a full-spectrum window is used. The apparent dead time also is longer when scattered radiation is present, because this also adds to the counting rate outside the photopeak window (Fig. 14-5). Therefore, when specifying gamma camera dead time, it is important to note the conditions of measurement. Dead time values as short as 1 to 2 μsec can be obtained in the absence of scattering material with a full-spectrum window; however, under clinically realistic conditions (^{99m}Tc source in scattering material, 15% photopeak window), system dead times of 4 to 8 μsec are more typical. For a dead time of 5 μsec, counting losses are approximately 20% for a counting rate of 4×10^4 counts per second (cps).

Dead time losses are not serious in most static imaging studies, but they can be important in certain high-counting-rate applications (e.g., first-pass cardiac studies) in which counting rates as high as 10^5 cps may occur. Pile-up rejection circuitry (see Chapter 8, Section B.3) is used to achieve higher usable counting rates in such situations. Another approach for shortening camera dead time is by the use of *analog buffers,* or *derandomizers.* These are electronic circuits that "hold" a

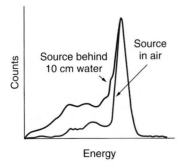
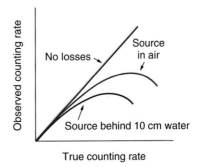

FIGURE 14-5 Effect of scattered radiation on counting losses. Scattered radiation decreases the window fraction recorded with a photopeak window (*left*), thus causing an apparent increase in dead time counting losses (*right*).

voltage level or pulse from one circuit component (e.g., an amplifier) until the next circuit in the pulse-processing sequence (e.g., the PHA) is ready to receive it.

Similarly, in digital gamma cameras, data can be buffered in memory until the computer is ready to process them. Both these approaches result in a decrease in the "apparent" dead time of the camera by effectively changing the arrival times of the pulses. This, however, means that the simple dead time models and corrections presented in Chapter 11, Section C can no longer be used, and more complex modeling of system dead time must be carried out to produce accurate correction at high counting rates.

It also is possible to physically shorten the dead time of a camera by shortening the charge integration time from the PM tubes and using electronic circuitry that returns the signal to baseline after the chosen integration time.[2] Clearly, this also decreases the amount of signal used for determining event location. For example, with a charge integration time of 0.4 μsec, only 81% of the scintillation light is collected, compared with 98% for a 1-μsec integration time. This causes a degradation of intrinsic spatial resolution and energy resolution. Some gamma cameras have a variable integration time, in which the charge integration is automatically shortened as the counting rate increases.

Other means for shortening dead time are to bypass altogether the pile-up rejection circuits and nonuniformity correction circuitry (see discussion on pile-up correction later in this section and on nonuniformity and its correction in Section B.3). The signal processing that occurs in these circuits slows down the rate at which the camera can handle individual events, and bypassing them can shorten system dead time from typical values of 4 to 8 μsec down to 1 to 3 μsec. Some cameras provide an optional "high count rate" mode of operation in which some or all of these corrections are turned off by software control. This mode is intended specifically for applications requiring high counting rates, such as first-pass cardiac studies. "Normal mode," in which all corrections are employed, is used for routine imaging to obtain the desired high-quality images. Obviously specifications for gamma camera dead time should indicate whether any circuits were bypassed to achieve the reported value.

The second undesirable effect of pulse pile-up is image distortion. Using standard pulse-positioning logic for gamma cameras

(see Chapter 13, Section B.2), two events detected simultaneously at different locations in the detector are recorded as a single event with energy equal to the sum of the two events, at a location somewhere between them (Fig. 14-6). If both are valid photopeak events, their total energy exceeds the value that would be accepted by the PHA window and both events are rejected, resulting in counting losses. On the other hand, it is possible for two Compton-scattered γ rays to have a total energy that falls within the selected energy window, so that two invalid events are accepted as a single valid event. The visible result at very high counting rates is to add a diffuse background to the image, as illustrated in Figure 14-7. Note as well the image in the upper right-hand corner of this figure, showing how contrast can be restored by shielding high-activity areas outside the imaging area of interest (e.g., with a thin sheet of lead).

Early pile-up rejection methods were based on measuring the length of a pulse. If the pulse did not return close to baseline level within the time expected given the decay time of NaI(Tl), it was assumed that pile-up of two pulses had occurred and the event was rejected, resulting in the loss of both γ rays. This improved image quality but resulted in an effective increase in system dead time, because many events were rejected at high counting rates.

Many gamma cameras now incorporate circuits that continuously monitor the decay of a pulse and use a method based on *pulse-tail extrapolation* for pile-up correction. Consider two γ-ray interactions that occur close together in time and create overlapping pulses. When the second γ ray arrives, the decay of the pulse created by the first γ ray immediately

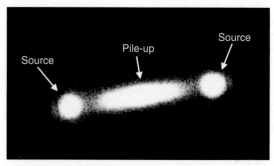

FIGURE 14-6 Images of two ⁹⁹ᵐTc point sources of relatively high activities (~370 MBq each). Events appearing in the band between the two point-source locations are mispositioned events caused by pulse pile-up.

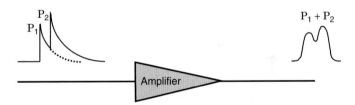

FIGURE 14-7 Demonstration of pile-up effects on images of a brain phantom. Times required to record 1.2×10^6 counts are indicated. At very high counting rates there is a noticeable loss of image contrast, which can be restored by shielding useless high-activity areas from the detector (*top right-hand image*).

deviates from the expected exponential decay and the gamma camera signal is switched to a second amplifier circuit. Estimator circuitry in the first amplifier circuit completes the signal from the first γ ray by extrapolating the remainder of the tail of the pulse with an exponential function based on the decay time of NaI(Tl). At the same time, this extrapolated tail is also sent to the second amplifier circuit and subtracted from the second pulse. This removes the contribution of the pulse generated by the first γ ray from that of the second γ ray. This process is summarized in Figure 14-8. The pulse-tail extrapolation

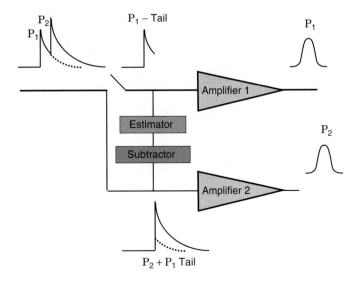

FIGURE 14-8 Illustration of pile-up correction using pulse-tail extrapolation techniques. See text for details. (*Adapted from Lewellen TK, Pollard KR, Bice AN, Zhu JB: A new clinical scintillation camera with pulse tail extrapolation electronics. IEEE Trans Nucl Sci 37:702-706, 1990.*)

technique results in both events being retained and allows them to contribute to the image, providing they also meet the PHA requirements. This method is very effective, unless the two pulses occur nearly simultaneously (within a few tens of nanoseconds of each other), in which case the extrapolation is of limited accuracy.

With modern digital gamma cameras, it also is possible to use the spatial distribution of PM tube signals to further reduce pile-up. For pile-up events occurring at different locations in the detector crystal, two distinct clusters of PM tubes will produce signals. If the light distributions produced by the two events on the PM tubes do not overlap, or only slightly overlap, the events can be clearly separated and retained.

B. DETECTOR LIMITATIONS: NONUNIFORMITY AND NONLINEARITY

1. Image Nonlinearity

A basic problem arising in the detector and electronics is image *nonlinearity*. Straight-line objects appear as curved-line images. An inward "bowing" of line images is called *pincushion distortion;* an outward bowing is called *barrel distortion* (Fig. 14-9). Nonlinearities result when the X- and Y-position signals do not change linearly with displacement distance of a radiation source across the face of the detector. For example, when a source is moved from the edge of one of the PM tubes toward its center, the light collection efficiency of that PM tube increases more rapidly than the distance the source is moved. This causes the image of a line source crossing in front of a PM tube to be bowed toward its center. The result is a characteristic pincushion distortion in areas of a gamma camera image lying directly in front of the PM tubes, and barrel distortion between them. Differences in sensitivity among the PM tubes, nonuniformities in optical light guides, as well as PM tube or electronic malfunctions, also can cause nonlinearities.

Figure 14-10A, shows an image of a straight-line "test pattern" recorded on a

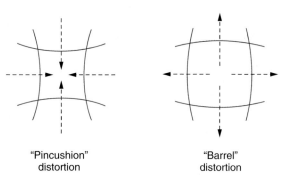

"Pincushion"
distortion

"Barrel"
distortion

FIGURE 14-9 Appearance of straight-line objects with "pincushion" and "barrel" distortions.

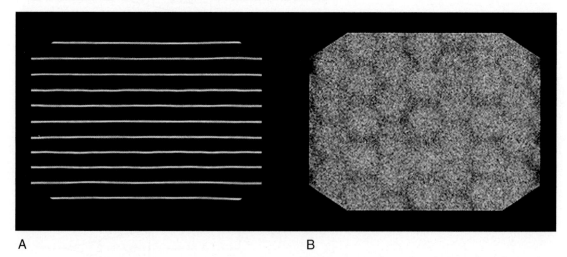

A   B

FIGURE 14-10 *A,* Illustration of nonlinearities in images of a straight-line test pattern obtained with a gamma camera. Image demonstrates subtle waviness in the lines. *B,* Flood-field image obtained by exposing the same camera to a uniform radiation field. This is the image obtained in the absence of any corrections for nonuniformity. Notice that the photomultiplier tube pattern can be seen. The gray levels in this image are confined to a narrow display window to improve visualization of the artifacts. (*Images courtesy Dr. Magnus Dahlbom, UCLA School of Medicine, Los Angeles, CA.*)

modern gamma camera to demonstrate the general appearance of nonlinearities. On close inspection, some waviness of the lines is apparent. On properly functioning cameras, including the one illustrated, the nonlinearities themselves (including the pincushion distortions in front of PM tubes) are barely perceptible and rarely interfere directly with image interpretation; however, they can have significant effects on image nonuniformities, as discussed in the following section.

2. Image Nonuniformity

A more noticeable problem is image nonuniformity. Exposing the detector crystal to a uniform flux of radiation produces a *flood-field image* with small but noticeable nonuniformities in intensity, even with a properly functioning camera. These variations may be equivalent to counting rate variations of ±10% or more. A flood-field image from a gamma camera demonstrating image nonuniformity is shown in Figure 14-10B. *Intrinsic* flood-field images are acquired with the collimator removed, using a point source placed far enough from the surface of the gamma camera to give uniform irradiation of the surface (distance equal to 4-5 times the camera diameter). *Extrinsic* flood-field images are acquired with the collimator in place using a disk or thin flood phantom that covers the area of the detector. ^{99m}Tc or ^{57}Co are the two most commonly used radionuclides for flood-field measurements.

There are two primary causes of gamma camera nonuniformities. The first is *nonuniform detection efficiency* arising from (1) small differences in the pulse-height spectrum for different PM tubes and (2) position-dependent collection efficiency of scintillation light, particularly for events located over the gaps and dead areas between the PM tubes compared with events located directly over the center of a PM tube. The differences in PM tube response can be minimized by careful selection and tuning of all of the PM tubes of a gamma camera; however, position-dependent effects on the pulse-height spectrum remain. If a fixed pulse-height window is used for all output pulses, the result is an apparent difference in detection efficiency owing to differences in the "window fraction" for different areas of the crystal (see Fig. 13-6, *top*).

The second cause of nonuniformities is image *nonlinearities* described in Section B.1. In areas of pincushion distortion events are crowded toward the center of the distortion, causing an apparent "hot spot," whereas in areas of barrel distortion events are pushed outward from the center, causing an apparent "cold spot." Because of the characteristic pincushion distortions occurring in front of PM tubes, it is common to see a pattern of hot spots at the locations of the PM tubes on an otherwise uniform gamma camera image. Other causes of nonlinearities (e.g., PM tube failure, crystal cracking, and collimator defects) also can result in nonuniformities.

Another characteristic nonuniformity is a bright ring around the edge of the image. This artifact, called *edge packing*, results from a somewhat greater light collection efficiency for events near the edge versus central regions of the detector crystal. This is the result of internal reflections of scintillation light from the sides of the detector crystal back into the PM tubes near the edge. Also, for events occurring toward the center of the crystal, there are always PM tubes on either side of the event location, whereas at the edges of the crystal there are PM tubes only to one side. Thus events at the very edges are not distributed uniformly across the edge, but are "pulled" toward the center, compounding the edge-packing artifact. The portion of the image demonstrating this artifact usually is masked on the image display and therefore is not a part of the useful field of view (UFOV). Typically, 5 cm or more of the detector width is eliminated by the mask. When specifying gamma camera detector dimensions, it is important to distinguish between the physical dimensions of the crystal and the dimensions of the useful imaging area.

Both nonuniformity and edge-packing artifacts are related to the pattern of the distribution of scintillation light falling on the PM tubes. For this reason, they also have an energy-dependent component. When the gamma camera is used to image higher-energy radionuclides, interactions, on average, occur deeper in the crystal, closer to the PM tubes. This produces a more narrow light spread distribution on the PM tubes and generally results in a worsening of detector nonuniformity.

3. Nonuniformity Correction Techniques

All modern gamma cameras incorporate techniques that attempt to correct the causes of nonuniformity described in the preceding section. All of these techniques begin with spatially varying *energy corrections*, normally derived from an intrinsic flood-field image. The flood-field image is divided into a matrix of small, square elements, typically 128×128 elements (or *pixels*). Using the PHA, the channel number (pulse amplitude) of the photopeak in the pulse-height spectrum is determined for each element. This information is stored in a 128×128 look-up table and used to set regionally varying PHA windows for subsequent studies on patients. For example, if a 20% window is chosen for a patient study, and the center of the photopeak is found in PHA channel 100 in a particular pixel in the flood-field image, then events at that location having Z-signal amplitudes between PHA channels 90 and 110 are accepted in patient imaging studies. If the center of the photopeak is in channel 110 at another location, events for which the Z-signal falls within the range of 99 to 121 are accepted at that location. The position-dependent PHA window corrects for variations in the pulse-height spectrum across the face of the camera detector. It also provides a partial correction for image nonuniformity.

The second step in the nonuniformity correction is to account for the remaining regional variations in image intensity, largely caused by detector nonlinearity. In one older method, the correction is based directly on variations in intensity of the energy-corrected flood-field image. The number of counts recorded within each pixel in that image is stored in a matrix and compared to the smallest number recorded in the pixel array. This is used to derive a matrix of normalized intensity values, which range from 100 for the "coldest" pixel to higher values for other pixels. In subsequent patient studies, a certain fraction of the counts recorded in each pixel are thrown out, depending on the relative value for that pixel in the energy-corrected flood-field image. For example, if the value in the normalized intensity-correction matrix is 110, then 1 of every 11 counts is subtracted from the patient image at that location. This process is sometimes called *count skimming*.

Most modern digital gamma cameras replace the second step described in the preceding paragraph with a correction for image nonlinearity, which more directly attacks the major underlying cause of image nonuniformity. For nonlinearity corrections, another flood-field image is obtained, this time with a sheet of lead having a uniformly spaced array of small holes (~1-mm diameter, ~4-mm separation) placed directly on the gamma camera face (no collimator). The locations of the images of these holes are compared with their known locations in the lead sheet to derive a matrix of offsets, Δx and Δy, for each (X,Y) location on the detectors, which is stored as another look-up table. When an event is detected, its X and Y coordinates are computed using conventional positional circuitry or algorithms. These values then are corrected using the positional offsets for that location stored in the look-up table. The offsets and the corresponding look-up table usually are measured and generated at the factory prior to shipment.

Figure 14-11 shows the same data as Figure 14-10, after the corrections for nonuniformity and nonlinearity described in the preceding paragraphs have been applied. Figure 14-12, showing intensity profiles across the flood-field images in Figures 14-10 and 14-11,

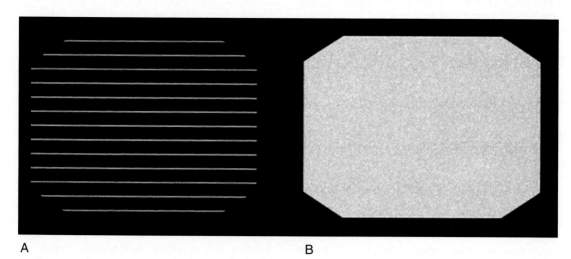

A   B

FIGURE 14-11 Straight-line test pattern (*A*) and uniform flood-field (*B*) images after nonuniformity corrections are applied. Compare with Figure 14-10. (*Images courtesy Dr. Magnus Dahlbom, UCLA School of Medicine, Los Angeles, CA.*)

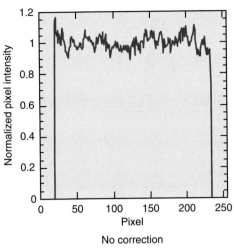

No correction

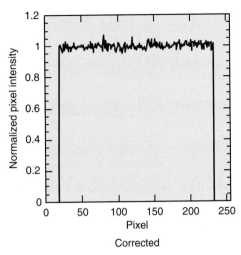

Corrected

FIGURE 14-12 Profiles through the uniform flood-field images in Figures 14-10 and 14-11 showing relative uniformity of flood-field image with and without nonuniformity correction. The standard deviation is improved from 3.4% to 1.9% after correction.

clearly illustrates the improvements. The examples in these figures are for ^{99m}Tc. Note that different correction matrices must be obtained for each radionuclide used, because the effects corrected for generally vary with γ-ray energy, for example, because of different average depths of interaction in the NaI(Tl) crystal.

Improvements in camera uniformity also have contributed to improvements in intrinsic resolution. Earlier cameras used thicker light guides and large-diameter PM tubes, in part to achieve satisfactory uniformity, at the expense of somewhat degraded spatial resolution. Because of effective uniformity corrections, newer gamma cameras can use thinner light guides (or eliminate the light guide entirely) and smaller PM tubes, both of which contribute to more accurate event localization and improved intrinsic spatial resolution.

4. Gamma Camera Tuning

The nonuniformity corrections described previously require that the gamma camera remain very stable over time. However, the gain of PM tubes invariably changes as the tubes age. The high-voltage supply and amplifier gain can also drift over time. A method to "tune" the PM tubes to ensure consistent performance over time is therefore necessary.

On many older systems, the tuning is done manually. One method involves irradiation of the gamma camera detector through a lead mask with holes centered over each of the PM tubes. The output of each PM tube is examined and the preamplifier gain is adjusted if the tube output has changed by more than 1% from the original reference value.

In newer gamma cameras, the large number of PM tubes makes the manual method impractical. Many digital gamma cameras therefore contain tuning circuitry that allows the output of each individual PM tube to be automatically adjusted to a set of reference outputs. One automated approach involves the use of light-emitting diodes (LEDs) that are coupled to the neck of each PM tube. These LEDs are pulsed to produce a light signal on the photocathode of the PM tube that does not vary with time. The PM tube signals are then monitored and the preamplifier adjusted electronically if the PM tube signal has drifted.

A second approach uses two narrow energy windows, placed just above the photopeak position to minimize the influence of scatter (Fig. 14-13). The count ratio between the two energy windows during flood-field irradiation by the radionuclide of interest is measured for each PM tube. This ratio remains constant, unless the PM tube signal drifts over time. If the count ratio changes, the PM tube preamplifier is adjusted electronically to restore the ratio to its original value.

Some of these tuning methods also can be adapted so that they are continuous, in the sense that the camera is tuned dynamically every few seconds during a patient study. This can be used to adjust the energy windows in real time, compensating for any drift that occurs during the course of a study. The major cause of drift on such short

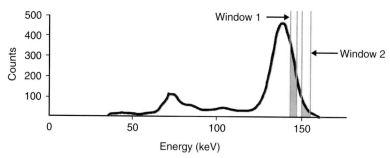

FIGURE 14-13 The ratio of counts detected in two narrow energy windows can be used to determine whether a photomultiplier tube is drifting. The windows are defined on the high side of the photopeak to avoid any contribution from scatter. The ratio is virtually independent of source distribution and the amount of scattering material present.

timescales usually is related to count-rate effects. At high counting rates, a small baseline shift can occur in the signal amplitudes owing to overlap of pulses, and continuous adjustment of the energy window minimizes such effects, keeping the energy window centered over the photopeak irrespective of the counting rate. Continuous tuning also is important in single photon emission computed tomographic imaging (Chapter 17), in which rotation of the gamma camera through the earth's magnetic field can result in changes of PM tube gain. A detailed discussion of automatic tuning methods can be found in reference 3.

C. DESIGN AND PERFORMANCE CHARACTERISTICS OF PARALLEL-HOLE COLLIMATORS

1. Basic Limitations in Collimator Performance

The collimator is a "weak link" for the performance of a gamma camera system, as indeed it is in any nuclear medicine imaging system employing the principles of *absorptive collimation*. *Collimator efficiency,* defined as the fraction of γ rays striking the collimator that actually pass through it to project the γ-ray image onto the detector, is typically only a few percent or less. *Collimator resolution,* which refers to the sharpness or detail of the γ-ray image projected onto the detector, also is rather poor, generally worse than the intrinsic resolution of the camera detector and electronics.

Because it is a limiting factor in camera system performance, it is important that

the collimator be designed carefully. Poor design can result only in poorer overall performance. Design considerations for parallel-hole collimators are discussed in this section. Design characteristics for converging and diverging collimators are similar to those of the parallel-hole type. Design characteristics of pinhole collimators are not discussed in detail here but are described in references 4 and 5. The analysis to be presented for parallel-hole collimators is similar to that presented by Anger in reference 1, which may be consulted for a more detailed discussion.

2. Septal Thickness

A primary consideration in collimator design is to ensure that *septal penetration* by γ rays crossing from one collimator hole into another is negligibly small. This is essential if an accurate γ-ray image is to be projected by the collimator onto the camera detector. No thickness of septal material is sufficient to stop *all* γ rays, so the usual criteria is to accept some reasonably small level of septal penetration (e.g., ~5%).

The required septal thickness can be determined by analysis of Figure 14-14. The shortest path length for γ rays to travel from one hole to the next is w. Septal thickness t is related to w, and to the length l and diameter d of the collimator holes, by

$$t \approx 2dw/(l - w) \qquad (14\text{-}1)$$

If septal penetration is to be less than 5%, the transmission factor for the thickness w must be

$$e^{-\mu w} \lesssim 0.05 \qquad (14\text{-}2)$$

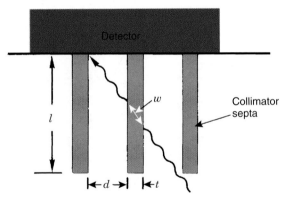

FIGURE 14-14 Minimum path length w for a γ ray passing through the collimator septa from one hole to the next depends on length l and diameter d of the collimator holes and on septal thicknesses t.

where μ is the linear attenuation coefficient of the septal material. Because e^{-3} is approximately 0.05, this implies

$$\mu w \gtrsim 3 \qquad (14\text{-}3)$$

$$w \gtrsim 3/\mu \qquad (14\text{-}4)$$

and thus

$$t \gtrsim \frac{6d/\mu}{l-(3/\mu)} \qquad (14\text{-}5)$$

It is desirable that septal thickness t be as small as possible so that the collimator septa obstruct the smallest possible area of detector surface and collimator efficiency is maximized. This objective is realized by using a material with a large value of μ for the collimator septa. Materials of high atomic number Z and high density ρ are preferred. Lead (Z = 82, ρ = 11.3 g/cm^3) is the material of choice for reasons of cost and availability; however, other materials, including tantalum (Z = 73, ρ = 16.6 g/cm^3), tungsten (Z = 74, ρ = 19.3 g/cm^3), gold (Z = 79, ρ = 19.3 g/cm^3) and even depleted uranium (Z = 92, ρ = 18.9 g/cm^3) have been employed in experimental applications.

As discussed in Chapter 6, Section D.1, attenuation coefficients of heavy elements depend strongly on γ-ray energy in the nuclear medicine energy range. Thus the required septal thickness also depends strongly on the γ-ray energy for which the collimator is designed to be used. Commercially available collimators are categorized according to the maximum γ-ray energy for which their septal thickness is considered to be adequate. *Low-energy collimators* generally have an upper limit of approximately 150 keV and *medium-energy collimators* of approximately 400 keV. *High-energy collimators* are used for imaging positron-emitting radionuclides at 511 keV.

EXAMPLE 14-1

Calculate the septal thickness required for low-energy (150 keV) and medium-energy (400 keV) lead collimators having hole diameters of 0.25 cm and lengths of 2.5 cm.

Answer
The linear attenuation coefficient of lead at 150 keV is $\mu_l = 1.91$ cm^2/g $\times$ 11.34 g/cm^3 = 21.66 cm^{-1} and at 400 keV is $\mu_l = 0.22$ cm^2/g $\times$ 11.34 g/cm^3 = 2.49 cm^{-1} (Appendix D). Therefore from Equation 14-5 for the low-energy collimator

$$t \gtrsim \frac{6 \times 0.25/21.66}{2.5-(3/21.66)}$$

$$\gtrsim 0.029 \text{ cm}$$

and for the medium-energy collimator

$$t \gtrsim \frac{6 \times 0.25/2.49}{2.5-(3/2.49)}$$

$$\gtrsim 0.465 \text{ cm}$$

As shown by this example, thicknesses needed for low-energy collimators are only a few tenths of a millimeter, which is in the range of lead "foil" thicknesses and approaches the limits of lead thicknesses that can be used without loss of necessary mechanical strength. Indeed, low-energy collimators generally are quite fragile, and their septa can be damaged easily by mechanical abuse (such as dropping or stacking on sharp objects). Medium-energy collimators require substantially greater septal thicknesses, typically a few millimeters of lead. Alternatively, medium-energy collimators can be made thicker (larger l in Equation 14-5).

Low-energy γ-ray emitters (e.g., ^{99m}Tc, 140 keV) can be imaged using medium-energy collimators. This is done, however, with an unnecessary sacrifice of collimator efficiency because the collimator septa are unnecessarily thick. (See Table 14-1 for comparative efficiencies of low- and medium-energy collimators.) Low-energy collimators are used whenever possible to obtain maximum collimator efficiency. When choosing a collimator, however, one must consider not only the energy of the γ rays to be imaged but also the energies of any other γ rays emitted by the radionuclide of interest or by other radionuclides that may be present as well (e.g., residual activity from another study or radionuclide impurities). Higher-energy γ rays may be recorded by Compton downscatter into a lower-energy analyzer window. If the collimator septa are too thin, the collimator may be virtually transparent to higher-energy γ rays, causing a relatively intense "foggy" background image to be superimposed on the desired image, with a resulting loss of image contrast. Whether a low-energy collimator can be used when higher-energy γ rays are present depends on the energy and intensity of those emissions and requires experimental evaluation in specific cases.

3. Geometry of Collimator Holes

Collimator performance also is affected by the geometry of the collimator holes, specifically, their *shape, length,* and *diameter.* The preferred hole shape, to maximize the exposed area of detector surface for a given septal thickness, is round or hexagonal, with the holes arranged in a close-packed hexagonal array, or square holes in a square array. Triangular holes also have been used.

Collimator hole length and diameter affect strongly both collimator resolution and collimator efficiency. Collimator resolution R_{coll} is defined as the FWHM of the radiation profile from a point or line source of radiation projected by the collimator onto the detector (Fig. 14-15). This profile is also called the *point-spread function* (PSF) or *line-spread function* (LSF). Collimator resolution R_{coll} is given by*

$$R_{coll} \approx d(l_{eff} + b)/l_{eff} \qquad (14\text{-}6)$$

where b is the distance from the radiation source to the collimator and d is the diameter and $l_{eff} = l - 2\mu^{-1}$ the "effective length" of the collimator holes. Here μ is the linear attenuation coefficient of the collimator material. The effective length of the collimator holes is somewhat less than their actual length owing

---

*Some versions of Equation 14-6 include additional correction terms involving the thickness of the detector crystal, reflecting the fact that the image actually is formed at some depth within the detector crystal. Because photons of different energies penetrate to different average depths within the crystal, the correction actually is photon-energy dependent, a point not noted in some texts. The correction is small and for simplicity is omitted from Equation 14-6, as well as from Equations 14-10 and 14-13 for the converging and diverging collimators presented later in this chapter.

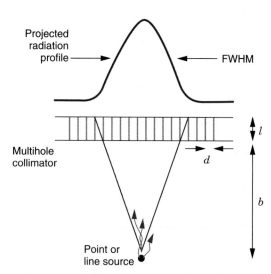

FIGURE 14-15   Radiation profile (point- or line-spread function) for a parallel-hole collimator. The full width at half maximum (FWHM) of the profile is used to characterize collimator resolution.

to septal penetration. For 2.5-cm thick low-energy collimators (150 keV), the difference between effective and actual length is approximately 0.1 cm, whereas for 2.5-cm thick medium-energy collimators (400 keV) it is approximately 0.8 cm.

EXAMPLE 14-2

Calculate the resolution (FWHM) of the low-energy collimator described in Example 14-1, at source depths $b = 0$ and $b = 10$ cm, assuming it has a septal thickness of 0.03 cm.

Answer
The effective length of the collimator is

$$l_{eff} = 2.5 \text{ cm} - (2/21.66) \text{ cm} \approx 2.4 \text{ cm}$$

Thus for $b = 0$

$$R_{coll} \approx 0.25 (2.4 + 0)/2.4 \text{ cm}$$
$$\approx 0.25 \text{ cm}$$

and at $b = 10$ cm

$$R_{coll} \approx 0.25 (2.4 + 10)/2.4 \text{ cm}$$
$$\approx 1.3 \text{ cm}$$

This example illustrates the strong dependence of collimator resolution on the distance of the source from the collimator.

Collimator efficiency g, defined as the fraction of γ rays passing through the collimator per γ ray emitted by the source is given by

$$g \approx K^2 (d/l_{\mathrm{eff}})^2 [d^2/(d+t)^2] \qquad (14\text{-}7)$$

where t is septal thickness and K is a constant that depends on hole shape (~0.24 for round holes in a hexagonal array, ~0.26 for hexagonal holes in a hexagonal array, ~0.28 for square holes in a square array[1]). Equation 14-7 applies to a source *in air* and assumes no attenuation of radiation by intervening body tissues.

Several aspects of Equations 14-6 and 14-7 should be noted. First, resolution improves as the ratio of hole diameter to effective length (d/l_{eff}) is made smaller. Long, narrow holes provide images with the best resolution; however, collimator efficiency decreases approximately as the square of the ratio of hole diameter to length $(d/l_{\mathrm{eff}})^2$. Thus an approximate relationship between collimator efficiency, g, and spatial resolution, R_{coll}, is

$$g \propto (R_{\mathrm{coll}})^2 \qquad (14\text{-}8)$$

Therefore for a given septal thickness, collimator resolution is improved only at the expense of decreased collimator efficiency, and vice versa.

EXAMPLE 14-3

Calculate the efficiency g of the collimator described in Examples 14-1 and 14-2, assuming it has hexagonal holes in a hexagonal array.

Answer

For hexagonal holes in a hexagonal array, $K = 0.26$. Thus,

$$g \approx (0.26)^2 (0.25/2.4)^2 \times [(0.25)^2/(0.25+0.03)^2]$$

$$\approx (0.0676) \times (0.0109) \times (0.797)$$

$$\approx 5.85 \times 10^{-4}$$
(photons transmitted/photons emitted)

This example illustrates the relatively small fraction of emitted γ rays that are transmitted by a typical gamma camera collimator.

Equation 14-7 also demonstrates the effect of septal thickness on efficiency. Medium-energy collimators have lower efficiencies than low-energy collimators because of their greater septal thicknesses.

In addition to providing low- and medium-energy collimators, manufacturers of gamma

camera systems also provide a selection of collimators with different combinations of resolution and efficiency. Those with good resolution but poor efficiency generally are described as "high-resolution" collimators, whereas those with the opposite characteristics are described as "high-sensitivity" collimators. Those with characteristics intermediate to the extremes are referred to as "general purpose," "all purpose," or by other similar names.

Equation 14-6 indicates that collimator resolution becomes poorer as source-to-collimator distance b increases. Thus structures closest to the collimator are imaged with sharpest detail. Figure 14-16 shows graphically the relationship between collimator resolution and source-to-collimator distance for three different collimators provided by one commercial manufacturer. Typically, collimator resolution deteriorates by a factor of 2 at a distance of 4-6 cm from the collimator.

On the other hand, according to Equation 14-7, collimator efficiency for a source in air is independent of source-to-collimator distance b. This rather surprising result is obtained provided the counting rate for the entire detector area is measured. The reason for this is illustrated by Figure 14-17. As the source is moved farther away from the collimator, the efficiency with which radiation is transmitted through any one collimator hole decreases in proportion to $1/b^2$ (inverse-square law), but the number of holes through which radiation can pass to reach the detector increases in proportion to b^2. The two effects cancel each other, with the result that total counting rate—and thus collimator efficiency—does not change with source-to-collimator distance. Another illustration of this effect is shown in Figure 14-18. As source-to-collimator distance increases, the maximum height of the PSF or LSF decreases, but the width increases (and resolution becomes poorer), so that the total area under the curve (total detector counting rate) does not change.

Invariance of collimator efficiency with source-to-collimator distance applies to point sources, line sources, and uniform sheet sources in air with parallel-hole collimators; however, it applies only to uniform sheet sources with converging, diverging, or pinhole collimators (Section D). When the source is embedded at different depths in the patient, attenuation effects also must be considered. Septal penetration and scatter of photons from the walls of the collimator holes also are not considered in the earlier analysis.

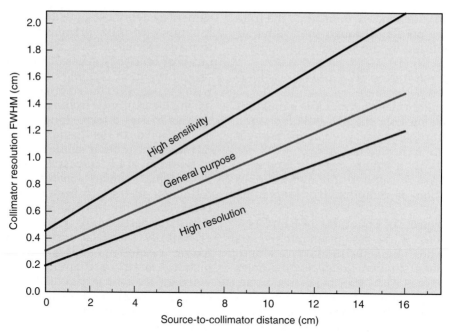

FIGURE 14-16 Collimator resolution versus source-to-collimator distance for three different collimators. (*Adapted from Hine GJ, Paras D, Warr CP: Recent advances in gamma-camera imaging. Proc SPIE 152:123, 1978.*)

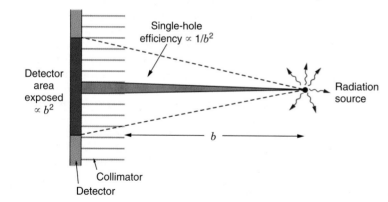

FIGURE 14-17 Explanation for constant counting rate (collimator efficiency) versus source-to-collimator distance for a point source in air and a parallel-hole collimator. Efficiency for a single hole decreases as $1/b^2$, but number of holes passing radiation (area of detector exposed) increases as b^2.

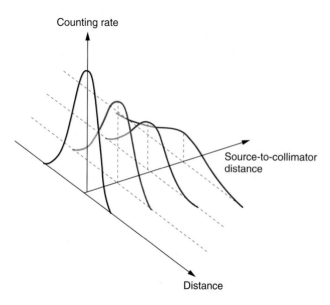

FIGURE 14-18 Point-spread functions versus distance for a parallel-hole collimator. Area under curve is proportional to collimator efficiency and does not change with distance.

TABLE 14-1

PERFORMANCE CHARACTERISTICS OF SOME TYPICAL COMMERCIALLY MANUFACTURED PARALLEL-HOLE COLLIMATORS

Collimator Type	Recommended Max. Energy (keV)	Efficiency, g	Resolution R_{coll} (FWHM at 10 cm)
Low-energy, high-resolution	150	1.84×10^{-4}	7.4 mm
Low-energy, general-purpose	150	2.68×10^{-4}	9.1 mm
Low-energy, high-sensitivity	150	5.74×10^{-4}	13.2 mm
Medium-energy, high-sensitivity	400	1.72×10^{-4}	13.4 mm

Adapted from Hine GJ, Erickson JJ: Advances in scintigraphic instruments. In Hine GJ, Sorenson JA (eds): *Instrumentation in Nuclear Medicine*, Vol 2. New York, 1974, Academic Press.
FWHM, full width at half maximum.

Table 14-1 summarizes the physical construction and typical performance characteristics of a number of collimators. Collimator resolution is the FWHM for a source at 10 cm from the face of the collimator. Collimator efficiency g refers to the relative number of γ rays transmitted by the collimator and reaching the detector per γ ray emitted by the source. Note that the approximate relationship between collimator efficiency and resolution given by Equation 14-8 is verified by these data. Note also the relatively small values for collimator efficiency.

4. System Resolution

The sharpness of images recorded with a gamma camera is limited by several factors, including intrinsic resolution, collimator resolution, scattered radiation, and septal penetration. In terms of the FWHM of a PSF or LSF, the most important factors are the intrinsic resolution R_{int} of the detector and electronics, and the collimator resolution R_{coll}. The combined effect of these two factors is to produce a *system resolution* R_{sys} that is somewhat worse than either one alone. System resolution R_{sys} (FWHM) is given by

$$R_{sys} = \sqrt{R_{int}^2 + R_{coll}^2} \qquad (14\text{-}9)$$

Because collimator resolution depends on source-to-collimator distance, system resolution also depends on this parameter. Figure 14-19 shows system resolution versus source-to-collimator distance for a typical parallel-hole collimator and different values of intrinsic resolution. At a distance of 5-10 cm (typical depth of organs inside the body), *system resolution* is much poorer than intrinsic resolution and is determined primarily by *collimator resolution*. There are significant

differences between system resolutions for cameras having substantially different intrinsic resolutions (e.g., 4 mm vs. 8 mm), but the difference in system resolutions for cameras having small differences in intrinsic resolutions (e.g., 4 mm vs. 5 mm) is minor and not clinically significant. Small differences in intrinsic resolution may be apparent on bar-pattern images or on images of very superficial structures in the patient, but they usually are not apparent on images of deeper-lying structures.

System resolution also is degraded by scattered radiation. This is discussed in Chapter 15, Section C. The method for combining component resolutions to determine system resolution also is discussed in Appendix G.

D. PERFORMANCE CHARACTERISTICS OF CONVERGING, DIVERGING, AND PINHOLE COLLIMATORS

Figure 14-20 illustrates the important design parameters for converging, diverging, and pinhole collimators. Equations for collimator resolution, R_{coll}, and efficiency, g, for these collimators are as follows:

Converging Collimator:

$$R_{coll} \approx [d(l'_{eff} + b)/l'_{eff}][1/\cos\theta] \\ \times [1 - (l'_{eff}/2)/(f + l'_{eff})] \qquad (14\text{-}10)$$

$$g \approx K^2(d/l'_{eff})^2[d^2/(d+t)^2][f^2/(f-b)^2] \qquad (14\text{-}11)$$

where

$$l'_{eff} \approx (l - 2\mu^{-1})/\cos\theta \approx l_{eff}/\cos\theta \qquad (14\text{-}12)$$

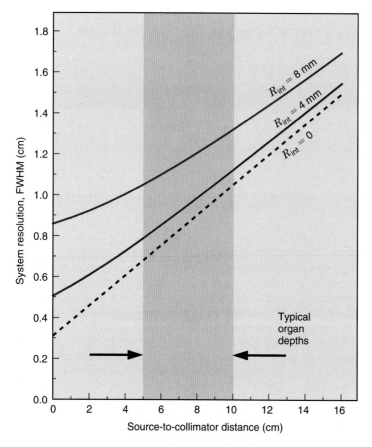

FIGURE 14-19 System resolution versus source-to-collimator distance for a typical parallel-hole collimator and different values of intrinsic resolution. At most typical organ depths, system resolution is determined primarily by collimator resolution.

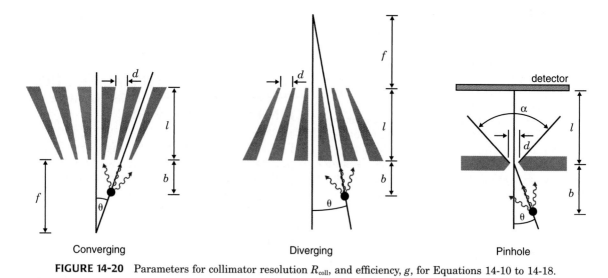

FIGURE 14-20 Parameters for collimator resolution R_{coll}, and efficiency, g, for Equations 14-10 to 14-18.

Diverging Collimator:

$$R_{coll} \approx [d(l'_{eff} + b)/l'_{eff}][1/\cos\theta][1 + (l'_{eff}/2f)] \tag{14-13}$$

$$g \approx K^2(d/l'_{eff})^2[d^2/(d+t)^2][(f+l)/(f+l+b)] \tag{14-14}$$

Pinhole Collimator:

$$R_{coll} \approx d_{eff,R}(l+b)/l \tag{14-15}$$

$$g \approx d_{eff,g}^2 \cos^3\theta/(16b^2) \tag{14-16}$$

where

$$d_{eff,R} = d + \frac{\ln(2)}{\mu} \tan\left(\frac{\alpha}{2}\right) \quad (14\text{-}17)$$

and

$$d_{eff,g} = \sqrt{d[d + (2/\mu)\tan(\alpha/2)] + [(2/\mu^2)\tan^2(\alpha/2)]}$$
$$(14\text{-}18)$$

l'_{eff} is the effective collimator length, accounting for septal penetration at different off-axis locations (see also Equation 14-6). For the pinhole collimator, $d_{eff,R}$ and $d_{eff,g}$ are the "effective" pinhole diameters, for resolution and sensitivity, respectively. d_{eff} takes into account the penetration of gamma rays through the edges of the pinhole aperture, but still assumes parallel rays, normally incident on the detector surface.[4,5] These expressions for d_{eff} also assume that the pinhole aperture has a "knife-edge" geometry, as illustrated in Figure 14-20.

The equations for collimator resolution R_{coll} refer to the equivalent FWHM of the PSF or LSF, corrected for magnification or minification of the image by the collimator described by Equations 13-3, 13-5, and 13-6. Thus, if the collimator projects a profile with a 2-cm FWHM measured on the detector and the

image magnification factor is ×2, the equivalent FWHM in the imaged plane is 1 cm.

These equations may be compared with Equations 14-6 and 14-7 for the parallel-hole collimator. They are similar except for the presence of additional terms involving collimator focal lengths f and, for off-axis sources, the angle θ between the source, the focal point (or pinhole), and the central axis of the collimator. The equations illustrate that for converging and diverging collimators, resolution is best at the center ($\theta = 0$, $\cos \theta = 1$).

The performance characteristics of different types of collimators are compared in Figure 14-21, which shows *system* resolution and efficiency versus distance, including effects of camera intrinsic resolution as well as collimator magnification. Equations 14-10, 14-13, and 14-15 show that resolution *always* is best with the source as close as possible to the collimator. Changes in collimator efficiency with distance depend on whether the radiation source is a point source or a uniform sheet source.

For a point source (Fig. 14-21, *right*), collimator efficiency increases with increasing source-to-collimator distance for the converging collimator. Maximum efficiency is obtained at the collimator convergence point

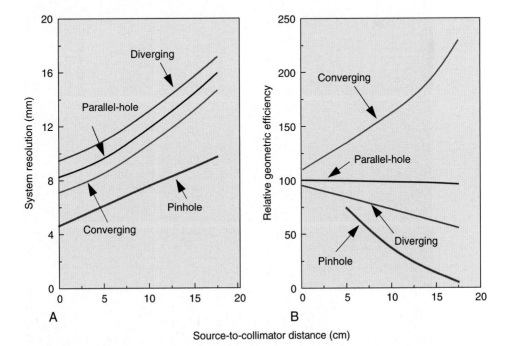

FIGURE 14-21 Performance characteristics (*A,* system resolution; *B,* point-source geometric efficiency in air) versus source-to-collimator distance for four different types of gamma camera collimators. (*Adapted with permission from Society of Nuclear Medicine from Moyer RA: A low-energy multihole converging collimator compared with a pinhole collimator.* J Nucl Med *15:59-64, 1974.*)

(~35 cm), where γ rays are transmitted through all of the collimator holes, and then decreases beyond that point. Point-source collimator efficiency decreases with distance for the diverging and pinhole collimators, more severely for the latter. For an extended, large-area sheet source, sufficiently large to cover the entire field of view of the collimator, efficiency does not change with source-to-collimator distance for all of these collimators. Again, for sources embedded within a patient, attenuation effects also must be accounted for.

Figure 14-21 illustrates that the converging collimator offers the best combination of resolution and efficiency at typical imaging distances (5 to 10 cm); however, the field-of-view is also somewhat limited at these distances (Equation 13-6 and Example 13-2), and for this reason converging collimators are most useful with cameras having relatively large-area detectors. Diverging collimators offer a larger imaging area (Example 13-1) but at the cost of both resolution and efficiency. Pinhole collimators offer very good resolution and reasonable efficiency at close distances but lose efficiency very rapidly with

distance; they also have a quite limited field of view because of magnification effects at typical imaging distances (Equation 13-3). Generally they are used for imaging smaller organs, such as the thyroid and heart, which can be positioned close to the collimator. They also are useful with high-energy γ-ray emitters because they can be designed to reduce septal penetration problems.

Differences between the resolution and field-of-view obtained at different source-to-collimator distances with parallel-hole, converging, diverging, and pinhole collimators are further illustrated by Figure 14-22. The distortions caused by changing magnification with depth for different structures inside the body sometimes make images obtained with the converging, diverging, and pinhole collimators difficult to interpret (see Fig. 13-9).

E. MEASUREMENTS OF GAMMA CAMERA PERFORMANCE

It is important to define standardized experimental protocols for measuring gamma camera performance that produce consistent

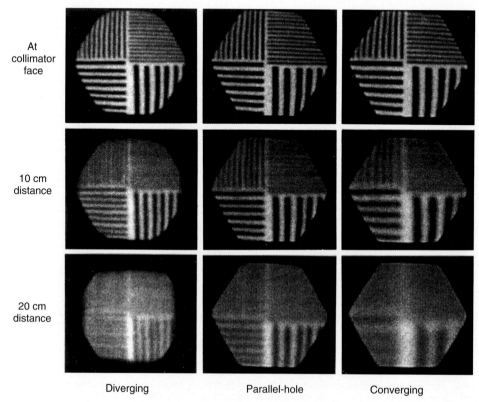

At collimator face

10 cm distance

20 cm distance

Diverging   Parallel-hole   Converging

FIGURE 14-22 Bar-pattern images demonstrating changing field size and resolution obtained versus distance for three collimator types.

results, are easily reproducible, and that do not require specialized equipment. Such protocols can then be used in comparing one gamma camera with another, in performing acceptance testing on a newly installed gamma camera, and as part of a quality assurance program to ensure that the camera is always performing to its specifications.

The exact regulations for gamma camera quality assurance and the guidelines for acceptance testing vary with locality. For example, in the United States, the Joint Commission requires that instruments be tested prior to initial use and that the performance of a gamma camera be tested at least once a year. Because of the rapidly changing regulatory environment, and differences between states and countries, a detailed review of the requirements of regulatory agencies is beyond the scope of this book. In this section, we therefore briefly summarize the more common measurements that are performed to assess gamma camera performance (whether they be for acceptance testing or for quality assurance). The protocols presented here are largely based on the recommendations of the National Electrical Manufacturers Association.[6] A typical quality assurance program might involve daily measures of flood-field uniformity, weekly checks of spatial resolution and spatial linearity, and semiannual checks of other performance parameters. It is important that all measurements be taken under the same conditions (pulse-height window width, correction algorithms, and correction circuitry on or off) as are used for routine clinical studies. More detailed information on performance measurements, quality assurance, and acceptance testing can be found in references 7 to 9.

1. Intrinsic Resolution

Intrinsic resolution is determined without a collimator using a linearity test pattern, such as the one shown in Figure 14-10 (left), placed directly on the surface of the NaI(Tl) crystal housing. The width of the strips in the pattern is approximately 1 mm, which is significantly smaller than the resolution expected in the measurement. A point source (usually ^{99m}Tc or ^{57}Co) is placed at a distance equal to five times the UFOV from the gamma camera face. The UFOV corresponds to the field of view of the gamma camera after masking off the portion of the camera face affected by edge-packing effects. Data are acquired with the system count rate below 30,000 cps

(<10,000 cps for a small field-of-view gamma camera) to avoid pile-up-related mispositioning. Two sets of images are taken, with a 90-degree rotation of the test pattern between acquisitions so that both X and Y resolution are measured. Data are acquired until the peak channel has at least 1000 counts. Images are acquired in a matrix with pixel sizes less than $\frac{1}{10}$ of the expected resolution (typically <0.35 mm). Profiles through the images of the line sources are taken at different locations across the gamma camera face and fitted to a Gaussian function (Chapter 9, Section B.3). The FWHM (Fig. 14-15) and full width at tenth maximum (FWTM) of the profiles are measured in both X and Y directions. The reported measurements usually are average measurements across the UFOV, and the average across the central field-of-view (CFOV) that has linear dimensions scaled by 75% with respect to the dimensions of the UFOV. Typical values of intrinsic spatial resolution are 2.5 to 3.5 mm.

2. System Resolution

This measurement is made with the collimator in place and should be repeated for each collimator of interest. The source consists of two 1-mm-diameter line sources, placed 5 cm apart at a distance of 10 cm from the front face of the collimator. The measurement also can be performed with the addition of a scattering medium by placing 10 cm of plastic between the sources and the collimator, and 5 cm of the same material behind the sources. Images are acquired (typically several million events, at a rate of <30,000 cps to avoid pile-up) and profiles taken through the image of the line sources are fitted to Gaussian functions to determine FWHM and FWTM as described for intrinsic resolution. The results vary widely depending on the exact type of collimator used but are typically in the range of 8 to 14 mm for ^{99m}Tc.

3. Spatial Linearity

This measurement uses the same slit pattern (Fig. 14-10A) and conditions as for the intrinsic resolution measurement. Once again, measurements are taken with two orientations of the test pattern, rotated by 90 degrees, to provide linearity measurements in both X and Y directions. Two measurements can be made from the resulting images. The *differential spatial linearity* is the deviation of the measured distance d_i between two slits from the actual distance D between them calculated for each row i in the image. The means and the standard deviations are reported for the X and Y directions across the UFOV and the CFOV and are defined as:

$$\text{Mean} = \frac{\sum_{i=1}^{n}(d_i - D)}{n} \qquad (14\text{-}19)$$

$$\text{Standard Deviation} = \sqrt{\frac{\sum_{i=1}^{n}(d_i - D)^2}{n-1}} \qquad (14\text{-}20)$$

In addition, the *absolute spatial linearity* is defined as the maximum deviation of the location of the slits from their true location. Once again this is assessed for the UFOV and the CFOV. It is not easy to detect small nonlinearities using these techniques, and tests of uniformity (discussed in the next section) usually are better at revealing the effects of small nonlinearities.

4. Uniformity

Intrinsic uniformity is determined from flood-field images acquired without a collimator. A ^{99m}Tc source is placed at a distance of approximately $5 \times$ the UFOV from the front face of the gamma camera. The source activity is such that the counting rate on the gamma camera is less than approximately 30,000 cps. Flood-field images are acquired so that there are a minimum of 4000 counts in each pixel of the image and then smoothed with a 9-point (3×3) smoothing filter with the following weightings:

1	2	1
2	4	2
1	2	1

Integral uniformity is based on the maximum and minimum pixel counts in the image and is defined as

Integral Uniformity (%)

$$= 100 \times \frac{\text{max. pixel count} - \text{min. pixel count}}{\text{max. pixel count} + \text{min. pixel count}} \qquad (14\text{-}21)$$

This is calculated for the UFOV and CFOV. Integral uniformity values are typically 2% to 4%.

Differential uniformity is based on the change in counts of five consecutive pixels across all rows and columns of the image. It is defined as

$$\text{Differential Uniformity } (\%) = 100 \times \frac{(\text{high} - \text{low})}{(\text{high} + \text{low})} \qquad (14\text{-}22)$$

where "high" refers to the maximum count difference for any five consecutive pixels (row or column) in the image and "low" refers to the minimum count different for any five consecutive pixels. This usually is reported for the UFOV.

For convenience, uniformity measurements often are made with the collimator in place (*extrinsic uniformity*). A thin flood-field source of ^{99m}Tc or a disk source of ^{57}Co that covers the active area of the gamma camera is placed on top of the collimator to provide uniform irradiation. This protocol is more practical for routine quality assurance because the measurement can be done without removing the collimator. Extrinsic uniformity measurements also have the advantage that they reveal any defects or problems caused by the collimator itself.

5. Counting Rate Performance

As described in Section A.4, most gamma cameras behave as paralyzable counting systems with the observed count rate described as a function of the true count rate by Equation 11-18. The basis for measurement of the dead time, τ, is the two-source method described in Chapter 11, Section C.4. Two ^{99m}Tc sources are placed approximately 1.5 m away from the camera face. The total activity should be sufficient to cause approximately a 20% loss in the observed counting rate relative to the true counting rate. Counting rates then are measured with both sources present, and then with each individual source present. Care must be taken that all measurements are performed with exactly the same source geometry, that pile-up rejection electronics or any other high counting rate correction circuitry is turned on, and that source decay is negligible (<1%) during the course of the measurement. The dead time can then be calculated from Equation 11-25. The observed count rate at which a 20% counting rate loss occurs, $R_{20\%}$, is also often quoted, and this can be computed from Equation 11-18 using the fact that $R_o = 0.8R_t$ as

$$R_{20\%} = -\frac{0.8}{\tau}\ln(0.8) \qquad (14\text{-}23)$$

6. Energy Resolution

Energy resolution is measured with a flood illumination of the gamma camera face, without a collimator, using a ^{99m}Tc source suspended $5 \times$ UFOV above the camera face. The resulting pulse-height spectrum is analyzed to determine the FWHM of the ^{99m}Tc photopeak. It usually is reported in keV or converted to a percent energy resolution based on the energy of the photopeak (see Equation 10-3). Typical values are in the range of 8% to 11% for ^{99m}Tc.

7. System Sensitivity

System sensitivity needs to be measured separately for each collimator. In general, the sensitivity of low-energy collimators is measured with ^{99m}Tc ($E_\gamma = 140$ keV), that of medium-energy collimators is measured with ^{111}In ($E_\gamma = 172$, 247 keV), and ^{131}I ($E_\gamma = 364$ keV) is used for high-energy collimators. A solution of the radionuclide (known total activity) is placed in a 10-cm diameter dish to a depth of 2 to 3 mm. The shallow depth minimizes self-absorption by the source. The source is placed 10 cm from the front face of the collimator and an image is acquired. The sensitivity is calculated by drawing a circular region of interest around the image of the dish and integrating all the counts in that region. A second image is recorded for an equal imaging time with the source removed to provide a measure of the background. The same region of interest is applied to this image. The sensitivity is given by

$$\text{Sensitivity(cps/Bq)}$$
$$= \frac{\text{counts in ROI} - \text{background counts in ROI}}{\text{time(sec)} \times \text{source activity(Bq)}}$$
$$(14\text{-}24)$$

A general-purpose collimator typically has a sensitivity on the order of 2 to 3×10^{-4} cps/Bq or 0.02% to 0.03%.

REFERENCES

1. Anger HO: Radioisotope cameras. In Hine GJ, editor: *Instrumentation in Nuclear Medicine*, Vol 1. New York, 1967, Academic Press, pp 485-552.
2. Muehllehner G, Karp JS: A positron camera using position-sensitive detectors: PENN-PET. *J Nucl Med* 27:90-98, 1986.
3. Graham LS: Automatic tuning of scintillation cameras: A review. *J Nucl Med Tech* 14:105-110, 1986.
4. Accorsi R, Metzler SD: Analytic determination of the resolution-equivalent effective diameter of a pinhole collimator. *IEEE Trans Med Imag* 23:750-763, 2004.
5. Smith MF, Jaszczak RJ: The effect of gamma ray penetration on angle-dependent sensitivity for pinhole collimation in nuclear medicine. *Med Phys* 24:1701-1709, 1997.
6. Performance measurements of scintillation cameras. Standards Publication No. NU-1-2007. Washington, DC, 2007, National Electrical Manufacturers Association.
7. Simmons GH, editor: *The Scintillation Camera*, New York, 1988, Society of Nuclear Medicine.
8. Graham LS: Scintillation camera imaging performance and quality control. In Henkin RE, Boles MA, Karesh SM, et al, editors: *Nuclear Medicine*, St. Louis, 1996, Mosby, pp 125-146.
9. Murphy PH: Acceptance testing and quality control of gamma cameras, including SPECT. *J Nucl Med* 28:1221-1227, 1987.

15

Image Quality in Nuclear Medicine

Image quality refers to the faithfulness with which an image represents the imaged object. The quality of nuclear medicine images is limited by several factors. Some of these factors, relating to performance limitations of the gamma camera, already have been discussed in Chapter 14. In this chapter, we discuss the essential elements of image quality in nuclear medicine and how it is measured and characterized. Because of its predominant role in nuclear medicine, the discussion focuses on planar imaging with the gamma camera; however, the general concepts are applicable as well to the tomographic imaging techniques that are discussed in Chapters 17 to 19.

A. BASIC METHODS FOR CHARACTERIZING AND EVALUATING IMAGE QUALITY

There are two basic methods for characterizing or evaluating image quality. The first is by means of *physical characteristics* that can be quantitatively measured or calculated for the image or imaging system. Three such characteristics that are used for nuclear medicine image quality are (1) *spatial resolution* (detail or sharpness), (2) *contrast* (difference in image density or intensity between areas of the imaged object containing different concentrations of radioactivity), and (3) *noise* (statistical noise caused by random fluctuations in radioactive decay, or structured noise, e.g., resulting from instrument artifacts). Although they describe three different aspects of image quality, these three factors cannot be treated as completely independent parameters because improvements in one of them frequently are obtained at the expense or deterioration of one or more of the others. For example, improved collimator resolution usually involves a tradeoff of decreased collimator efficiency (see Chapter 14, Section C) and, hence, decreased counting rates and increased image statistical noise.

The second method for characterizing or evaluating image quality is by means of *human observer performance studies* using images obtained with different imaging systems or under different imaging conditions. Although observer performance can be characterized objectively, and certainly is related to the physical measures of image quality described earlier, the relationships are not well established because of the complexity of the human visual system and other complicating factors, such as observer experience. Hence, the two methods, though related, are somewhat independent.

A related approach, known as *computer observer performance studies,* uses a mathematical model that under appropriate conditions predicts the performance of a human observer and can be used as a surrogate for actual human observer studies. Because human observer studies require large numbers of images and therefore are very time consuming, computer observers often are more practical. Details regarding computer observer models are beyond the scope of this text and the interested reader is referred to reference 1 for further information.

B. SPATIAL RESOLUTION

1. Factors Affecting Spatial Resolution

Spatial resolution refers to the sharpness or detail of the image, or to the ability of the imaging instrument to provide such sharpness or detail. The sample images presented in Chapters 13 and 14 already have demonstrated that nuclear medicine images have somewhat limited spatial resolution, at least in comparison with photographic or radiographic images. A number of factors contribute to the lack of sharpness in these images.

Collimator resolution is perhaps the principal limiting factor when absorptive collimators are used for spatial localization (Chapter 14, Section C). Because collimator hole diameters must be relatively large (to obtain reasonable collimator efficiency), there is blurring of the image by an amount at least as great as the hole diameters (Equation 14-6). Collimator resolution also depends on source-to-detector distance (Figs. 14-16 through 14-22). Note that collimator resolution is not a factor in positron emission tomography (PET) imaging, which uses annihilation coincidence detection for spatial localization (Chapter 18, Section A).

A second factor is *intrinsic resolution* of the imaging detector. With the gamma camera, this limitation arises primarily because of statistical variations in the distribution of light photons among the photomultiplier tubes (Chapter 14, Section A.1). Intrinsic resolution is a function of γ-ray energy with the gamma camera, becoming poorer with decreasing γ-ray energy (Fig. 14-2). For imaging devices with discrete detector elements, such as many PET systems (see Chapter 18, Section A.3), the size of the individual detector elements largely determines the intrinsic resolution of the device.

Image sharpness also can be affected by patient motion. Figure 15-1 shows images of a brain phantom obtained with and without motion. Respiratory and cardiac motion can be especially troublesome because of the lengthy imaging times required in nuclear medicine and the relatively great excursions in distance (2-3 cm) that are possible in these instances. Gated-imaging techniques (see Chapter 20, Section A.4) have been employed to minimize motion blurring, especially in cardiac studies. Breath-holding also has been used to minimize blurring caused by respiratory motion.

Nuclear medicine imaging systems acquire data on a discrete matrix of locations, or pixels, which leads to *pixelation effects* in the image. As discussed in Chapter 20, the size of the discrete pixels sets a limit on the spatial resolution of the image. In general, it is desirable to have at least two pixels per full width at half maximum (FWHM) of system resolution to avoid creating distracting pixelation effects and possible loss of image detail.

2. Methods for Evaluating Spatial Resolution

Spatial resolution may be evaluated by subjective or objective means. A subjective evaluation can be obtained by visual inspection of images of organ phantoms that are meant to simulate clinical images (e.g., the brain phantom in Fig. 15-1). Although they attempt to project "what the physician wants to see," organ phantoms are not useful for quantitative comparisons of resolution between different imaging systems or techniques. Also, because of the subjective nature of the evaluation, different observers might give different interpretations of comparative image quality.

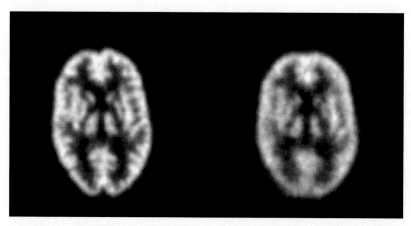

Stationary   With motion

FIGURE 15-1 Images of a brain phantom obtained with phantom stationary (*left*) and with random translations (several mm) and rotations (several degrees) (*right*) during the imaging procedure, demonstrating motion-blurring effects. (*Adapted from Fulton R et al: Accuracy of motion correction methods for PET brain imaging. 2004 IEEE Nuclear Science Symposium Conference Record, 4226-4230.*)

A phantom that can be used for more objective testing of spatial resolution is shown in Figure 15-2. *Bar phantoms* are constructed of lead or tungsten strips, generally encased in a plastic holder. Strips having widths equal to the spaces between them are used. For example, a "5-mm bar pattern" consists of 5-mm-wide strips separated edge to edge by 5-mm spaces. The four-quadrant bar phantom shown in Figure 15-2 has four different strip widths and spacings. To evaluate the intrinsic resolution of a gamma camera, the bar phantom is placed directly on the uncollimated detector and irradiated with a uniform radiation field, typically a point source of radioactivity at several meters distance from the detector. To evaluate the resolution with a collimator, the phantom is placed directly on the collimated detector and irradiated with a point source at several meters distance, or with a sheet source of radioactivity placed directly behind the bar phantom. Spatial resolution is expressed in terms of the smallest bar pattern visible on the image. There is a certain amount of subjectivity to the evaluation, but not so much as with organ phantoms.

To properly evaluate spatial resolution with bar phantoms, one must ensure that the thickness of lead strips is sufficient so that they are virtually opaque to the γ rays being imaged. Otherwise, poor visualization may be due to poor contrast of the test image rather than poor spatial resolution of the imaging device. For ^{99m}Tc (140 keV) and similar low-energy γ-ray emitters, tenth-value thicknesses in lead are approximately 1 mm or less, whereas for ^{131}I (364 keV), annihilation photons (511 keV), and so on, they are on the order of 1 cm (see Table 6-4). Most commercially available bar phantoms are designed for ^{99m}Tc and are not suitable for use with higher-energy γ-ray emitters.

A still more quantitative approach to evaluating spatial resolution is by means of the point-spread function (PSF) or line-spread function (LSF). General methods for recording these functions were described in Chapter 14, Section E.2. Examples of LSFs are shown in Figure 17-8 for a single-photon emission computed tomography (SPECT) camera and in Figure 18-5 for a PET system. Although the complete profile is needed to fully characterize spatial resolution, a partial specification is provided by its FWHM (Fig. 14-15). The FWHM is not a complete specification because PSFs or LSFs of different shapes can have the same FWHM. (Compare, for example, the different shapes in Figs. 18-5 and 18-7). However, the FWHM is useful for general comparisons of imaging devices and techniques. Roughly speaking, the FWHM of the PSF or LSF of an imaging instrument is approximately 1.4-2 times the width of the smallest resolvable bar pattern (Fig. 15-3). Thus an instrument having an FWHM of 1 cm should be able to resolve 5- to 7-mm bar patterns.

In most cases, multiple factors contribute to spatial resolution and image blurring. The method for combining FWHMs for intrinsic and collimator resolutions to obtain the overall system FWHM is discussed in Chapter 14, Section C.4 and in Appendix G. In general, if a system has n factors or components that each contribute independently to blurring, individually characterized by FWHM_1, FWHM_2, . . ., FWHM_n, the FWHM for the system is given by

$$\text{FWHM}_{\text{sys}} \approx \sqrt{\text{FWHM}_1^2 + \text{FWHM}_2^2 + \ldots + \text{FWHM}_n^2} \tag{15-1}$$

This equation provides an exact result when all of the components have gaussian-shaped

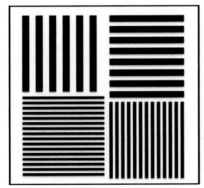

FIGURE 15-2   Design (*left*) and gamma camera image (*right*) of a four-quadrant bar phantom used for evaluation of spatial resolution.

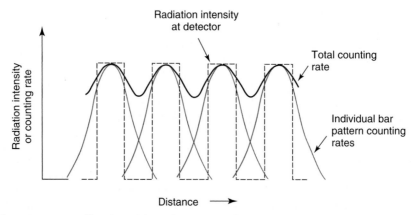

FIGURE 15-3   Counting-rate profiles obtained on a bar pattern phantom with an imaging system having FWHM resolution approximately 1.6 times the width of individual bars and spaces.

blurring functions, but it is an approximation when nongaussian shapes are involved. Note that if the FWHM for any one factor is significantly larger than the others, it becomes the dominating factor for system FWHM. Thus, for example, if $FWHM_1 \gg FWHM_2$, it makes little sense to expend substantial effort toward improving $FWHM_2$.

The most detailed specification of spatial resolution is provided by the *modulation transfer function* (MTF). The MTF is the imaging analog of the frequency response curve used for evaluating audio equipment. In audio equipment evaluations, pure tones of various frequencies are fed to the input of the amplifier or other component to be tested, and the relative amplitude of the output signal is recorded. A graph of relative output amplitude versus frequency is the frequency response curve for that component (Fig. 15-4). A system with a "flat" curve from lowest to highest frequencies provides the most faithful sound reproduction.

By analogy, one could evaluate the fidelity of an imaging system by replacing the audio tone with a "sine-wave" distribution of activity (Fig. 15-5). Instead of varying in time (cycles per second), the activity distribution

varies with distance (cycles per centimeter or cycles per millimeter). This is called the *spatial frequency* of the test pattern, customarily symbolized by k.* The modulation of the test pattern, which is a measure of its contrast, is defined by

$$M_{in} = (I_{max} - I_{min})/(I_{max} + I_{min}) \quad (15\text{-}2)$$

where I_{max} and I_{min} are the maximum and minimum radiation intensities emitted by the test pattern. M_{in} is the input modulation for the test pattern and ranges from zero ($I_{max} = I_{min}$, no contrast) to unity ($I_{min} = 0$, maximum contrast). Similarly, output modulation M_{out} is defined in terms of the modulation of output image (e.g., image density or counting rate recorded from the test pattern).

$$M_{out} = (O_{max} - O_{min})/(O_{max} + O_{min}) \quad (15\text{-}3)$$

---

*Technically speaking, the notation k is used in physics to denote "cycles per radian," and the notation $\bar{k}$ or "k-bar" is used to denote "cycles per distance." Mathematically, $\bar{k} = k/2\pi$, because there are 2π radians per cycle. For notational simplicity, we use k for cycles per distance in this text.

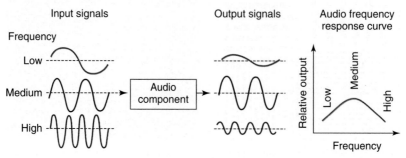

FIGURE 15-4   Basic principles for generating frequency response curves for an audio system.

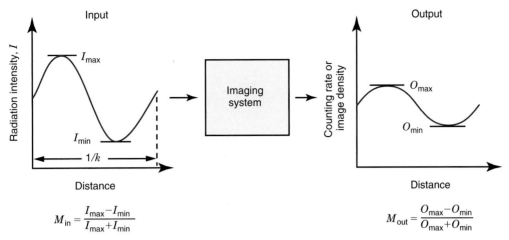

FIGURE 15-5 Basic principles for determining the modulation transfer function of an imaging instrument. Input contrast is measured in terms of object radioactivity or emission rate. Output contrast is measured in terms of counting rate, image intensity, etc. Spatial frequency is k.

The ratio of output to input modulation is the MTF for the spatial frequency k of the test pattern,

$$\mathrm{MTF}(k) = M_{\mathrm{out}}(k)/M_{\mathrm{in}}(k) \qquad (15\text{-}4)$$

The usefulness of the MTF (or frequency response curve) derives from the fact that any image (or audio signal) can be described as a summation of sine waves of different frequencies. For audio signals, the sound "pitch" is determined by its basic sine-wave frequency, whereas superimposed higher frequencies create the unique sound characteristics of the instrument or human voice producing it. An audio system with a flat frequency response curve over a wide frequency range generates an output that matches faithfully the sound of the instrument or voice producing it. Inexpensive audio systems generally reproduce the midrange audio frequencies accurately but have poor response at low and high frequencies. Thus they have poor bass response (low frequencies) and poor sound "quality" (high frequencies).

An imaging system with a flat MTF curve having a value near unity produces an image that is a faithful reproduction of the imaged object. Good low-frequency response is needed to outline the coarse details of the image and is important for the presentation and detection of relatively large but low-contrast lesions. Good high-frequency response is necessary to portray fine details and sharp edges. This is of obvious importance for small objects but sometimes also for larger objects because of the importance of edges and sharp borders for detection of low-contrast objects and for accurate assessment of their size and shape.

Figure 15-6 illustrates some typical MTF curves for a gamma camera collimator. The MTF curves have values near unity for low frequencies but decrease rapidly to zero at higher frequencies. Thus the images of a radionuclide distribution obtained with this collimator show the coarser details of the distribution faithfully but not the fine details. Edge sharpness, which is a function of the high-frequency MTF values, also is degraded. This type of performance is characteristic of virtually all nuclear medicine imaging systems. Note also that the MTF curve at higher frequencies decreases more rapidly with increasing source-to-collimator distance.

The MTF curve characterizes completely and in a quantitative way the spatial resolution of an imaging system for both coarse and fine details. Images of bar patterns and similar test objects are quantitative only for specifying the limiting resolution of the imaging system, for example, the minimum resolvable bar pattern spacing. Bar-pattern images and MTF curves can be related semiquantitatively by noting that the spatial frequency of a bar pattern having bar widths and spaces of x cm is one cycle per $2x$ cm. Thus a "5-mm bar pattern" has a basic spatial frequency of one cycle per centimeter (one bar and one space per centimeter). Roughly speaking, bar patterns are no longer visible when the MTF for

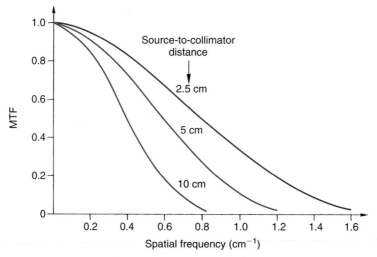

FIGURE 15-6 Modulation transfer function curves for a typical parallel-hole collimator for different source-to-collimator distances. (*Data from Ehrhardt JC, Oberly LW, Cuevas JM:* Imaging Ability of Collimators in Nuclear Medicine. *Publication No. [FDA] 79-8077. Rockville, MD, U.S. Department of Health, Education, and Welfare, 1978.*)

their basic spatial frequency drops below a value of approximately 0.1. MTF curves thus can be used to estimate the minimum resolvable bar pattern for an imaging system.

In practice, MTFs are not determined using sinusoidal activity distributions, as illustrated in Figure 15-5, which would be difficult to construct. Instead, they are obtained by mathematical analysis of the LSF or PSF. Specifically, the MTF of an imaging system can be derived from the Fourier transform (FT) of the LSF or PSF.* The one-dimensional (1-D) FT of the LSF is the MTF of the system measured in the direction of the profile, that is, perpendicular to the line source. Similarly, the 1-D FT of a profile recorded through the center of the PSF gives the MTF of the system in the direction of the profile. Alternatively, a 2-D FT of the 2-D PSF provides a 2-D MTF that can be used to determine the frequency response of the system at any angle relative to the imaging detector. This sometimes is useful for imaging systems that have asymmetrical spatial resolution characteristics, such as detector arrays with rectangular elements. Some PET detector arrays have this property (see Chapter 18, Section B).

It also is possible to obtain a 3-D representation of the MTF from a complete 3-D data set for the PSF. This is potentially useful for characterizing the spatial resolution of tomographic instruments in all three spatial directions. Note that in all cases the diameter or width of the source should be much smaller than resolution capability of the imaging device ($d \leq \text{FWHM}/4$). Additional discussions about the measurement of MTFs and their properties can be found in references 2 and 3.

Another useful feature of MTFs is that they can be determined for different components of an imaging system and then combined to determine the system MTF. This feature allows one to predict the effects of the individual components of the system on the MTF of the total system. For example, one can obtain the MTF for the intrinsic resolution of the Anger camera detector, $\text{MTF}_{int}(k)$, and another for the collimator, $\text{MTF}_{coll}(k)$. The system MTF then is obtained by point-by-point multiplication of the intrinsic and collimator MTFs at each value of k:

$$\text{MTF}_{sys}(k) = \text{MTF}_{int}(k) \times \text{MTF}_{coll}(k) \quad (15\text{-}5)$$

In general, the MTF of a system is the product of the MTFs of its components.

If two systems have MTF curves of the same general shape, one can predict confidently that the system with the higher MTF values will have superior spatial resolution; however, the situation is more complicated when comparing two systems having MTF curves of different shapes. For example, Figure 15-7 shows MTF curves for two collimators, one of which would be better for visualizing large low-contrast structures (low frequencies), the other for fine details (high

*More specifically, the MTF is the modulus, or amplitude, of the FT, the latter generally being a complex number. See Appendix F for a more detailed discussion of FTs.

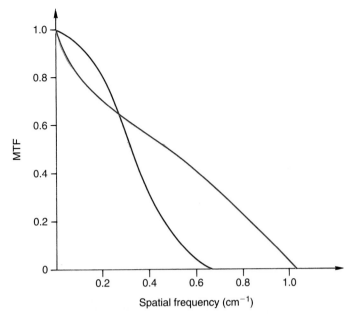

FIGURE 15-7 Modulation transfer function (MTF) curves for two different collimators. One has better low-frequency resolution for coarse details (*blue line*), whereas the other is better for fine details (*orange line*). (*Data from Ehrhardt JC, Oberly LW, Cuevas JM:* Imaging ability of collimators in nuclear medicine. *Rockville, MD, U.S. Department of Health, Education, and Welfare, Publ. No. [FDA] 79-8077, 1978, p 20.*)

frequencies). To gain an impression of comparative image quality in this situation, one would probably have to evaluate organ phantoms or actual patient images obtained with these collimators.

C. CONTRAST

Image contrast refers to differences in intensity in parts of the image corresponding to different levels of radioactive uptake in the patient. In nuclear medicine, a major component of image contrast is determined by the properties of the radiopharmaceutical. In general, it is desirable to use an agent having the highest lesion-to-background uptake or concentration ratio. Some aspects of radiopharmaceutical design that affect this issue were discussed in Chapter 5, Section F. Physical factors involved in image formation also can affect contrast. In general, factors that affect contrast in nuclear medicine also affect the statistical noise levels in the image. More specifically, they affect the contrast-to-noise ratio (CNR), which is discussed in detail in the next section. Here we focus only on some factors that affect contrast.

A general definition of contrast is that it is the ratio of signal change of an object of

interest, such as a lesion, relative to the signal level in surrounding parts of the image. Thus if R_o is the counting rate over normal tissue and R_ℓ is the counting rate over a lesion, the contrast of the lesion is defined as

$$C_\ell = \frac{R_\ell - R_o}{R_o}$$
$$= \frac{\Delta R_\ell}{R_o} \qquad (15\text{-}6)$$

where ΔR_ℓ is the change in counting rate over the lesion relative to the surrounding background.* Contrast sometimes is expressed as a percentage, for example, $C_\ell = 0.1 =$ "10% contrast."

Perhaps the major factor affecting contrast is *added background counting rates* that are superimposed more or less uniformly over the activity distribution of interest. For example, suppose that in the absence of background counts a certain object (e.g., a lesion) has

---

*This equation is related to, but not the same as, the equations for modulation given in Equations 15-2 and. 15-3. The definition used here has the disadvantage that it does not apply when $R_o = 0$. However, this situation rarely, if ever, applies in nuclear medicine, and the definition in Equation 15-6 is more straightforward for the analysis of contrast and CNR.

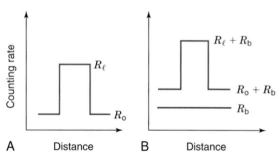

FIGURE 15-8 Effect on image contrast of adding a background counting rate R_b.

intrinsic contrast as defined by Equation 15-6. Suppose then that a uniform background counting rate R_b is superimposed on the image (Fig. 15-8). Then the lesion contrast becomes

$$
\begin{aligned}
C'_\ell &= \frac{(R_\ell + R_b) - (R_o + R_b)}{R_o + R_b} \\
&= \frac{\Delta R_\ell}{R_o + R_b} \\
&= \frac{\Delta R_\ell}{R_o} \times \left[\frac{1}{1 + (R_b / R_o)} \right] \quad (15\text{-}7) \\
&= C_\ell \times \left[\frac{1}{1 + (R_b / R_o)} \right]
\end{aligned}
$$

Comparing the last line of Equation 15-7 with Equation 15-6, it can be seen that contrast is decreased by the additional factor R_b/R_o in the denominator.

EXAMPLE 15-1

Suppose that under ideal conditions, a certain radiopharmaceutical produces lesion and normal tissue counting rates given by R_o and $R_\ell = 1.2R_o$, respectively. Suppose further that a background counting rate $R_b = R_o$ then is added to the image. Calculate the image contrast with and without the added background counting rate.

Answer

Using Equation 15-6 for the intrinsic contrast without the added background

$$
\begin{aligned}
C_\ell &= \frac{(1.2R_o - R_o)}{R_o} \\
&= 0.2 \ (20\%)
\end{aligned}
$$

When background amounting to $R_b = R_o$ is added, according to Equation 15-7

$$
\begin{aligned}
C'_\ell &= 0.2 \times \left[\frac{1}{1 + (R_b / R_o)} \right] \\
&= 0.2 \times \left[\frac{1}{1 + 1} \right] \\
&= 0.2 \times \frac{1}{2} = 0.1 \ (10\%)
\end{aligned}
$$

Thus contrast is reduced by 50% by the added background.

Example 15-1 illustrates that added background can reduce image contrast substantially. It should be noted again that background counting rates also add to the noise levels in the image, just as they add to the noise levels in counting measurements (see Chapter 9, Section D.4). This is discussed in more detail in Section D.

Background counting rates can arise from a number of sources. Septal penetration and scattered radiation are two examples. Another would be inadequately shielded radiation sources elsewhere in the imaging environment. Septal penetration is avoided by using a collimator that is appropriately designed for the radionuclide of interest (Chapter 14, Section C.2). Scattered radiation can be minimized by pulse-height analysis; however, sodium iodide [NaI(Tl)] systems cannot reject all scatter, and rejection becomes especially difficult for γ-ray energies below approximately 200 keV, as illustrated by Figure 10-10. Using a narrower analyzer window for scatter rejection also decreases the recorded counting rate and increases the statistical noise in the image. A reasonable tradeoff between counting efficiency and scatter rejection for imaging systems using NaI(Tl) detectors is obtained with a 15% energy window centered on the γ-ray photopeak. There has been continuing interest in applying semiconductor detectors to nuclear medicine imaging to take advantage of their superior energy resolution for discrimination against scattered radiation by pulse-height analysis (see Figs. 10-14 and 10-15).

Figure 15-9 shows the effect of scattered radiation on images of a phantom. With a very wide analyzer window, there is virtually no rejection of scattered radiation and a noticeable loss of image contrast. The loss of contrast can result in degraded visibility of both large low-contrast objects and fine details in the image. Figure 15-10, for example, illustrates the effects of scattered radiation (or

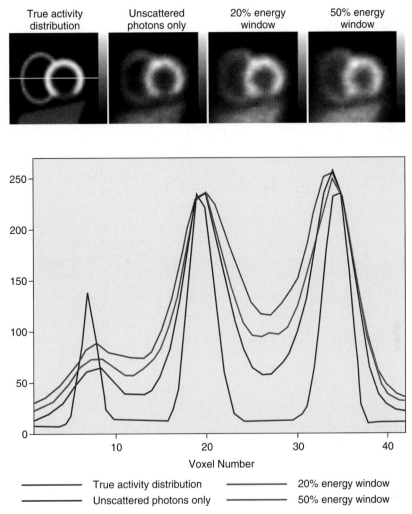

FIGURE 15-9 Effect of scatter and pulse-height analysis on image contrast. The images were generated by Monte Carlo simulations mimicking a clinical study of myocardial function using the radiotracer ^{201}TlCl. Count profiles through the images also are shown. These profiles are taken along the line shown in the image of the true activity distribution. The images also demonstrate blurring of the activity distribution caused by the finite camera resolution. (*Courtesy Dr. Hendrik Pretorius and Dr. Michael King, University of Massachusetts Medical School, Worcester, MA*).

septal penetration, which has similar effects) on the LSF and MTF of an imaging system. The addition of long "tails" to the LSF results first in the suppression of the MTF curve at low frequencies. This is reflected in poorer contrast of large objects that would make large low-contrast objects more difficult to detect or characterize. The high-frequency portion of the MTF curve also is suppressed, which has the effect of shifting the limiting frequency for detection of high-contrast objects (e.g., bar patterns) to lower frequencies. Thus the contrast-degrading effects of added background decrease the visibility of *all* structures in the image, particularly those

that may already be near the borderline of detectability. These effects are apparent in Figure 15-9, which demonstrates a perceptible loss of image sharpness as well as overall image contrast when the added background is present.

An important contributor to background radiation in conventional planar imaging is radioactivity above and below the object of interest. Image contrast is improved in emission computed tomography (SPECT and PET) (see Chapters 17 and 18) because it permits imaging of an isolated slice without the superimposed activities in overlying and underlying structures. Tomographic techniques offer

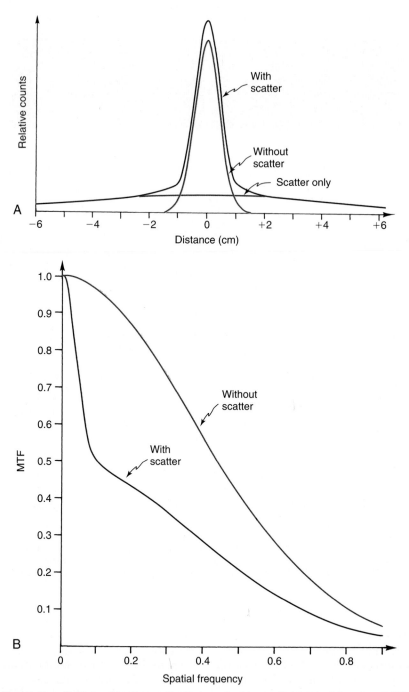

FIGURE 15-10 Illustration of effects of scatter and septal penetration on line-spread function (LSF) (*A*) and modulation transfer function (MTF) (*B*) of an imaging system. The long "tails" on the LSF have the effect of suppressing the MTF curve at both low and high spatial frequencies.

significant improvements for the detection of low-contrast lesions. Figure 15-11 illustrates this effect. Details of emission computed tomographic imaging are presented in Chapters 16 to 18; however, even at this point, the benefits of removing the interfering effects of overlying and underlying activity should be evident.

The preceding discussion relates to the effects of various types of background radiation on input contrast to the imaging system. It is possible with computers to apply "background subtraction" or "contrast enhancement" algorithms and thereby restore the original contrast, at least in terms of the

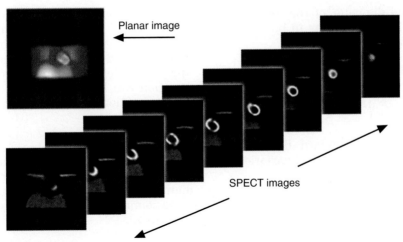

FIGURE 15-11 Planar (*upper left*) and single-photon emission computed tomographic (SPECT) (*center*) images of a thoracic phantom. Note the improved contrast and visibility of the voids in the cardiac portion of the phantom when overlying and underlying activity are removed in the SPECT images. (*Courtesy Dr. Freek Beekman, Delft University of Technology, Netherlands.*)

relative brightness levels between a lesion and its surrounding area. However, these techniques also enhance the statistical noise levels in the image as well as the contrast of any underlying artifacts, such as gamma camera image nonuniformities. Thus the critical parameter to consider regarding computer enhancement techniques is their effect on CNR. This concept is discussed in the following section.

D. NOISE

1. Types of Image Noise

Image noise generally can be characterized as either random or structured. *Random noise* refers to the mottled appearance of nuclear medicine images caused by random statistical variations in counting rate (Chapter 9). This is a very important factor in nuclear medicine imaging and is discussed in detail in this section.

Structured noise refers to nonrandom variations in counting rate that are superimposed on and interfere with perception of the object structures of interest. Some types of structured noise arise from the radionuclide distribution itself. For example, in planar imaging, uptake in the ribs may be superimposed over the image of the heart in studies to detect myocardial infarction with ^{99m}Tc-labeled pyrophosphates. Bowel uptake presents a type of structured noise in studies to detect inflammation or abscess with ^{67}Ga.

Structured noise also can arise from imaging system artifacts. Nonuniformities in gamma camera images (see Fig. 14-10) are one example. Various "ring" or "streak" artifacts generated during reconstruction tomography are another (e.g., see Figs. 16-11 and 16-13).

2. Random Noise and Contrast-to-Noise Ratio

Random noise, also called *statistical noise* or *quantum mottle*, is present everywhere in a nuclear medicine image. Even when the size of an object is substantially larger than the limiting spatial resolution of the image, statistical noise can impair detectability, especially if the object has low contrast. The critical parameter for detectability is the CNR of the object in the image. In the following discussion, we present an analysis and some illustrations of the effects of CNR on detectability of objects in 2-D planar nuclear images.

Suppose that a 2-D image contains a circular lesion of area A_ℓ having contrast C_ℓ (Equation 15-6) against a uniform background counting rate, R_o (cps/cm^2). The number of counts recorded in a background area of the same size as the lesion during an imaging time, t, is

$$N_o = R_o \times A_\ell \times t$$
$$= R_o \times \frac{\pi}{4} d_\ell^2 \times t \qquad (15\text{-}8)$$

where d_ℓ is the diameter of the lesion. The statistical variation of counts in background areas of size A_ℓ is

$$\sigma_{N_o} = \sqrt{N_o}$$

$$= \sqrt{R_o \times \frac{\pi d_\ell^2}{4} \times t} \qquad (15\text{-}9)$$

Thus the fractional standard deviation of counts due to random statistical variations is

$$C_{\text{noise}} = \frac{\sigma_{N_o}}{N_o}$$

$$= \frac{1}{\sqrt{R_o \times \frac{\pi d_\ell^2}{4} \times t}} \qquad (15\text{-}10)$$

As indicated by the notation in Equation 15-10, this factor can be considered as the "noise contrast" for a circular area of diameter d_ℓ in background areas of the image. The ratio of lesion-contrast to noise-contrast is defined as its CNR_ℓ.

$$\text{CNR}_\ell = \frac{|C_\ell|}{C_{\text{noise}}}$$

$$\approx |C_\ell| \times d_\ell \times \sqrt{R_o \times t} \qquad (15\text{-}11)$$

$$\approx |C_\ell| \times d_\ell \times \sqrt{\text{ID}_o}$$

where we have used the approximation $\sqrt{\pi/4} \approx 1$. The quantity $\text{ID}_o = (R_o \times t)$ is the background *information density* of the image and has units (counts/cm^2). The absolute value of contrast, $|C_\ell|$, is used in Equation 15-11 to indicate that it applies to either positive or negative contrast.

To detect a lesion or other object in an image, the observer must be able to distinguish between the lesion or object and noise-generated contrast patterns in background areas of the same size in the image. A substantial amount of research has gone into this subject. The conclusion is that, to be detectable, an object's CNR must exceed 3-5. This factor is known as the *Rose criterion,* after the individual who did basic studies on this subject.[4] The actual value depends on object size and shape, edge sharpness, viewing distance, observer experience, and so forth. Choosing a factor of 4, the requirement for detectability becomes $\text{CNR}_\ell \geq 4$, and Equation 15-11 can be written as

$$|C_\ell| \times d_\ell \times \sqrt{R_o \times t} \geq 4$$

$$|C_\ell| \times d_\ell \times \sqrt{\text{ID}_o} \geq 4 \qquad (15\text{-}12)$$

Equation 15-12 applies to somewhat idealized conditions of more or less circular objects against a relatively uniform background of nonstructured noise. Such conditions rarely apply in nuclear medicine. Nevertheless, this equation can be used to gain some insights into lesion detectability and the factors that affect it.

EXAMPLE 15-2

Estimate the minimum contrast for detection of circular objects of 1-cm and 2-cm diameter in an area of an image where the background information density is $\text{ID}_o = 400$ counts/cm^2.

Answer

Rearranging Equation 15-12 and inserting the specified information density,

$$|C_\ell| \geq \frac{4}{d_\ell \sqrt{\text{ID}_o}} = \frac{4}{\sqrt{400}d_\ell} = \frac{0.2}{d_\ell}$$

Thus for a 1-cm diameter object, the minimum contrast required for detectability is approximately 0.2 (20%), whereas for a 2-cm diameter object it is approximately 0.1 (10%).

Example 15-2 shows that, all other factors being the same, the contrast required for detectability is inversely proportional to object size.

EXAMPLE 15-3

Estimate the minimum diameter for detection of an object that has 10% contrast, $|C_\ell| = 0.1$, in an area of the image where the background information density is 100 counts/cm^2.

Answer

Rearranging Equation 15-12 and inserting the specified parameters,

$$d_\ell \geq \frac{4}{|C_\ell|\sqrt{\text{ID}_o}} = \frac{4}{0.1\sqrt{400}} = 4 \text{ cm}$$

Examples 15-2 and 15-3 illustrate that the minimum size requirement for object detectability decreases inversely with the square root of information density, from 2 cm with $\text{ID}_o = 400$ counts/cm^2 in Example 15-2 to 4 cm with $\text{ID}_o = 100$ counts/cm^2 in Example 15-3.

At first glance, it would seem that adding background radiation to an image would improve lesion detectability, by increasing information density, ID_o. However, as illustrated by Example 15-1, background radiation also degrades lesion contrast. The following example illustrates the overall effect of background radiation on object detectability.

EXAMPLE 15-4

Example 15-3 indicates that a 4-cm diameter object with 10% contrast should be detectable against a background information density of $ID_o = 100$ counts/cm^2. Suppose that background radiation with the same information density, ($ID_b = 100$ counts/cm^2) is added to the image. Estimate the minimum detectable lesion size after this is done.

Answer

According to Example 15-1, the addition of background radiation with $ID_b = ID_o$ decreases contrast by a factor of 2, from $C_\ell = 0.1$ to $C'_\ell = 0.05$. At the same time, the background information density increases from 100 counts/cm^2 to 200 counts/cm^2. Rearranging Equation 15-12 and inserting these values, one obtains

$$d_\ell \geq \frac{4}{|C'_\ell|\sqrt{ID_o}} = \frac{4}{0.05\sqrt{200}} \approx 5.7 \text{ cm}$$

Example 15-4 illustrates that the minimum detectable object size becomes larger (from 4 cm to 5.7 cm in the example) when background radiation is added. This example illustrates that the effect of background on degradation of object contrast more than offsets its effect toward increasing the information density of the image.

These examples and analyses assume that object size (d_ℓ) and contrast (C_ℓ) are independent variables. This may be true for computer-generated test images; however, in planar nuclear medicine imaging these parameters often are intimately linked. In many cases, lesions are somewhat spherical in shape so that their thickness varies linearly with their diameter. Thus a larger lesion not only has a larger diameter but generates greater contrast as well. The following example illustrates how these two factors operating together affect lesion CNR and detectability.

EXAMPLE 15-5

Suppose that a certain radionuclide concentrates in normal tissue to a level that provides a counting rate of 10 cpm/cm^2 per cm of tissue thickness. Suppose further that it concentrates in a certain type of lesion to a level that is twice this value, that is, 20 cpm/cm^2 per centimeter of lesion thickness. Compare the contrast, CNR and detectability of 1-cm diameter versus 2-cm diameter lesions embedded in normal tissue of total thickness 10 cm and

an imaging time of 1 min. Ignore the effects of attenuation and source-to-detector distance for this comparison.

Answer

In both cases, the uptake in normal tissues would generate a background counting rate of $R_o = 10$ cm $\times$ 10 cpm/cm^2 per centimeter thickness = 100 cpm/cm^2. For the 1-cm diameter lesion, the count rate over the center of the lesion is

$$(9 \text{ cm} \times 10 \text{ cpm/cm}^2 \text{ per cm})$$
$$+ (1 \text{ cm} \times 20 \text{ cpm/cm}^2 \text{ per cm})$$
$$= 110 \text{ cpm/cm}^2$$

Thus the contrast of the 1-cm diameter lesion is $(110 - 100)/100 = 0.1$ (10%). For a 1-min imaging time, its CNR (Equation 15-11) is

$$CNR_{1\text{ cm}} = 0.1 \times 1 \times \sqrt{100 \times 1} = 1$$

For the 2-cm diameter lesion, the counting rate over the center of the lesion is

$$(8 \text{ cm} \times 10 \text{ cpm/cm}^2 \text{ per cm})$$
$$+ (2 \text{ cm} \times 20 \text{ cpm/cm}^2 \text{ per cm})$$
$$= 120 \text{ cpm/cm}^2$$

Thus the contrast of the 2-cm diameter lesion is $(120 - 100)/100 = 0.2$ (20%). For a 1-min imaging time, its CNR is

$$CNR_{2\text{ cm}} = 0.2 \times 2 \times \sqrt{100 \times 1} = 4$$

According to Equation 15-11, when the diameter of a *planar* object is doubled, its CNR increases by a factor of 2 as well. However, Example 15-5 shows that when the object is spherical, so that its thickness is doubled as well, its CNR increases by another factor of 2, that is, the total change in CNR is a factor of 4. This example illustrates the strong dependence of lesion detectability on lesion size when its contrast increases with its size. In essence, CNR increases as the *square* of spherical lesion diameter, not as the first power as implied by Equation 15-11. Going in the opposite direction, this factor becomes a significant impediment for the detection of smaller and smaller lesions in nuclear medicine.

Finally, Equations 15-11 and 15-12 and the discussion thus far assume that detectability is the same for positive ("hot spot") and negative ("cold spot") contrast. Indeed, if two objects generate identical levels of contrast, this is a valid assumption. Again, however, additional factors come into play in nuclear medicine imaging. Specifically, the intrinsic contrast of a lesion can depend on whether its contrast is generated by preferential uptake or by preferential suppression of uptake relative to surrounding normal tissues. The following example provides an illustration.

EXAMPLE 15-6

Suppose that two radiopharmaceuticals are available for a study. Radiopharmaceutical A generates contrast by selective uptake in a lesion that is 10 times higher than the uptake in surrounding normal tissue, whereas radiopharmaceutical B generates contrast by suppression of uptake in the same lesion, to a level that is 1/10 (10%) of the uptake in surrounding tissue. Thus the uptake ratio is 10:1 in both cases. Assume that a 1-cm thick lesion is present in a total thickness of 10 cm of tissue. Ignoring the effects of attenuation and source-to-detector distance, calculate the CNR generated by the two radiopharmaceuticals. For both radiopharmaceuticals, assume that the uptake in normal tissue generates a counting rate of 10 cpm/cm^2 per centimeter thickness of tissue and that imaging time is 1 min in both cases.

Answer

For both radiopharmaceuticals, the uptake in normal tissues generates a background counting rate of $R_o = 10$ cm $\times$ 10 cpm/cm^2 per centimeter thickness = 100 cpm/cm^2. For radiopharmaceutical A ("hot" lesion), the uptake of the lesion is 10 times greater and the counting rate over the lesion is

$$(9 \text{ cm} \times 10 \text{ cpm/cm}^2 \text{ per cm})$$
$$+ (1 \text{ cm} \times 100 \text{ cpm/cm}^2 \text{ per cm})$$
$$= 190 \text{ cpm/cm}^2$$

Thus the contrast of the hot lesion is (190 − 100)/100 = 0.9 (90%). For a 1-min imaging time, its CNR (Equation 15-11) is

$$\text{CNR}_A = 0.9 \times 1 \times \sqrt{100 \times 1} = 9$$

This value easily exceeds the requirement for detectability given by the Rose criterion.

For the "cold" lesion, the uptake by the lesion is 1/10 of the uptake in normal tissue. Thus the counting rate over the lesion is
$$(9 \text{ cm} \times 10 \text{ cpm/cm}^2 \text{ per cm})$$
$$+ (1 \text{ cm} \times 1 \text{ cpm/cm}^2 \text{ per cm})$$
$$= 91 \text{ cpm/cm}^2$$

Thus the contrast of the "cold" lesion is (91 − 100)/100 = −0.09 (−9%), in which the minus sign indicates "negative" contrast. For a 1-min imaging time, the CNR for radiopharmaceutical B is

$$\text{CNR}_B = 0.09 \times 1 \times \sqrt{100 \times 1} = 0.9$$

which is well below the threshold of detectability specified by the Rose criterion.

Example 15-6 illustrates the basis for the generally held (and generally accurate) belief that "cold" lesions are more difficult to detect than "hot" ones. One way to overcome this deficit is to inject more radioactivity for "cold" lesions; however, the specified levels of uptake in normal tissue in this example leads to comparable radiation doses in both cases and thus the higher level of radioactivity required for radiopharmaceutical B would presumably lead to greater radiation dose. Specific comparisons of radiopharmaceuticals vary, depending on details of the uptake distribution and properties of the radionuclides involved.

These examples illustrate that contrast and information density can be limiting factors for lesion detection, even when the size of the lesion easily exceeds the spatial resolution limits of the imaging system. Figure 15-12 further illustrates this point for images of a heart phantom. Although spatial resolution and contrast are the same for all the images shown in this figure, there are marked differences in lesion visibility because of differences in information density and noise.

Although not specifically included in the analysis of CNR presented earlier, spatial resolution of the imaging system also affects the detectability of small, low-contrast objects. As shown in Figure 15-13, high-resolution collimators (or imaging detectors) provide better image contrast and improved visibility for fine details, even for smaller numbers of counts in the image. In essence, "sharpening" the edges of lesions lowers the CNR required for detectability (Rose criterion) specified in Equation 15-12.

Nevertheless, the tradeoff between improved collimator resolution and decreased collimator sensitivity (see Equations 14-7 and 14-8), as well as the requirement for greater information density, eventually establishes a point of diminishing returns in the effort to detect smaller and smaller lesions by improvements in imaging resolution. In the end, detectability in nuclear medicine is limited by

True activity distribution 1×10^6 counts 4×10^6 counts (~clinical level) 16×10^6 counts

FIGURE 15-12 Example of effects of information density on visibility of a low-contrast lesion (*arrow*) in a computer-simulated cardiac phantom. The simulation assumes a ^{99m}Tc radiotracer imaged on a gamma camera with a 15% energy window. Note the region of reduced radiotracer uptake in the myocardium (*arrow*) that can be clearly visualized only with the highest information density. (*Courtesy Dr. Hendrik Pretorius and Dr. Michael King, University of Massachusetts Medical School, Worcester, MA*).

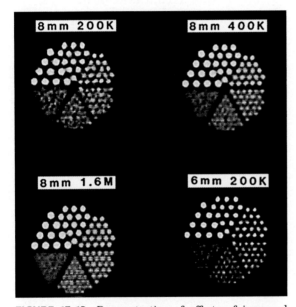

FIGURE 15-13 Demonstration of effects of improved resolution on contrast and detectability of small objects. Improved spatial resolution results in improved contrast (*lower right*), providing improved visibility in spite of fewer counts in comparison with the other images. Decreased sensitivity of high-resolution collimators ultimately sets practical limits for high-resolution imaging in nuclear medicine. (*From Muehllehner G: Effect of resolution improvement on required count density in ECT imaging: A computer simulation. Phys Med Biol 30:163-173, 1985.*)

information density rather than image resolution. Information densities in planar nuclear medicine images are typically in the range of 100 to 3000 counts/cm². This is well below the levels encountered in radiography and photography, in which information densities (x-ray or visible light photons detected to form the image) are on the order of 10^6 events per mm². Practical limitations on imaging time and the amount of activity that can be administered safely to patients are serious impediments to improvements in nuclear medicine information densities and are the reason why photographic or radiographic image quality are unlikely to ever be achieved in nuclear medicine.

In general, the rules regarding image CNR and object detectability are the same for planar images and tomographic images; however, the approaches for calculating CNR are different. This is discussed further in Chapter 16, Section C.3.

E. OBSERVER PERFORMANCE STUDIES

The physical measures of image quality discussed in the preceding sections are helpful for comparing different imaging systems, as well as for preparing purchase specifications, establishing quality assurance parameters, and so forth. They also can in some cases provide useful estimates of minimum detectable object size and contrast, as in Examples 15-2 through 15-6. In most cases, however, object detectability is determined more accurately by direct evaluation, using human observers. The general name for such evaluations is *observer performance studies*. They test both the ability of an imaging device to produce detectable objects as well as the ability of individual observers to detect them. Two types of experiments commonly used for this purpose in nuclear medicine imaging are *contrast-detail* (C-D) and *receiver operating characteristic* (ROC) studies.

1. Contrast-Detail Studies

A contrast-detail, or C-D study is performed using images of a phantom having a set of objects of different sizes and contrasts. Typically, the objects are graded in size along one axis of the display and in contrast along the other. An example is the Rollo phantom, shown in Figure 15-14A. This phantom consists of solid spheres of four different diameters immersed in four different thicknesses of a radioactive solution of uniform concentration. Images of this phantom thus contain cold lesions of different sizes and contrasts (Fig. 15-14B).

To perform a C-D study with this or a similar phantom, images are obtained using the different imaging systems or techniques to be evaluated. An observer then is given the images, usually without identification and in random order to avoid possible bias, and asked to indicate the smallest diameter of sphere that is visible at each level of contrast. Borderline visibility may be indicated by selecting a diameter between two of the diameters actually present in the image. The results then are presented on a C-D diagram as illustrated by Figure 15-15. A C-D study can be helpful for comparing detectability of both large low-contrast lesions as well as small high-contrast lesions. For example, in Figure 15-15, system A would be preferred for the former and system B for the latter. Because of the subjective nature of C-D studies, the use of multiple observers is recommended. Also, because observers may change their detection threshold from one study to the next or as they gain familiarity with the images, it usually is helpful to repeat the readings for verification of results.

C-D studies have a number of limitations. Because they are subjective, they are susceptible to bias and other sources of differences in the observer's detection thresholds in different experiments. This is especially true for phantoms having a design similar to the one illustrated in Figure 15-14, because the observer has a priori knowledge of the locations of the simulated lesions. Thus such a phantom does not test for the possibility of false-positive results, that is, the mistaken detection of objects that actually are not present in the image. This is particularly important for noisy images in which noise not only can mask the presence of real objects but also can create apparent structures that masquerade as real objects. Finally, C-D phantoms generally are lacking in clinical realism.

2. Receiver Operating Characteristic Studies

Some of the deficiencies of the C-D method outlined earlier are overcome by the ROC method. For an ROC study, a set of images is obtained with the different imaging systems or techniques to be tested. Phantoms containing simulated lesions can be used, but it also is possible to use actual clinical images. In the simplest approach, each image contains either one or no lesions. The former are called *positive images* and the latter are called *negative images*. The images are given to the observer, who is asked to indicate whether a lesion is present or absent in each image, as well as where it is and his or her confidence that it actually is present. Usually the confidence levels are numbered and four different levels are permitted; for example, 1 = definitely present,

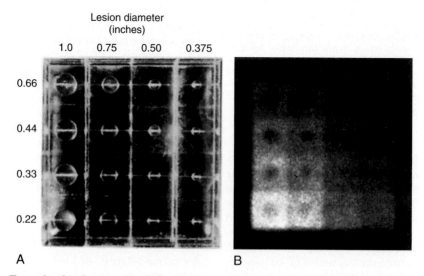

FIGURE 15-14 Example of a phantom, the Rollo phantom, which can be used to obtain images for a contrast-detail study. *A,* Phantom. *B,* Example image. (*From Rollo FD, Harris CC: Factors affecting image formation. In Rollo FD [ed]: Nuclear Medicine Physics, Instrumentation, and Agents. St. Louis, 1977, CV Mosby, p 397.*)

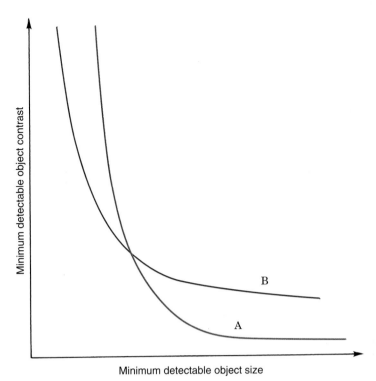

FIGURE 15-15 Hypothetical results of a contrast-detail study comparing two imaging systems, A and B. Reading horizontally, one can estimate the minimum size of an object that can be detected for a specified level of object contrast. Reading vertically, one can estimate the minimum contrast required for detection of an object of a specified size. In this example, system A provides better detectability for large low-contrast objects, suggesting perhaps a better lesion-to-noise contrast ratio, whereas system B is better for small high-contrast lesions, suggesting perhaps better spatial resolution.

2 = probably present, 3 = probably not present, and 4 = definitely not present. Then the following results are calculated for each confidence level:

True-positive fraction (TPF) = fraction of positive images correctly identified as positive by the observer

False-positive fraction (FPF) = fraction of negative images incorrectly identified as positive by the observer

Two other parameters that are calculated are the *true-negative fraction* (TNF) = (1 − FPF), and the *false-negative fraction* (FNF) = (1 − TPF). The TPF is sometimes called the *sensitivity* and TNF the *specificity* of the test or the observer.

The ROC curve then is generated by plotting TPF versus FPF for progressively relaxed degrees of confidence, that is, highest confidence = level 1 only, then confidence levels 1 + 2, then confidence levels 1 + 2 + 3, and so forth. An example of data and the resulting

ROC curves are shown in Figure 15-16. The ROC curve should lie above the ascending 45-degree diagonal, which would represent "guessing." The farther the curve lies above the 45-degree line, the better the performance of the imaging system and observer.

An ROC curve shows not only the true-positive detection rate for an observer or an imaging system or technique but also its relationship to the false-positive detection rate. Thus it is relatively immune to the sources of observer bias that can occur in C-D studies, for example, a tendency to "over-read" to avoid missing a possible lesion or test object. It also is applicable to other types of detection questions, such as the presence or absence of disease, which might be indicated by a general pattern of uptake within an organ, as opposed to the simple detection of individual lesions.

As with C-D studies, the interpretation of ROC results sometimes can be challenging. For example, the ROC curves for two different imaging systems can "cross," leading to some ambiguity in the results. One approach to simplifying the interpretation is to report the

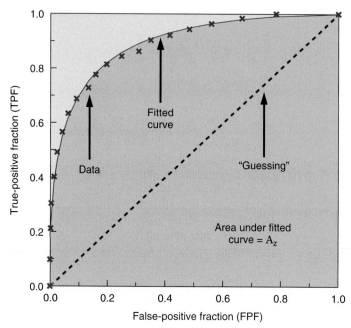

FIGURE 15-16   Example of results from an ROC study. ✖, data points; *orange line*, fitted curve; *blue line,* 45-degree line, which is equivalent to "guessing." Area under the curve (*shaded in darker blue*) is A_z, which is one measure of detection accuracy.

results of an ROC study as a single number. Most commonly the parameter calculated is the area under the ROC curve, usually denoted by A_z. This number can range from zero (all readings wrong) to 1 (all readings correct). A value of 0.5 indicates an overall accuracy of 50%, which is equivalent to "guessing."

An extensive amount of theoretical and experimental work has been done on the properties of A_z, including such issues as statistical comparisons of values obtained from different ROC studies. References 5 to 7 present detailed analyses of these and other practical issues in ROC studies. A_z also has an interesting practical interpretation: It is the probability that, given a side-by-side pair of images, one of which has a lesion or test object and the other does not, the observer will correctly identify the image with the lesion.[8]

Despite their power and potential usefulness, ROC studies also have a number of limitations. Perhaps the most challenging is the verification of absolute "truth" for images obtained from clinical studies. Ideally, the outcome of the ROC study itself (i.e., the tested images) should not be used for this determination. This means that other equally or even more reliable information about the presence or absence of disease in the patient must be available. Often nonimaging tests (e.g., surgical results) must be obtained for

verification of "truth" when clinical images are used.

Another potential problem is the possible presence of multiple lesions on a single image. Conventional ROC methodology allows only for a single "yes-no" interpretation of each image. This allows straightforward calculations of false-positive rates. However, if multiple lesions are possible, as in many clinical images, the potential number of false positives is virtually infinite, making the calculation of false-positive rates difficult, if not impossible. Alternative methods, called the *free-response operating characteristic*, that allow for the presence of multiple lesions have been developed and are discussed in reference 9.

Finally, even a "perfect" image evaluation technique with a clearly defined outcome might not provide the final answer regarding the merit or value of an imaging device or technique. Even after the physician or scientist has demonstrated that he or she has developed a truly "better" device or technique in terms of lesion or disease detectability, there is still the bottom-line question: "So what?" Does the improved detectability alter the care of the patient or the outcome of that care? Does it improve the patient's quality of life? In an age of cost-consciousness, what are the cost-benefit tradeoffs? For example, from a public health perspective, is it really worth spending a small fortune to detect the next smaller size of lesion, as compared with directing those funds toward simpler health measures, such as education and behavior modification? These are difficult questions to answer, but efforts are being made to develop methodology for answering them in a quantitative and objective manner. The general term for these investigations is *efficacy studies*. Additional discussion of this topic can be found in reference 10.

REFERENCES

1. Barrett HH, Yao J, Rolland JP, Myers KJ: Model observers for assessment of image quality. *Proc Natl Acad Sci* 90:9758-9765, 1993.
2. Cunningham IA: Introduction to linear systems theory. In Beutel J, Kundel HL, Nan Metter RL, editors: *Handbook of Medical Imaging*, Bellingham, WA, 2000, SPIE, Chapter 2.
3. Vayrynen T, Pitkanen U, Kiviniitty K: Methods for measuring the modulation transfer function of gamma camera systems. *Eur J Nucl Med* 5:19-22, 1980.
4. Rose A: *Vision: Human and Electronic*. New York, 1973, Plenum Press, pp 21-23.
5. Swets JA, Pickett RM: *Evaluation of Diagnostic Systems: Methods from Signal Detection Theory*, New York, 1982, Academic Press.
6. Metz CE: ROC methodology in radiologic imaging. *Invest Radiol* 21:720-733, 1986.
7. Metz CE: Fundamental ROC analysis. In Beutel J, Kundel HL, Nan Metter RL, editors: *Handbook of Medical Imaging*, Bellingham, WA, 2000, SPIE, Chapter 15.
8. Hanley JA, McNeil BJ: The meaning and use of the area under the receiver operating characteristic (ROC) curve. *Radiology* 143:29-36, 1982.
9. Chakraborty DP: The FROC, AFROC, and DROC variants of the ROC analysis. In Beutel J, Kundel HL, Nan Metter RL, editors: *Handbook of Medical Imaging*, Bellingham, WA, 2000, SPIE, Chapter 16.
10. Fryback DG, Thornbury JR: The efficacy of diagnostic studies. *Med Decis Making* 11:88-94, 1991.

Tomographic Reconstruction in Nuclear Medicine

A basic problem in conventional radionuclide imaging is that the images obtained are two-dimensional (2-D) projections of three-dimensional (3-D) source distributions. Images of structures at one depth in the patient thus are obscured by superimposed images of overlying and underlying structures. One solution is to obtain projection images from different angles around the body (e.g., posterior, anterior, lateral, and oblique views). The person interpreting the images then must sort out the structures from the different views mentally to decide the true 3-D nature of the distribution. This approach is only partially successful; it is difficult to apply to complex distributions with many overlapping structures. Also, deep-lying organs may have overlying structures from all projection angles.

An alternative approach is *tomographic imaging*. Tomographic images are 2-D representations of structures lying within a selected plane in a 3-D object. Modern *computed tomography* (CT) techniques, including positron emission tomography (PET), single photon emission tomography (SPECT), and x-ray CT, use detector systems placed or rotated around the object so that many different angular views (also known as *projections*) of the object are obtained. Mathematical algorithms then are used to *reconstruct* images of selected planes within the object from these projection data. Reconstruction of images from multiple projections of the detected emissions from radionuclides within the body is known as *emission computed tomography* (ECT). Reconstruction of images from transmitted emissions from an external source (e.g., an x-ray tube) is known as *transmission computed tomography* (TCT or, usually, just

CT, See Chapter 19, Section B). The mathematical basis is the same for ECT and TCT, although there are obviously differences in details of implementation.

ECT produces images in which the activity from overlying (or adjacent) cross-sectional planes is eliminated from the image. This results in a significant improvement in contrast-to-noise ratio (CNR), as already has been illustrated in Figure 15-11. Another advantage of SPECT and PET over planar nuclear medicine imaging is that they are capable of providing more accurate quantitation of activity at specific locations within the body. This is put to advantage in tracer kinetic studies (Chapter 21).

The mathematics underlying reconstruction tomography was first published by Johann Radon in 1917, but it was not until the 1950s and 1960s that work in radioastronomy and chemistry resulted in practical applications. The development of x-ray CT in the early 1970s initiated application of these principles for image reconstruction in medical imaging. An interesting historical perspective on the origins and development of tomographic image reconstruction techniques is presented in reference 1.

Instrumentation for SPECT imaging is discussed in Chapter 17 and instrumentation for PET is discussed in Chapter 18. Although the instruments differ, the same mathematics can be used to reconstruct SPECT or PET images. In this chapter, we focus on the basic principles of reconstructing tomographic images from multiple projections. A detailed mathematical treatment of image reconstruction is beyond the scope of this text. The reader is referred to references 2 to 4 for more detailed accounts.

A. GENERAL CONCEPTS, NOTATION, AND TERMINOLOGY

We assume initially that data are collected with a standard gamma camera fitted with a conventional parallel-hole collimator. (Applications involving other types of collimators are discussed in Section E.) To simplify the analysis, several assumptions are made. We consider only a narrow cross-section across the detector. The collimated detector is assumed to accept radiation only from a thin slice directly perpendicular to the face of the detector. This reduces the analysis to that of a 1-D detector, as shown in Figure 16-1. Each collimator hole is assumed to accept radiation only from a narrow cylinder defined by the geometric extension of the hole in front of the collimator. This cylinder defines the *line of response* for the collimator hole. For further simplification, we ignore the effects of attenuation and scatter and assume that the counts recorded for each collimator hole are proportional to the total radioactivity contained within its line of response. The measured quantity (in this case, counts recorded or radioactive content) sometimes is referred to as the *line integral* for the line of response. A full set of line integrals recorded across the detector is called a *projection,* or a *projection profile,* as illustrated in Figure 16-1.

Obviously, the assumptions noted earlier are not totally valid. Some of the effects of the inaccuracies of these assumptions are discussed in Chapter 17, Section B and in Chapter 18, Section D.

A typical SPECT camera is mounted on a gantry so that the detector can record projections from many angles around the body. PET systems generally use stationary arrays of detector elements arranged in a ring or hexagonal pattern around the body. In either case, the detectors acquire a set of projections at equally spaced angular intervals. In reconstruction tomography, mathematical algorithms are used to relate the projection data to the 2-D distribution of activity within the projected slice. A schematic illustration of the data acquisition process is shown in

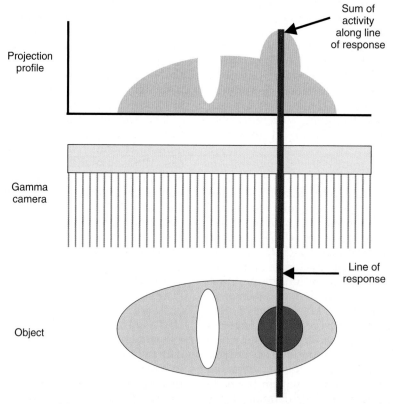

FIGURE 16-1   Cross-section of the response characteristics of an idealized gamma camera. Each collimator hole views the radioactivity within a cylinder perpendicular to the face of the gamma camera, called its *line of response.* Under idealized conditions (such as no attenuation or scatter) the signal recorded by the detector at that point reflects the sum of activity within the line of response. For a row of holes across the detector, the gamma camera generates a projection profile as shown. The projection profiles provide the data from which the image is reconstructed.

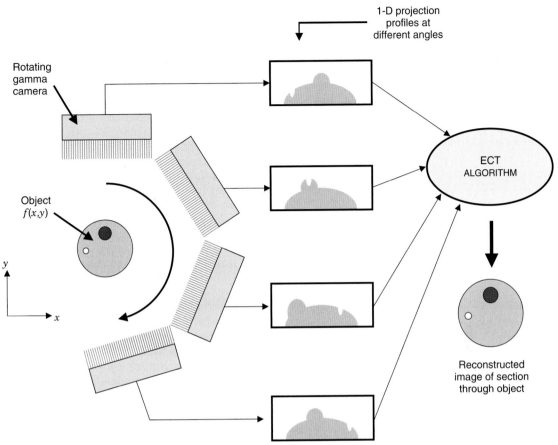

FIGURE 16-2 Rotating the gamma camera around the object provides a set of one-dimensional projection profiles for a two-dimensional object, which are used to calculate the two-dimensional distribution of radioactivity in the object. ECT, emission computed tomography.

Figure 16-2. Note that the data collected correspond to a slice through the object perpendicular to the bed and that this is called the *transverse* or *transaxial direction*. The direction along the axis of the bed, which defines the location of the slice, is known as the axial direction.

We assume that N projections are recorded at equally spaced angles between 0 and 180 degrees. Under the idealized conditions assumed here, the projection profile recorded at a rotation angle of $(180 + \phi)$ degrees would be the same (apart from a left-right reversal) as the profile recorded at ϕ degrees. Thus the data recorded between 180 and 360 degrees would be redundant; however, for practical reasons (e.g., attenuation), SPECT data often are acquired for a full 360-degree rotation. This is discussed further in Chapter 17.

For purposes of analysis, it is convenient to introduce a new coordinate system that is stationary with respect to the gamma camera detector. This is denoted as the (r,s)

coordinate system and is illustrated in Figure 16-3. If the camera is rotated by an angle ϕ with respect to the (x,y) coordinate system of the scanned object, the equations for transformation from (x,y) to (r,s) coordinates can be derived from the principle of similar triangles and are given by

$$r = x \cos\phi + y \sin\phi \qquad (16\text{-}1)$$

and

$$s = y \cos\phi - x \sin\phi \qquad (16\text{-}2)$$

These equations can be used to determine how radioactivity at a location (x,y) in the object contributes to the signal recorded at location r in the projection acquired at rotation angle ϕ.

One commonly used way to display a full set of projection data is in the form of a 2-D matrix $p(r,\phi)$. A representation of this matrix, generically known as a *sinogram,* is shown for

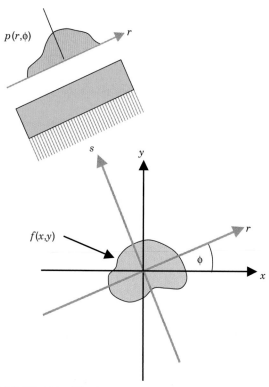

FIGURE 16-3   The (r,s) coordinate system is rotated by projection angle ϕ with respect to the (x,y) coordinate system of the object and is fixed with respect to the gamma camera.

a simple point-source object in Figure 16-4. Each row across the matrix represents an intensity display across a single projection. The successive rows from top to bottom represent successive projection angles. The name *sinogram* arises from the fact that the path of a point object located at a specific (x,y) location in the object traces out a sinusoidal path down the matrix. (This also can be deduced from Equations 16-1 and 16-2.) The sinogram provides a convenient way to represent the full set of data acquired during a scan and can be useful for determining the causes of artifacts in SPECT or PET images.

B. BACKPROJECTION AND FOURIER-BASED TECHNIQUES

1. Simple Backprojection

The general goal of reconstruction tomography is to generate a 2-D cross-sectional image of activity from a slice within the object, $f(x,y)$, using the sinogram, or set of projection profiles, obtained for that slice. In practice, a set of projection profiles, $p(r,\phi_i)$, is acquired at discrete angles, ϕ_i, and each profile is sampled at discrete intervals along r. The image is reconstructed on a 2-D matrix of discrete

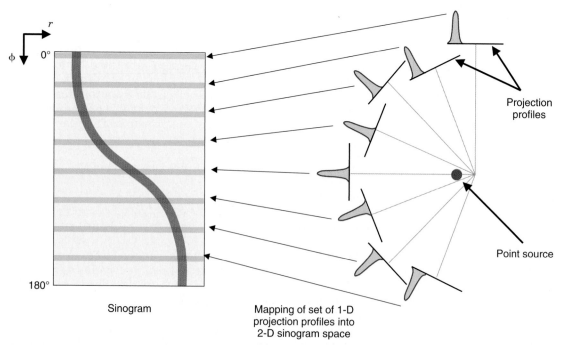

FIGURE 16-4   Two-dimensional (2-D) intensity display of a set of projection profiles, known as a *sinogram*. Each row in the display corresponds to an individual projection profile, sequentially displayed from top to bottom. A point source of radioactivity traces out a sinusoidal path in the sinogram.

pixels in the (x,y) coordinate system. For mathematical convenience, the image matrix size usually is a power of 2 (e.g., 64 × 64 or 128 × 128 pixels). Pixel dimensions Δx and Δy can be defined somewhat arbitrarily, but usually they are related to the number of profiles recorded and the width of the sampling interval along r.

The most basic approach for reconstructing an image from the profiles is by *simple backprojection*. The concepts will be illustrated for a point source object. Figure 16-5A shows projection profiles acquired from different angles around the source. An approximation for the source distribution within the plane is obtained by projecting (or distributing) the data from each element in a profile back across the entire image grid (Fig. 16-5B). The counts recorded in a particular projection profile element are divided uniformly amongst the pixels that fall within its projection path.* This operation is called *backprojection*. When the backprojections for all profiles are added

together, an approximation of the distribution of radioactivity within the scanned slice is obtained. Mathematically, the backprojection of N profiles is described by

$$f'(x, y) = \frac{1}{N} \sum_{i=1}^{N} p(x \cos \phi_i + y \sin \phi_i, \phi_i) \quad (16\text{-}3)$$

where ϕ_i denotes the ith projection angle and $f'(x,y)$ denotes an approximation to the true radioactivity distribution, $f(x,y)$.

As illustrated in Figure 16-5B, the image built up by simple backprojection resembles the true source distribution. However, there is an obvious artifact in that counts inevitably are projected outside the true location of the object, resulting in a blurring of its image. The quality of the image can be improved by increasing the number of projection angles and the number of samples along the profile. This suppresses the "spokelike" appearance of the image but, even with an infinite number of views, the final image still is blurred. No matter how finely the data are sampled, simple backprojection always results in some apparent activity outside the true location for the point source. Figure 16-6 shows an image reconstructed by simple backprojection for a somewhat more complex object and more clearly illustrates the blurring effect.

---

*In practice, counts are assigned to a pixel in proportion to the fraction of the pixel area contained within the line of response for the projection element. However, owing to the complexity of the notation, this part of the algorithm is not included here.

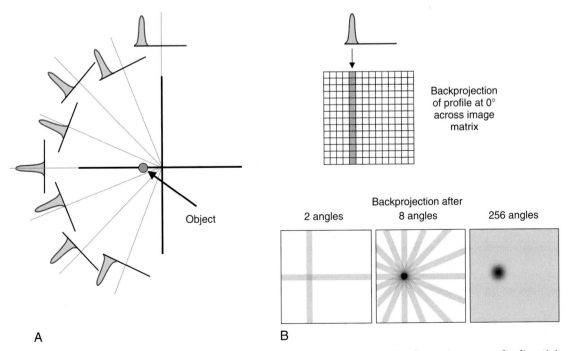

A   **B**

FIGURE 16-5 Illustration of the steps in simple backprojection. *A*, Projection profiles for a point source of radioactivity for different projection angles. *B*, Backprojection of one intensity profile across the image at the angle corresponding to the profile. This is repeated for all projection profiles to build up the backprojected image.

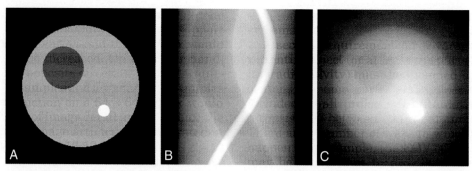

FIGURE 16-6 *A*, Computer-simulation phantom used for testing reconstruction algorithms. *B*, Sinogram of simulated data for a scan of the phantom. *C*, Image of simulation phantom for simple backprojection of data from 256 projection angles. 1/r blurring is apparent in the object, and edge details are lost. (*Computer simulations performed by Dr. Andrew Goertzen, University of Manitoba, Canada*)

Mathematically, the relationship between the true image and the image reconstructed by simple backprojection is described by

$$f'(x, y) = f(x, y) * (1/r) \qquad (16\text{-}4)$$

where the symbol $*$ represents the process of convolution described in Appendix G. A profile taken through the reconstructed image for a point source that is reconstructed from finely sampled data decreases in proportion to $(1/r)$, in which r is the distance from the center of the point-source location. Because of this behavior, the effect is known as *1/r blurring*. Simple backprojection is potentially useful only for very simple situations involving isolated objects of very high contrast relative to surrounding tissues, such as a tumor with avid uptake of a radiopharmaceutical that in turn has very low uptake in normal tissues. For more complicated objects, more sophisticated reconstruction techniques are required.

2. Direct Fourier Transform Reconstruction

One approach that avoids 1/r blurring is Fourier transform (FT) reconstruction, sometimes called *direct Fourier transform reconstruction* or *direct FT*. Although direct FT is not really a backprojection technique, it is presented here as background for introducing the filtered backprojection (FBP) technique in the next section.

Basic concepts of FTs are discussed in Appendix F. Briefly, in the context of nuclear medicine imaging, the FT is an alternative method for representing spatially varying data. For example, instead of representing a 1-D image profile as a spatially varying function, $f(x)$, the profile is represented as a summation of sine and cosine functions of different spatial frequencies, k. The amplitudes for different spatial frequencies are represented in the FT of $f(x)$, which is denoted by $F(k)$. The operation of computing the FT is symbolized by

$$F(k) = \mathscr{F}[f(x)] \qquad (16\text{-}5)$$

The function $f(x)$ is a representation of the image profile in *image space* (or "object space"), whereas $F(k)$ represents the profile in "spatial frequency space," also called *k-space*. FTs can be extended to 2-D functions, $f(x,y)$, such as a 2-D image. In this case, the FT also is 2-D and represents spatial frequencies along the x- and y-axes, $F(k_x, k_y)$, in which k_x and k_y represent orthogonal axes in 2-D k-space. Symbolically, the 2-D FT is represented as

$$F(k_x, k_y) = \mathscr{F}[f(x, y)] \qquad (16\text{-}6)$$

Mathematically, a function and its FT are equivalent in the sense that either one can be derived from the other. The operation of converting the FT of a function back into the original function is called an *inverse FT* and is denoted by

$$\mathscr{F}^{-1}[F(k_x, k_y)] = f(x, y) \qquad (16\text{-}7)$$

FTs can be calculated quickly and conveniently on personal computers, and many image and signal-processing software packages contain FT routines. The reader is referred to Appendix F for additional information about FTs.

The concept of k-space will be familiar to readers who have studied magnetic resonance imaging (MRI), because this is the coordinate system in which MRI data are acquired. To reconstruct an image from its 2-D FT, the full 2-D set of k-space data must be available (Equation 16-7). In MRI, data are acquired point-by-point for different (k_x, k_y) locations in a process known as "scanning in k-space." There is no immediately obvious way to directly acquire k-space data in nuclear medicine imaging. Instead, nuclear medicine CT relies on the *projection slice theorem*, or *Fourier slice theorem*. In words, this theorem says that the FT of the projection of a 2-D object along a projection angle ϕ [in other words, the FT of a profile, $p(r,\phi)$], is equal to the value of the FT of the object measured through the origin and along the same angle, ϕ, in k-space (note, the *value* of the FT, not the *projection* of the FT). Figure 16-7 illustrates this concept. Mathematically, the general expression for the projection slice theorem is

$$\mathscr{F}[p(r,\phi)] = F(k_r, \phi) \qquad (16\text{-}8)$$

where $F(k_r,\phi)$ denotes the value of the FT measured at a radial distance k_r along a line at angle ϕ in k-space.

The projection slice theorem provides a means for obtaining 2-D k-space data for an object from a series of 1-D measurements in object space. Figure 16-7 and Equation 16-8 provide the basis for reconstructing an object from its projection profiles as follows:
1. Acquire projection profiles in object space at N projection angles, ϕ_i, $i = 1, 2, ..., N$ as previously described.
2. Compute the 1-D FT of each profile.
3. Insert the values of these FTs at the appropriate coordinate locations in k-space. Note that values are inserted in polar coordinates along radial lines through the origin in k-space. For a specific value of k_r in the FT of the projection acquired at a rotational angle ϕ, the data are inserted at rectangular coordinates given by

$$\begin{aligned} k_x' &= k_r \cos\phi \\ k_y' &= k_r \sin\phi \end{aligned} \qquad (16\text{-}9)$$

where primed notation is used to indicate that the coordinate locations do not correspond exactly to points on a rectangular grid. The inserted values are closely spaced near the origin and more widely spaced farther away from the origin. This "over-representation" of data near the origin in k-space is one explanation for the 1/r

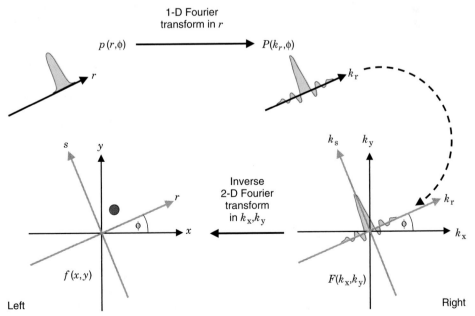

FIGURE 16-7 Concepts of the projection slice theorem. (*Left*) $p(r,\phi)$ is a one-dimensional (1-D) profile of the 2-D object $f(x,y)$ at projection angle ϕ. The theorem states that the 1-D Fourier transform of this projection profile (*right*) is equal to the values of the 2-D Fourier transform of the object, $F(k_x, k_y)$, along a line through the origin of k-space at the same angle ϕ.

blurring that occurs in simple backprojection, as was discussed in Section B.1.

4. Using the values inserted in polar coordinates, interpolate values for k_x and k_y on a rectangular grid in k-space.

5. Use the interpolated values in k-space and a standard 2-D (inverse) FT (Equation 16-7) to compute the image of the object.

With noise-free data, perfect projection profiles (i.e., line integrals that represent precisely the sum of activity along a line measured through the object) and perfect interpolation, the direct FT reconstruction technique is capable of producing an exact representation of the object. Additional criteria regarding the required numbers of projection profiles and sampled points across each profile are discussed in Section C.

A major drawback of direct Fourier reconstruction is that the interpolation from polar to rectangular coordinates in k-space is computationally intensive. As well, it can lead to artifacts in the image, if not done carefully. A more elegant (and practical) approach, called *filtered backprojection (FBP),* is described in the next section.

3. Filtered Backprojection

Like the direct FT algorithm, FBP employs the projection slice theorem but uses the theorem in combination with backprojection in a manner that eliminates 1/r blurring. The steps are as follows:

1. Acquire projection profiles at N projection angles (same as direct FT).

2. Compute the 1-D FT of each profile (same as direct FT). In accordance with the projection slice theorem (see Fig. 16-7 and Equation 16-8), this provides values of the FT for a line across k-space.

3. Apply a "ramp filter" to each k-space profile. Mathematically, this involves multiplying each projection FT by $|k_r|$, the absolute value of the radial k-space coordinate at each point in the FT. Thus the value of the FT is increased (amplified) linearly in proportion to its distance from the origin of k-space. Figure 16-8 illustrates the profile of a ramp filter, with filter amplitude denoted by $H(k_r)$. Applying the ramp filter produces a modified FT for each projection, given by

$$P'(k_r, \phi) = |k_r| P(k_r, \phi) \qquad (16\text{-}10)$$

where $P(k_r, \phi)$ is the unfiltered FT.

4. Compute the inverse FT of each filtered FT profile to obtain a modified (filtered) projection profile. This is given by

$$\begin{aligned} p'(r, \phi) &= \mathscr{F}^{-1}[P'(k_r, \phi)] \\ &= \mathscr{F}^{-1}[|k_r| P(k_r, \phi)] \end{aligned} \qquad (16\text{-}11)$$

5. Perform conventional backprojection using the filtered profiles. Mathematically, the result is

$$f(x, y) = \frac{1}{N} \sum_{i=1}^{N} p'(x \cos \phi_i + y \sin \phi_i, \phi_i) \qquad (16\text{-}12)$$

Step 5 is essentially the same as simple backprojection, but with filtered profiles. However, unlike Equation 16-3, in which $f'(x,y)$ is only an approximation of the true distribution, FBP, when applied with perfectly measured noise-free data, yields the exact value of the true distribution, $f(x,y)$. Figure 16-9 schematically illustrates the process of FBP for a pointlike object.

The only difference between simple and filtered backprojection is that in the latter method, the profiles are modified by a *reconstruction filter* applied in k-space before they are backprojected across the image. The effect

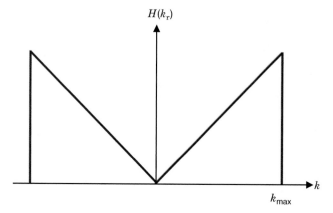

FIGURE 16-8 Ramp filter in the spatial-frequency (k-space) domain. The filter selectively amplifies high-frequency components relative to low-frequency components. The filter removes the 1/r blurring present in simple backprojection and sharpens image detail, but it also amplifies high-frequency noise components in the image.

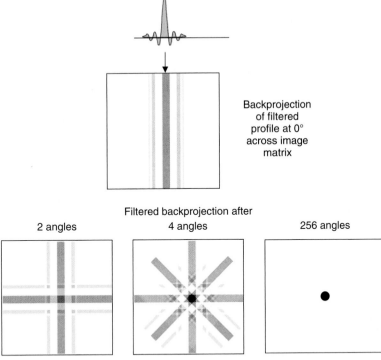

FIGURE 16-9 Illustration of the steps in filtered backprojection. The one-dimensional Fourier transforms of projection profiles recorded at different projection angles are multiplied by the ramp filter. After taking the inverse Fourier transform of the filtered transforms, the filtered profiles are backprojected across the image, as in simple backprojection.

of the ramp filter is to enhance high spatial frequencies (large k_r) and to suppress low spatial frequencies (small k_r). The result of the filtering is to eliminate 1/r blurring.* One way to visualize the effect is to note that, unlike unfiltered profiles (see Fig. 16-5), the filtered profiles have both positive and negative values (see Fig. 16-9). The negative portions of the filtered profiles near the central peak "subtract out" some of the projected intensity next to the peak that otherwise would create 1/r blurring.

Amplification of high spatial frequencies in FBP also leads to amplification of high-frequency noise. Because there usually is little signal in the very highest frequencies of a nuclear medicine image, whereas statistical noise is "white noise" with no preferred

frequency, this also leads to degradation of signal-to-noise ratio (SNR). For this reason, images reconstructed by FBP appear noisier than images reconstructed by simple backprojection. (This is a general result of any image filtering process that enhances high frequencies to "sharpen" images.) In addition, filters that enhance high frequencies sometimes have edge-sharpening effects that lead to "ringing" at sharp edges. This is an unwanted byproduct of the positive-negative oscillations introduced by the filter, illustrated in the filtered profile at the top of Figure 16-9.

To minimize these effects on SNR and artifacts at sharp edges, the ramp filter usually is modified so as to have a rounded shape to somewhat suppress the enhancement of high spatial frequencies. Figure 16-10 illustrates a ramp filter and two other commonly used reconstruction filters. Also shown are the equations describing these filters. A variety of reconstruction filters have been developed, each with its own theoretical rationale. Filters also play a role in the suppression of artifacts caused by aliasing in FT-based reconstruction techniques, as discussed in Section C. Additional discussions of these filters can be found in reference 4.

---

*More precisely, 1/r blurring is the convolution of the true image with a blurring function, $b(r) = 1/r$ (Equation 16-4). As discussed in Appendix G, convolution in image space is equivalent to multiplying by the FT of the blurring function in k-space, which for $b(r)$ is $B(k_r) = 1/|k_r|$. Thus multiplying by $|k_r|$ in k-space is equivalent to *deconvolving* the blurring function in image space, thereby eliminating the blurring effect.

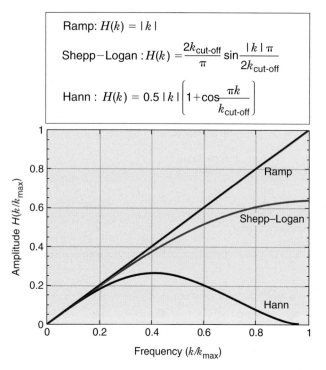

$$\text{Ramp: } H(k) = |k|$$

$$\text{Shepp–Logan : } H(k) = \frac{2k_{\text{cut-off}}}{\pi} \sin \frac{|k|\,\pi}{2k_{\text{cut-off}}}$$

$$\text{Hann : } H(k) = 0.5\,|k|\left[1+\cos\frac{\pi k}{k_{\text{cut-off}}}\right]$$

FIGURE 16-10 Ramp filter and two other reconstruction filters that are designed to prevent artifacts and noise amplification caused by the sharp cut-off of the ramp filter at the maximum frequency k_{max}. Note that all of the filters shown have the same response at lower frequencies and that cut-off frequencies are set so that $k_{\text{cut-off}} = k_{\text{max}}$. However, the Hann and Shepp-Logan filters roll off gradually at higher frequencies, thereby minimizing artifacts and noise amplification.

Because of its speed and relative ease of implementation, FBP became a widely used reconstruction method in nuclear medicine. A single 2-D image slice can be reconstructed in a fraction of a second on a standard computer. Under idealized conditions (noise-free data acquisition, completely sampled data, and so forth), FBP produces an accurate representation of the distribution of radioactivity within the slice. However, FBP is not without its limitations. First, it is susceptible to major artifacts if data from the object are measured incompletely (possibly because of collimator defects, portions of the object outside the FOV of the camera for some projections, etc.). Second, in datasets that have poor counting statistics or random "noise" spikes (perhaps caused by instrument malfunction), FBP produces annoying "streak artifacts." These artifacts can be suppressed by employing a k-space filter with a strong roll-off at high spatial frequencies, but this results in loss of image resolution as well.

Finally, the FBP algorithm cannot readily be modified to take into account various physical aspects of the imaging system and data acquisition, such as limited spatial resolution of the detector, scattered radiation, and the fact that the sensitive volume of the detector collimator holes actually is a cone rather than a cylinder, as assumed for the reconstruction process. These factors require additional preprocessing or postprocessing data manipulations that work with varying degrees of success. They are discussed further in Chapter 17, Section B.

By contrast, another set of reconstruction methods, known as *iterative reconstruction techniques,* can build these steps directly into the reconstruction algorithm and are less prone to the artifacts described in the preceding paragraph. These techniques are described in Section D.

4. Multislice Imaging

The analysis presented earlier for backprojection and Fourier-based reconstruction techniques applies to single-slice images. In practice, as described in Chapters 17 and 18, both SPECT and PET imaging are performed with detectors that acquire data simultaneously for multiple sections through the body. Projection data originating from each individual section through the object can be reconstructed as described. Individual image slices then are "stacked" to form a 3-D dataset, which in turn can be "resliced" using computer techniques to obtain images of planes other than those that are directly imaged. Thus 2-D multislice imaging of contiguous slices can be used to generate 3-D volumetric images. In many SPECT systems, the distance between image slices and slice thickness can be adjusted to achieve different axial resolutions, much the same as the sampling interval, Δr, across the image profile can be adjusted to vary the in-plane resolution. In PET systems, the distance between image slices and the slice thickness often is fixed by the axial dimensions of the segmented scintillator crystals typically used in the detectors (See Chapter 18, Sections B and C).

C. IMAGE QUALITY IN FOURIER TRANSFORM AND FILTERED BACKPROJECTION TECHNIQUES

In this section, we discuss some general issues involving image quality in reconstruction tomography based on the direct FT and FBP techniques. These issues affect all reconstruction tomography based on these techniques, including both x-ray CT and ECT. Additional aspects that are specifically relevant to SPECT and PET image quality are discussed in Chapters 17 and 18, respectively. The issues discussed here do not pertain directly to iterative reconstruction techniques, which are discussed separately in Section D.

1. Effects of Sampling on Image Quality

Projection data are not continuous functions but discrete point-by-point samples of projection profiles. The distance between the sample points is the *linear sampling distance*. In addition, projection profiles are acquired only at a finite number of *angular sampling intervals* around the object. The choice of linear and angular sampling intervals and the cut-off frequency of the reconstruction filter (see Fig. 16-10), in conjunction with the spatial resolution of the detector system, determine the spatial resolution of the reconstructed image. The effects of the imaging system depend on the type of detector, collimator, and so forth and are discussed in Chapters 17 and 18. Here we discuss briefly those aspects that are related to the reconstruction process, which are applicable to all types of imaging devices.

The sampling theorem[5] states that to recover spatial frequencies in a signal up to a maximum frequency k_{max} requires a linear sampling distance given by

$$\Delta r \leq 1/(2k_{max}) \qquad (16\text{-}13)$$

This means that the highest spatial frequency to be recovered from the data must be sampled at least twice per cycle. Coarser sampling does not allow higher spatial frequencies to be recovered and leads to image artifacts known as *aliasing*. Mathematical aspects of aliasing are discussed in detail in Appendix F, Section C.

Thus the linear sampling distance sets a limit on spatial resolution for the imaging system. This limit (k_{max} in Equation 16-13) also is known as the *Nyquist frequency*, $k_{Nyquist}$ (see also Equation F-9). The highest spatial frequency that is present in an image profile depends on the spatial resolution of the collimator-detector system. Higher resolution implies higher frequency content. As a rule of thumb, the sampling requirement for an imaging detector is

$$\Delta r \leq \text{FWHM}/3 \qquad (16\text{-}14)$$

where FWHM is the full width at half maximum of its point-spread function (see Chapter 15, Section B.2).

Figure 16-11 shows images of a computer-simulation phantom that were reconstructed with progressively coarser sampling of the image profiles. Undersampling not only results in image blurring but also creates image artifacts resulting from the effects of aliasing.

The Nyquist frequency is the highest spatial frequency represented in k-space and thus defines an upper frequency limit for the reconstruction filter. However, a lower-frequency filter cut-off, $k_{cut\text{-}off} < k_{Nyquist}$ can be used in the reconstruction. This improves SNR by suppressing the high-frequency end of the spatial frequency spectrum, where a large fraction of the signal is statistical noise. Lowering the cut-off frequency also degrades spatial resolution, because the higher frequencies also contain the fine details of the image. Thus the choice of the reconstruction filter and its cut-off frequency involve a tradeoff between spatial resolution and SNR in the image. This is illustrated in Figure 16-12, which shows images of a computer-simulation phantom reconstructed with a Shepp-Logan filter with different cut-off frequencies.

The angular sampling interval (angle between projections) should provide sampling around the periphery at approximately the same intervals as the linear sampling distance. Thus if projections are acquired around a FOV of diameter D, the minimum number of angular views, N_{views}, should be approximately the length of the 180-degree arc over

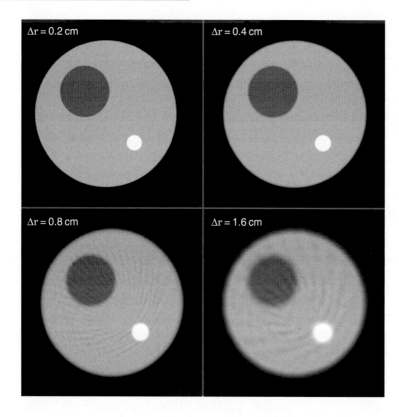

FIGURE 16-11 Images of a computer-simulation phantom reconstructed with progressively coarser sampling of the image profiles. Linear under-sampling results both in loss of resolution and image artifacts. (*Computer simulations performed by Dr. Andrew Goertzen, University of Manitoba, Canada.*)

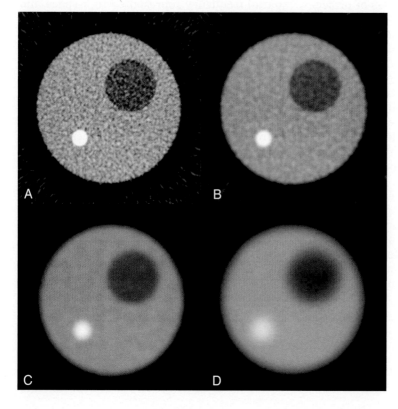

FIGURE 16-12 Filtered backprojection reconstructions of the computer-simulation phantom shown in Figure 16-6, using a Shepp-Logan filter with different cut-off frequencies. *A*, $k_{cut\text{-}off} = k_{max}$; *B*, $k_{cut\text{-}off} = 0.8\ k_{max}$; *C*, $k_{cut\text{-}off} = 0.6\ k_{max}$; and *D*, $k_{cut\text{-}off} = 0.2\ k_{max}$. Note the tradeoff between image detail and signal-to-noise ratio. (*Computer simulations performed by Dr. Andrew Goertzen, University of Manitoba, Canada.*)

which projections are acquired ($\pi D/2$) divided by the linear sampling distance, Δr:

$$N_{\text{views}} \geq \pi D/2\Delta r \qquad (16\text{-}15)$$

Figure 16-13 illustrates the effect of angular sampling interval on images of a computer-simulation phantom. Spokelike artifacts are evident around high-intensity objects when the number of angular samples is inadequate.

EXAMPLE 16-1

Suppose you are working with an ECT system that has spatial resolution FWHM ≈ 1 cm and FOV = 30 cm. Estimate the sampling interval, Δr, and the number of angular views, N_{views}, that would support the available spatial resolution of the system.

Answer

From Equation 16-14, the sampling interval should be

$$\Delta r \leq 1\,\text{cm}/3 \approx 0.33\,\text{cm}$$

For FOV = 30 cm, this amounts to

$$N_{\text{samp}} \geq 30/(1/3) \approx 90 \text{ samples per profile}$$

According to Equation 16-15, the number of views should be such that

$$N_{\text{views}} \geq (\pi \times 30)/[2 \times (1/3)] \approx 140 \text{ views}$$

Thus 140 views over a 180-degree degree arc, with linear sampling at approximately 0.33-cm intervals, would fully support the available system resolution.

The closest power-of-two image reconstruction and display matrix that would meet the sampling requirements in Example 16-1 is 128 × 128. One possibility would be to interpolate the sampled profiles from 90 samples to 128 samples. A more practical option, however, is to acquire 128 samples over 30 cm, which would somewhat exceed the linear sampling requirement. If this were done, Equation 16-15 would suggest that additional views would be needed to support the smaller value of Δr;

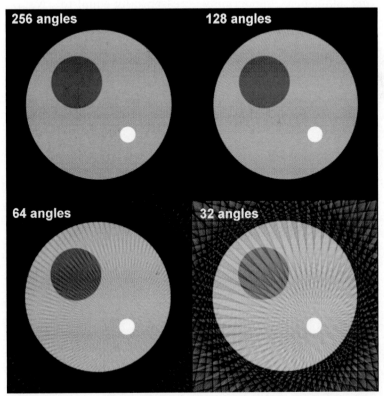

FIGURE 16-13 Effect of the number of angular samples recorded on the reconstructed image of a computer-simulation phantom. Spokelike streak artifacts are evident when an inadequate number of projections are used. (*Computer simulations performed by Dr. Andrew Goertzen, University of Manitoba, Canada.*)

however, 140 views still would provide the number of angular views needed to support the *system resolution* and this number would not have to be increased. On the other hand, going in the opposite direction, that is, acquiring only 64 samples and fewer angular samples for reconstruction on a 64 × 64 matrix would lead to a loss of image detail and introduce the possibility of image artifacts, as illustrated in Figures 16-11 and 16-13.

2. Sampling Coverage and Consistency Requirements

In addition to meeting the requirements described in the preceding section regarding linear and angular sampling intervals, the data acquired must provide full coverage of the object. Thus it is necessary that data be acquired over a full 180-degree arc. If an arc less than 180 degrees is used, geometric distortions are produced. Figure 16-14 demonstrates that an inadequate angular-sampling range causes data to flare out past the true objects and produces geometric distortions perpendicular to the direction of the absent projections. This is a problem for a number of systems developed in nuclear medicine that are classified as "limited-angle tomography" (e.g., rotating slant-hole tomography

and some positron emission mammography systems).

A second requirement for coverage is that the entire object (or at least the parts containing radioactivity) must be included in all projections. If some parts of the object are not included in all projections, the data will be *inconsistent* between different projections. There are a number of ways in which this can happen. For example, the FOV of the detector may be insufficient to provide full coverage from all directions. Figure 16-15 illustrates the effect of incomplete coverage of the object during some parts of the scan.

Two other possible sources of inconsistency between projections are patient movement and missing or distorted values in individual profiles caused by instrumentation failures, such as an unstable element in a detector array. Figure 16-16 illustrates some effects of these types of inconsistencies.

3. Noise Propagation, Signal-to-Noise Ratio, and Contrast-to-Noise Ratio

Noise propagation, SNR, and CNR differ in ECT from their behavior in conventional planar imaging. In conventional planar imaging, the SNR for an individual pixel is essentially equal to $\sqrt{N_{pixel}}$, in which N_{pixel} is

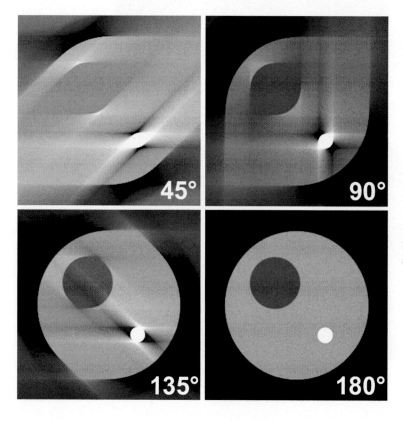

FIGURE 16-14 Effects of angular sampling range on images of a computer-simulation phantom. Images obtained by sampling over 45 degrees, 90 degrees, 135 degrees, and 180 degrees. Sampling over an interval of less than 180 degrees distorts the shape of the objects and creates artifacts. (*Computer simulations performed by Dr. Andrew Goertzen, University of Manitoba, Canada.*)

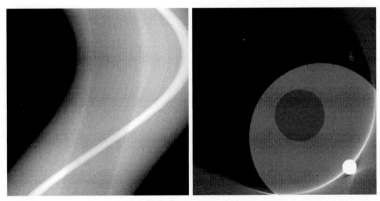

FIGURE 16-15 Effects of having some profiles that do not cover the entire object. *Left,* Sinogram of computer-simulation phantom. *Right,* Reconstructed image. (*Computer simulations performed by Dr. Andrew Goertzen, University of Manitoba, Canada.*)

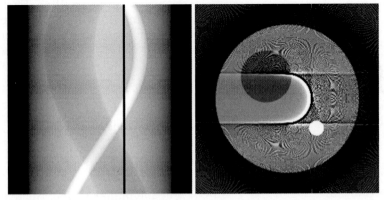

FIGURE 16-16 Effects of missing projection elements on reconstructed image. *Left,* Sinogram of computer-simulation phantom. *Right,* Reconstructed image. This simulation would apply to a SPECT image reconstructed from profiles acquired over a 180-degree sampling range with a single-headed camera, with one region of the detector "dead." (*Computer simulations performed by Dr. Andrew Goertzen, University of Manitoba, Canada.*)

the number of counts recorded for that pixel. In ECT, the computation of noise and SNR is much more complicated because the intensity level for each pixel is derived by computations involving different views and many other pixels in the image. In addition, a variety of mathematical manipulations, such as filtering operations, are performed along the way. As a result, although SNR still depends on the square root of the total number of counts recorded during the imaging procedure, the relationship between those counts and the SNR of individual pixels is more complicated.

Suppose that an ECT image is acquired of a cylindrical object of diameter D containing a uniform concentration of radioactivity. Suppose further that projection data are acquired with a linear sampling interval Δr across all projection profiles, that a total of N_{image} counts are recorded during the imaging procedure, and that the image is reconstructed by FBP with a ramp filter on a

square matrix of size $D \times D = D^2$ with pixel size $\Delta r \times \Delta r = \Delta r^2$. It can be shown that the SNR for an individual pixel in the resulting image of the object is given by[6]

$$\text{SNR}_{\text{pixel}} \approx \sqrt{\frac{12 N_{\text{image}}}{\pi^2 (D/\Delta r)^3}} \qquad (16\text{-}16)$$

Equation 16-16 indicates that SNR decreases when pixel size, Δr, is made smaller, that is, as spatial resolution is improved.* The

---

*Note that Equation 16-16 specifically assumes that pixel width is the same as the sampling interval, Δr. Often in nuclear medicine, interpolation techniques are used to generate images with pixels that are smaller than the sampling interval. Equation 16-16 is valid in these situations provided that the sampling interval rather than pixel size is used in the equation. Some texts describe Δr as the "resolution element" to avoid confusing it with pixel size.

dependence is relatively strong, as illustrated by the following example.

EXAMPLE 16-2

Suppose that an image of a 20-cm diameter cylinder containing a uniform distribution of activity is generated by FBP reconstruction. The image is reconstructed on a square matrix, 20×20 cm in size with 1×1 cm pixel size. A total of 1 million counts are acquired for the image. Calculate the number of counts required to generate an image with the same SNR per pixel if both the sampling interval and pixel size are reduced to 0.5 cm.

Answer
For the image with 1 cm resolution, from Equation 16-16

$$\mathrm{SNR}_{\mathrm{pixel}} \approx \sqrt{\frac{12 \times 10^6}{\pi^2 \times (20/1)^3}}$$

$$\approx 12.33$$

To maintain the same value of $\mathrm{SNR}_{\mathrm{pixel}}$ with 0.5-cm pixels, the required number of counts in the image, $N_{0.5\,\mathrm{cm}}$, must be such that

$$\sqrt{\frac{12 \times N_{0.5\,\mathrm{cm}}}{\pi^2 \times (20/0.5)^3}} \approx 12.33$$

Solving this equation yields the requirement of $N_{0.5\,\mathrm{cm}} = 8$ million counts.

Example 16-2 indicates that if Δr is decreased by a factor of 2 (to $\Delta r/2$), the total number of counts required to keep the SNR per pixel constant increases by a factor of 8, that is, as the inverse cube of the size of the pixel.

The total number of pixels in the reconstructed image is $n_{\mathrm{pixels}} = (D/\Delta r)^2$. Equation 16-16 can be rewritten as

$$\mathrm{SNR}_{\mathrm{pixel}} \approx \sqrt{12/\pi^2} \times \frac{\sqrt{<N_{\mathrm{pixel}}>}}{\sqrt[4]{n_{\mathrm{pixels}}}} \quad (16\text{-}17)$$

where $<N_{\mathrm{pixel}}>$ is the average number of counts recorded per reconstructed pixel in the object. This can be simplified even further by noting that $\sqrt{12/\pi^2} = 1.103 \approx 1$, for more approximate work.

Equation 16-17 indicates that SNR per pixel improves in proportion to the square root of the average number of counts recorded per pixel. This part of the equation is consistent with conventional counting statistics. However, as compared with conventional planar imaging (or photon counting), there is an additional factor equal to the fourth root of the total number of pixels (or resolution elements) in the denominator of the equation. This places stronger requirements on counting statistics for reconstruction tomography as compared with planar imaging, to achieve a specified level of $\mathrm{SNR}_{\mathrm{pixel}}$, as illustrated in the following example.

EXAMPLE 16-3

Consider two images of a cylindrical cross-section 20 cm in diameter with a uniform concentration of radioactivity, one a planar image and the other generated by FBP reconstruction with a ramp filter. Each image is 20×20 cm in size with 1×1 cm pixel size, and a total of 1 million counts are acquired for each image. What is the percent noise level, relative to signal level, in each image?

Answer
For both images,

percent noise level
$$= (\mathrm{noise/signal}) \times 100\%$$
$$= 100\%/\mathrm{SNR}_{\mathrm{pixel}}$$

For the planar image, the average number of counts per pixel within the area occupied by the 20-cm diameter object is

$$n_{\mathrm{pixel}} = [10^6/(\pi \times 10^2)] \text{ counts/cm}^2 \times 1 \text{ cm}^2/\text{pixel}$$

$$\approx 3180 \text{ counts/pixel}$$

This yields an SNR given by

$$\mathrm{SNR}_{\mathrm{pixel}} = \frac{n_{\mathrm{pixel}}}{\sqrt{n_{\mathrm{pixel}}}} = \sqrt{n_{\mathrm{pixel}}}$$

$$= \sqrt{3180} = 56$$

from which the percent noise level is $100\%/56 \approx 1.8\%$.

For the image reconstructed by FBP, the result is as given in Example 16-2:

$$\mathrm{SNR}_{\mathrm{pixel}} \approx 12.33$$

from which the percent noise level is $100\%/12.33 \approx 8.1\%$.

The "noise enhancement" factor for reconstruction tomography illustrated in Example 16-3 is the result of noise propagation from pixels at many locations in the imaged object into the pixel of interest in the backprojection process, as well as the ramp filtering operation.

Example 16-3 applies to the SNR of a single pixel in images of a uniform object. The result would seem to imply a statistical disadvantage for the detection of low-contrast objects by ECT. However, for purposes of applying the Rose criterion for detectability of lesions and other objects (see Chapter 15, Section D.2) this must be converted to CNR for the object of interest. Using the definitions given in Chapter 15, it can be shown that the CNR for a lesion that occupies n_ℓ pixels in an ECT image is

$$\text{CNR}_\ell \approx |C_\ell| \times \sqrt{n_\ell} \times \text{SNR}_{\text{pixel}} \quad (16\text{-}18)$$

where the absolute value indicates that CNR always is a positive quantity. Although the noise characteristics in ECT differ somewhat from those of planar imaging (particularly regarding possible artifacts), the same general rules for detectability apply for ECT and planar images, that is, $\text{CNR}_\ell \geq 4$.

EXAMPLE 16-4

Consider the situation described for radiopharmaceutical B in Example 15-6. In that example, the radiopharmaceutical produced "cold" lesions with uptake that was 10% of the surrounding normal tissue and a CNR of only 0.9 for a 1-cm diameter lesion. Using the same parameters, estimate the CNR that would be achieved using the same radiopharmaceutical, spatial resolution, and total imaging time with ECT. Assume that the normal tissue fills a volume of (10 × 10 × 10 cm) and, as in Example 15-6, ignore the effects of attenuation and source-to-detector distance.

Answer

The planar image described for Example 15-6 could be obtained by facing the detector toward any face of the cubic volume of tissue and acquiring counts for a 1-min imaging time. For purposes of computing the SNR of an ECT image, many projection views would be required (e.g., 60 1-sec views), but the total number of counts recorded in 1 min of imaging time, in the absence of attenuation and distance effects, would be the same as for planar imaging. The total number of pixels in the ECT image is $n_{\text{pixels}} = 10 \times 10 = 100$, and the average number of counts per pixel in the image is $<N_{\text{pixel}}> \approx 10$ counts per minute (cpm) × 1 min = 10. Thus from Equation 16-17, the SNR per pixel for the ECT image would be

$$\text{SNR}_{\text{pixel}} \approx \frac{\sqrt{10}}{\sqrt[4]{100}} \approx \frac{\sqrt{10}}{\sqrt{10}} \approx 1$$

Using the definition given in Equation 15-6, the contrast of the lesion in the ECT image is

$$C_\ell = (1 - 10)/10 = -0.9$$

Substituting these values into Equation 16-18, one obtains

$$\text{CNR}_\ell \approx |-0.9| \times \sqrt{1} \times 1 \approx 0.9$$

which is the same result as obtained for planar imaging. In neither case would the lesion be detectable using the Rose criterion.

Example 16-4 shows that, for the same level of object contrast and total number of counts in the image, and in the absence of attenuation and distance effects, *there is no intrinsic difference in CNR between ECT and planar imaging.* This result is obtained, in spite of the apparent statistical disadvantage of ECT illustrated in Example 16-3, because of the increased contrast of the low-contrast lesion in an ECT image as compared with a projection image. On the other hand, the sophisticated data manipulations of ECT do not improve the detectability of the lesion. This is not too surprising, because it should not be possible to improve CNR when noise is generated by counting statistics by applying mathematical manipulations (e.g., reconstruction tomography and contrast enhancement) of otherwise comparable data.

Thus it is inaccurate to conclude that ECT improves detectability of lesions or other objects by improving CNR. Rather, *the primary advantage of ECT for detecting low-contrast lesions derives from its ability to remove confusing overlying structures that may interfere with detectability of those lesions,* such as ribs overlying a lesion in the lungs. Not only does an object become more detectable when overlying clutter is removed by ECT, but its shape and borders become more clear.

An additional advantage of ECT is the ability to determine more accurately the concentration of radioactivity in a particular volume of tissue. For example, in Example 16-4, the same planar image would be obtained if the lesion were twice as thick along the viewing direction, but with half the uptake suppression and thus twice the concentration as originally specified in Example 15-6. However, such a difference would be readily evident on the ECT image (assuming the CNR requirements for detectability were met). Some appreciation for all of these advantages can be gained by inspection of Figure 15-11.

D. ITERATIVE RECONSTRUCTION ALGORITHMS

A viable and increasingly used alternative to FBP is a class of methods known as *iterative reconstruction*. These methods are computationally more intensive than FBP and for this reason have been more slowly adopted in the clinical setting. However, as computer speeds continue to improve, and with a combination of computer acceleration techniques (e.g., parallel processors), and intelligent coding (e.g., exploiting symmetries

and precomputing factors), reconstruction times have become practical and iterative methods are finding their way into more general use.

1. General Concepts of Iterative Reconstruction

The general concepts of iterative reconstruction are outlined in Figure 16-17. In essence, the algorithm approaches the true image, $f(x,y)$, by means of successive approximations, or estimates, denoted by $f^*(x,y)$. Often the initial estimate is very simple, such as a blank or uniform image. The next step is to compute the projections that would have been measured for the estimated image, using a process called *forward projection*. This process is exactly the inverse of backprojection. It is performed by summing up the intensities along the potential ray paths for all projections through the estimated image. The set of projections (or sinogram) generated from the estimated image then is compared with

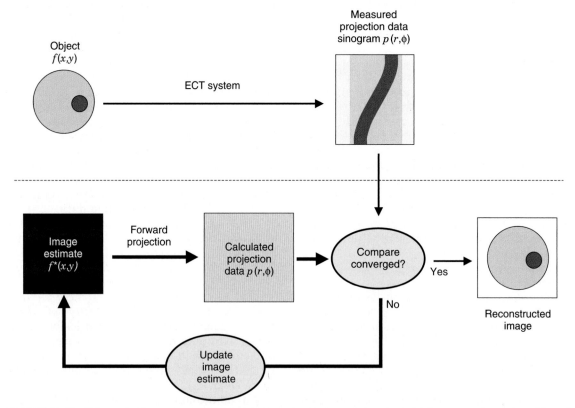

FIGURE 16-17 Schematic illustration of the steps in iterative reconstruction. An initial image estimate is made and projections that would have been recorded from the initial estimate then are calculated by forward projection. The calculated forward projection profiles for the estimated image are compared to the profiles actually recorded from the object and the difference is used to modify the estimated image to provide a closer match. The process is repeated until the difference between the calculated profiles for successively estimated images and the actually observed profiles reaches some acceptably small level.

the actually recorded projections (or sinogram). Most likely, they will not agree, because it is unlikely that the initial estimate of $f^*(x,y)$ closely resembles the true image. However, the difference between the estimated and actual projections can be used to adjust the estimated image to achieve closer agreement.

The update-and-compare process is repeated until the difference between the forward-projected profiles for the estimated image and the actually recorded profiles falls below some specified level. With proper design of the image updating procedure, the estimated image progressively converges toward the true image. Figure 16-18 shows the progress of the estimated image during iterative reconstruction with an increasing number of iterations.

The two basic components of iterative reconstruction algorithms are (1) the method for comparing the estimated and actual profiles and (2) the method by which the image is updated on the basis of this comparison. In generic terms, the first component is performed by the *cost function*, which measures the difference between the profiles generated by forward projections through the estimated

image and the profiles actually recorded from the scanned object. The second component is performed by the *search* or *update function*, which uses the output of the cost function to update the estimated image. A general goal of algorithm development is to devise versions of these functions that produce convergence of the estimated image toward the true image as rapidly and accurately as possible. One area of algorithmic differences is the method for dealing with statistical noise. For example, some algorithms give more weight to portions of projections (or sinograms) that contain the highest number of counts, and thus the lowest percentage levels of statistical noise (see Chapter 9, Section B.1). Another approach is to incorporate some sort of "prior information," such as the expected shape or smoothness of the image. Some algorithms also "force" the reconstructed image to be nonnegative. A concise history and review of iterative reconstruction methods are presented in reference 7.

Two factors make iterative reconstruction computationally more intensive than FBP. First, most iterative algorithms require several iterations to converge to an acceptable

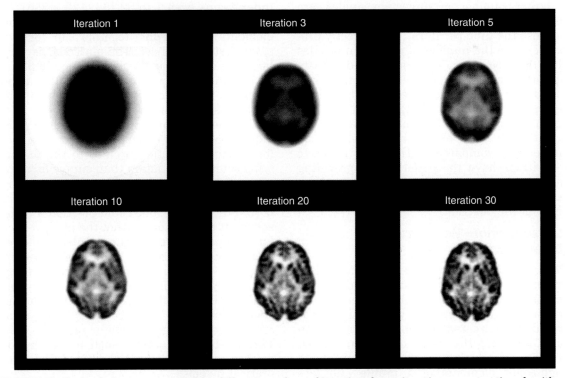

FIGURE 16-18 Brain images generated for different numbers of iterations by an iterative reconstruction algorithm. Image resolution progressively improves as the number of iterations increases. In practice, the iterations are performed until an acceptable level of detail is achieved or until further iterations produce negligible improvement. (*Courtesy Dr. Richard Leahy, University of Southern California, Los Angeles, CA.*)

image, and each of these iterations is essentially equivalent to a separate backprojection procedure. Backprojection is the most time-consuming part of the FBP algorithm but only needs to be done once for FBP. Forward projection is similarly time-consuming in iterative reconstruction algorithms.

Second, iterative algorithms often incorporate factors that account for the specific characteristics of the imaging device, such as collimator and object scatter, system geometry, and finite detector resolution. Simple forward projection along a single ray path no longer is used to calculate the projection profiles for the estimated image. Instead, all image pixels are considered to have a finite probability of contributing data to virtually all ray paths. In practice, very distant pixels might not be considered. Nevertheless, this adds to the computing time, because the reconstruction must include effects not only from pixels directly along a ray path but from pixels outside that ray path as well.

A number of methods have been developed to speed up these advanced algorithms. One of the most popular is called *ordered subsets*. In this method only a small number (or subset) of projection angles are used in the initial iterations. As the image is refined, a larger number of projection angles are included. This speeds up the algorithm, because the time per iteration is directly proportional to the number of projection profiles that must be computed. The ordered-subsets approach can be used to speed up both simple forward projection-based iterative algorithms as well as the advanced algorithms that use complex modeling of the imaging system.

Although they are more challenging to implement compared with FBP, iterative algorithms have the potential for providing quantitatively more accurate reconstructions. An example of one algorithm is presented in the following section.

2. Expectation-Maximization Reconstruction

The *expectation-maximization* (EM) algorithm incorporates statistical considerations to compute the "most likely," or *maximum-likelihood* (ML), source distribution that would have created the observed projection data, including the effects of counting statistics. Specifically, it assigns greater weight to high-count elements of a profile and less weight to low-count regions. (By comparison, backprojection algorithms assign a uniform statistical weighting to all elements of a

profile.) Because of the statistical weighting factor, the algorithm often is referred to as the *ML-EM method*. A detailed discussion of this algorithm and its theoretical underpinnings are beyond the scope of this text but can be found in references 8 and 9. Here we present only a description of how it is implemented.

In the EM algorithm, the reconstruction process is formulated as follows

$$p_j = \sum_i M_{i,j} f_i \qquad (16\text{-}19)$$

where f_i is the intensity (or activity) in the ith pixel in the image, p_j is the measured intensity in the jth projection element, and $M_{i,j}$ is the probability that radiation emitted from the ith pixel will be detected in the jth projection element. Note that, unlike previous uses of i and j to represent different (x,y) locations in a 2-D image (or a set of projections), the indices here each apply to the full set of the subscripted quantities. Thus, if the image is reconstructed on a grid of 128×128 pixels, the subscript i runs from 1 to 16,348 (128×128). If the imaging system records projections at 128 different angles around the object, and each projection has 256 elements, the index j runs from 1 to 32,768 (128×256). In essence, all of the image pixels and projection elements are "strung together" to form a single list for each set. The matrix M is very large, even for a single-slice image (16,384 × 32,768 in the previous example). It can be extended to three dimensions as well, in which case it becomes even larger.

The matrix approach described above provides a potentially much more accurate model for relating projection profiles to the underlying source distribution than simple forward projection. The matrix could be determined by calculations, simulations, or a combination of both. For example, one could position a point source at all locations within the imaged slice (or volume) and record the counts in all elements of all possible projection profiles. However, this would be very time consuming. Symmetry considerations could somewhat shorten the project. In practice, many of the geometric effects can be calculated from simple models (e.g., collimator response—see Fig. 14–15) and others, such as collimator scatter, can be simulated or derived from theoretical models.

Once the matrix M has been determined and projection profiles have been recorded, the operating equation for computing the estimated intensity value f of pixel i in the $(k + 1)^{st}$ iteration of the EM algorithm is as follows:

$$f_i^{k+1} = \frac{f_i^k}{\sum\limits_j M_{i,j}} \times \sum\limits_j M_{i,j} \frac{p_j}{\left(\sum\limits_l M_{l,j} f_l^k\right)} \qquad (16\text{-}20)$$

where k refers to the immediately preceding k^{th} iteration. The term in parentheses in the denominator on the right hand side of Eq. 16-20 represents a summation over all image pixels. This term must be evaluated first before the summation over the j projection elements can be computed. Therefore it is given a different pixel index, l, instead of i, to avoid confusion.

The number of iterations can be fixed, or the iteration process can be terminated when some measure of the difference between images from one iteration to the next (e.g., the sum of the squares of differences for all pixels in the reconstructed image) falls below some predetermined value. In theory, with perfectly measured noise-free data and an exact matrix M, the algorithm eventually would converge to the point where the estimated projection data, $\sum\limits_l M_{l,j} f_l^k$ exactly equals the measured projection data, p_j, for each profile. At that point

$$f_i^{k+1} = f_i^k \qquad (16\text{-}21)$$

that is, there is no further change in the estimated image and the estimated activity image is identical to the true activity distribution. In practice, this never happens, owing to inaccuracies or simplifications in M and statistical noise. Therefore some practical limit must be set for an acceptable difference that will be used to terminate the reconstruction process.

The computational issues relating to iterative reconstruction techniques already have been mentioned. Equation 16-20 (which represents only a single-slice version of the algorithm) illustrates this point. Nevertheless, the ML-EL algorithm can produce high-quality images with good quantitative accuracy and is now a selectable option on many PET and SPECT cameras.

The sampling and noise-propagation rules summarized in Section C do not apply to iterative reconstruction. Although insufficient sampling also has consequences for iterative algorithms, the aliasing and streaking artifacts associated with FBP are not seen. More typically, undersampling in the linear sampling distance or in the number of angular views results in a more or less uniform loss of spatial resolution across the reconstructed image. If only partial angular coverage of the object is obtained (e.g., from 0 to 120 degrees instead of over the full 180 degrees), the resolution is likely to be degraded along the direction of the missing data. Because iterative algorithms are non-linear in nature, the exact effects of under-sampling are object and algorithm dependent.

E. RECONSTRUCTION OF FAN-BEAM, CONE-BEAM AND PINHOLE SPECT DATA, AND 3-D PET DATA

The discussion thus far has focused on reconstructing projection data in which the acquired rays for a given projection angle are parallel and the projection data arises from parallel sections through the body. This is the situation when a parallel-hole collimator is used. Tomographic reconstruction also can be performed using data acquired with fan-beam, cone-beam, or pinhole collimators. The rationale for using these collimators is that they can provide higher spatial resolution (converging-hole or pinhole collimators) or greater coverage (diverging-hole collimators— see Chapter 14, Section D). However, these collimators introduce an added degree of complexity for reconstruction tomography of SPECT data, because they do not provide simple parallel-ray line integral projections such as were illustrated in Figure 16-2. Similar issues arise in PET scanning. In addition to acquiring projection data for transverse sections through the body, PET scanners, as discussed in Chapter 18, Section C, also are capable of acquiring additional projection data at oblique angles with respect to these transverse slices. Accurately incorporating this additional projection data requires 3-D reconstruction algorithms.

1. Reconstruction of Fan-Beam Data

We should first distinguish between fan-beam versus cone-beam collimators. Figure 16-19 schematically illustrates the difference. Consider first the *fan-beam collimator* shown at the top of the figure. In this collimator, each row of holes across the collimator has its own focal point. Sequential rows of collimator holes are stacked and evenly spaced, parallel to each other, along the z-axis of the object. Apart from overlapping coverage resulting from the finite diameters of the collimator

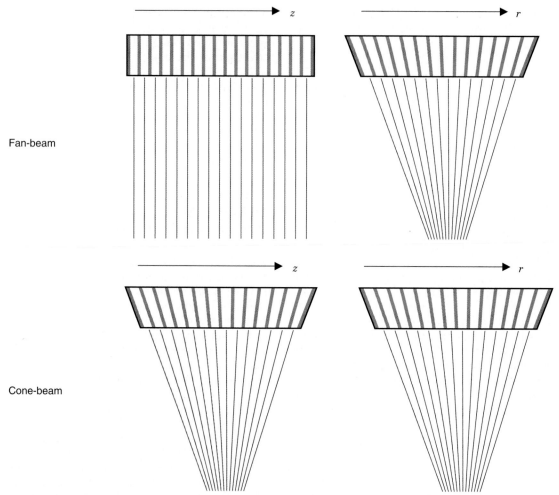

FIGURE 16-19 Schematic illustrations of fan-beam and cone-beam collimators. Cross-sections are shown for perpendicular viewing angles.

holes, each row of holes provides its own independent and nonoverlapping projection profile.

Data from a fan-beam collimator cannot be inserted directly into algorithms used for reconstructing data acquired with a parallel-beam collimator. However, the data can be rearranged so that these algorithms can be used. One approach is to re-sort the fan-beam data into parallel-beam data. Figure 16-20 illustrates how this is done for a few elements of adjacent projection profiles. Once the data have been re-sorted, any of the algorithms discussed in the preceding sections for parallel-beam collimators can be used. Alternative, and more elegant, approaches reformulate the FBP algorithm itself to handle fan-beam data. These are discussed in references 2 and 7.

A fan-beam collimator provides complete 3-D coverage of a volume of tissue in a single rotation around the object. However, whereas complete coverage can be obtained with a 180-degree rotation using a parallel-hole collimator, the required rotation for a converging-beam collimator is $(180 + \theta)$ degrees, in which θ is half the fan angle for the collimator (see Fig. 14-20). Conversely, for a diverging collimator, the required angle of rotation is $(180 - \theta)$ degrees.

2. Reconstruction of Cone-Beam and Pinhole Data

In a *cone-beam collimator* (Fig. 16-19, *bottom*), all of the holes are directed toward (or away from) a common focal point. Each row of holes across the center of the collimator provides a projection profile, but the profiles all intersect at the center. (This also applies to the pinhole collimator.) It is not possible to re-sort the data acquired from a single rotation of a

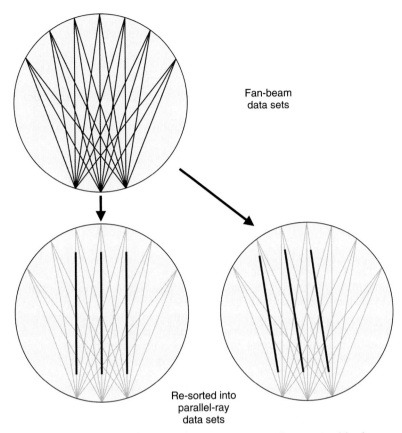

FIGURE 16-20 Procedure for creating parallel-beam projections from a set of fan-beam projections.

cone-beam collimator around the object into a full set of parallel-ray projections. Only one set (corresponding to the projections acquired from a single slice across the center of collimator, oriented perpendicular to the axis of rotation) can be re-sorted in this way. Therefore to obtain complete projection coverage of a volume of tissue to allow accurate reconstruction of multiple slices, a more complex rotation is required.

One approach is to perform a helical scan around the object, translating the collimator along the z-axis as it rotates about that axis. This provides a dataset that can be re-sorted into a complete set of parallel projections for multiple slices through the object. An alternative approach is to use approximations and interpolations to convert the cone-beam data into fan-beam data. The most popular of these methods is called the *Feldkamp algorithm,* described in reference 10. These methods work best when the cone angle is small.

Finally, iterative algorithms, conceptually similar to those described in Section C, have been developed for direct reconstruction of

3-D cone-beam data; however, computing time increases dramatically as compared with already time-consuming single-slice iterative algorithms. The matrix M (Equation 16-19) becomes very large for a full 3-D algorithm and, even with accelerated approaches and specialized computer hardware, full 3-D image reconstructions are typically at least an order of magnitude slower than multislice 2-D image reconstructions.

3. 3-D PET Reconstruction

PET scanners typically consist of multiple detector rings (see Chapter 18, Section B). Projection data acquired within a given detector ring can be reconstructed into a transverse image with the methods described previously in Sections B and D. However, PET scanners also can acquire projection data at oblique angles between detector rings (see Chapter 18, Section C and Fig. 18-24). To incorporate these additional projection angles requires some form of 3-D reconstruction algorithm. 3-D algorithms have been developed based on both FBP and iterative

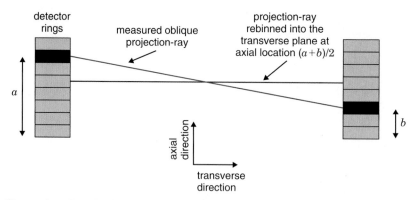

FIGURE 16-21 Illustration of single-slice re-binning in which an oblique projection-ray between the detector pair a and b is "re-assigned" to the projection data for the non-oblique slice corresponding to a transverse detector pair at axial location $(a + b)/2$.

reconstruction methods. A detailed description of these algorithms is beyond the scope of this text. However, some general concepts will be presented.

One common approach is to "re-bin" the 3-D dataset, such that each oblique projection ray is placed within the projection data for a particular nonoblique 2-D transverse slice. In effect, the 3-D dataset is collapsed back into a multislice 2-D dataset. The most simple method to accomplish this is to assign each ray to its average axial location.[11] Thus an oblique projection-ray between a detector at location a and a detector at location b would be positioned as if it were a projection from a directly opposed pair located halfway between them, i.e., at location $(a+b)/2$ (Fig. 16-21). Processing of all the projection rays in this manner results in a series of sinograms of parallel-ray projections, each corresponding to different axial locations through the object. Each sinogram can then be reconstructed using the 2-D FBP or iterative algorithms described previously. This method is known as single-slice rebinning.

Figure 16-21 illustrates that for events originating close to the center of the field of view (FOV) of the scanner, only small errors in positioning are made by this approximation. However, for events originating close to the edge of the scanner, and for projections at a large angle with respect to the transverse plane, significant mispositioning errors are made in the axial direction using this method. To overcome this, more accurate methods that include more sophisticated re-binning algorithms and an axial filtering step[12] or that use Fourier re-binning techniques[13] have been developed.

An alternative approach is to formulate the iterative reconstruction equations (Eqs. 16-19 to 16-21), for fully 3-D reconstruction, thus implicitly accounting for the exact orientation of each line of response. Because of the additional dimensionality of the projection data, and the fact that the images are being reconstructed into a 3-D volume rather than a 2-D slice, the matrix M in Equation 16-19 becomes very large. The number of elements can be in the range of 10^{13} to 10^{15} with modern PET scanners that have large numbers of detector rings. Furthermore, the backprojection and forward-projection steps must now be performed in 3-D, tracing each ray though a volume rather than across a 2-D slice.

Thus the computational challenges are formidable, although great progress has now been made in reducing the matrix size using sparse storage techniques and symmetry arguments, and multiprocessor hardware and efficient coding have produced fast methods for 3-D backprojection and forward projection. Fully 3-D iterative algorithms are now available on some systems, especially small-animal imaging systems in which the small FOV typically leads to more manageable projection dataset sizes.

REFERENCES

1. Webb S: *From the Watching of Shadows: The Origins of Radiological Tomography*, Bristol, England, 1990, Adam Hilger.
2. Kak AC, Slaney M: *Principles of Computerized Tomographic Imaging*, Philadelphia, 2001, SIAM.
3. Herman GT: *Fundamentals of Computerized Tomography: Image Reconstruction from Projections*, ed 2, London, 2009, Springer-Verlag.
4. Natterer F: *The Mathematics of Computerized Tomography*, New York, 1986, Wiley.

5. Oppenheim AV, Wilsky AS: *Signals and Systems*, Englewood Cliffs, NJ, 1983, Prentice-Hall, pp 513-555.

6. Hoffman EJ, Phelps ME: Positron emission tomography: Principles and quantitation. In Phelps ME, Mazziotta JC, Schelbert HR, editors: *Positron Emission Tomography and Autoradiography: Principles and Applications for the Brain and Heart*, New York, 1986, Raven Press, pp 237-286.

7. Leahy RM, Clackdoyle R: Computed tomography. In Bovik A, editor: *Handbook of Image and Video Processing*, Burlington, MA, 2005, Elsevier Academic Press, pp 1155-1174.

8. Shepp LA, Vardi Y: Maximum likelihood reconstruction for emission tomography. *IEEE Trans Med Imag* 1:113-122, 1982.

9. Lange K, Carson R: EM reconstruction algorithms for emission and transmission tomography. *J Comput Assist Tomogr* 8:306-316, 1984.

10. Feldkamp LA, Davis, LC, Dress JW: Practical cone-beam algorithm. *J Opt Soc Am* 1:612-619, 1984.

11. Daube-Witherspoon ME, Muehllehner G: Treatment of axial data in three-dimensional PET. *J Nucl Med* 28:1717-1724, 1987.

12. Lewitt RM, Muehllehner G, Karp JS: Three-dimensional image reconstruction for PET by multi-slice rebinning and axial image filtering. *Phys Med Biol* 39:321-339, 1994.

13. Defrise M, Kinahan PE, Townsend DW, Michel C, Sibomana M, Newport DF: Exact and approximate rebinning algorithms for 3D PET data. *IEEE Trans Med Imag* 16:145-148, 1997.

chapter 17

Single Photon Emission Computed Tomography

As discussed in Chapter 16, a rotating gamma camera can be used to acquire data for computed tomographic (CT) images. This approach to tomography, which is employed with radionuclides that emit single γ rays or multiple γ rays with no angular correlations, is known as *single photon emission computed tomography* (*SPECT*). In this chapter, we describe the design features and performance characteristics of SPECT systems. We also discuss some practical aspects of SPECT imaging and some of its major clinical applications. A second form of tomographic nuclear medicine imaging, *positron emission tomography* (*PET*), uses radionuclides that decay by positron emission. PET imaging systems and their characteristics are discussed in Chapter 18. Multimodality systems that combine SPECT or PET with x-ray CT are discussed in Chapter 19.

A. SPECT SYSTEMS

1. Gamma Camera SPECT Systems

Almost all commercially available SPECT systems are based on the gamma camera detector that was described in detail in Chapters 13 and 14. A single gamma camera head, mounted on a rotating gantry, is sufficient to acquire the data needed for tomographic images. The gamma camera acquires two-dimensional (2-D) projection images at equally spaced angular intervals around the patient. These images provide the 1-D projection data needed for reconstructing cross-sectional images using the techniques described in Chapter 16. Typically, clinical SPECT images are reconstructed on a matrix of 64 × 64 or 128 × 128 pixels. Cross-sectional images are produced for all axial locations (slices) covered by the field of view (FOV) of the gamma camera, resulting in a stack of contiguous 2-D images that form a 3-D image volume.

The number of angular projections (or views) needed when using a standard parallel-hole collimator can be calculated using Equation 16-15. Because the resolution of a general-purpose parallel-hole collimator is approximately 1 cm at a distance of 10 cm from the collimator (see Fig. 14-16), the number of views required generally is between 64 and 128, for a FOV ranging from 20 to 60 cm in diameter. Although data acquired over an arc of 180 degrees are sufficient for tomographic reconstruction in SPECT, there are advantages in terms of resolution uniformity and correction for γ-ray attenuation in acquiring data over a full 360-degree arc. This is discussed in Section B.1.

The sensitivity of a SPECT system can be improved by incorporating multiple detector heads in the system. Both dual-headed and triple-headed SPECT systems are available, with dual-headed systems being the most commonly encountered. These systems allow two or three angular projections to be acquired simultaneously. For the same total data acquisition time, each projection can be recorded two or three times, leading to a twofold or threefold increase in the total number of counts acquired for the image. Alternatively, a multihead system can be used to acquire the same number of counts in one half or one third the time needed with a single-head system. This can be useful for dynamic SPECT imaging to observe changes

279

in the distribution of a radiopharmaceutical as a function of time (Section D). One also could replace the parallel-hole collimator with a converging collimator to obtain improved sensitivity (see Fig. 14-21); however, this results in a smaller FOV (see Fig. 13-7).

Photographs of single-headed and dual-headed gamma cameras that are capable of SPECT imaging are shown in Figures 13-11 and 13-12. In dual-headed SPECT systems, the detector heads are typically placed at 180 degrees relative to each other for whole-body SPECT imaging, and at 90 degrees relative to each other for cardiac imaging. In some systems the location of the detector heads is fixed, whereas in others, they can be adjusted by the operator. Figure 13-12 shows a system in which the detector head orientations can be changed. In addition to mechanical capabilities for rotating the detector heads, gamma camera systems intended for SPECT imaging must be provided with computer capabilities and software for image reconstruction, for attenuation and scatter corrections, and for display and analysis of 3-D image volumes. The ability to perform conventional planar imaging as well as tomographic imaging is a very useful feature of these cameras.

Many SPECT systems have more sophisticated gantries that allow the detector heads to trace out elliptical rather than circular orbits. Some even allow orbits that follow the contours of the patient. The body contour can be determined by an initial scout scan that, for example, uses an infrared light source and camera to trace the outline of the patient and the bed as a function of angle. The importance of this feature is evident from Figure 14-16, which shows the rapid degradation in spatial resolution with increasing distance of the object from the collimator. As shown in Figure 17-1, elliptical orbits or orbits that follow the contours of the patient allow the detector to pass closer to the patient than would be the case with circular orbits, which can lead to significant improvements in spatial resolution.

2. SPECT Systems for Brain Imaging

A disadvantage common to all of the SPECT systems described in the preceding section is that the detector heads must be rotated around the patient to record the multiple projections required for tomographic reconstruction. Because of the mechanical motions involved and the bulk of the detector head or heads, the shortest time in which a complete set of projections can be recorded generally is

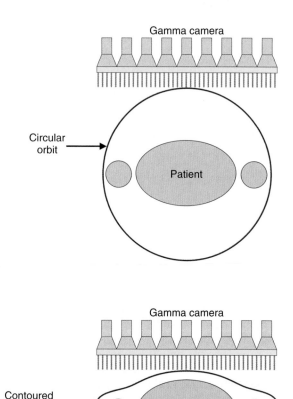

FIGURE 17-1   Illustration of circular (*top*) and contoured (*bottom*) orbits for SPECT imaging. The contoured orbit minimizes the distance of the detector from the patient, thus optimizing spatial resolution.

several minutes. For imaging fast biologic processes, it would be desirable to acquire a full set of projections in a few seconds. Furthermore, traditional collimator design limits reconstructed spatial resolution to 1 cm or greater, which is inadequate for some applications in the human brain. A number of efforts have been made toward addressing these limitations. Although they involve specialized systems that are not currently commercially available, we describe briefly some of the concepts involved to illustrate possible alternatives to performing SPECT with a rotating gamma camera.

The University of Michigan *SPRINT II* system[1] employs 11 γ-ray detectors in a polygonal arrangement, each detector consisting of 44 thin bars of sodium iodide [NaI(Tl)] scintillator (dimensions 3-mm-wide × 13-mm-deep × 15-cm-long) coupled to an array of twenty 38-mm-diameter photomultiplier (PM) tubes arranged behind the NaI(Tl) bars in a close-packed hexagonal array. In-slice collimation is achieved using a lead aperture ring, with 12 equally spaced 2.4-mm-wide slits, that rotates in front of the stationary detector array (Fig. 17-2A). Axial collimation is achieved using a set of stacked stationary lead foils. As the collimator ring rotates through one slit interval (30 degrees), the system acquires a complete set of fan-beam projection data.

A system based on similar principles is the *CERASPECT* system.[2] In this case, the detector is a single annular NaI(Tl) crystal (31 cm inner diameter × 8 mm thick × 13 cm wide) coupled to 63 5-cm diameter PM tubes via glass light guides (Fig. 17-2B). A parallel-hole collimator with six segments rotates in front of the detector, simultaneously providing six angular views. Each collimator segment has a different FOV. This gives a higher weighting to activity at the center of the object (which is viewed by all six collimator segments) in comparison to activity toward the periphery of the object (which is seen by a smaller number of collimator segments). This nonuniform weighting helps compensate for the effects of photon attenuation (see Sections B.1 and B.2) and provides more uniform signal-to-noise ratio across the image.

Both of these systems were designed primarily for brain imaging applications, and both provide better image resolution than a conventional SPECT system by placing the collimated detector relatively close to the head. The reconstructed spatial resolution is approximately 8 mm at the center of the brain, improving to approximately 5 mm at the edge of the brain. By comparison, a typical single-head SPECT system operated with a radius of rotation of 12.5 cm (appropriate for brain imaging) would have a resolution of approximately 12.5 mm full width at half maximum (FWHM) at the center of the brain (see Fig. 14-16). These systems also have roughly twofold to threefold higher sensitivity than a single-headed gamma camera with a general-purpose parallel-hole collimator, because multiple sets of projection data can be acquired simultaneously. This enables higher resolution to be achieved without injecting more radioactivity or lengthening the imaging time.

3. SPECT Systems for Cardiac Imaging

One of the most common uses for SPECT is to image myocardial function in patients with a range of cardiovascular diseases. Because of the relatively small FOV required for this application, and the increasing use of cardiac

SPECT in smaller clinics, dedicated and compact cardiac SPECT systems with high sensitivity have been developed. In contrast to the SPECT systems discussed so far, the majority of these cardiac systems involve the patient sitting upright in a chair during the examination.

Figure 17-3A shows a triple-headed SPECT system designed for cardiac applications.[3] Note the much smaller size of the detector heads compared with general-purpose SPECT systems. The detector heads use pixelated cesium iodide [CsI(Tl)] scintillator crystals, read out by avalanche photodiodes (Chapter 7, Section C.3). The size of each CsI(Tl) pixel is 6 mm, and there are 768 pixels in each detector head, providing a detector FOV of approximately 16 cm × 20 cm. In this system the detectors are fixed and the patient chair rotates to provide the necessary angular sampling for tomographic reconstruction. The distance between the detectors and the chair can be adjusted to accommodate patients of different size. With a low-energy, high-resolution (LEHR) collimator (see Table 14-1), the reported spatial resolution of the SPECT images is 11 mm (for a 20-cm detector-object separation) and the sensitivity for each detector head is ~72 cps/MBq (160 cpm/μCi). Because of the vertical orientation of the patient, the entire system can fit in a room as small as 2.4 m × 2.4 m.

Another type of cardiac SPECT system replaces standard scintillation Anger camera designs with detector heads made up of pixels of the dense semiconductor cadmium zinc telluride (CZT) (Chapter 7, Section B).[4] This system contains 10 small detector heads, distributed in an arc over the chest of the patient (Fig. 17-3B). Each detector head [2-mm CZT pixels tiled on a 40-mm (transaxial) by 160-mm (axial) area] is equipped with a parallel-hole collimator and can rotate independently to sample different projection angles, enabling a complete projection dataset to be acquired for tomographic reconstruction. The patient chair is stationary. The reconstructed spatial resolution is quite dependent on the location in the FOV, and is nonisotropic, ranging between approximately 8 and 14 mm. The sensitivity of the system is on the order of ~400 cps/MBq (900 cpm/μCi) for each detector head. The high sensitivity derives from the use of a shorter collimator with larger holes compared with the LEHR collimator.

A third design employs principles similar to the brain system shown in Figure 17-2A and described in Section 2. A multislit collimator

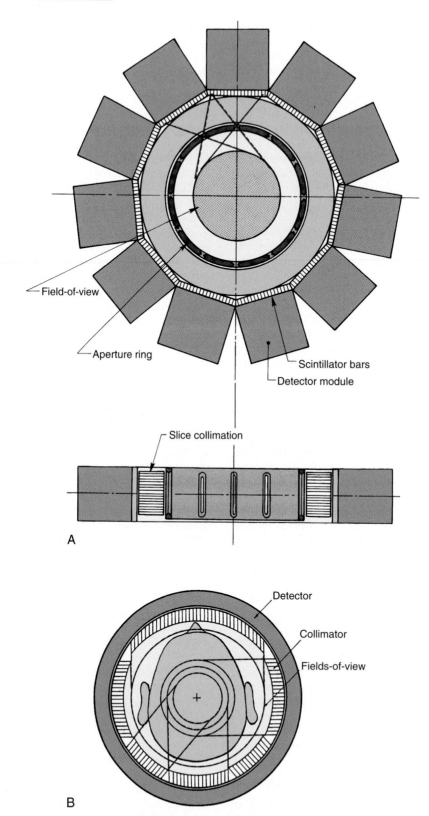

FIGURE 17-2   Cross-sectional views showing the design of two SPECT systems designed for brain imaging. *A,* The SPRINT system developed at the University of Michigan. This system employs a rotating collimator with 12 axial slits. Transverse and axial views are shown. *B,* The CERASPECT system developed by Digital Scintigraphics, Inc., Cambridge, MA. In this system, each collimator segment has a different field-of-view diameter. (*A, From Rogers WL, Clinthorne NH, Shao L, et al: SPRINT II: A second-generation single-photon ring tomograph. IEEE Trans Med Imag 7:291-297, 1988; B, From Genna S, Smith AP: The development of ASPECT, an annular single-crystal brain camera for high-efficiency SPECT. IEEE Trans Nucl Sci 35: 654-658, 1988.*)

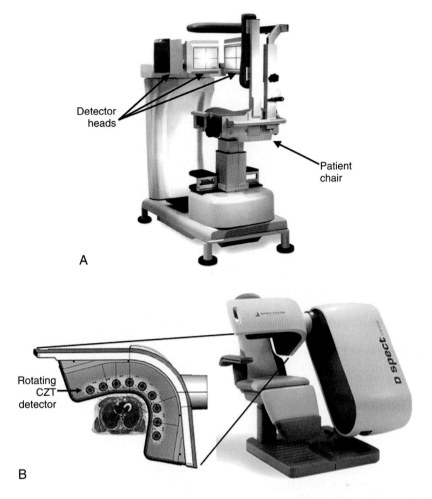

FIGURE 17-3 Photographs of dedicated cardiac SPECT scanners: *A,* A system with three detector heads. The patient sits upright in the chair and the chair rotates in front of the detectors to acquire the projection angles necessary for tomographic reconstruction. *B,* A system comprising nine cadmium zinc telluride detector heads arranged in an arc around the patient. Each detector head rotates independently to provide different angular views. The patient remains stationary in the chair. (*A, Courtesy Digirad Corp., Poway, CA; B, Courtesy Spectrum Dynamics Ltd. Caesarea, Israel.*)

rotates in front of an arc-shaped gamma camera that employs traditional continuous NaI(Tl) crystals read out by PM tubes.[5] This rotating collimator gives rise to the different projection angles, with slice collimation provided by a stationary stack of thin concentric lead rings. For a source-to-aperture distance of 20 cm, this system can provide reconstructed spatial resolution in the range of 6 to 7 mm.

4. SPECT Systems for Small-Animal Imaging

In addition to their widespread clinical role, SPECT systems also are used for applications in biomedical research involving small animals. Typical applications are the

evaluation of new radiopharmaceuticals being developed for diagnostic purposes, or the use of established radiopharmaceuticals to measure functional, physiologic, or metabolic processes in an animal model to monitor or understand the response to a new therapeutic approach.

The challenge is the small size of the organs (a few millimeters in diameter) in commonly used experimental animals such as mice and rats, relative to the spatial resolution typically obtained with SPECT systems. However, two key factors can be exploited to obtain much higher spatial resolution with SPECT in small animals compared with what can be achieved in human imaging. The first is the small volume of tissue to be imaged, which, depending on the detector size used, can permit high magnification of the object onto the detector with pinhole or converging hole collimators (see Fig. 13-7). The second related factor is that the organ of interest can always be positioned within a few millimeters (rather than many centimeters in humans) of the collimator. Because of the strong dependence of spatial resolution on source-to-collimator distance (see Fig. 14-16), much higher resolution can therefore be obtained. Furthermore, if pinhole collimation is used, the sensitivity also increases rapidly as objects of interest are moved close to the pinhole aperture (see Fig. 14-21B and Equation 14-16).

Thus the most common approach to small-animal imaging with SPECT has been to use pinhole collimation, with some magnification. Indeed, standard clinical SPECT systems have been used to great effect in small-animal imaging using a pinhole or multipinhole collimator. However, to achieve optimal resolution and sensitivity performance, in a compact device suited to a laboratory environment, dedicated small-animal SPECT systems have been developed. Although there are many different designs, these systems commonly consist of a series of compact detector heads, with interchangeable pinhole collimators that have apertures ranging from approximately 0.3 mm to 2 mm in diameter, allowing the operator to trade off between improved spatial resolution (smaller pinholes) or improved sensitivity (larger pinholes) (see Equations 14-15 and 14-16). The detector heads often use pixelated NaI(Tl) or CsI(Tl) scintillator arrays (similar to those shown in Fig. 13-15B) or arrays of CZT semiconductor elements to achieve high intrinsic spatial resolution. The simplest systems consist of two opposing detector heads, each with a single pinhole collimator, that rotate around the animal and translate along the animal to produce the angular projections required for reconstruction tomography.

More advanced systems employ collimators with multiple pinholes to improve sensitivity. Such collimators also can increase the FOV along the axial direction without the need to translate the animal. In the most straightforward implementation, the pinholes are arranged with sufficient distance between them such that the image of the animal projected through adjacent pinholes does not overlap on the detector (Fig. 17-4, *left*). In some systems, the projections are allowed to overlap to a certain degree, which enables more pinholes to be used for a given detector area and magnification (Fig. 17-4, *right*). However, this leads to ambiguity in the projection data in the region of overlap, and tomographic reconstruction into SPECT images must use algorithms that properly model this ambiguity.

Some multipinhole animal SPECT systems are completely stationary and designed such that they simultaneously acquire sufficient angular data for tomographic reconstruction with no moving parts. This type of system is particularly suited for rapid dynamic studies, often of interest when evaluating the biodistribution of a novel radiotracer in the first seconds and minutes after injection. These systems consist of an annular collimator sleeve containing many pinholes each projecting a different angular view of the radionuclide distribution onto a detector that sits behind the collimator. Figure 17-5 shows one such collimator and a drawing showing how

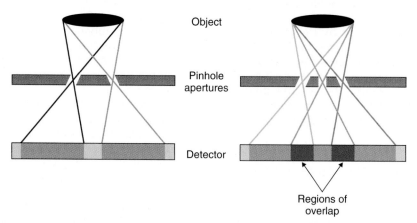

FIGURE 17-4 Illustration of multiple pinhole systems used for small-animal SPECT. *Left,* Two pinholes spaced far enough apart to avoid overlap of projections. *Right,* Increased number of pinholes provides increased sensitivity, but comes at the expense of a partial overlap in the projection data viewed through adjacent pinholes.

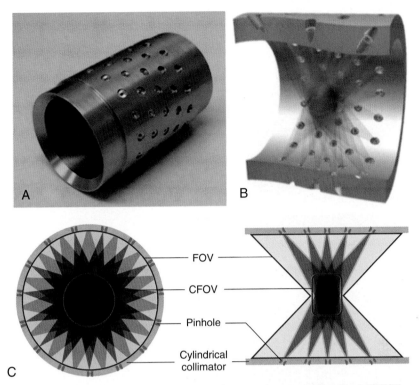

FIGURE 17-5 *A,* Photograph of multiple-pinhole collimator for a stationary small-animal SPECT system. There are 75 pinholes viewing the object from different angles. *B,* Cut-away view of the pinhole collimator showing angulation of pinholes and dense sampling in the central region. *C,* Transverse and axial sections through the center of the collimator showing sensitive region for each pinhole and extent of field of view. (*Courtesy MILabs, Utrecht, The Netherlands*).

the pinholes are angled to cover a large FOV without the need for any movement.

Using these approaches, small-animal SPECT systems routinely reach a resolution in the range of 1 mm, and in some instances are able to produce images with a reconstructed resolution much smaller than 1 mm. Sensitivity can be high as well, because of the large number of pinholes used. Numbers are commonly in the 0.5 to 2 × 10³ cps/MBq (~1000-4000 cpm/μCi) range, but depend strongly on the number and diameter of the pinhole apertures used, and the source-to-pinhole distance. For systems in which the projections overlap, the measured sensitivity must be interpreted with caution, as the information content of an event detected in a region of overlap is not as high as for an event in a system in which the projections do not overlap. Almost all small-animal systems use some form of iterative reconstruction algorithm with detailed modeling of the collimator apertures for reconstructing images at the highest possible resolution. Figures 17-6A and B show photographs of two SPECT systems designed for small-animal imaging

applications. A useful review of small-animal SPECT systems is given in reference 6.

B. PRACTICAL IMPLEMENTATION OF SPECT

Ideally, the signal level for a voxel in a SPECT image is linearly proportional to the amount of activity contained within the volume of tissue in the patient that corresponded to the location of that voxel. This would be useful not only for quantitative applications, such as perfusion studies, but also for visual interpretations of the image. In practice, this ideal result is not achieved because the realities of data acquisition do not match the idealized assumptions made for the development of reconstruction algorithms. As shown in Figure 16-1, it was assumed that the line of response (or a projection element) for a single hole in a parallel-hole collimator is an extended cylinder, but the actual response resembles a diverging cone. It was further assumed that the signal recorded was proportional to the total activity within the line of response, but

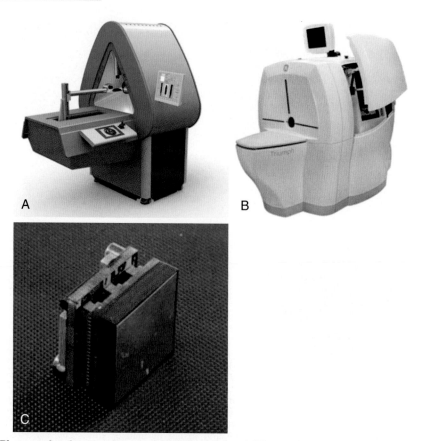

FIGURE 17-6 Photographs of two small-animal SPECT systems: *A,* This stationary system uses the cylindrical multipinhole collimator shown in Figure 17-5, and views are projected onto three large-area Anger cameras with no overlap between projections. With no need to move the detectors, fast dynamic studies can be performed. *B,* This system employs four 12.7 cm × 12.7 cm pixelated cadmium zinc telluride (CZT) detector heads and multipinhole collimators. Some rotation of the detectors is necessary to obtain all projection angles. The use of CZT provides excellent energy resolution. *C,* Photograph of one detector module from the system shown in B. The detector measures 2.54 cm × 2.54 cm and has a 16 × 16 array of CZT elements on a 1.59-mm pitch. The CZT thickness is 5 mm. Twenty-five of these modules are tiled together to form a 12.7 cm × 12.7 cm detector head. (*A, Courtesy MILabs, Utrecht, The Netherlands; B, Copyright Gamma Medica, Northridge, CA and GE Healthcare, Waukesha, WI; used with permission of GE Healthcare; C, Copyright Gamma Medica, Northridge, CA.*)

in fact the signal from activity closest to the detector is more heavily weighted than from deeper-lying activity, because of attenuation by overlying tissues. Finally, it was assumed that activity outside the line of response did not contribute to the signal for the projection element, whereas there may be crosstalk between elements resulting from scattered radiation or septal penetration through the collimator. To further complicate matters, most of the discrepancies vary with the energy of the γ rays involved.

Some of the discrepancies between the idealized assumptions and the actual situation in SPECT are illustrated in Figure 17-7. The discrepancies distort the desired linear relationship between signal level and amount of activity present. They also can lead to artifacts and seriously degraded image quality. To avoid this, one must use somewhat modified approaches to data acquisition or apply postprocessing of the acquired data. This is always the case when backprojection algorithms are used, because they are rigorously grounded in the idealized assumptions noted earlier. Some of the discrepancies can be accounted for with iterative algorithms, such as the maximum likelihood-expectation maximization algorithm (see Chapter 16, Section D.2), which can incorporate these factors in its probability matrix. In this section, we describe some general approaches that are valid and potentially useful for all reconstruction algorithms.

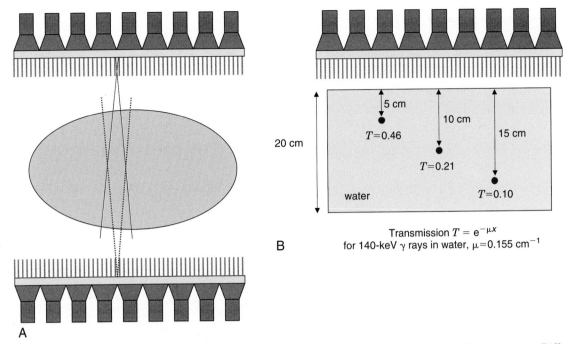

FIGURE 17-7 *A,* Volumes of tissue viewed by a collimator hole at two different angles separated by 180 degrees. Differences in the volumes viewed results in different projections from the two viewing angles. *B,* Attenuation leads to further differences in these two projections, emphasizing activity that is close to the gamma camera compared with activity further away that has to penetrate more tissue to reach the gamma camera. Values are shown for the attenuation of the 140-keV γ rays from ^{99m}Tc in water.

1. Attenuation Effects and Conjugate Counting

The attenuation of γ rays in SPECT imaging depends on the distance the γ rays have to travel through the tissue to reach the detector. Figure 17-7B illustrates the depth-dependent nature of this attenuation for point sources located at different positions within the body. The transmission factor for a source at a certain depth can be calculated using Equation 6-22. For 140-keV γ rays, the linear attenuation coefficient of tissue is 0.155 cm^{-1}; therefore γ rays that are emitted from a depth of 10 cm in the body would only have a probability of 0.21 ($e^{-10 \times 0.155}$) of emerging from the body in their original direction. The attenuation of γ rays is even more severe in parts of the body containing significant amounts of bone, because the linear attenuation coefficient of bone is ~0.25 cm^{-1} at 140 keV.

One approach to reducing both the divergence of the response profile (Fig. 17-7A) and the effects of tissue attenuation (Fig. 17-7B) is conjugate counting. *Conjugate counting* refers to acquiring data (or image profiles) for directly opposing views and then combining these data into a single dataset or line of response. A source that is located relatively close to the detector from one view will be relatively far away in the opposing view. Hence, the response profile will be narrower and attenuation by overlying tissues will be smaller in the first view and larger in the second, with partially offsetting effects.

Conjugate counting in SPECT requires that views be obtained over a full 360-degree range around the object. Data from opposing views then are combined to yield the equivalent of a single 180-degree scan. Conjugate counts (or views) generally are combined in one of two ways. The first is to use the *arithmetic mean.* If I_1 and I_2 are the counts recorded from opposing directions for a particular line of response through the object, the arithmetic mean is given by

$$\bar{I}_A = \frac{(I_1 + I_2)}{2} \qquad (17\text{-}1)$$

An alternative is the *geometric mean,* given by

$$\bar{I}_G = \sqrt{I_1 \times I_2} \qquad (17\text{-}2)$$

Figure 17-8 shows response profiles versus source depth for a ^{99m}Tc line source for a single-view projection and for projections

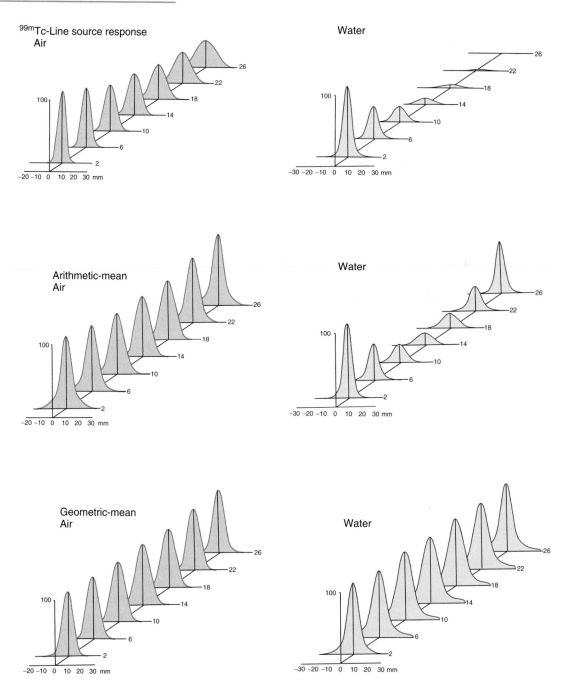

FIGURE 17-8   Line-spread functions versus distance in air (*left*) and in water (*right*) for high-resolution parallel-hole collimator on a gamma camera. The line source measured 2.5 mm in diameter and was mounted inside a tank measuring 410 mm in length, 310 mm in width, and 300 mm in thickness. Measurements were made either with the tank empty (in air) or filled with water. *Top,* single detector only; *middle,* arithmetic mean of opposing detector profiles; *bottom,* geometric mean of opposed detector profiles. (*From Larsson SA: Gamma camera emission tomography: Development and properties of a multisectional emission computed tomography system.* Acta Radiol Suppl *363:1-75, 1980.*)

created from the arithmetic and geometric means of opposing views. The profiles across the top are for a single view with the source in air and for the same view with the source in water. The profile for the source in air illustrates the degradation of spatial resolution with distance from the collimator that is characteristic for a parallel-hole collimator (see also Figs. 14-18 and 14-19). The profile for the source in water shows similar degradation of spatial resolution with increasing distance but, in addition, shows decreasing amplitude of response owing to attenuation of photons by the overlying thickness of water.

The middle row of Figure 17-8 shows response profiles for the arithmetic mean. The profiles for the source in air show significantly improved uniformity of spatial resolution with depth, as compared with the single-view profile directly above it. The profiles for the source in water show similar improvement in uniformity of spatial resolution with depth; however, there still is marked variation in the amplitude of the profile versus distance (and depth in water), indicating that the effects of photon

attenuation are only partially corrected for by the arithmetic mean.

The response profile for the arithmetic mean has its minimum amplitude when the source is near the center of the water phantom. Figure 17-9 shows simulated SPECT images of a water-filled cylinder containing a solution of uniform concentration of activity for different γ-ray energies, using the arithmetic mean of opposing views. Also shown are profiles of the images across the center of the phantom. As one might expect from the profiles illustrated in Figure 17-8, there is a marked decrease in intensity at the center of the

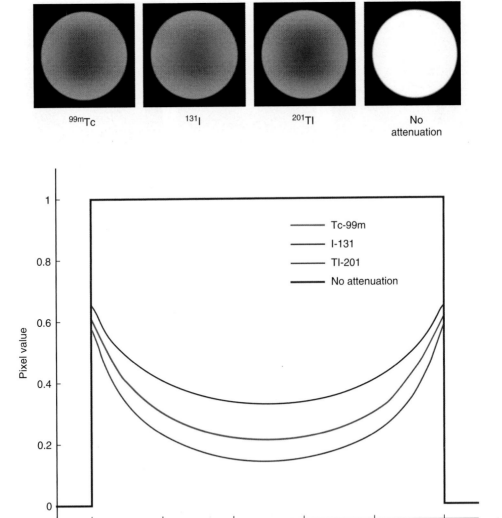

FIGURE 17-9 *Top,* Simulated SPECT images of 20-cm diameter water-filled cylinders containing uniform concentrations of ^{99m}Tc (140 keV), ^{131}I (364 keV), and ^{201}Tl (70 keV). *Bottom,* Arithmetic mean of count profiles through the center of the simulated images. Note the reduction in image intensity at all points in the image caused by attenuation, with the largest reduction occurring at the center of the cylinder. The amount of attenuation is energy dependent, with greatest attenuation occurring at the lower γ-ray energies.

phantom. The strongest effect occurs for the lowest energy γ rays. It is apparent from Figure 17-9 that relatively strong attenuation effects are present with the arithmetic mean for all of the photon energies commonly used in SPECT imaging.

Figure 17-10 shows simulated images and profiles for phantoms of different sizes with $\mu = 0.155$ cm^{-1}, corresponding to the attenuation of 140-keV photons of ^{99m}Tc in water. As the diameter of the phantom increases, the images and count profiles demonstrate progressively greater suppression at the center of the image.

The bottom row of Figure 17-8 shows profiles in air and in water for the geometric mean of opposing views. In this case, both the amplitude and the width of the profile remains nearly constant at all distances and depths in the water phantom.

Tables 17-1 and 17-2 summarize numerical data derived from the profiles in Figure 17-8. From this summary, it can be seen that attenuation has much stronger effects than distance. The combined effects of distance and attenuation result in a 100-fold range in counts recorded with a single detector (see Fig. 17-8, *top right*). With the arithmetic

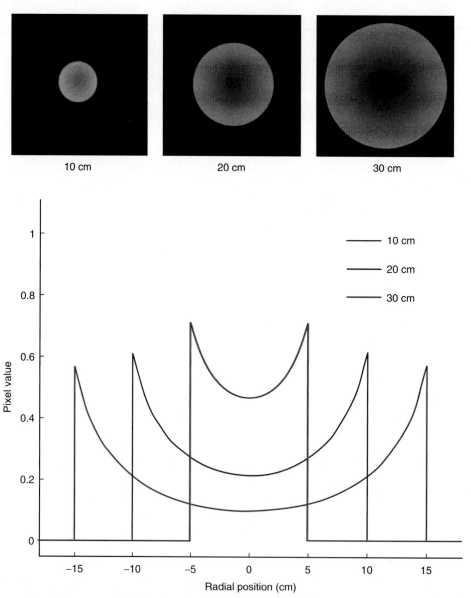

FIGURE 17-10 *Top,* Simulated SPECT images of water-filled cylinders of different diameters containing uniform concentrations of ^{99m}Tc (140 keV). *Bottom,* Count profiles through the centers of the images. The strong dependence of attenuation on cylinder size is evident.

TABLE 17-1
NUMERICAL DATA FOR FIGURE 17-8 *(LEFT COLUMN)* ^{99m}Tc LINE SOURCE IN AIR

	Distance from Collimator (cm)	Relative Maximum Response* in Air (%)	Resolution FWHM (mm)
^{99m}Tc line-source response	2	100	7.4
	6	84	8.6
	10	66	10.0
	14	58	12.5
	18	49	14.7
	22	44	17.2
	26	38	20.0
Arithmetic mean	2-28[†]	100	8.5
	6-24	93	10.6
	10-20	84	12.3
	14-16	84	13.2
	18-12	82	12.7
	22-8	90	11.2
	26-4	99	9.8
Geometric mean	2-28[†]	100	9.4
	6-24	98	10.8
	10-20	94	12.4
	14-16	94	13.2
	18-12	93	12.7
	22-8	97	11.7
	26-4	100	10.4

Adapted from Larsson SA: Gamma camera emission tomography: Development and properties of a multisectional emission computed tomography system. *Acta Radiol Suppl* 363:1-75, 1980.

FWHM, full width at half maximum.

*Value at peak of profile.

[†]For arithmetic and geometric means, paired values refer to the distance between the source and collimator for the two conjugate views.

mean (see Fig. 17-8, *middle*), the range is reduced to a factor of five. With the geometric mean (see Fig. 17-8, *bottom*), the variations with depth are virtually eliminated, although there still is a small reduction in relative counts when the source is at the deepest location in the phantom.

The constancy of amplitude for the geometric mean can be understood from the following analysis. Consider the arrangement of the radioactive source and detectors shown in Figure 17-11. The attenuation of photons directed toward detector 1 is given by

$$I_1 = I_{01}(e^{-\mu a}) \qquad (17\text{-}3)$$

and for those directed toward detector 2 by

$$I_2 = I_{02}(e^{-\mu b}) \qquad (17\text{-}4)$$

where a and b are the source depths. Note that $a + b = D$. Taking the geometric mean of I_1 and I_2, one obtains

$$\sqrt{I_1 \times I_2} = \sqrt{(I_{01} \times I_{02})e^{-\mu \times (a+b)}}$$
$$= \sqrt{(I_{01} \times I_{02})}e^{-\mu D/2} \qquad (17\text{-}5)$$

Thus the geometric mean of counts from opposed detectors depends on total tissue thickness, D, but not on source depths, a and b. This result is exact only for a point or plane source, but corrections can be applied as approximations for simple extended sources (e.g., uniform volume sources).[7] The geometric mean also depends on the unattenuated counts, I_{01}, and I_{02}, which may change with a and b because of distance effects; however, for systems using parallel-hole collimators, such as SPECT gamma camera systems, unattenuated counts do not change with distance (see Chapter 14, Section C.3 and Fig. 14-17). In this case, $I_{01} = I_{02} = I_0$ and Equation 17-5 reduces to

$$\sqrt{I_1 \times I_2} = I_0 e^{-\mu D/2} \qquad (17\text{-}6)$$

TABLE 17-2
NUMERICAL DATA FOR FIGURE 17-8 *(RIGHT COLUMN)* ^{99m}Tc LINE SOURCE IN WATER

	Distance from Collimator (cm)	Relative Maximum Response* in Water (%)	Resolution FWHM (mm)
^{99m}Tc line-source response	2	100	7.5
	6	47	9.4
	10	23	11.2
	14	10	13.9
	18	8.1	16.3
	22	1.9	19.1
	26	0.8	21.6
Arithmetic mean	2-28[†]	100	7.8
	6-24	50	8.6
	10-20	25	11.5
	14-16	17	14.3
	18-12	21	13.0
	22-8	34	10.4
	26-4	71	8.1
Geometric mean	2-28	100	9.9
	6-24	93	11.7
	10-20	95	13.3
	14-16	93	14.3
	18-12	95	14.1
	22-8	93	13.2
	26-4	96	11.9

Adapted from Larsson SA: Gamma camera emission tomography: Development and properties of a multisectional emission computed tomography system. *Acta Radiol Suppl* 363:1-75, 1980.

FWHM, full width at half maximum.

*Value at peak of profile.

[†]For arithmetic and geometric means, paired values refer to the distance between the source and collimator for the two conjugate views.

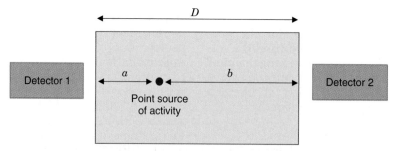

FIGURE 17-11 Point source of activity within an attenuating medium of thickness D. The attenuation can be compensated for by using the geometric or arithmetic mean and a correction for total tissue thickness, D.

These analyses and equations are accurate for a single radioactive source. When multiple sources are present, the situation is more complicated, as shown in Example 17-1.

EXAMPLE 17-1

Derive the equation for the geometric mean of counts from two point sources located along a line between two detectors, and show why it cannot be described only in terms of the unattenuated counts, I_{01}, and I_{02}, and μ and D (distance between the detectors) as can be done for a single point source (Equation 17-5). Assume that a parallel-hole collimator is being used and therefore that the unattenuated counts do not depend on distance.

Answer

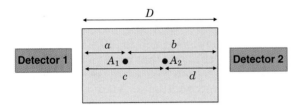

Referring to this figure:

$$a + b = D$$

$$c + d = D$$

We represent the unattenuated counts from source A_1 as I_{01} and from source A_2 as I_{02}. The measured counts at detector 1 will be

$$I_1 = I_{01}e^{-\mu a} + I_{02}e^{-\mu c}$$

The measured counts at detector 2 will be

$$I_2 = I_{01}e^{-\mu b} + I_{02}e^{-\mu d}$$

The geometric mean is

$$
\sqrt{I_1 \times I_2} = \left[I_{01}^2 e^{-\mu(a+b)} + I_{02}^2 e^{-\mu(c+d)} + I_{01}I_{02}e^{-\mu(a+d)} \right.
$$
$$
\left. + I_{01}I_{02}e^{-\mu(c+b)} \right]^{1/2}
$$
$$
= \left[\left(I_{01}^2 + I_{02}^2 \right) e^{-\mu D} + I_{01}I_{02}e^{-\mu(a+d)} \right.
$$
$$
\left. + I_{01}I_{02}e^{-\mu(c+b)} \right]^{1/2}
$$

Only the first term in the last expression depends solely on I_{01}, I_{02}, μ, and D. The other two terms contain exponential terms $e^{-\mu(c+b)}$ and $e^{-\mu(a+d)}$ that depend on the relative locations of the two sources between the detectors. Therefore attenuation effects depend on the source distribution, and the simple correction scheme for point sources and line sources

must be modified for more complicated source distributions.[7]

---

2. Attenuation Correction

Conjugate-counting techniques, especially using the geometric mean, can substantially reduce the variation of width and amplitude of counting rate profiles that are present in single-view profiles. However, even with the geometric mean, there are residual scaling factors caused by attenuation [exp($-\mu D/2$) in Equation 17-6]. Thus, for quantitative accuracy, attenuation corrections are required.

A relatively simple method for attenuation correction is to correct projection profiles generated with the geometric or arithmetic mean before reconstruction using an estimate for tissue thickness, D. The attenuation correction is particularly simple for the geometric mean (Equation 17-6) and is given by multiplying the projection profiles by an attenuation correction factor (ACF) of

$$ACF = \frac{1}{e^{-\mu D/2}} = e^{\mu D/2} \qquad (17\text{-}7)$$

A constant value for μ, the linear attenuation coefficient of tissue, is assumed. An estimate for tissue thickness D can be derived from a preliminary uncorrected image or by assuming a standard body size and shape.

As demonstrated in Example 17-1, simply generating profiles using the geometric mean does not correctly deal with attenuation in the general case in which γ rays are emitted at different locations in the FOV. An alternative approach is to calculate an ACF for each pixel after image reconstruction. In this method, an initial image, $f'(x,y)$, is reconstructed by filtered back-projection without any attenuation correction. The contours of this image are used to obtain an estimate of the attenuation path length through the tissue for all projection views. Once again, it is assumed that the linear attenuation coefficient at a given energy is constant for all body tissues. The ACF for each pixel (x,y) in the reconstructed image then is calculated by

$$ACF(x, y) = \frac{1}{\dfrac{1}{N}\displaystyle\sum_{i=1}^{N} e^{-\mu d_i}} \qquad (17\text{-}8)$$

where d_i is the attenuation path length for the pixel at projection view i and μ is the assumed constant value for the attenuation coefficient. The reconstructed image $f'(x,y)$ is corrected on a pixel-by-pixel basis by multiplying it by the ACF

$$f(x, y) = f'(x, y) \times ACF(x, y) \quad (17\text{-}9)$$

This technique is known as *Chang's multiplicative method.*[8]

There also is a more involved implementation of the Chang method in which the image obtained with the first-order correction described by Equation 17-8 is forward-projected (see Chapter 16, Section D.1), with appropriate attenuation of the image counts corresponding to the path length through the tissue.[9] The forward-projected profiles $p_{fp}(r, \phi)$ are subtracted from the original measured projection profiles $p(r, \phi)$ to form an "error projection" $P_{error}(r, \phi)$

$$P_{error}(r, \phi) = p(r, \phi) - p_{fp}(r, \phi) \quad (17\text{-}10)$$

This error projection is itself reconstructed with filtered backprojection to form an error image, $f_{error}(x, y)$. The final attenuation corrected image is given by

$$f(x, y) = f'(x, y) \times ACF(x, y)$$
$$+ f_{error}(x, y) \times ACF(x, y)$$

$$(17\text{-}11)$$

Adding the two images together is made possible by the fact that filtered backprojection is a linear algorithm. Figure 17-12 shows

No attenuation correction

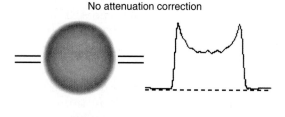

Chang correction ($\mu = 0.15$ cm^{-1})

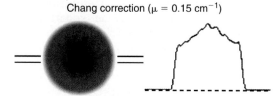

FIGURE 17-12 SPECT images of a 20-cm diameter cylinder containing a uniform concentration of ^{99m}Tc with and without attenuation correction (Chang method with narrow-beam attenuation coefficient of $\mu = 0.15$ cm^{-1}). Profiles are through the center of the images. The apparent overcorrection of attenuation is due to scattered events in the dataset. (*Courtesy Dr. Freek Beekman, Delft University of Technology, The Netherlands.*)

SPECT images of a cylinder that contains a uniform concentration of radionuclide before and after attenuation correction with the Chang method.

Methods based on the Chang approach are used in most commercial SPECT systems and yield reasonable results in the brain and abdomen, where the assumption of a uniform attenuation coefficient is not unreasonable (the amount of bone and any air spaces are small). These methods do not work well in the thorax or in the pelvic region, where the presence of the lungs and significant amounts of bone, respectively, can lead to significant errors. SPECT projection profiles must be acquired over a full 360 degrees to use these methods.

3. Transmission Scans and Attenuation Maps

An alternative approach for SPECT imaging in regions of the body that have variable attenuation is to actually measure tissue attenuation using an additional scan known as a *transmission scan.*[10,11] This scan can be performed using the same detector system as is used for acquiring the emission data. An external source of radiation is used to acquire transmission profiles that can be used to reconstruct cross-sectional images reflecting the linear attenuation coefficient of the tissue, often referred to as an *attenuation map*. This is equivalent to an x-ray CT scan, although the quality of the image is poorer, because of the limited resolution of the gamma camera and the low-photon flux used to obtain the transmission image. As well, the higher energy of the γ rays, as compared with most diagnostic x-ray beams, leads to lower contrast. Approaches that use x-ray CT scans for attenuation correction are discussed in Chapter 19, Section E.

Transmission data can be acquired using a collimated flood source, line source, multiple line sources, or a moving line source. Several possible acquisition geometries for transmission scans are shown in Figure 17-13. The data usually are acquired with a parallel-hole collimator on the detector, although on triple-headed cameras, a single line source sometimes is used in conjunction with a fan-beam collimator. The radionuclide chosen for the transmission sources usually has an emission energy that is different from ^{99m}Tc, to allow for simultaneous emission imaging of ^{99m}Tc, as described later. A radionuclide with a long half-life is convenient so that the source does

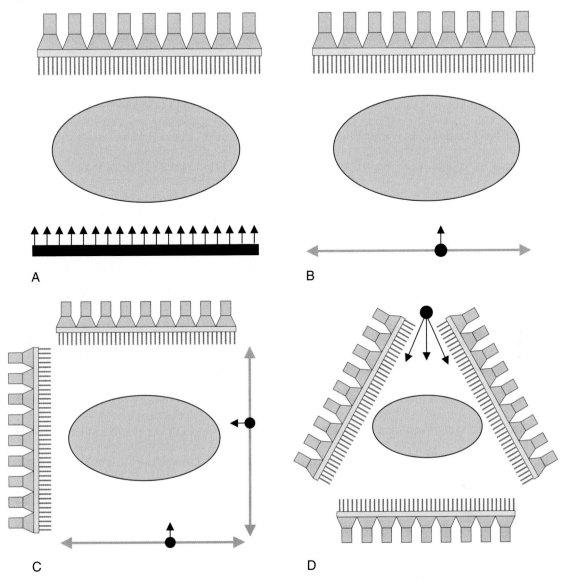

FIGURE 17-13 Examples of transmission source geometries that are being employed for attenuation correction in SPECT. The *black arrows* show the direction of γ rays emitted from collimated transmission source; the *gray arrows* show the direction of motion of moving line sources. *A,* Flood source. *B,* Collimated moving line source. *C,* Two orthogonal moving collimated line sources on dual-headed gamma camera. *D,* Stationary line source (collimated in axial direction irradiating opposite detector head (fan-beam collimator) in a triple-headed gamma camera.

not require frequent replacement. ^{153}Gd ($T_{1/2}$ = 242 days, E_γ = 97 and 103 keV) and ^{123m}Te ($T_{1/2}$ = 120 days, E_γ = 159 keV) are among the radionuclides suitable for this purpose.

To obtain an attenuation map, two separate scans are acquired with the transmission source. Typically one of the geometries shown in Figure 17-13 is used and the gamma camera system is rotated through 360 degrees to acquire a full set of projection views. The first scan is acquired with no object in the FOV of the SPECT camera. This is referred to as the *blank* or *reference scan*. The second

scan is acquired with the object of interest in the FOV. This is the transmission scan. The relationship between the reference (I_{ref}) and transmission (I_{trans}) counts in any particular projection element is given by the usual exponential relationship for γ-ray attenuation

$$I_{trans} = I_{ref} e^{-\mu x} \qquad (17\text{-}12)$$

Taking the natural logarithm of the ratio of the two scans results in

$$\ln(I_{ref}/I_{trans}) = \mu x \qquad (17\text{-}13)$$

Projection profiles of μx represent the sum of the attenuation coefficients along each line of response

$$\mu x = \sum_i \mu_i \Delta x_i \qquad (17\text{-}14)$$

where μ_i is the linear attenuation coefficient for the ith pixel and Δx_i is the pathlength of the line of response through the ith pixel. This is analogous to the standard emission projection profiles that represent the sum of the radioactivity along each line of response. Using the methods described in Chapter 16, Section B, the projection profiles of μx (calculated from the transmission scan profiles using Equation 17-13) are reconstructed, resulting in images of μ_i. Figure 17-14 shows a SPECT attenuation map reconstructed from transmission and reference scans.

The attenuation map can be used to more accurately compute the ACFs in the Chang algorithm (Section B.2) by taking into account the nonuniform attenuation at each source location in Equation 17-8. It can also be incorporated in the forward-projection step of the modified Chang algorithm to more accurately compute the error term. More commonly, tissue attenuation information is directly incorporated into iterative reconstruction algorithms (Chapter 16, Section D) in which it becomes another factor in the calculation of the probability matrix M in Equation 16-19. The probability for γ rays emitted from a given pixel i reaching projection element j is reduced by the probability of attenuation in the tissue lying between the point of emission and the detection point in the gamma camera. For both filtered backprojection and iterative algorithms, the difference in energy (and

therefore attenuation) between the transmission and emission photons must be taken into account in applying the information from the transmission map.

Emission and transmission scans can be acquired simultaneously if the separation between the photon energies of the transmission source and the emission radionuclide is sufficient to allow them to be acquired in two separate energy windows. This is shown schematically in Figure 17-15. Even if two different windows are used, however, some events from the higher-energy radionuclide will be recorded in the lower-energy window. This effect, known as *downscatter*, arises from two causes. The first is spillover from higher-energy events into lower-energy regions of the spectrum (e.g., see Fig. 10-3). These events may arise from partial absorption of higher-energy photons in the detector or from natural broadening of the photopeak.

A second cause is γ rays that have experienced a partial loss of energy in Compton-scattering interactions in the body. For emission scans using ^{99m}Tc (140 keV), downscatter from a ^{123m}Te transmission source (159 keV) can appear in the emission source window. Conversely, when ^{153}Gd is used as the transmission source, downscatter from ^{99m}Tc can appear in the transmission source window. Even if the emission and transmission scans are acquired sequentially, rather than simultaneously, the patient normally is injected with a ^{99m}Tc-labeled radiopharmaceutical prior to transmission imaging, and hence downscatter is still an issue with ^{153}Gd transmission sources.

Downscatter can be estimated and corrections can be applied for it in ways that are similar to the methods used for correcting for scattered radiation in the emission scan. For example, one can use an energy window between the transmission and emission windows to estimate the level of downscatter. These methods are similar to those employed for scatter corrections, which are discussed in the following section.

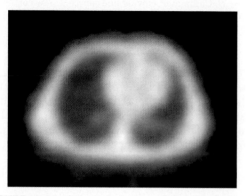

FIGURE 17-14 Attenuation map of the thorax reconstructed from the reference and transmission scans obtained with a moving line transmission source. (*Data courtesy Dr. Freek Beekman, Delft University of Technology, The Netherlands.*)

4. Scatter Correction

The idealized model used for developing the filtered backprojection reconstruction algorithms described in Chapter 16 assumed that only radioactivity within the line of response for a projection element contributed to the signal for that element. In practice, the signal can include events that have been scattered into the line of response from radioactivity elsewhere in the body. With the typical 20%

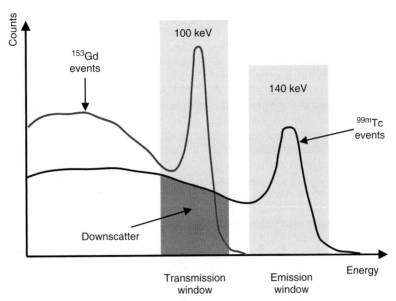

FIGURE 17-15 Dual-energy windows used to simultaneously acquire SPECT emission (⁹⁹ᵐTc) and transmission (¹⁵³Gd) data. Note the presence of downscatter from the ⁹⁹ᵐTc activity in the ¹⁵³Gd window. The magnitude of the downscatter contamination depends on the relative amounts of ⁹⁹ᵐTc activity in the body and ¹⁵³Gd activity in the transmission source, the amount of scattering material in the field of view, the details of how the transmission source is collimated, and the precise energy windows that are used.

pulse-height analyzer (PHA) window that is used for ⁹⁹ᵐTc, events that have scattered through angles as large as 50 degrees still have a 50% probability of being accepted.

Compared with the effects of attenuation, the effects of scattering are of lesser magnitude. Nevertheless, Compton scattering, and at low γ-ray energies ($\lesssim$100 keV) coherent scattering (see Chapter 6, Section C.5), still can have a significant effect on image quality and on the quantitative relationship between the reconstructed image intensity and source activity. In a typical patient study with a ⁹⁹ᵐTc-labeled radiopharmaceutical, even using a narrow 15% PHA window, the ratio of the number of detected scattered photons to the number of nonscattered photons may be as large as 40%. The presence of scattered events results in reduced image contrast (the tails of the point-spread function [PSF] are elevated with respect to the peak) and leads to an overestimation of the concentration of radioactivity in the pixel (see Fig. 17-12, *bottom*). The loss of image contrast may obscure clinically important details, particularly "cold" areas in the images, for example, areas of low radiopharmaceutical accumulation in the heart caused by coronary artery disease or infarction.

A first-order correction for scatter can be made by recognizing that scatter and attenuation are part of the same phenomenon. Attenuation is caused by the scattering (and

only rarely for energies $\gtrsim$100 keV, the total absorption) of γ rays. Because of the broad-beam geometry (Chapter 6, Sections D.2 and D.3) of SPECT imaging systems, some of the scattered γ rays are detected, leading, on average, to a reduction in the "apparent" attenuation coefficient that is measured relative to narrow-beam attenuation coefficients. One can provide an averaged correction for scatter by using the apparent or broad-beam value for μ in Equations 17-7 and 17-8. For example, for the 140-keV γ rays from ⁹⁹ᵐTc in a typical patient, the broad-beam attenuation coefficient is ~0.12 cm⁻¹ as compared with a narrow-beam value of 0.155 cm⁻¹. Although this works well in objects with uniform radioactivity distributions in a uniform attenuation medium (Fig. 17-16), it does not properly

Chang correction (μ = 0.12 cm⁻¹)

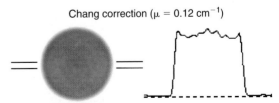

FIGURE 17-16 Effect of Chang attenuation correction on 20-cm uniform cylinder data in Figure 17-12 using a broad-beam attenuation coefficient of μ = 0.12 cm⁻¹. Note the improvement in the uniformity of the profile, which is due to compensation for scattered events. (*Data courtesy Dr. Freek Beekman, Delft University of Technology, The Netherlands.*)

take into account the spatial distribution of the scattered events and is therefore of limited accuracy in more realistic imaging situations.

A second simple method that has been used to correct for scattered events involves measuring the scatter component in the projection profiles using a line source (or point source) immersed in a scattering medium that is representative of the dimensions of the body. By measuring projection profiles of the source with and without the scattering medium, the distribution of scattered events in the projection profile can be determined. This can be considered to be the line-spread function (LSF) of the scattered events and can be deconvolved (see Appendix G) from the measured projection profiles measured in patient studies to correct for scatter. The accuracy of this correction is limited by differences in radioactivity distribution and attenuation distribution between the phantom in which the scatter response is measured, and the patient, and by the spatially invariant nature of the correction.

One of the most commonly used methods to correct for scattered γ rays is to simultaneously acquire counts with a photopeak window and a lower-energy *scatter window*. For example, the photopeak window for ^{99m}Tc might be set to 127-153 keV and the scatter window to 92-125 keV (Fig. 17-17). The resulting scatter projection profiles then are multiplied by a weighting factor and subtracted from the photopeak profiles to obtain scatter-corrected projection data. The weighting factor applied to the counts in the scatter window for the subtraction process must be determined experimentally and depends in general on the size of the source, the exact settings of the energy windows, and the energy resolution of the gamma camera detector. The accuracy of this method is limited by the fact that γ rays in the scatter window are more likely to have undergone multiple Compton interactions than scattered events in the photopeak window; therefore the spatial distributions of the scatter recorded in the two energy windows may differ.

Many variants on the use of multiple-energy windows for scatter correction have been developed. Some SPECT systems use as many as 32 separate energy windows to more accurately model the scatter distribution. Accurate scatter corrections require very good spatial linearity and uniformity of the gamma camera detector (Chapter 14, Section B) to avoid creating artifacts in the scatter correction process. Scatter corrections also increase the statistical noise in the reconstructed image because of the inevitable propagation of noise in the subtraction process (Chapter 9, Section C.1).

Note that for all the corrections described here, if scattered events are not "removed" prior to applying attenuation corrections, the scattered events also are amplified during the attenuation correction procedure. Therefore it is important that scatter corrections precede attenuation corrections.

Attenuation maps, described in the preceding section, also can be used in conjunction with iterative reconstruction algorithms to correct for scattered radiation. In essence, the matrix M in Equation 16-19 can be modified to account for the probability of scatter from a source at a location (x,y) into a specified detector element. This probability can be calculated by combining knowledge of the

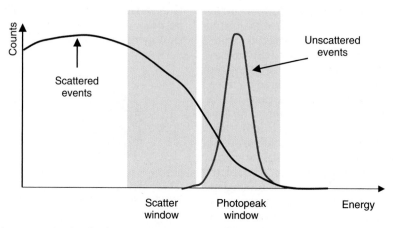

FIGURE 17-17 Diagrammatic sketch showing dual-energy windows superimposed on the spectral distribution of unscattered and scattered events for a patient-sized phantom filled with ^{99m}Tc.

distribution of the attenuation coefficients from the attenuation map, along with the probability of Compton scattering at different angles (see Fig. 6-13), and a model of the camera and collimator geometry, as well as the energy resolution of the camera.

Because computing the scatter probability from every source location in the subject, to every possible detector location, is computationally expensive, simplifying approximations often are made. Nonetheless, iterative algorithms in conjunction with patient-specific attenuation maps offer a powerful approach to quantitative SPECT imaging in the presence of nonuniform attenuating and scattering media. This is of particular value in cardiac imaging, in which the mixture of soft tissue and lungs in a cross-section of the thorax can cause major artifacts and quantitative errors if accurate corrections for attenuation and scatter are not employed. Scatter correction methods are discussed in more detail in reference 11.

5. Partial-Volume Effects

Ideally, the intensity of each pixel in a SPECT image would be proportional to the amount of radioactivity within the corresponding volume of tissue in the patient. The methods described in the preceding sections described how this can be facilitated by corrections for attenuation and scatter. Even with these corrections, however, there still may be errors in assigning activity and concentration values to small sources and small tissue volumes.

As described in Section C, a SPECT system has a characteristic "resolution volume" that is determined by the combination of its in-plane (x-y) and axial (z) resolutions. For systems that produce a stack of contiguous 2-D images, this volume has an approximately cylindrical shape of height = $2 \times$ FWHM axial resolution and diameter = $2 \times$ FWHM in-plane resolution. For sources or measurement volumes of the size of the resolution volume or larger, the intensity of images produced by the SPECT system reflect both the amount and concentration of activity within that volume. For smaller objects that only partially fill a resolution-volume element, the sum of the intensities of all the pixels that are attributable to that object still reflects the total amount of activity within it. However, the intensities of the individual pixels no longer accurately reflect the concentration of activity contained within them, because the signal is distributed over a volume that is larger than the actual size of the source. This

effect is illustrated in Figure 17-18, in which objects of identical concentration are seen to decrease in intensity, and thus in apparent concentration, with decreasing size.

This *partial-volume effect* is important for both qualitative and quantitative interpretation of SPECT images. Although they may be visible in the image, small objects near the resolution limits of the device appear to contain smaller concentrations of radioactivity than they actually do. The ratio of apparent concentration to true concentration is called the *recovery coefficient* (RC). Figure 17-19 illustrates RC versus object size for the cylinders in Figure 17-18. In principle, if a SPECT system has a known and uniform spatial resolution and if the size of the object is known, an RC correction factor can be applied to correct for the partial-volume underestimation of concentrations for small objects. Although this approach works well in phantom studies in which object sizes are well characterized, the sizes of in vivo objects usually are too poorly defined for this method to be useful, unless high-resolution anatomic information is available from another modality such as CT or magnetic resonance imaging.

In some situations, RC can be greater than one. This occurs when the object of interest has low radiotracer accumulation relative to surrounding structures (e.g., an area of reduced blood flow in the heart) and activity from these surrounding areas "spills over" into the structure of interest as a result of the same resolution effects described previously. This is commonly known as *spillover*. The net effect of partial-volume effects in all cases is to reduce the contrast between areas of high radiotracer uptake and those of low uptake and to lead to underestimation or overestimation of radiotracer concentrations. This can be the dominant source of error in quantitative SPECT studies of small structures and must be carefully considered when comparing images of different-sized objects, or sequential images of objects that are changing size (e.g., tumor shrinkage caused by therapy).

C. PERFORMANCE CHARACTERISTICS OF SPECT SYSTEMS

The following sections provide representative methods for characterizing the performance and for quality assurance of SPECT systems. A complete description of these measurements can be found in reference 11.

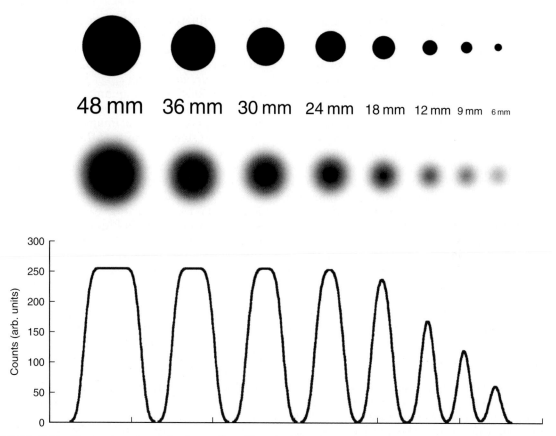

FIGURE 17-18   Illustration of partial-volume effect. The cylinders shown in the *top row* have diameters ranging from 48 mm down to 6 mm, and each contains the same concentration of radionuclide. The *middle row* shows a simulation of the images that would result from scanning these cylinders on a SPECT system with an in-plane spatial resolution of 12-mm full width at half maximum. The cylinders are assumed to have a height much greater than the axial resolution. The *bottom row* shows count profiles through the center of the images. Although each cylinder contains the same concentration of radionuclide, the intensity, and therefore the apparent concentration, appears to decrease when the cylinder size approaches and then becomes smaller than the resolution of the SPECT system. The integrated area under the count profiles does, however, accurately reflect the total amount of activity in the cylinders.

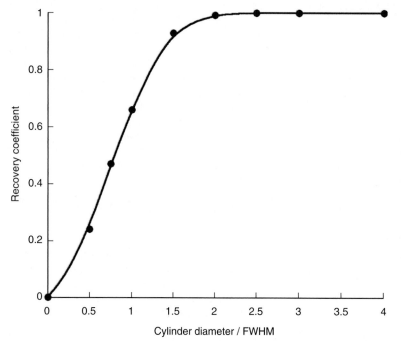

FIGURE 17-19   Recovery coefficient versus object size [in units of size/full width at half maximum (FWHM)] using the data from Figure 17-18. A recovery coefficient of 1 is obtained when the size of the object is $\gtrsim 2 \times$ FWHM. Note that the recovery coefficient depends on the radionuclide concentration in the objects of interest and in the surrounding tissue. For example, if the cylinders in Figure 17-18 contained no radionuclide and were placed in a background containing a uniform level of radionuclide, the recovery coefficients would be the reciprocal of those shown here.

1. Spatial Resolution

Spatial resolution in SPECT is characterized using profiles through the reconstructed image of a line or point source. Because the reconstructed images usually are stacked into a 3-D volume, two components of spatial resolution should be reported. *In-plane* or *transaxial resolution* refers to the component within the plane of a reconstructed slice (x- and y-direction). *Axial resolution* refers to the component that is perpendicular to the slice, along the axis of rotation of the SPECT scanner (z-direction). The latter sometimes is referred to as *slice thickness*.

In-plane resolution is determined by the intrinsic resolution of the gamma camera, the resolution of the collimator, the angular and linear sampling intervals used for acquiring the projection data, and the shape and cut-off frequency of the reconstruction filter. Of these, the dominant factor almost always is the collimator resolution (Chapter 14, Section C). Axial resolution depends on the intrinsic resolution of the gamma camera, the resolution of the collimator, and the linear-sampling interval along the axis of rotation of the gamma camera. Reconstructed spatial resolution is typically measured as follows:

A small ^{57}Co or ^{99m}Tc point source is placed in the FOV of the SPECT camera at the location of interest. To measure spatial resolution accurately, the diameter of the test source should be much smaller than the resolution capability of the imaging instrument (e.g., source diameter <1/4 of the expected FWHM resolution). Generally, measurements are made with the source at the center of the FOV and at several peripheral locations in both the axial and transaxial directions. SPECT images of the point source then are acquired using linear and angular sampling intervals normally employed in clinical practice. The image is reconstructed using a ramp filter and a pixel size that is less than 20% of the anticipated spatial resolution (FWHM). Thus, for a SPECT system with 12-mm FWHM resolution, the point-source size should be smaller than ~3 mm and the pixel size should be smaller than ~2.4 mm.

The reconstructed image of the point source is a nearly spherical dot within the 3-D image volume. A profile drawn directly through the center of the image of the point source yields the PSF. Usually, spatial resolution is characterized by the FWHM of the PSF. The FWHM is not a complete specification because PSFs of different shapes can have the same FWHM;

however, it is useful for general comparisons of imaging systems. A more complete specification is provided by the modulation transfer function, which can be obtained by taking the Fourier transform of the PSF (Chapter 15, Section B.2). PSFs and FWHMs usually are measured along the three orthogonal directions x, y, and z through the image of the point source. The average FWHM for the x- and y-axis profiles gives the transaxial resolution, whereas the FWHM of the z-axis profile gives the axial resolution.

The ramp filter provides the best achievable resolution for a specific collimator. However, as described in Chapter 16, Section B.3, it also enhances noise and often is not used clinically for this reason. If so, the measurement should be repeated for the reconstruction filters that are used in clinical practice. The measurement is repeated for different locations in the camera FOV to provide information about the uniformity of the reconstructed image resolution.

The PSF provides resolution values along three orthogonal directions. However, point sources can be difficult to manufacture, and PSFs provide values only for one image plane. Line sources and LSFs are an alternative solution for both of these limitations. If the line source is aligned parallel to the axis of the SPECT scanner, the in-plane resolution components can be measured simultaneously for all image planes. However, a second measurement with the line source rotated so that it lies perpendicular to the scanner axis then is necessary to obtain the axial resolution.

2. Volume Sensitivity

The *volume sensitivity* measurement gives the number of events detected by the SPECT system per second per unit of concentration of radionuclide uniformly distributed in a 20-cm-diameter cylinder. The result depends, among other things, on the efficiency of collimator, the energy window setting, the energy resolution of the detector, the thickness of the NaI(Tl) crystal, the radionuclide, and the number of gamma camera heads in the SPECT system.

The measurement uses a 20-cm-diameter × 20-cm-tall (outer dimensions) acrylic fillable phantom with a wall thickness between 8 and 12 mm. The phantom is filled with a known and uniform concentration A (becquerels per milliliter) of the radionuclide of interest (determined by using a dose calibrator or other calibrated radiation detector). The activity used in the cylinder is chosen such that it leads to a gamma camera counting rate of ~10,000 cps. At this counting rate, dead time and pile-up effects are negligible.

The well-mixed phantom is placed at the center of the axis of rotation of the SPECT system. A 360-degree circular orbit SPECT scan, with a radius of rotation of 150 mm, is performed. Typically 128 projection views are acquired, with each projection view containing on the order of 100,000 counts. Uniformity corrections, or any other corrections that may alter the number of counts in the projection images, are turned off.

The total time T required to complete the acquisition and the total number of counts N recorded across all the projection views are determined. The system volume sensitivity (SVS) is then given by

$$SVS(\text{cps/Bq/mL}) = \frac{N(\text{counts})/T(\text{sec})}{A(\text{Bq/mL})}$$

(17-15)

Sometimes, the measurement is reported as the volume sensitivity per axial centimeter, in which case Equation 17-15 is divided by the axial length of the phantom, in this case 20 cm.

3. Other Measurements of Performance

Other important performance parameters of a SPECT system are energy resolution, count-rate performance, and dead time. These parameters are the same, irrespective of whether a gamma camera is used for planar imaging or for SPECT; therefore the measurements described in Chapter 14, Sections E.5 and E.6, are equally applicable to SPECT.

4. Quality Assurance in SPECT

As described in Chapter 14, Section E, quality assurance programs are important to ensure that an imaging system is functioning correctly. Many of the quality assurance procedures used for gamma cameras also serve to ensure high-quality images when the gamma camera is used for SPECT. However, there are certain differences in the requirements and additional measurements that should be made when a gamma camera is used for SPECT imaging.

One difference relates to the specifications of flood-field uniformity discussed in Chapter 14, Section E.4. In planar imaging, nonuniformities of a few percent are acceptable for producing images of diagnostic quality. In SPECT,

even smaller nonuniformities can lead to major artifacts in reconstructed images. The artifacts often appear as rings in images acquired with single-headed SPECT systems or as arcs in images acquired with multi-headed systems. The intensity of the artifact is inversely proportional to the distance of the nonuniformity from the axis of rotation. This is because nonuniformities in peripheral areas are spread across a larger area of the image during reconstruction, whereas a nonuniformity near the axis of rotation affects a concentrated area near the center of the image, which can lead to strong artifacts.

Flood-field uniformities of 1% or better are desirable for gamma camera detectors used for SPECT. To detect nonuniformities at this level, it is necessary that sufficient counts be acquired to ensure that the measurement is not limited by counting statistics. Poisson counting statistics dictate that 10,000 counts are required per image element to reach an uncertainty of 1% in a planar image (Equation 9-6). If an image matrix of 64×64 is used, a total of ~41 million counts will be required in the uniformity image.

In many cases, it also is instructive to measure the reconstructed image uniformity of a SPECT system. At the present time, there is no specific National Electrical Manufacturers Association procedure for this measurement, but a reasonable approach would be to use the data acquired using the methods described in Section C.2 for measuring volume sensitivity. An image of the uniform cylinder would be reconstructed with appropriate attenuation and scatter corrections and with the use of clinically relevant reconstruction algorithms and filters. The reconstructed images are visually inspected for any noticeable artifacts or structured noise. Care must be taken to ensure that the scatter and attenuation corrections themselves are not the cause of any artifacts.

A test that is specific to SPECT systems is the measurement of system alignment. It is critical that the mechanical center of rotation (COR) coincide with the COR defined for the projection data used for reconstruction. If the camera detector sags or wobbles as it rotates around the patient, additional blurring or ring artifacts may be introduced into the image. Because it is extremely difficult to make a mechanically or rotationally perfect gantry, most manufacturers measure the alignment of their systems prior to shipment and incorporate software that corrects on a projection-by-projection basis for any small deviations from the COR. An additional requirement for multiheaded systems is that all the heads be accurately aligned in the axial direction. Otherwise, each head records data from a different slice, which leads to additional blurring or artifacts in the axial direction.

System alignment errors can be measured by recording profiles from different projection angles for a point source placed off-center in the FOV of the SPECT system. A typical protocol involves recording an even number of projection profiles N at equal angular intervals over 360 degrees. For example, projections could be acquired at 0, 90, 180, and 270 degrees. For each projection profile, the centroid (r_{cen}, z_{cen}) of the image of the point source on the gamma camera face is determined, in which r is the radial coordinate (as defined in Fig. 16-3) and z is the axial coordinate. The average COR error (Err_{COR}) is then given by

$$Err_{COR} = \frac{1}{N}\sum_{n=1}^{N} r_{cen} \qquad (17\text{-}16)$$

The individual x and y components of the COR error are given by averaging r_{cen} for the projection data at $\phi = 0$ degrees and 180 degrees for x and $\phi = 90$ degrees and 270 degrees for y, in which the projection angle is defined as shown in Figure 16-3. The average axial deviation Err_{AX} of the detector heads (caused by detector tilt) can be calculated from

$$Err_{AX} = \frac{1}{N}\sum_{n=1}^{N} |\bar{z} - z_{cen}| \qquad (17\text{-}17)$$

in which z_{cen} is the PSF centroid in the z direction and $\bar{z}$ is the mean value of z_{cen}. In multiheaded SPECT systems, differences in Err_{COR} and Err_{AX} between detector heads must also be assessed to determine any relative misalignment of the heads. The recommended methods for doing this can be found in reference 12.

D. APPLICATIONS OF SPECT

It is estimated that there are more than 5,000 SPECT-capable systems worldwide. Not surprisingly, the major uses of SPECT as opposed to planar imaging are for situations involving organs with complex geometry or structure, for which accurate 3-D localization of signals is critical for patient diagnosis and management.

The most frequent use of SPECT is for studies of myocardial perfusion for assessing coronary artery disease and heart muscle damage following infarction. Figure 17-20 shows a series of SPECT images that reflect perfusion of the heart. It is common to perform cardiac perfusion studies both under resting conditions and also following a "stress" to the heart created by exercise or by the injection of a drug that causes vasodilation (e.g., adenosine). These are called *rest/stress studies*.

In some cases, SPECT studies are gated to the electrocardiogram signal from the heart, and data from specific portions of the cardiac cycle (e.g., end-systolic or end-diastolic) can be isolated. This reduces blurring caused by motion of the heart and leads to improved image sharpness and contrast. However, the number of events contributing to the image are reduced (data are acquired only for a fraction of the cardiac cycle), leading to poorer signal-to-noise ratio for a fixed data

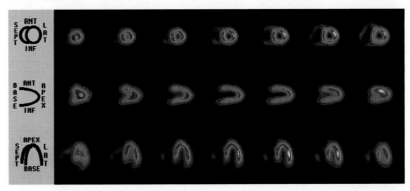

FIGURE 17-20 SPECT images showing perfusion in the heart muscle of a normal adult using ^{99m}Tc-sestamibi as the radiopharmaceutical. The image volume has been resliced into three different orientations as indicated by the schematics on the left of each image row. SPECT data were acquired over 64 views with a data acquisition time of 20 sec/view. Images were reconstructed with filtered backprojection onto a 128 × 128 image array. (*Images courtesy Siemens Medical Systems USA, Inc., Hoffman Estates, IL.*)

acquisition time. Figure 17-21 shows a comparison between gated and ungated SPECT images of the heart. Myocardial perfusion studies usually are carried out using ^{201}Tl, ^{99m}Tc-sestamibi, or ^{99m}Tc-tetrofosmin.

Cerebral perfusion studies with SPECT also are widespread, with applications including cerebrovascular disease, dementia, seizure disorders, brain tumors, and psychiatric disease (Fig. 17-22). Commonly used radiotracers are ^{99m}Tc-hexamethylpro pyleneamine oxime (^{99m}Tc-HMPAO) and ^{99m}Tc-ethyl cysteinate dimer (^{99m}Tc-ECD). A number of research groups also are developing novel SPECT radiotracers that bind to specific receptor populations in the brain, enabling the imaging of

Gated

Ungated

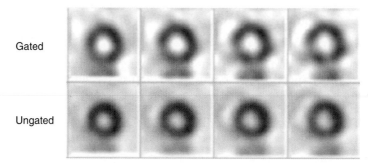

FIGURE 17-21 Comparison of gated (*top row*) and ungated (*bottom row*) SPECT images showing perfusion in the heart muscle of a normal adult following the injection of 1100 MBq of ^{99m}Tc-sestamibi. Images were acquired on a triple-headed gamma camera with low-energy, high-resolution parallel-hole collimators. Ninety projection views were acquired in 4-degree steps (30 projections per head), and the total imaging time was approximately 20 minutes. Data acquisition started 30 minutes after radiotracer injection. Short-axis views of the heart are shown. In the gated study, the cardiac cycle is divided into eight equal time intervals. Based on the electrocardiographic trigger, events are histogrammed into the appropriate interval. The gated images in this figure correspond to the time interval that represents end-diastole. Cardiac gating leads to a small improvement in apparent image contrast by reducing blurring caused by cardiac motion. (*Images courtesy Dr. Steve Meikle, University of Sydney, Australia.*)

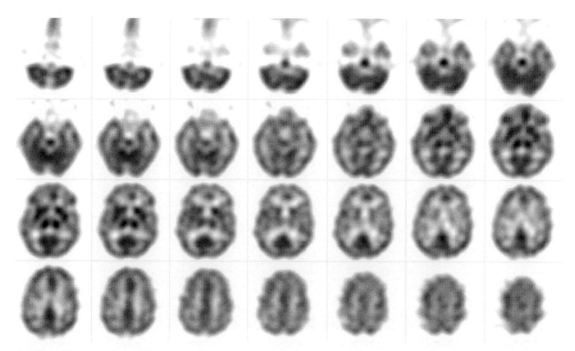

FIGURE 17-22 Transaxial SPECT images showing perfusion in the brain of a normal adult following injection of 890 MBq of ^{99m}Tc-HMPAO. Data were acquired on a triple-headed gamma camera with low-energy, high-resolution fan-beam collimators. One hundred twenty projection views were collected in 3-degree increments (40 views per camera head) with an imaging time of 40 sec/view. Total imaging time was approximately 30 minutes, with acquisition commencing 50 minutes after radiotracer injection. (*Courtesy Dr. Steve Meikle, University of Sydney, Australia.*)

important receptors such as dopamine receptors and serotonin receptors.

A third important application of SPECT is in oncology. Radiotracers such as ^{67}Ga, ^{201}Tl, and ^{99m}Tc-sestamibi often show accumulation in cancerous cells, and both primary and metastatic lesions can be visualized in SPECT images. SPECT (in comparison with planar imaging) is particularly useful for tumor imaging in the thorax, abdomen, or brain, where the tomographic information aids in tumor detection and localization against a complex background of heterogeneous tissues. New radiopharmaceuticals such as ^{99m}Tc-depreotide and ^{111}In-pentetreotide, which are specifically targeted to somatostatin receptors that are overexpressed on many malignant cells (as well as in areas of inflammation), have become available and appear promising. There also is interest in using radiolabeled

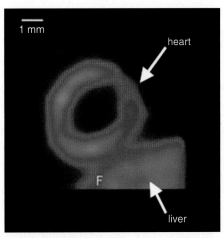

FIGURE 17-23 SPECT image showing myocardial perfusion in a mouse obtained with a small-animal system. The injected radiotracer is ^{99m}Tc-sestamibi and the study was electrocardiogram-gated. (*Courtesy Bioscan Inc., Washington D.C.*).

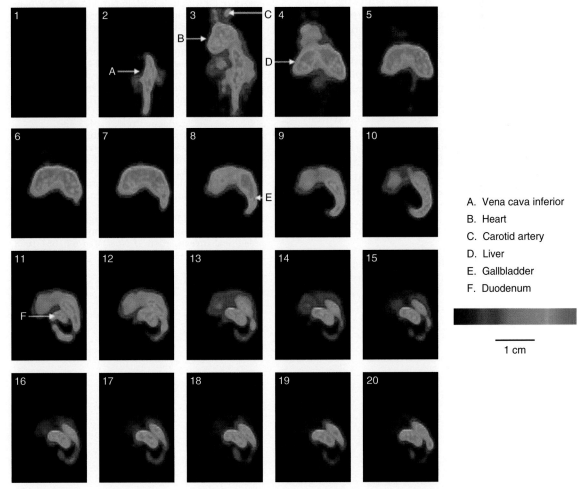

A. Vena cava inferior
B. Heart
C. Carotid artery
D. Liver
E. Gallbladder
F. Duodenum

1 cm

FIGURE 17-24 Dynamic SPECT images showing a single reconstructed coronal image slice over time following the intravenous injection of 34 MBq of ^{99m}Tc-HIDA into a mouse. Each image represents a 15-second acquisition over a period of 300 seconds starting at the time of injection. The dynamic sequence of images shows the temporal sequence of uptake and clearance of the radiopharmaceutical in the liver. (*Courtesy Félicie Sherer, Université Libre de Bruxelles, Belgium; Steven Staehlens, Ghent University, Belgium; and Freek Beekman, Delft University of Technology, The Netherlands*).

antibodies and peptides for tumor localization, and ultimately also for tumor therapy by substituting the imaging radionuclide with a β emitter that can cause localized cell death.

Other areas in which SPECT is used include imaging of infection and inflammation and measurement of liver and kidney function. An extensive discussion of the clinical applications of both planar gamma camera imaging and SPECT can be found in reference 13.

SPECT also has seen growing use for preclinical studies and basic biomedical research, often with dedicated small-animal SPECT systems such as those described in Section A.4. Studies may be performed in small animals to study the biodistribution over time of a potential new therapeutic entity, for example peptides, antibodies, nanoparticles, or cells that have been radiolabeled to allow them to be imaged. A second common use is to measure the effect of a new therapy. For example, the ability of implanted stem cells to improve myocardial perfusion could be monitored by SPECT in an animal model using the same radiotracers for perfusion that are used in the clinic. Figure 17-23 shows perfusion images of the myocardium in a mouse obtained with a small-animal SPECT system. Figure 17-24 shows a single image slice from a fast dynamic SPECT study showing the uptake of the radiopharmaceutical ^{99m}Tc-HIDA (see Table 5-5) in the liver following injection into a mouse. Some of the major applications for small-animal SPECT systems are discussed in reference 14.

REFERENCES

1. Rogers WL, et al: SPRINT II: A second-generation single-photon ring tomograph. *IEEE Trans Med Imag* 7:291-297, 1988.

2. Genna SG, Smith AP: The development of ASPECT, an annular single-crystal brain camera for high-efficiency SPECT. *IEEE Trans Nucl Sci* 35:654-658, 1988.

3. Babla H, Bai C, Conwell R: A triple-head solid state camera for cardiac single photon emission tomography (SPECT). *Proc SPIE* 6319:63190M, 2006. DOI: 10.1117/12.683765.

4. Erlandsson K, Kacperski K, van Gramberg D, Hutton BF: Performance evaluation of D-SPECT: a novel SPECT system for nuclear cardiology. *Phys Med Biol* 54:2635-2649, 2009.

5. CardiArc. Accessed October 11, 2011 from http://www.cardiarc.com.

6. Meikle SR, Kench P, Kassiou M, Banati RB: Small-animal SPECT and its place in the matrix of molecular imaging technologies. *Phys Med Biol* 50:R45-61, 2005.

7. Sorenson JA: Quantitative measurement of radioactivity in vivo by whole-body counting. In Hine GJ, Sorenson JA, editors: *Instrumentation of Nuclear Medicine*, Vol 2, New York, 1974, Academic Press, pp 311-348.

8. Chang LT: A method for attenuation correction in radionuclide computed tomography. *IEEE Trans Nucl Sci* 25:638-643, 1978.

9. Jaszczak RJ, Chang LT, Stein NA, Moore FE: Whole-body single-photon emission computed tomography using dual, large-field-of-view scintillation cameras. *Phys Med Biol* 24:1123-1142, 1979.

10. Bailey DL: Transmission scanning in emission tomography. *Eur J Nucl Med* 25:774-787, 1998.

11. King MA, et al: Attenuation, scatter and spatial resolution compensation in SPECT. In Wernick MN, Aarsvold JN, editors: *Emission Tomography: The Fundamentals of PET and SPECT*, Amsterdam, 2004, Elsevier Academic Press, pp 473-498.

12. Performance Measurements of Scintillation Cameras: *National Electrical Manufacturers Association (NEMA) Standards Publication NU 1-2007*. Rosslyn, VA, 2007, NEMA.

13. Sandler MP, et al, editors: *Diagnostic Nuclear Medicine*, ed 4, Baltimore, 2002, Lippincott Williams & Wilkins.

14. Franc BL, Acton PD, Mari C, Hasegawa BH: Small-animal SPECT and SPECT/CT: Important tools for preclinical investigation. *J Nucl Med* 49:1651-1663, 2008.

Positron Emission Tomography

The second major method for tomographic imaging in nuclear medicine is *positron emission tomography* (PET). This mode can be used only with positron-emitting radionuclides (see Chapter 3, Section G). PET detectors detect the "back-to-back" annihilation photons that are produced when a positron interacts with an ordinary electron. Although the annihilation photons could be detected using single photon emission computed tomography (SPECT) systems operating in conventional single-photon counting mode, these systems are not optimally designed for the relatively high energy of annihilation photons (511 keV). They have relatively low detection efficiencies at these energies and require relatively inefficient high-energy collimators. As well, SPECT systems do not take advantage of the back-to-back directional characteristics of annihilation photons. This unique feature is exploited advantageously with special annihilation-coincidence detector systems for PET.

PET has gained widespread clinical acceptance and now is firmly established alongside planar imaging and SPECT in clinical nuclear medicine. In this chapter, we describe the basic features of annihilation coincidence detection, the design and performance characteristics of PET detectors and scanners, and some of the important clinical applications of PET.

A. BASIC PRINCIPLES OF PET IMAGING

1. Annihilation Coincidence Detection

When a positron undergoes mutual annihilation with a negative electron, their rest masses are converted into a pair of annihilation photons (see Fig. 3-7). The photons have identical energies (511 keV) and are emitted simultaneously, in 180-degree opposing directions, usually within a few tenths of a mm to a few mm of the location where the positron was emitted, depending on the energy and range of the positrons. Near-simultaneous detection of the two annihilation photons allows PET to localize their origin along a line between the two detectors, without the use of absorptive collimators. This mechanism is called *annihilation coincidence detection* (ACD). Detection of a pair of annihilation photons in opposing detectors actually defines the volume from which they were emitted. Most ACD detectors have square or rectangular cross sections. Thus the volume is essentially a box of square or rectangular cross section, with dimensions equal to those of the detectors (Fig. 18-1).

Coincidence logic (Chapter 8, Section F and Fig. 8-15) is employed to analyze the signals from the opposing detectors. For many PET scanners, this is accomplished by having the electronics attach a digital "time stamp" to the record for each detected event. Typically, this is done with a precision of approximately 1 or 2 nanoseconds (1 nsec = 10^{-9} sec). The coincidence processor examines the time stamp for each event in comparison with events recorded in the opposing detectors. A coincidence event is assumed to have occurred when a pair of events are recorded within a specified *coincidence timing window*, which typically is 6 to 12 nanoseconds.

Although annihilation photons are emitted simultaneously, a small but finite coincidence window width is needed to allow for differences in signal transit times through the cables and electronics, as well as different distances of travel by the two photons from the annihilation event to the detectors (see Section A.2). In addition, the detectors in a PET scanner do not have perfect timing precision and therefore have a finite *timing*

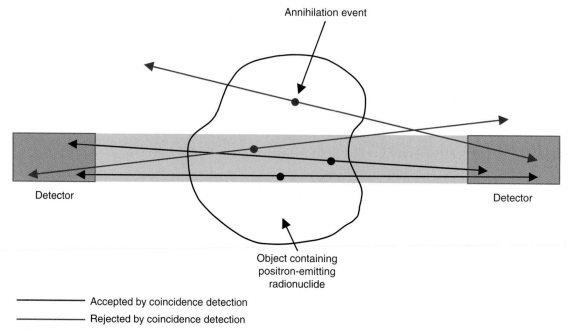

FIGURE 18-1 Volume (*green shaded area*) from which a pair of simultaneously emitted annihilation photons can be detected in coincidence by a pair of detectors. Not all decays in this volume will lead to recorded events, because it is necessary that both photons strike the detectors. Outside the shaded volume, it is impossible to detect annihilation photons in coincidence unless one or both undergo a Compton scatter in the tissue and change direction.

resolution. Uncertainties that govern the timing resolution can arise from the statistical nature of the signal (which is produced by the conversion of 511-keV photons into light, electrons, or electron-hole pairs in the detector) and from electronic noise in the detector and associated circuits. Uncertainties also can arise from the electronic method used to determine the time at which the interaction occurred (see Chapter 8, Section F). For a pair of similar detectors, the timing uncertainties typically are well described by a gaussian distribution, and the timing resolution is defined as the full width at half maximum (FWHM) of this distribution. For scintillation detectors, the timing uncertainty is reduced, and the timing resolution improved, by using brighter and faster scintillators that produce a large number of light photons over a short time interval immediately after an interaction occurs. Timing resolution is typically in the range of 0.5 to 5 nsec, depending on which scintillator and photodetector is used.

The need for a finite window width permits other types of events to occur in coincidence, as discussed in Section A.9. Also, as discussed in Section A.4, the annihilation photons are not always emitted in precise back-to-back directions. The effects of these deviations from the ideal are discussed in the sections indicated.

The ability of ACD to localize events on the basis of coincidence timing, without the need for absorptive collimation, is referred to as *electronic collimation.* As was discussed in Chapter 14, Section C, the lead septa in standard parallel-hole collimators, which are necessary to obtain adequate spatial localization, also are responsible for the relatively low sensitivity of these collimators. Because ACD does not require a collimator to define spatial location, its sensitivity (number of events detected per unit of activity in the object) is much higher than is obtainable with the absorptive collimators used for conventional planar imaging and for SPECT. For comparable midplane resolution, the sensitivity of PET is many times higher than for SPECT.

In addition, by incorporating multiple opposing detectors in a complete ring or other geometric array around the patient, and operating each detector in the array in coincidence with multiple detectors on the other side of the array, data for multiple projection angles can be acquired simultaneously (Fig. 18-2). Indeed, with a stationary ring or geometric array that completely surrounds the patient, it is possible to acquire data for all projection angles simultaneously. This allows the performance of relatively fast dynamic studies and the reduction of artifacts caused by patient motion.

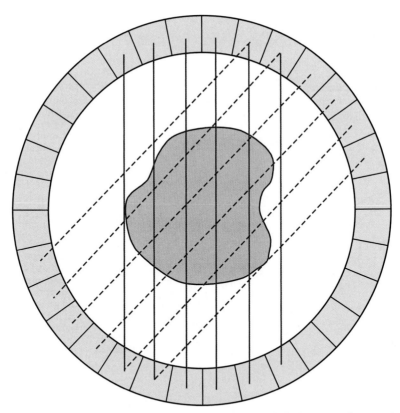

FIGURE 18-2 Array of detectors operating in electronic coincidence with detectors on the opposite side of the ring. This allows simultaneous acquisition of projection views from many different angles. Solid and dotted lines illustrate two simultaneously acquired projection views.

2. Time-of-Flight PET

In theory, it is possible to determine the location along a line between the two ACD detectors at which the annihilation photons originated by determining the difference in the time at which they are detected by the two detectors. This technique, which would allow the formation of tomographic images without mathematical reconstruction algorithms, is called *time-of-flight* PET. If the difference in the arrival times of the photons is Δt, the location of the annihilation event, with respect to the midpoint between the two detectors, is given by

$$\Delta d = \frac{\Delta t \times c}{2} \qquad (18\text{-}1)$$

where c is the velocity of light (3×10^{10} cm/sec). According to this equation, to achieve 1-cm depth resolution would require timing resolution of approximately 66 picoseconds (1 psec = 10^{-12} sec = 0.001 nsec). Although electronic circuits are capable of measuring this timing difference, the rise times of light output from scintillators currently available for PET imaging are too slow to provide this level of timing resolution. As well, the finite number of photoelectrons generated when an annihilation photon is detected gives rise to a "time jitter" during the rise time that adds to the uncertainty in event timing. This effect becomes more severe with detectors that have relatively low light output.

With the fastest available scintillators and careful design of electronic components and connections, it is possible to achieve timing accuracy at the level of a few hundred picoseconds. Although this is adequate to achieve localization only to within a few centimeters, images reconstructed from data acquired at this level of timing resolution have a higher signal-to-noise ratio than images reconstructed without time-of-flight information. This is because individual events can be constrained to lie within a smaller volume in the image reconstruction process. Figure 18-3 illustrates how the backprojection of data along one particular line of response is constrained to a smaller region of the reconstructed image matrix by the addition of time-of-flight information. To provide practically useful levels of time-of-flight information, only the fastest and brightest scintillators,

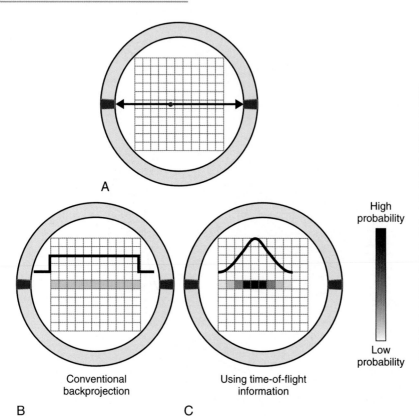

A

B Conventional
backprojection

C Using time-of-flight
information

High
probability

Low
probability

FIGURE 18-3 *A,* A pair of annihilation photons are emitted from a source (*red dot*) and detected in coincidence by opposing detectors. *B,* In the absence of time-of-flight information, there is no information about the location of the source along the line joining the two detectors. During reconstruction, the event is backprojected with equal probability of having occurred in all pixels along that line. *C,* With time-of-flight information, some limited localization of the event is possible and events are backprojected with probabilities that follow a Gaussian distribution, centered on pixel Δd (Equation 18-1) from the center of the scanner and with a full width at half maximum equal to the timing resolution of the detector pair.

such as LSO and LaBr$_3$, can be used (see Tables 7-2 and 18-2). Several commercially built systems incorporate some level of time-of-flight information using these materials. Although there are scintillators with even faster decay components, such as BaF$_2$, these are not favored because the signal-to-noise improvements that can be realized from time-of-flight information is typically more than offset by their lower density and therefore lower efficiency for detecting 511-keV annihilation photons.

3. Spatial Resolution: Detectors

The spatial resolution of ACD with discrete detector elements is determined primarily by the size of the individual detector elements. As shown in Figure 18-4A, for elements of width d, a one-dimensional (1-D) slice through the ACD point-source response profile at midplane between the detector pair is a triangle. The detector resolution, R_{det} has a FWHM = $d/2$. As the source moves toward either detector, the response profile becomes trapezoidal, eventually becoming a box of width d at the face of either detector. Considering a 2-D detector with width d and height h, the ACD response profile becomes a 3-D function, which is a pyramid at midplane and

a rectangular box of area $d \times h$ at the face of either detector. Between these extremes, it is the frustum of a pyramid with lower base size equal to the size of the detectors and upper base size increasing linearly from zero at midplane to the size of the detectors at their face.

Alternatively, consider an uncollimated pair of gamma camera detectors, also operating in coincidence mode (Fig. 18-4B). If their intrinsic spatial resolution (see Chapter 14, Section A.1) is a gaussian function with FWHM = R_{int}, then the spatial resolution of the detector pair for ACD also is a gaussian function with FWHM = $R_{int}/\sqrt{2}$ at midplane. The ACD response profile becomes wider as the source moves toward either detector, with its FWHM eventually becoming equal to R_{int} at the face of either detector. Assuming that the resolution of the imaging detector is the same in all directions, the 2-D ACD response profile is obtained by rotating the 1-D gaussian function around its center.

For both discrete or gamma camera–type detectors, the spatial resolution of ACD varies by only approximately 30% in the central 60% of the space between the detectors (Fig. 18-5). By comparison, the resolution of a parallel-hole collimator can vary by several hundred percent over a comparable range (see Fig.

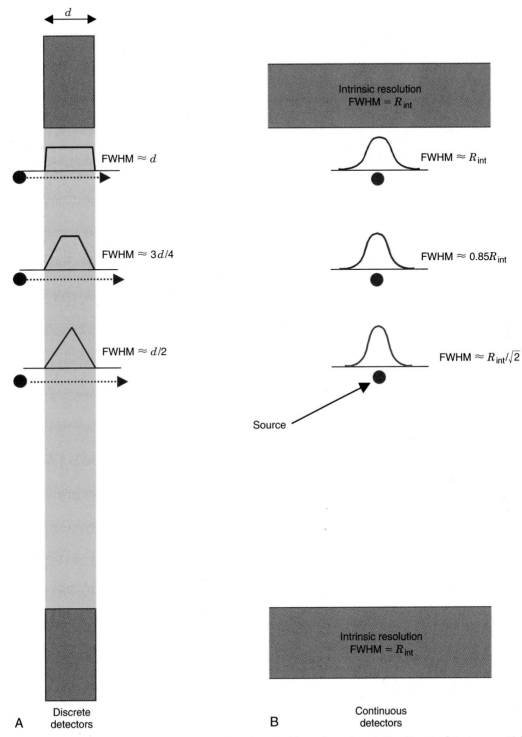

FIGURE 18-4   Spatial resolution of detector pair (R_{det}) for coincidence detection. *A,* For discrete detectors, spatial resolution is determined by the width of the detector element, d. At midplane, the coincidence response function is a triangle with full width at half maximum (FWHM) = $d/2$. As the source is moved closer to one of the detectors, the response function becomes trapezoidal in shape, eventually becoming a rectangle of width, d. *B,* For continuous detectors, spatial resolution is determined by the intrinsic resolution of the detector, R_{int}. At the midplane, the coincidence response function is approximately gaussian, with FWHM = $R_{int}/\sqrt{2}$. Near the face of a detector, it becomes FWHM = R_{int}.

14-19). When profiles obtained from opposing views in SPECT are combined using the geometric mean, the variation in resolution within the space between the opposing detectors is reduced to a level comparable to ACD (see Fig. 17-8). The key difference is that the resolution of ACD is determined primarily by the size of the detector element or the intrinsic resolution of the camera detector, whereas for SPECT, the resolution is primarily determined by the collimator resolution at the *midpoint* between the two detectors. The latter is substantially degraded from its value at the face of the detector. This means that the collimator must have very high resolution at its face to achieve even moderately good resolution at midplane between the detectors. In turn, the requirement for high spatial resolution leads to relatively low detection efficiency with absorptive collimators (see Chapter 14, Section C). As discussed in Section A.8, this results in relatively low sensitivity for a SPECT system as compared with a PET system with comparable spatial resolution.

4. Spatial Resolution: Positron Physics

The spatial resolution of an ACD system is degraded from the values derived from simple geometry indicated in Figure 18-4 by two factors relating to the basic physics of positron emission and annihilation. The first is the *finite range of positron travel* before it undergoes annihilation. ACD defines the line along which the annihilation event took place, which is not precisely the location from which the decaying radioactive nucleus emitted the positron. The range of travel for a positron before it undergoes annihilation is essentially the same as the range of travel of an ordinary electron (or β particle) of similar energy (see Chapter 6, Section B.2). Figure 6-10 and Table 6-1 show the extrapolated range versus maximum energy for β particles, E_β^{max}. The maximum energies of the positrons emitted from radionuclides used for nuclear medicine are in the range of 0.5 to 5 MeV (Table 18-1; see also Appendix C). Thus their extrapolated ranges are in the 0.1- to 2-cm range.

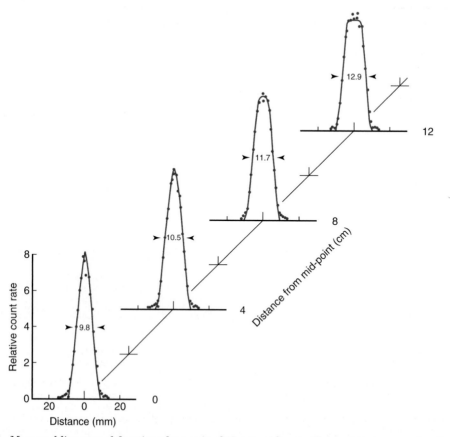

FIGURE 18-5 Measured line-spread functions for a pair of 17-mm-wide coincidence detectors as a function of source position between the two detectors. The detector separation was 42 cm. The FWHM varies by only 30% within the central 24 cm (57%) of the space between the two detectors. (*From Hoffman EJ, Huang S-C, Plummer D, Phelps ME: Quantitation in positron emission computed tomography: VI. Effect of nonuniform resolution. J Comput Assist Tomogr 6:987-999, 1982.*)

TABLE 18-1
SOME POSITRON-EMITTING NUCLIDES USED FOR IN VIVO IMAGING

Radionuclide	Half-Life	β⁺ fraction	Maximum β⁺ Energy	How Produced
^{11}C	20.4 min	0.99	960 keV	Cyclotron
^{13}N	9.96 min	1.00	1.19 MeV	Cyclotron
^{15}O	123 sec	1.00	1.72 MeV	Cyclotron
^{18}F	110 min	0.97	635 keV	Cyclotron
^{62}Cu	9.74 min	0.98	2.94 MeV	Generator (from ^{62}Zn)
^{64}Cu	12.7 hr	0.19	580 keV	Cyclotron
^{68}Ga	68.3 min	0.88	1.9 MeV	Generator (from ^{68}Ge)
^{76}Br	16.1 hr	0.54	3.7 MeV	Cyclotron
^{82}Rb	78 sec	0.95	3.35 MeV	Generator (from ^{82}Sr)
^{124}I	4.18 days	0.22	1.5 MeV	Cyclotron

The extrapolated range applies to the highest-energy positrons emitted by a radionuclide. However, positrons, like β particles, are emitted with a spectrum of energies. Only a small fraction have the full amount of energy available from the decay (see Fig. 3-2). In addition, the extrapolated range is the maximum distance that the electron would travel if it were not significantly deflected in any of its interactions and traveled in essentially a straight line to the end of its range. In reality, most electrons (and positrons) travel a tortuous path, often with multiple large-angle deflections (see Fig. 6-4). The result is that the average distance measured from the origin of the positrons to the end of their path is significantly smaller than their extrapolated range.

For purposes of defining the spatial resolution of ACD, the distance of interest is the *effective positron range*. This is the average distance from the emitting nucleus to the end of the positron range, measured *perpendicular* to a line defined by the direction of the annihilation photons (Fig. 18-6). This distance always is smaller than the extrapolated range for the positrons emitted by the radionuclide.

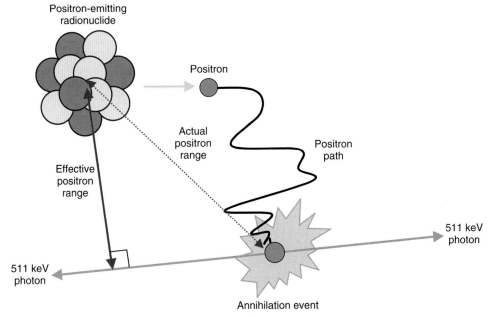

FIGURE 18-6 Blurring caused by positron range effects. The perpendicular distance from the decaying atom to the line defined by the two 511-keV annihilation photons is referred to as the *effective positron range*.

Figure 18-7 shows the positron range distribution for point sources of ^{18}F ($\bar{E}_\beta^{max} = 0.635$ MeV) and ^{15}O ($E_\beta^{max} = 1.72$ MeV). According to Figure 6-10, the extrapolated ranges for these positrons in water would be approximately 2 mm and 8 mm, respectively; however, the FWHMs of their distribution profiles are only 0.1 mm and 0.5 mm.

Note as well that the positron range distributions shown in Figure 18-7 have long tails and thus are not well described by gaussian functions. Therefore the FWHM is not the best indicator of the effect of positron range on ACD spatial resolution. Instead, the *root mean square (rms) effective range* often is used. Figure 18-8 shows the general relationship between rms effective range and maximum positron energy. Typical rms effective ranges (and thus the blurring caused by positron ranges) are on the order of 0.5 to 3 mm. Note that positron range is inversely proportional to the density of the absorber. Thus rms ranges would be proportionately higher in lung tissue and airways ($\rho \sim$ 0.1-0.5 g/cm^3) and lower in dense tissues such as bone ($\rho \sim$ 1.3-2 g/cm^3).

A second factor involving the physics of positrons is that the annihilation photons almost never are emitted at *exactly* 180-degree directions from each other (Fig. 18-9). This effect, which is due to small residual momentum of the positron when it reaches the end of its range, is known as *noncolinearity*. The angular distribution is approximately gaussian with FWHM approximately 0.5 degree. The effect on spatial resolution, expressed in terms of FWHM, is linearly dependent on the separation of the ACD detectors, *D*, and is given by

$$R_{180°} = 0.0022 \times D \qquad (18\text{-}2)$$

A typical value of *D* for a whole-body PET scanner is 80 cm. Thus the FWHM for blurring caused by noncolinearity is approximately 2 mm.

The *system resolution* of an ACD or PET detector system is obtained by combining the individual resolution components, in the same manner as the component resolutions are combined to determine the system resolution

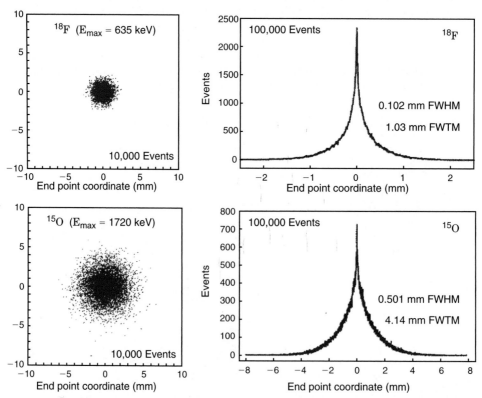

FIGURE 18-7 Results of Monte Carlo simulations showing the distribution of annihilation sites for positron-emitting point sources in water for ^{18}F ($E_\beta^{max} = 0.635$ MeV) and ^{15}O ($E_\beta^{max} = 1.72$ MeV). The profile of the distribution is broader for ^{15}O because of its higher average positron energy, which leads to a longer positron range prior to annihilation. (*From Levin CS, Hoffman EJ: Calculation of positron range and its effect on the fundamental limit of positron emission tomography system spatial resolution. Phys Med Biol 44:781-799, 1999.*)

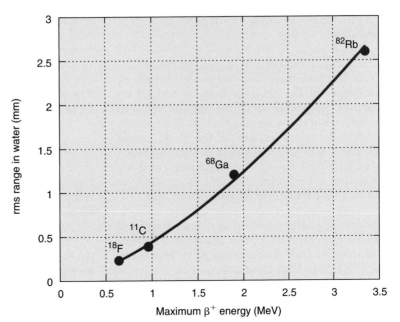

FIGURE 18-8 Root mean square range for positrons in water versus $E_\beta^{\max}$. (*Data from Derenzo SE: Mathematical removal of positron range blurring in high-resolution tomography.* IEEE Trans Nucl Sci *33:565-569, 1986.*)

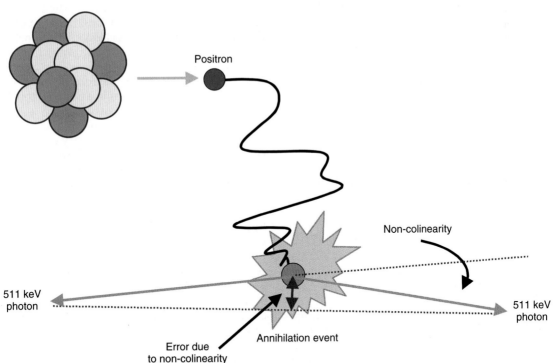

FIGURE 18-9 Noncolinearity of annihilation photons resulting from residual momentum of the electron and positron at annihilation. Noncolinearity leads to positioning errors. Angles are exaggerated in this example for purposes of illustration. Actual range of angles is about ±0.25 degree, centered at 180 degrees.

for a gamma camera system (see Chapter 14, Section C.4). Thus

$$R_{\text{sys}} \approx \sqrt{R_{\text{det}}^2 + R_{\text{range}}^2 + R_{180°}^2} \qquad (18\text{-}3)$$

where R_{det} is the spatial resolution of the detector system, as determined by the size of

the discrete detector elements or the intrinsic resolution of continuous detectors (see Section A.3 and Fig. 18-4). For typical whole-body PET scanners, with either discrete detector elements or gamma camera detectors, the effects of positron range and noncolinearity combine to add anywhere from a few tenths of a millimeter to a few millimeters to system resolution.

EXAMPLE 18-1

What fraction of PET system resolution at the center of the scanner bore is caused by positron range and noncolinearity blurring in the following three situations? (1) 6-mm-wide discrete detectors, separated by 80 cm, using ^{18}F; (2) 2-mm-wide discrete detectors separated by 60 cm, using ^{68}Ga; (3) gamma camera detectors with 3-mm intrinsic resolution at 511 keV, separated by 60 cm, using ^{68}Ga.

Assume that the positron range distribution can be approximated by a gaussian function and use the rms effective range to represent the FWHM of that function.

Answer

For *situation 1*, the spatial resolution at the midpoint between 6-mm-wide discrete detectors is given by

$$R_{det} = d/2 = (6 \text{ mm})/2 = 3 \text{ mm}$$

The rms range for ^{18}F is $R_{range} \approx 0.2$ mm (see Fig. 18-8), and the noncolinearity for an 80-cm separation is

$$R_{180°} = 0.0022 \times 800 \text{ mm} = 1.76 \text{ mm}$$

Using Equation 18-3, the system resolution is

$$R_{sys} \approx \sqrt{3^2 + 0.2^2 + 1.76^2} \approx 3.5 \text{ mm}$$

Thus resolution blurring caused by positron range and noncolinearity for situation 1 adds approximately 17% to the system resolution, relative to the detector resolution.

For *situation 2*, using the same equations and Figure 18-8, the spatial resolution of 2-mm-wide discrete detectors is $R_{det} = 1$ mm, the rms range for ^{68}Ga is $R_{range} \approx 1.2$ mm and the noncolinearity blurring for the 60-cm separation is 1.32 mm. Thus the system resolution is

$$R_{sys} \approx \sqrt{1^2 + 1.2^2 + 1.32^2} \approx 2.05 \text{ mm}$$

In this situation, the additional blurring caused by positron range and noncolinearity approximately doubles the system resolution relative to detector resolution. Note that if ^{68}Ga is replaced by the lower-energy ^{18}F in situation 2, the system resolution becomes ~1.7 mm, so the effects of positron range and noncolinearity still are substantial (~70%).

For *situation 3*, $R_{det} = 3 / \sqrt{2} \approx 2.12$ mm (see Fig. 18-4) while other factors remain the same as in situation 2. Thus

$$R_{sys} \approx \sqrt{3^2/2 + 1.2^2 + 1.32^2} \approx 2.77 \text{ mm}$$

Positron range and noncolinearity increase the system resolution by approximately 30% relative to detector resolution in this situation.

From Example 18-1, one can conclude that positron range and noncolinearity have a relatively small effect on system resolution for whole-body systems, which usually have larger detector elements (situation 1) or only moderately high spatial resolution (situation 3). This is especially true for ^{18}F, the radionuclide most commonly used for clinical applications. However, their effects can be important limitations for high-resolution brain imaging or small-animal imaging devices (see Section B.5) that employ small discrete detector elements, especially for applications involving higher-energy positrons (situation 2).

Because they do not depend on technology, positron range and non colinearity create barriers for improving the spatial resolution of PET systems that cannot be overcome simply by using smaller detector elements. For example, independent of the detectors used, the blurring caused by noncolinearity will limit the achievable spatial resolution for a whole-body PET scanner to approximately 2 mm, because a bore size of 80-90 cm is required to accommodate the human body. Example 18-1 also demonstrates that when system resolution is dominated by one component, the gains achieved by improving other components of resolution may be small.

5. Spatial Resolution: Depth-of-Interaction Effect

A substantial thickness of scintillator material is required to efficiently stop 511-keV annihilation photons. In a gamma camera, typical NaI(Tl) detector crystal thicknesses are 1.25 cm or less. PET systems generally employ 2- to 3-cm-thick scintillators with greater stopping power, such as BGO or LSO. For PET systems using arrays of detectors in multiple coincidence mode around the object, the relatively thick detector elements lead to a degradation of resolution known as the *depth of interaction* (DOI) effect. Although the effect also occurs in the axial direction in scanners that use cross-plane coincidence detection for 3-D data acquisition (see Section C.2), the primary effect is in the radial direction, and the discussion focuses on this aspect of the problem.

Figure 18-10 illustrates the cause of the problem for a detector system that uses a circular array of elements, all or most elements of which operate in multicoincidence mode with other elements in the array. For a source located near the center of the scanner, spatial resolution is determined by the width of the detector element, $R_{det} = d/2$, as described in Section A.3 and illustrated in Figure 18-4. However, for a source located away from the center, the apparent width of the detector element becomes

$$d' = d\cos\theta + x\sin\theta \qquad (18\text{-}4)$$

where d, x, and θ are as indicated in Figure 18-10. The apparent change in width results from the angulation between the detectors and from lack of knowledge about the depth

at which an interaction has occurred within the detector crystal. The spatial resolution (FWHM) then becomes $R'_{det} = d'/2$. Using Equation 18-4, this can be written as

$$\begin{aligned} R'_{det} &\approx (d/2) \times [\cos\theta + (x/d)\sin\theta] \\ &\approx R_{det} \times [\cos\theta + (x/d)\sin\theta] \end{aligned} \qquad (18\text{-}5)$$

From this equation it can be seen that the DOI effect is described by a multiplicative factor applied to the value of detector resolution at the midpoint between a pair of directly opposed detectors.

Equation 18-5 is only an approximation because the thickness of detector material is not constant across the width of the detector element seen by the source. Note as well that, for thin detector elements [$(x/d) \ll 1$], it is possible that $R'_{det} < R_{det}$. The same would be true for a very efficient detector material (or a very thin detector) that would stop most of the annihilation photons in a thin layer at the entrance to the detector. However, these conditions never apply in practice. Typically, $x \sim$ 2 to 3 cm and $d \sim$ 0.3 to 0.6 cm. For a whole-body PET scanner with 4-mm-wide detectors on a diameter of 80 cm, the DOI effect causes

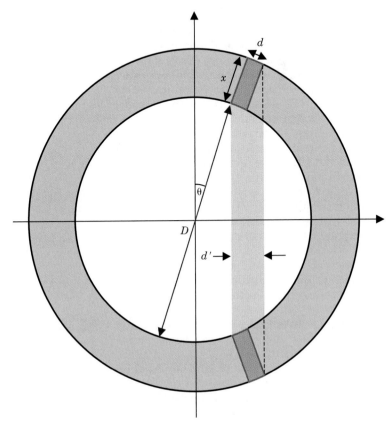

FIGURE 18-10 Apparent width of a detector element, d', increases with increasing radial offset in a PET scanner consisting of a circular array of detector elements. Because the depths at which the γ rays interact within the scintillation crystal are unknown, the annihilation event for a pair of photons recorded in coincidence could have occurred anywhere within the shaded volume. The magnitude of the effect depends on the source location, the diameter of the scanner, D, the length of the crystal elements, x, and the width of the detector elements, d.

approximately a 40% degradation of resolution at a distance of 10 cm from the center of the field of view (FOV).

The DOI effect is somewhat different for systems that use hexagonal or octagonal arrays as opposed to circular arrays of detector elements. With hexagonal or octagonal arrays, as the source moves away from the center of the scanner, some portion of an opposing array still remains perpendicular to it, at least over a distance comparable to the width of a segment of the array. Consequently, there is less variation in the DOI effect across the FOV. At the center of the FOV, the effect is somewhat larger than at the center for a circular array of the same diameter, whereas at the periphery it is somewhat smaller. On average, the DOI effect is comparable for both segmented and circular arrays.

Note also in Equations 18-4 and 18-5 that, for a given radial distance away from the center of the scanner, θ and $\sin \theta$ become smaller (and thus R'_{det} becomes smaller) as the diameter of the detector ring becomes larger. Because of the DOI effect, PET scanners often are built with detector arrays that are of larger diameter than would be necessary to fit the patient, which in turn increases detector costs.

6. Spatial Resolution: Sampling

In ACD, the FWHM of the detector resolution is one-half the width of a detector element (see Fig. 18-4). For a stationary array of detector elements, each of width d, the sampling interval between parallel projection lines also is d (Fig. 18-11A). Considering only detector resolution, it can be shown that this leads to undersampling of object profiles, which in turn leads to a distortion of the high-frequency content (i.e., the fine details) of the reconstructed image (see Fig. 16-11 and Section C). According to Equation 16-14, three samples should be acquired per FWHM of spatial resolution. In theory, this would translate into a requirement for *six* samples over an interval equal to the width of a detector element. However, as described elsewhere in this section, system resolution is degraded from the theoretical limits established by intrinsic resolution by other factors, so that coarser sampling is acceptable. Nonetheless, some additional sampling is required beyond that illustrated in Figure 18-11A.

In practice, two samples usually are acquired per detector element width. Although less than the theoretical ideal, this results in little noticeable distortion of clinical images. Some

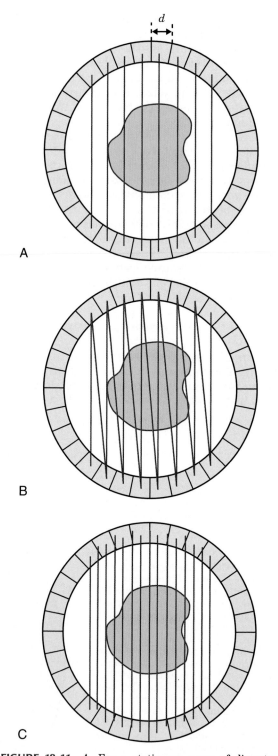

FIGURE 18-11 *A,* For a stationary array of discrete detector elements, the linear sampling distance is the same as the detector element width, d, which is insufficient to support the resolution of the detectors ($\sim d/2$). *B,* Linear sampling distance is reduced to $d/2$ when coincidences with immediately adjacent detectors are allowed. *C,* For image reconstruction, these samples are treated as if they came from a set of virtual detectors offset by half the detector width relative to the actual detectors.

earlier scanners actually incorporated a mechanical shift to acquire two sets of data with the detector elements shifted by half the width of a detector element between acquisitions. This approach was mechanically cumbersome. The modern approach is to use coincidence events in adjacent pairs of detector elements, thereby creating additional samples between the detector elements, as illustrated in Figure 18-11B. These are treated as samples acquired with a virtual set of detectors located between the actual detector elements, as shown in Figure 18-11C. The ray paths for the additional samples are not quite parallel to those for directly opposed detectors, but this seems to have little effect on the reconstructed image across most of the useful FOV.

Note that combining data for adjacent pairs of detectors into a single projection view as illustrated in Figure 18-11B and C, reduces the number of projection angles available from a stationary ring of detectors by half. Note also that PET systems that use continuous (i.e., gamma camera type) detectors (see Section B.3) can use arbitrarily chosen sampling intervals and thus can avoid some of the sampling problems associated with discrete detector arrays.

7. Spatial Resolution: Reconstruction Filters

The discussions in the preceding sections describe the spatial resolution achievable with PET systems, as determined by the physical characteristics of the imaging device and the basic physics of positron decay. However, as discussed in Chapter 16, Sections B.3 and C.3, spatial filters are applied to the recorded projection profiles to suppress noise in the reconstructed image (see Fig. 16-12). Inevitably, this results in some degradation of spatial resolution. In general, the fewer the number of counts recorded in an image, the lower the filter cut-off frequency ($k_{\text{cut-off}}$) and the greater the loss of spatial resolution.

The selection of cut-off frequency depends in part on the type of study. Thus a brain scan might be reconstructed with a cut-off frequency yielding 6-mm spatial resolution, whereas an abdominal scan, with more tissue attenuation and generally lower count densities, might be reconstructed with a filter yielding 10-mm spatial resolution. As well, the sensitivity (number of counts recorded per unit of activity in the patient) affects the statistical quality of the image and the degree of

spatial filtering required to achieve an acceptable noise level in the image. As discussed in the following section and in Section B, PET systems, especially those employing multiple detector rings and multi-ring coincidence detection, have substantially higher detection efficiencies (by orders of magnitude) than is achievable with typical SPECT systems. Thus PET images usually can be reconstructed with higher cut-off frequencies, and their final spatial resolution generally is superior to SPECT images.

8. Sensitivity

The sensitivity of PET, like that of all imaging devices, is determined primarily by the absorption efficiency of the detector system and its solid angle of coverage of the imaged object. The true coincidence rate, R_{true}, for a positron-emitting source located in an absorbing medium between a pair of coincidence detectors is given by

$$R_{\text{true}} = E\varepsilon^2 g_{\text{ACD}} e^{-\mu T} \qquad (18\text{-}6)$$

where E is the source emission rate (positrons/sec); ε is the intrinsic efficiency of each detector, that is, the fraction of incident photons detected (Equation 11-9, assumed to be the same for both detectors); and μ and T are the linear attenuation coefficient and total thickness of the object, respectively. g_{ACD} is the geometric efficiency of the detector pair, that is, the fraction of annihilation events in which both photons are emitted in a direction to be intercepted by the detectors.

As shown in Figure 18-4, the shape as well as the amplitude of the point-source response profile (i.e., g_{ACD} for a point source) varies with the location of the source between the two detectors. In one dimension, the profile is triangular at the midpoint between the detectors, rectangular at the face of either detector, and trapezoidal at locations between. In *two dimensions,* the ACD response at the midpoint is a pyramid, whereas at the surface of either of the detectors it is a rectangular box. At points between, it is the frustum of a pyramid. In all cases, the lower base is equal to the area of the detectors.

Maximum geometric efficiency for ACD is obtained for a point source located precisely at the midpoint of the centerline between the two detectors. However, as illustrated in Figure 18-4, this value does not apply when the point source is moved even slightly away from the centerline, or from the midpoint of that line. Thus a more appropriate measure for distributed sources is the *average* geometric efficiency within the sensitive volume for ACD. Midway between the two detectors, this is given by

$$\bar{g}_{ACD} \approx 2 \times \frac{1}{3} \times \left[A_{det} / \pi D^2 \right]$$

$$\approx \frac{2 A_{det}}{3 \pi D^2} \qquad (18\text{-}7)$$

where D is the distance between the detectors and A_{det} is the area of the detector facing the source.

The term in brackets in Equation 18-7 is the geometric efficiency for a single detector for a point source located at the midpoint of the centerline between the detectors. (As discussed in Chapter 11, Section A.2, this expression is valid when the detector dimensions are "small" in comparison with the source-to-detector distance.) The factor of 2 accounts for the fact that two detectors are used and that if one photon is emitted in a proper direction toward one detector, the other photon is virtually assured of being emitted in the proper direction toward the other. The factor of 1/3 is the average geometric efficiency across the sensitive volume at midplane, that is, the average height of a pyramid.

Actual PET systems typically employ many small detector elements arranged in circular, hexagonal, or octagonal arrays. Each detector element is operated in coincidence with many detectors on the opposing side of the ring, as shown in Figure 18-12. This multicoincidence operation has useful and important consequences for both the magnitude and uniformity of geometric efficiency.

The simplest way to visualize its effects on geometric efficiency is to consider a complete ring of detectors on a diameter D, with detector height h in the axial dimension and detector width $d \ll D$ in the plane of the ring. Assume that any interdetector gaps are "very small," so that the individual elements form a virtually continuous ring of detector material. For a point source located precisely at the center of the ring, the geometric efficiency would be equal to the solid angle subtended by the ring, because if either annihilation photon is intercepted by the ring, it is virtually assured that the second photon is traveling the proper direction to be intercepted as well. From simple geometric considerations, if $h \ll D$, it can be shown that the solid angle, and thus the geometric efficiency, for a point source precisely at the center of the ring is given by

$$g_{ACD,RING} \approx h/D \qquad (18\text{-}8)$$

Under the conditions described, geometric efficiency is relatively constant as the source is moved away from the center of the ring but still in its center plane; however, as the source

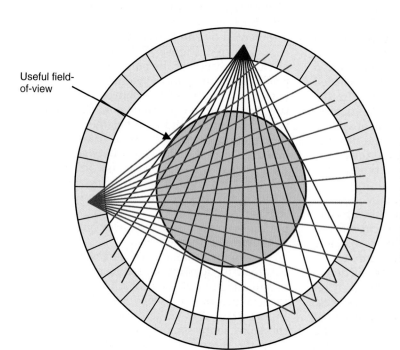

Useful field-of-view

FIGURE 18-12 Ring of detectors in which individual elements are operated in coincidence with multiple elements on the opposite side of the ring. In multicoincidence operation, each detector element is associated with a fan-beam acquisition, shown for two individual detector elements in the drawing. Data are recorded simultaneously for all possible fan beams. The inner circle formed by edges of all such fan beams defines the useful field-of-view.

is moved axially toward the ends of the ring, the geometric efficiency still has a triangular shape. Thus the *average* geometric efficiency for a source distributed within the sensitive volume for ACD across the width of the ring is half the value given by Equation 18-8, that is,

$$\bar{g}_{ACD,RING} \approx \frac{h}{2D} \qquad (18\text{-}9)$$

Equations 18-8 and 18-9 also are valid for polygonal arrays, with D representing the diameter of a circle drawn tangential to the surface of the array. As long as h << D, they also apply to continuous detectors that use gamma camera electronics, rather than discrete detector elements, to determine event locations.

In addition to increasing geometric efficiency and improving its uniformity, multicoincidence detection with a ring or polygonal array of detectors also allows simultaneous acquisition of multiple projection views without moving the detectors. Suppose the ring consists of N individual detector elements. When each detector in the ring or array is operated in coincidence with a bank of detectors on the opposite side, as illustrated in Figure 18-12, a total of $N/2$ fan-beam projections are acquired. These fan-beam projections typically are arranged to form parallel-beam profiles, as illustrated in Figures 16-20 and 18-2. However, as illustrated in Figure 18-11, data from adjacent pairs of detector usually are assigned to the same projection profile, thereby decreasing the number of views to $N/4$.

The number of detectors that are enabled for multicoincidence detection determines the width of the fan-beam projections and thus the diameter of the useful FOV. Sources located within the circle illustrated in Figure 18-12 are seen in all projections. Once a source is included within the useful FOV, a further increase in fan-beam width does not increase the counts recorded from that source. Sources outside the circle are not seen in some views, which could be a cause of image artifacts (see Chapter 16, Section C.2). Typical PET systems operate with each detector in coincidence with approximately two-thirds the total number of detectors in the ring.

Geometric efficiency varies somewhat across the useful FOV of the detector ring. In part, this is because the solid-angle for coincidence detection changes with source position. There also are geometric effects caused by differences in angle of incidence of the

photons onto the detectors and by gaps between detector elements. Corrections for this and other image nonuniformities are described in Section D.1.

It is noteworthy that, by segmenting large detectors into smaller elements and operating them in coincidence with multiple elements in the opposing array, it is possible to improve the spatial resolution in PET with only a modest loss of geometric efficiency. This effect is seen in Equation 18-9, in which geometric efficiency depends on the diameter of the ring, D, but not on the width d of the individual detector elements. Most of the loss of sensitivity that does occur is due to the requirement for interelement spacing and shielding, which is only approximately 0.2 to 0.3 mm in practical systems. For comparison, from fundamental principles, the geometric efficiency of absorptive collimators is degraded approximately as the square of spatial resolution (Equation 14-8). This presents a formidable challenge for improving spatial resolution in imaging applications based on single-photon counting, including SPECT.

The benefits of multicoincidence operation extend as well to the third (axial) dimension in multi-ring PET systems. This is discussed further in Section C.

As is the case for any imaging system, the sensitivity of a PET system also depends critically on the detection efficiency of the detector, which enters as a squared term in Equation 18-6. As was discussed in Chapter 11, Section A.3, detection efficiency is given by

$$\varepsilon = 1 - e^{-\mu_l x} \qquad (18\text{-}10)$$

where μ_l is the linear attenuation coefficient of the detector material and x is the detector thickness. Values of μ_l for several detector materials of interest for PET are given in Table 18-2. Also indicated are values of ε for 2-cm-thick detectors of each material, without a low-energy threshold and with an energy threshold that eliminates 50% of the detected pulses. Values for ε^2 in this table are useful for calculating scanner sensitivity and ACD counting rates (Equation 18-6). These values illustrate why materials such as BGO, LSO, and LYSO (Chapter 7, Section C.4) are preferred over NaI(Tl) for PET imaging.

Overall sensitivities for PET systems for a small-volume source of activity located near the center of the scanner range from 0.2% to 0.5% (0.002-0.005 cps/Bq) for single-ring systems or for multi-ring systems operated in

TABLE 18-2

**LINEAR ATTENUATION COEFFICIENTS AND DETECTION EFFICIENCIES FOR SOME
SCINTILLATORS AT 511 keV***

Scintillator	μ_ℓ(511 keV) cm^{-1}	ε (2 cm)†	ε^2 (2 cm)	ε_{50} (2 cm)‡	ε_{50}^2 (2 cm)
NaI(Tl)	0.34	0.49	0.24	0.25	0.061
BGO	0.95	0.85	0.72	0.43	0.18
LSO, LYSO	0.88	0.83	0.69	0.41	0.17
GSO	0.70	0.75	0.57	0.38	0.14
BaF$_2$	0.44	0.58	0.34	0.29	0.086

*Efficiency values are for 2-cm thick crystals.
†Detection efficiency (see Equation 18-10), assuming no low-energy threshold (all pulses counted).
‡Detection efficiency, assuming low-energy threshold is used, with 50% of pulses counted ($f = 0.5$ in Equation 11-4).

2-D acquisition mode (one slice per ring; Section C.1). For multi-ring systems in which coincidences between rings are allowed for 3-D data acquisition (see Section C.2), the sensitivity typically is 2% to 10% (0.02-0.10 cps/Bq). For comparison, the sensitivities for SPECT systems with a general-purpose parallel-hole collimator are in the range of 0.01% to 0.03% (0.0001-0.0003 cps/Bq), depending on the number of detector heads (see Chapter 14, Section E.7). The substantially greater sensitivity of PET versus SPECT systems is due primarily to their ability to achieve a high degree of spatial resolution without the use of absorptive collimators.

9. Event Types in Annihilation Coincidence Detection

ACD produces an output whenever two events are recorded within a specified coincidence timing window. Generically, any such events are called *prompt coincidences*. The discussion and analysis presented thus far assumes that all prompt coincidences arise from a pair of photons produced from the same annihilation event and that the annihilation event occurs somewhere within the coincidence volume between the detectors (see Fig. 18-1). These events are called *true coincidences*. Equation 18-6 describes the sensitivity of the system for these events. However, other prompt coincidence events also can occur within the resolving time of the detector system.

Two examples are shown in Figure 18-13. *Random coincidences* (also called *accidental coincidences*) occur when annihilation photons from two unrelated positron annihilation events are detected in two different detectors, within the coincidence timing window, and recorded as a single coincidence event. This can happen if one photon from each annihilation event is detected in each detector element.

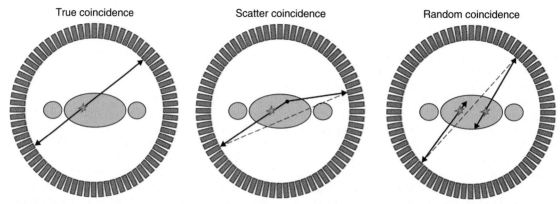

FIGURE 18-13 True coincidence event (*left*), scatter coincidence event (*center*), and random or accidental coincidence (*right*). Scatter and accidental coincidences yield incorrect positional information and contribute a relatively uniform background to the image that results in a loss of contrast. (*Courtesy Dr. Magnus Dahlbom, University of California—Los Angeles.*)

Random coincidences are not rare events, because the volume of tissue from which the photons for a random coincidence event could arise generally is much larger than the potential volume for true coincidence events.

The random coincidence counting rate in a detector pair is given by

$$R_{\text{random}} = \Delta T \times R_{\text{single,1}} \times R_{\text{single,2}} \quad (18\text{-}11)$$

where ΔT is the coincidence timing window used by the system* and $R_{\text{single,1}}$ and $R_{\text{single,2}}$ are the single-channel counting rates in the two detectors of the pair. Unlike true coincidence events, which can occur only when the source is located within the volume that is geometrically defined by the sides of the detector pair (see Fig. 18-1), random coincidences can arise from activity anywhere in the region between the detectors, including activity outside the useful FOV for a ring or array of detectors. Thus the single and random coincidence counting rates depend in a complicated way on both the source and detector geometry. A detailed analysis of these factors is beyond the scope of this text. Reference 1 provides a more comprehensive analysis. Nonetheless, some general observations can be made.

In general, the greater the total amount of activity used in a study, the higher the ratio of random-to-true coincidence rates. This is because the random coincidence rate increases as the square of the amount of activity present (product of single-channel counting rates in Equation 18-11), whereas the true coincidence rate increases only linearly with the amount of activity administered (Equation 18-6). A second general observation is that the ratio of random-to-true coincidence rates decreases in proportion to the width of the coincidence timing window. However, as noted in Sections A.1 and A.2, there are lower limits for this value, because of electronic and time-of-flight considerations. Finally, a general way to reduce the single-channel counting rate from activity outside the true coincidence volume is to use tungsten septa to restrict the FOV of individual detectors (see Section B.3). In turn, this reduces the random coincidence rate.

In actual PET scanners, the ratio of random-to-true coincidence counting rates typically ranges from approximately 0.1 to 0.2 for brain imaging to greater than 1 for applications where large amounts of activity may be nearby, but outside the true coincidence volume of the scanner. The latter could apply, for example, to some types of abdominal imaging when large amounts of activity are excreted into the bladder. Random coincidences occur more or less uniformly across the FOV of the scanner, causing a loss of image contrast as well as inaccuracies in quantification of activity within the patient. Methods for correcting for random coincidences are discussed in Section D.2.

A second category of nonvalid prompt coincidences are *scatter coincidences*. These occur when one (or both) of the photons from an annihilation event outside the sensitive volume for true coincidence events undergoes scattering and is detected in a detector other than the one that would be appropriate for a true coincidence event. The scattering event shown in Figure 18-13 occurs within the patient, but it also can occur within components of the scanner. Because the two annihilation photons were emitted simultaneously, they reach the detectors virtually simultaneously, apart from small time-of-flight differences (see Section A.2). Because these differences are very small, the detector system and its associated coincidence logic cannot typically distinguish them from valid events.

As is the case for the random-to-true coincidence ratio, the ratio of scatter-to-true coincidence counting rates depends in a complicated way on the source distribution and detector geometry. Placement of lead shielding on either side of the detector ring, or of thin tungsten septa between detector rings in a multi-ring PET system, reduces the likelihood of accepting scattered photons. However, unlike the random-to-true ratio, the scatter-to-true ratio does not depend on the amount of activity administered, because both the scatter and true coincidence rates increase linearly with this parameter. It also does not depend on the width of the coincidence timing window because scatter coincidences arise from the same positron annihilation event, and the two photons actually do arrive almost simultaneously at the two detectors. In clinical studies, the scatter-to-true coincidence ratio ranges from 0.2 to 0.5 for brain imaging and from 0.4 to 2 for abdominal imaging. The higher end of these ranges applies for 3-D

---

*The value of ΔT used here refers to the time separation between any two events that is determined by the electronics to indicate a prompt coincidence event (see Chapter 8, Section F). This differs from the definitions used in some articles and texts and leads to an additional factor of 2 in the version of Equation 18-11 in those publications.

acquisitions, which do not use interplane septa (see Section C.2).

Scatter coincidences provide incorrect localization of the positron annihilation event. The degree of position error depends on the scattering angle and location of the scatter event. Scatter coincidences lead to a broad distribution of mispositioned events, generally peaked toward the center of the object. Methods for minimizing the acceptance of scattered photons and for correcting for residual scatter coincidences are discussed in Section D.3.

B. PET DETECTOR AND SCANNER DESIGNS

As discussed in Section A.6, detection efficiency (Equation 18-10) is an important parameter in PET scanner sensitivity and performance. Sodium iodide detectors, which are the "workhorse" for many nuclear medicine applications, also have been used for PET scanners. Indeed, as discussed in Section B.3, it is possible to use appropriately modified dual-headed SPECT systems for PET imaging. However, because of the relatively high energy of the 511-keV annihilation photons, sodium iodide generally is not the detector material of choice for PET imaging. For these reasons, most PET scanners use denser higher-Z scintillation detectors arranged in rings or banks of discrete elements around the scanned object. These systems not only provide a high detection efficiency but they allow the simultaneous collection of data for all projection angles with a completely stationary set of detectors. In this section, we discuss the design of modern PET detector systems and scanners. Reference 2 is a useful review describing emerging detector technologies for PET.

1. Block Detectors

Early PET systems used individual detector units consisting of a piece of scintillator coupled to a photomultiplier tube (PMT). The individual detectors were arranged in a ring or in multiple rings around the subject. As illustrated in Figure 18-4, the response profile at midplane of a pair of coincidence detectors is a triangle with FWHM equal to one half the width of the detector. Thus to improve the intrinsic resolution of a PET scanner, the detectors must be made smaller. However, the cost increases rapidly if each detector element requires its own PMT.

The *block detector,* designed in the mid-1980s by Casey and Nutt,[3] allows small detector elements to be used (improving spatial resolution) while reducing the number of PMTs required to read them out (controlling cost). Figure 18-14 shows a typical block detector. A large piece of scintillator (most commonly BGO, LSO, or LYSO), is segmented into an array of many elements by making partial cuts through the crystal with a fine saw. The cuts between the elements are filled with a reflective material that serves to reduce and control optical cross-talk between scintillator elements. The array of crystals is read out by four individual PMTs. The depth of the saw cuts is determined empirically to control the light distribution to the four PMTs in a fairly linear fashion.

To determine the segment of the crystal in which an annihilation photon is detected, the signals from a four-PMT array are combined as follows:

$$X = \frac{(PMT_A + PMT_B) - (PMT_C + PMT_D)}{PMT_A + PMT_B + PMT_C + PMT_D}$$

$$Y = \frac{(PMT_A + PMT_C) - (PMT_B + PMT_D)}{PMT_A + PMT_B + PMT_C + PMT_D}$$

$$(18\text{-}12)$$

where PMT_A, PMT_B, and so forth are the signals from different PMTs. It will be recognized that these are essentially identical to Equations 13-1 and 13-2 for position localization for an Anger camera, except that only four PMTs are used here. The X and Y signals then are used to determine the subelement of the array in which the annihilation photon was detected.

Figure 18-15 shows the image obtained from uniform irradiation of a block detector. The image is not uniform. Rather, the calculated locations for recorded events are clustered in small localized areas corresponding to the individual detector elements. There is a small amount of overlap, but the individual elements are clearly resolved. Although the array pattern is nonlinear, the separation is sufficiently clear to allow each (x,y) location in the image to be assigned to a specific detector element in the array, for example, by using a look-up table.

The major advantage of the block detector is that it enables many detector elements (e.g., $8 \times 8 = 64$) to be decoded using only four PMTs. This dramatically lowers the cost per detector element while providing high spatial

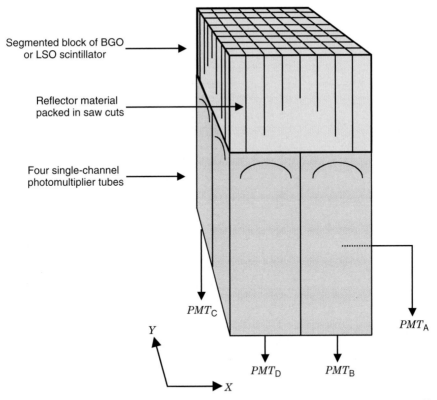

Segmented block of BGO
or LSO scintillator

Reflector material
packed in saw cuts

Four single-channel
photomultiplier tubes

PMT_C

Y

PMT_A

PMT_D

PMT_B

X

FIGURE 18-14 Block detector commonly used in clinical PET scanners. A piece of BGO or LSO scintillator is cut into an array of smaller elements that are read out using four single-channel photomultiplier tubes (PMTs). The cuts in the material are filled with an opaque reflective material that, along with the depths of the cuts, helps control the distribution of scintillation light reaching the PMTs.

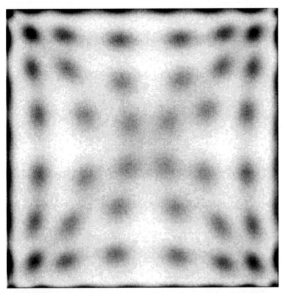

FIGURE 18-15 Flood-field image obtained by uniformly irradiating a block detector with 511-keV annihilation photons. Individual block detector elements appear as distinct "blobs" in the image, allowing separation of events recorded within individual detector elements.

resolution. Typical block detectors are made from 20- to 30-mm-thick BGO, LSO, or LYSO scintillator crystals (see Chapter 7, Section C.4), with 4- to 6-mm-wide sub-elements.

2. Modified Block Detectors

Two important modifications have been made to the basic design of the block detector. The first is to use proportionately larger PMTs positioned so they overlap portions of adjacent blocks (Fig. 18-16). Thus each block still is monitored by four PMTs, but each PMT also monitors the corners of four different blocks. This approach, known as *quadrant sharing,* reduces the total number of PMTs required for the array by approximately a factor of four as compared with the basic block design described in the preceding section. (The actual reduction is slightly smaller, as a result of edge effects.)

Quadrant sharing is used to create large planar detector panels that can be combined in hexagonal or other polygonal arrangements in a PET system. These detector panels closely resemble a standard gamma camera detector

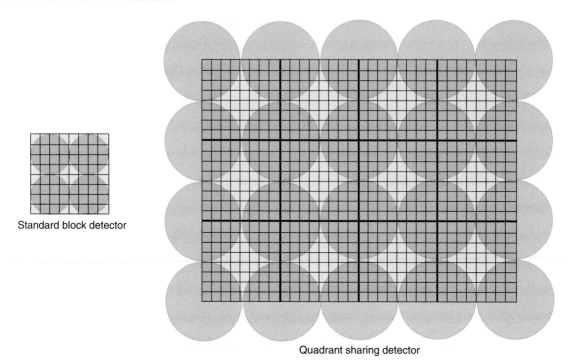

Standard block detector

Quadrant sharing detector

FIGURE 18-16 *Right,* Quadrant-sharing detector, in which each scintillator block straddles the corners of four photomultiplier tubes (PMTs). This allows larger PMTs to be used as compared to a standard block detector (*left*). For large panels, this leads to almost a fourfold reduction in the number of PMTs required to read out a given number of scintillator elements.

(Chapter 13, Section B), with the continuous scintillator plate replaced by an array of discrete scintillator elements. Although this approach reduces the cost per detector element of a PET system (by reducing the number of PMTs and electronic channels required compared with a block detector), it has the disadvantage of higher dead time losses, because each PMT views signals from a larger volume of scintillator.

The second modification of the basic block detector design is to use layers of two different scintillator materials, creating what is known as a *phoswich* (Fig. 18-17). This approach makes use of the difference in decay times of the two scintillators. By monitoring the decay time of the pulse, the event can be localized into either the upper or lower layer. For example, combinations of LSO (decay time ~40 nsec) and GSO (decay time ~60 nsec) scintillators can be used. Because the location of photon interaction can be determined to within half the total scintillator thickness, this reduces the DOI effects (described in Section A.5) by approximately a factor of 2. The disadvantage of this approach is that manufacturing of the detectors is more involved and that the light output and stopping power of GSO are worse than LSO. Thus

the overall detector performance is slightly degraded as compared with a detector of the same dimensions made purely from LSO. The phoswich design also can be combined with the quadrant-sharing approach.

3. Whole-Body PET Systems

Figure 18-18 illustrates several different whole-body PET scanner designs that have been developed, some using block detectors comprising discrete scintillator elements as introduced previously (*A, B, C*) and others that use continuous large-area gamma camera detectors (*D, E, F*). Systems that use a stationary ring or polygonal array of detectors (*A, C, E,* and *F*), with the detectors operating in multicoincidence mode, can acquire data for all projection angles simultaneously and these designs have been the basis for most commercial systems. Others (*B, D*) use only a few opposing banks of detectors, which must be rotated to get full tomographic information. Most PET systems use a ring diameter of 80 to 90 cm. After inserting scatter shielding and a shroud to cover the detectors and other components, the clear bore of the scanner typically is 55 to 60 cm, which is sufficient to comfortably accommodate most patients. The FOV in the axial direction is determined by the axial

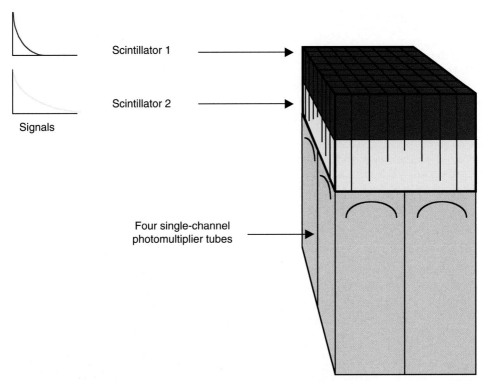

FIGURE 18-17 Phoswich detector constructed from two scintillator materials with different decay times. By analyzing pulse decay time, an event can be assigned to either the upper or lower layer, reducing the effective thickness of the detector by one half and providing some depth of interaction information.

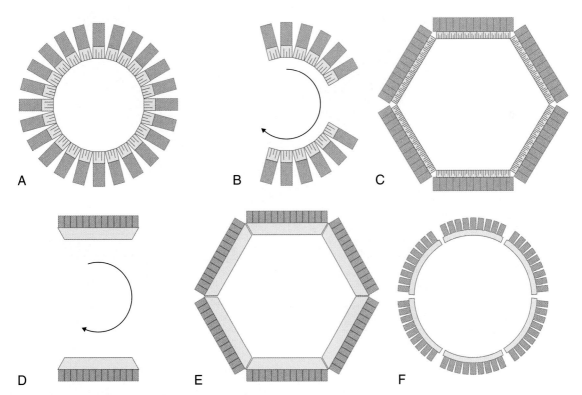

FIGURE 18-18 PET scanner geometries based on discrete scintillator elements (*top row*) or continuous scintillator plates (*bottom row*). *A,* Full ring of modular block detectors. *B,* Partial ring of modular block detectors. *C,* Hexagonal array of quadrant-sharing panel detectors. *D,* Dual-headed gamma camera with coincidence circuitry. *E,* Hexagonal array of gamma camera detectors. *F,* Continuous detectors using curved plates of NaI(Tl). A complete set of profiles can be acquired without motion with systems shown in A. C, E, and F, whereas detector motion is required with systems shown in B and D.

extent of the detectors and typically is in the range of 15 to 40 cm.

Figure 18-19 shows schematically the design of a representative whole-body PET scanner based on block detectors. This scanner employs 336 BGO block detectors, arranged in three rings of 112 blocks per ring.[4] Each block is cut into a 6 × 6 array of elements, with element sizes of 4 mm (transaxial) × 8.1 mm (axial) × 30 mm (thickness). The inside diameter of the detector ring is 92.7 cm and the clear bore of the scanner is 59 cm. The 18 crystals (three rings of three blocks) in the axial direction cover an axial FOV of 15.2 cm. The gantry can be tilted ± 20 degrees from the vertical, which can be useful for aligning the scan planes with the optimal viewing angle for an organ of interest. The system contains a set of tungsten interplane

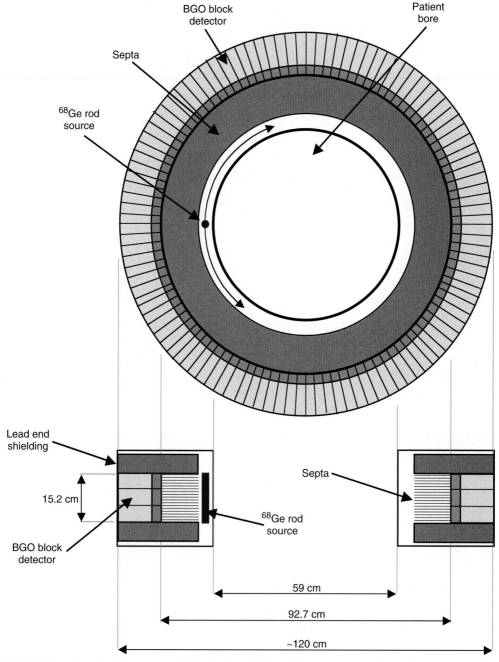

FIGURE 18-19 Drawings showing transaxial (*top*) and axial (*bottom*) cross-sections through a representative whole-body PET scanner.

septa of 1-mm thickness and 12-cm length between the crystal rings. The septa can be extended or retracted to provide varying levels of scatter rejection, as described in more detail in Section C.2. They also provide shielding from potential high concentrations of activity outside the scanning volume of interest, which helps control the random and scatter coincidence rates, as described in Section A.9.

The scanner also incorporates a rod source made from ^{68}Ge ($T_{1/2}$ = 273 days) to perform transmission scans for attenuation corrections. The source is permanently mounted in the system and is retracted into a lead shield when not in use. This is discussed further in Section D.4. Typically, the system uses a coincidence timing window of 12.5 nanoseconds and an energy window of 300 to 650 keV. The intrinsic spatial resolution of the detectors is approximately 3 mm, whereas the system resolution is approximately 4.5 mm near the center of the FOV and approximately 6.2 mm near the periphery of the scanner bore, the difference being due primarily to DOI effects (see Section A.5). The axial (slice thickness) resolution is approximately 4.2 mm at the center of the scanner, whereas near the periphery of the FOV, approximately 20-cm distance axially from the center, it is approximately 6.9 mm. The scanner simultaneously acquires data for 35 slices, separated center-to-center by 4.25 mm, in 2-D acquisition mode. In 3-D acquisition mode (see Section C.2), the number of slices and slice thickness in the axial direction can be chosen arbitrarily.

Figure 18-20 shows photographs for another whole-body PET scanner, which is described in reference 5. This system has smaller detector dimensions in the axial direction and uses the scintillator LYSO. Its design and performance capabilities are broadly similar to those of the scanner described previously. The trend in recent years has been to improve spatial resolution by further reducing the dimensions of the detector elements (4 mm × 4 mm is current state-of-the-art), and to increase the number of detectors along the axial direction to improve axial coverage of the body, increase sensitivity and thus reduce imaging times for studies that cover the whole body. A second trend has been to improve the detectors and electronics to provide sufficient timing resolution that time-of-flight information (see Section A.2) can be extracted. Commercial systems with a coincidence timing resolution of better than 600 psecs (corresponding to a spatial localization of ~9 cm) are now available.

Gamma camera technology similar to that used for conventional planar imaging and SPECT (see Chapters 13, 14, and 17) also has been employed for PET imaging. In one approach, coincidence timing circuitry has been installed between the heads of dual-headed scanners and the collimators removed for PET imaging. The spatial localization provided by the detector heads allows many coincidence lines to be acquired simultaneously. The basic concept is illustrated in Figure 18-18D. These systems can still be used for planar or SPECT imaging, by replacing the collimators.

The performance of standard gamma cameras for PET suffers from a number of limitations. Chief among these is the relatively low detection efficiency of the camera detectors for 511-keV annihilation photons (see Figs. 11-4 and 11-5). As well, although removing the collimator allows simultaneous

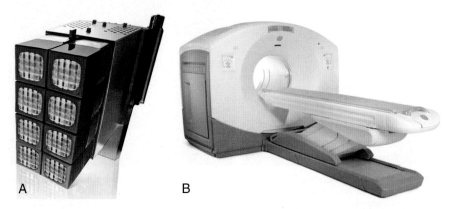

FIGURE 18-20 *A,* Modular cassette from a PET scanner containing eight block detectors. These cassettes are mounted on the PET scanner gantry to form complete rings of detector blocks that surround the patient. *B,* Clinical PET scanner based on rings of these block detectors. (*Courtesy GE Healthcare, Waukesha, WI.*)

data acquisition for many projection angles, the resulting high counting rates can lead to significant dead time losses and pile-up effects. An event detected anywhere in the detector can affect all other events detected at the same time. By contrast, dedicated PET systems use blocks of detectors that operate essentially independently from each other. Random and scatter coincidence rates, both of which increase with the geometric efficiency for detecting events outside the true-coincidence volume (see Section A.9), also tend to be high when the collimators are removed from the camera heads.

Some manufacturers addressed these limitations by incorporating thicker NaI(Tl) crystals (up to 2.5 cm) into systems intended for PET usage, and by employing more sophisticated circuitry in their gamma cameras to minimize dead time and suppress pile-up effects (see Chapter 14, Section A.4). Other manufacturers have developed gamma camera detectors specifically for use in PET. One such scanner used six curved gamma camera detector plates, arranged in a ring around the object (Fig. 18-18F). At the present time, PET scanners based on continuous gamma camera detectors are not widely used.

PET systems have been integrated with x-ray computed tomography (CT) technology to create combined PET/CT scanners in a single gantry. Almost all PET scanners sold today are combined with CT. These multi-modality systems are discussed in Chapter 19.

4. Specialized PET Scanners

Specialized PET systems also have been developed for high-resolution brain imaging and for breast imaging. These systems have smaller-diameter detector rings or arrays for improved geometric efficiency. They also generally have smaller detector elements for higher spatial resolution. Figure 18-21 shows a scanner designed for brain imaging that incorporates 2.1-mm×2.1-mm dimension scintillator elements in a phoswich configuration (Fig. 18-17) to limit DOI degradation of spatial resolution.[6] The detectors are based on the quadrant sharing design (Fig. 18-16) and eight detector panels are arranged in an octagonal configuration around the head. The reconstructed spatial resolution of images from this system is ~2.5 mm, superior to the 4- to 6-mm resolution that can be obtained on whole-body PET systems. Blurring caused by noncolinearity is reduced, as the detector separation in this brain imaging system is only 47 cm.

Figure 18-22 shows an example of a PET system developed for breast imaging.[7] This system uses two scanning detector panels (area 16.4 cm × 6 cm) to image the breast under mild compression. With scanning motion, the FOV can be adjusted up to 16.4 cm × 24 cm. The system uses LYSO scintillator elements with dimensions of 2 mm × 2 mm × 13 mm and these are read out using position-sensitive PMTs. Not all projection angles are measured, as the detectors do not rotate about the breast. However, as all points on one

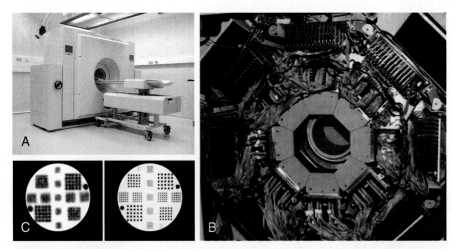

FIGURE 18-21 *A,* Photograph of a high-resolution brain imaging system. *B,* Interior of the scanner, showing the eight panels of detectors arranged in an octagonal geometry that are made up of phoswich detectors read out by photomultiplier tubes in a quadrant-sharing configuration. *C,* Phantom images from this system (*right*) compared with those obtained from a typical whole-body PET scanner (*left*). The improvement in spatial resolution arising from the smaller detector dimensions is apparent. (*Courtesy Dr. Adriaan Lammertsma VU Medical Center, Amsterdam, Netherlands, and Siemens Medical Solutions, Knoxville, TN.*)

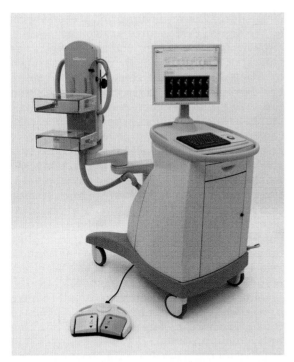

FIGURE 18-22 A PET system designed for breast imaging. Two detector panels scan back and forth to acquire an image of the breast under mild compression. (*Courtesy Naviscan Inc., San Diego, CA.*)

detector are in coincidence with all points on the opposing detector, there is sufficient angular information for approximate tomographic reconstruction using iterative reconstruction algorithms. The spatial resolution achievable with such systems is on the order of 2 to 2.5 mm for image planes parallel to the detector plates. In addition to improving geometric efficiency, the cost of dedicated breast PET systems is lower compared with whole-body scanners, because the volume of detector material is much smaller. Attenuation effects also are reduced, as the annihilation photons only need traverse the breast tissue, and not the entire cross-section of the body, for detection.

5. Small-Animal PET Scanners

PET scanners that are designed for small-animal imaging studies also are available. These are typically being used to evaluate and optimize new diagnostic and therapeutic agents destined for human use. The challenge is to obtain sufficiently high spatial resolution and sensitivity for imaging in mice and rats. Unlike SPECT imaging of small animals with pinhole collimators (Chapter 17, Section A.4), PET cannot readily improve the image resolution through magnification techniques, and

also has additional constraints because of positron physics and the DOI effects discussed in Sections A.4 and A.5. Fortunately, for small diameter systems, the blurring caused by noncolinearity (see Section A.4) is significantly reduced compared with whole-body clinical scanners. For a scanner with a 15-cm detector separation, this blurring is only 0.33 mm.

Most small-animal PET scanners use small scintillator elements decoded with position-sensitive or multichannel PMTs. The dimensions of the scintillator elements are typically on the order of 1 to 2 mm to achieve high spatial resolution. To avoid major DOI blurring, most systems only use a thickness of 10 to 15 mm of scintillator, and keep the ring diameter quite large (12 to 20 cm) compared with the size of the subject. Despite this, good sensitivity can be obtained, as high geometric efficiency can be achieved with relatively small amounts of detector material and the FOV required to cover the animal is quite small.

A typical small-animal PET scanner based on such technology is shown in Figure 18-23. The detector in this system consists of a block of LSO scintillator segmented into 1.5-mm ×

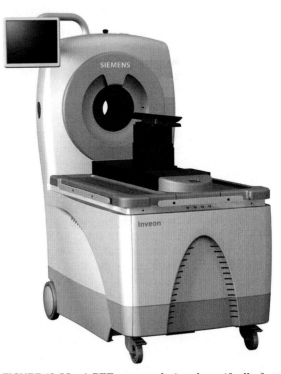

FIGURE 18-23 A PET scanner designed specifically for small-animal imaging. Such systems can achieve a spatial resolution on the order of 1.5 mm. (*Courtesy Siemens Preclinical Solutions, Knoxville, TN.*)

1.5-mm × 10-mm elements in a 20 × 20 matrix, and read out by a position-sensitive PMT, which has six anodes in the X-direction and six anodes in the Y-direction. The scanner uses 64 of these detector blocks arranged in four rings of 16 detector blocks per ring. The transaxial FOV of the system is 10 cm and the axial FOV is 12.7 cm. The system operates exclusively in 3-D acquisition mode (see Section C.2). The reconstructed spatial resolution of this system is approximately 1.4 mm, and the sensitivity is 7.4% for an energy window of 250-750 keV.[8]

Some small-animal PET systems incorporate alternative detector technologies. Successful devices have been constructed based on avalanche photodiode detectors (Chapter 7, Section C.3), multiwire proportional chamber detectors with high-density converters to improve the efficiency of these detectors for 511-keV photons (Chapter 7, Section A.3), or direct detection using the semiconductor cadmium zinc telluride (Chapter 7, Section B). A review of selected small-animal PET technology is presented in references 9 and 10.

C. DATA ACQUISITION FOR PET

1. Two-Dimensional Data Acquisition

Originally, most PET scanners were designed with axial collimators or septa between each ring of detectors (see Fig. 18-19). As shown in Figure 18-24A, the septa allow only those photons that are emitted parallel to the plane of the detector ring to be detected. This is known as *2-D data acquisition*. The septa provide efficient rejection of annihilation photons that have been scattered in the body. They also reduce the single-channel counting rate, thereby lowering the random coincidence rate (Equation 18-11) and minimizing dead time losses.

Because each crystal ring collects data from a single slice (oblique lines of response are not allowed because of the septa), 2-D projection data are analogous to the data obtained with a rotating gamma camera with a parallel-hole collimator used for SPECT imaging. Thus the images can be reconstructed using filtered backprojection (Chapter 16, Section B) or iterative approaches (Chapter 16, Section D). Using a scanner that employs multiple detector rings, one obtains a series of contiguous 2-D transaxial image planes that can be stacked together to form an image volume.

With slight or no modification of the lengths of the septa, PET scanners also can acquire data from immediately adjacent rings, as shown in Figure 18-24, B. These are known as *cross planes*. At the center of the scanner, the cross planes fall exactly halfway between the direct planes that are defined by individual crystal rings. For purposes of analysis, the cross-plane data can be assumed to have been acquired with a virtual ring of detectors shifted by half the detector width along the axial direction relative to the direct planes. For a scanner with n detector rings, this leads to a total of $(2n - 1)$ image planes in the axial direction. Because the cross planes receive data from two different lines of response, they have roughly twice the sensitivity (and therefore twice the counting rates) as the direct planes. They also are "x-shaped," but the amount of this distortion is too small to have a practical effect, except at the periphery of the FOV, where it leads to additional blurring in the axial direction.

Cross-plane data are reconstructed in the same manner as direct-plane data. In PET scanners with very small detector elements, the number of cross planes can be increased even further to include crystal ring differences of ± 2, ± 3, and so forth (Fig. 18-24C). As larger ring differences are accepted, the sensitivity increases; however, there is a loss of spatial resolution in the axial direction, because of the superposition of data that come from axially disparate locations.

2. Three-Dimensional Data Acquisition

Multi-slice 2-D data acquisition as described earlier rejects any photons that have an obliquity of more than the maximum accepted ring difference (typically ± 2 or 3 rings). This is very wasteful because the annihilation photons from many potentially valid coincidence events are absorbed by the septa. In *3-D acquisition mode,* the interplane septa are removed from the PET scanner and data are obtained for all possible lines of response, as shown in Figure 18-24D. Typically, this leads to a fourfold to eightfold improvement in sensitivity; however, the number of scattered photons and the single-channel counting rates also are increased. In brain scans using 3-D acquisition mode, 30% to 40% of the detected photons will have been scattered in the head prior to reaching the detectors. The axial sensitivity profile for 3-D acquisition is determined geometrically and is a triangular function, peaked at the center of the FOV. In 3-D mode, it is important to place the

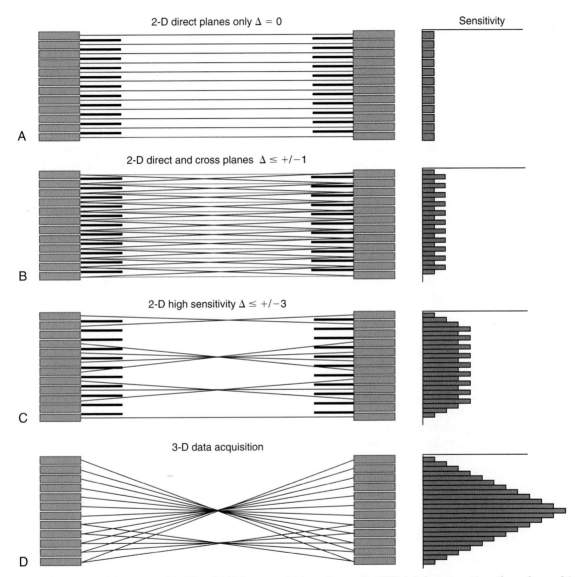

FIGURE 18-24 Two-dimensional (2-D) and 3-D data acquisition schemes for PET. Axial cross-sections through a multi-ring scanner are shown on the *left* and corresponding axial sensitivity profiles on the *right*. *A*, 2-D direct plane data acquisition. *B*, 2-D direct and cross-plane data acquisition. *C*, 2-D high-sensitivity data acquisition. *D*, Full 3-D data acquisition. For clarity, lines of response are shown only for selected axial positions in *C* and *D*.

structures of interest as close to the center of the axial FOV as possible. Multi-ring PET systems have relatively high overall sensitivity, as shown by the following example.

EXAMPLE 18-2

Consider a 32-ring PET scanner with BGO detector elements that are 6-mm-wide × 6-mm axial height × 2-cm-thick. The crystals are tightly packed on a 73.3-cm-diameter ring, such that each ring contains 384 crystals. In single-slice mode, each ring is operated in coincidence only with other detector elements within the same ring. In 3-D acquisition mode, each ring is operated in coincidence with all other rings. Estimate the sensitivity for a source located at the center of scanner bore for single-slice and 3-D modes. Assume that the source is comparable in size to the axial length of the detector crystals, and that a low-energy threshold that passes 50% of detected events is used.

Answer

For a small-volume source, comparable in size to the axial thickness of a single detector ring, Equation 18-9 applies. For such a source, the average geometric efficiency at the center of a single-ring is

$$\bar{g}_{RING} \approx \frac{1}{2} \times \frac{6 \text{ mm}}{733 \text{ mm}} \approx 0.00409 = 0.409\%$$

From Table 18-2, the intrinsic efficiency (squared) for coincidence detection with 2-cm-thick BGO crystals and an energy threshold that passes 50% of detected pulses is 0.18. When combined with geometric efficiency (Equation 18-6), this gives a total detection efficiency (sensitivity) for single-slice acquisition of

$$\text{Sensitivity}_{RING} \approx 0.18 \times 0.00409$$
$$\approx 0.00074 = 0.074\%$$

In 3-D mode, all 32 rings are operated in coincidence with each other (Fig. 18-24D). Because of the increased solid-angle of coverage, this immediately increases the geometric efficiency for a source located at the center of the axial FOV by approximately a factor of 32 (ignoring small geometric effects). As well, a small-volume source at the center of the axial FOV more closely approximates a true point source, for which Equation 18-8 applies. As

compared with Equation 18-9, this adds another factor of 2 to the sensitivity at the center of the FOV. The final result is

$$\text{Sensitivity}_{3D} \approx 32 \times 2 \times 0.18 \times 0.00409$$
$$\approx 0.047 = 4.7\%$$

For an extended source, for example, a line source that has length comparable to the total thickness of the 32-ring array, the average sensitivity of the scanner across the axial FOV is given by Equation 18-9, and the 3-D result given previously would be reduced by a factor of 2.

The estimated sensitivity in Example 18-2 for 3-D operation is 2-3 orders of magnitude greater than the sensitivity achieved with a gamma camera for single-photon imaging with absorptive collimation (see Chapter 14, Section E.7). Note as well that the additional rings extend the volume of coverage, so that a volume of tissue can be imaged in less time with a multi-ring scanner, as compared with a single-ring device.

Reconstruction of 3-D PET data also is more complex, because the projection data arises not only from transverse slices used for 2-D reconstruction, but also from many oblique angles through the subject. Thus the full 3-D image volume must be considered during the reconstruction process. Fully 3-D Fourier-based and iterative reconstruction algorithms are both available; however, computation times are roughly an order of magnitude longer than for 2-D reconstructions, because they involve backprojections and computations in three dimensions rather than two. Approximate 3-D reconstruction algorithms have been developed in which the 3-D dataset is reduced to a 2-D dataset using rebinning methods (see Chapter 16, Section E.3). In many situations, any loss in accuracy resulting from the approximations made in these algorithms is small when compared with the benefit of enabling 3-D PET data to be reconstructed in clinically acceptable timeframes.

Despite the increased computational and data storage requirements for 3-D PET, the large increases in sensitivity that it produces has resulted in it being offered as an option on all commercial whole-body PET systems. In some systems, interplane septa have been completely eliminated and only 3-D acquisition is possible. All small-animal and breast imaging systems currently operate in 3-D mode. Reference 11 provides further details on 3-D data acquisition and reconstruction in PET.

3. Data Acquisition for Dynamic Studies and Whole-Body Scans

PET scanners having detectors that surround the patient can acquire profiles simultaneously for all of the projection angles required for reconstruction. This allows dynamic studies to be performed with frame times of just a few seconds. For a dynamic scan, the number of time frames required and the length of each frame typically are entered into the computer. The scan starts as the tracer is being injected and the location to which the data are sent in the computer memory is incremented at the end of each frame.

Some systems allow what is called *list-mode acquisition*. In this mode, each coincidence event is written sequentially to a computer disk, along with a time stamp that indicates when the even occurred. After the scan is completed, the result is a single file with a list of coincidence events in the order that they were received. The events in this list can then be integrated over any time interval, allowing the number and duration of frames to be chosen and altered as necessary, after a scan has been completed. For example, this may allow the elimination of segments of data where a patient moved.

Once the projection data have been organized into frames, each corresponding to a certain time interval, the frames are individually reconstructed into tomographic images using the methods described in Chapter 16. In many studies, the dynamics of tracer delivery, accumulation, and clearance are important indicators of tissue function. The data often are analyzed using mathematical models described in Chapter 21.

Whole-body studies are performed by translating the patient through the scanner and acquiring data at multiple axial locations. This is achieved by using a computer-controlled bed. In 3-D PET, to improve the uniformity in sensitivity along the axial direction, bed positions are typically overlapped by 1/4 or 1/3 of the axial FOV. The data from different bed positions then are "stitched" together to form a single whole-body image.

D. DATA CORRECTIONS AND QUANTITATIVE ASPECTS OF PET

One goal of tomographic imaging is that the intensity of the reconstructed image should be proportional to the amount or concentration of activity at the corresponding location in the object. This is desirable for accurate comparisons of activity levels in different organs or in diseased versus normal tissues. It is essential for some types of dynamic studies (see Chapter 21). A number of corrections are required to achieve this goal in SPECT, as described in Chapter 17, Section B. In this section, we describe similar corrections that are needed in PET.

1. Normalization

PET scanners that are based on gamma camera detectors require corrections for nonlinearity and nonuniformity similar to those that are employed in SPECT (see Chapter 17, Section C.4). Inaccurate correction of these factors can lead to rings or other artifacts in reconstructed images.

A typical PET scanner may have 10,000 to 20,000 individual detector elements, which may have small variations in dimensions or in the fraction of scintillation light that is coupled to the PMTs. There also may be differences in the effective thickness of crystal seen by photons traveling along different angles of incidence for different cross planes. Correction for these variations is known as *normalization*.

Conceptually, the most straightforward approach to normalization would be to record the number of counts detected by each coincidence detector pair while exposing all pairs to the same radiation source. This could be accomplished, for example, using a rod source that extends through the axial FOV and rotating it around the periphery of the FOV (see Fig. 18-19). (This could be the same source that is used for attenuation correction, described in Section D.4.) One revolution of the rod source around the FOV would expose all detector pairs to the same number of annihilation photon pairs.

In an ideal scanner, each detector pair (i,j) would record the same number of counts N (within statistical limits) in the rod source scan. In practice, some detector pairs record more counts and some record less counts because of efficiency variations. The normalization factor for a specific pair of detectors is computed from

$$Norm_{i,j} = \frac{N_{i,j}}{<N>} \qquad (18\text{-}13)$$

where $<N>$ is the average value of $N(i,j)$ for all of the coincidence detector pairs in the scanner. Hence, the average normalization factor is equal to 1.

The normalization factor then is used to correct the counts recorded for each detector pair in a scan of a patient, $C(i,j)$, as follows:

$$C_{\text{Norm}_{i,j}} = \frac{C_{i,j}}{Norm_{i,j}} \qquad (18\text{-}14)$$

where $C_{\text{Norm}_{i,j}}$ are the corrected counts. This correction is applied to the projection (sinogram) data prior to image reconstruction.

Statistical errors caused by the finite number of counts in the normalization scan will increase the noise levels in the corrected data, which is undesirable. In 3-D mode, PET scanners can have on the order of 10^8 lines of response. To achieve a statistical uncertainty of ~3% in the normalization factor would therefore require a normalization scan with a total of approximately $1000 \times 10^8 = 10^{11}$ counts. Even at relatively high total counting rates of approximately 500,000 cps, this would require approximately 55 hours of scanning time. Thus the straightforward approach outlined previously must be modified to reduce the number of counts required without increasing statistical noise. Most of the modified methods are based on computing the efficiencies of the individual detector elements (rather than all possible detector pairs) and then combining them to estimate the efficiency of the detector pairs. Details of these methods are beyond the scope of this text but are discussed in reference 12.

2. Correction for Random Coincidences

As discussed in Section A.9, random coincidences add a relatively uniform background across the reconstructed image, suppressing contrast and distorting the relationship between image intensity and the actual amount of activity in the image. There are two approaches to estimating the random coincidences so that they can be subtracted from the measured projection data: the *delayed window* method and the *singles* method.

In most PET scanners, the arrival time of each photon is recorded and "tagged" with an accuracy of approximately 2 nanoseconds. At the end of each clock cycle (typically 256 nanoseconds), the computer checks to see if any events have occurred, and if so, whether they occurred with arrival times within ΔT nanoseconds of each other, where ΔT is width of the coincidence timing window (typically 4 to 12 nanoseconds). If two photons arrive within this time interval, they are recorded as a valid event and the appropriate memory location corresponding to that particular detector pair is incremented by +1.

An estimate of the random coincidence rate can be obtained by delaying the coincidence timing window by a time that is much greater than its width. For example, the coincidence timing window might be delayed by 64 nanoseconds, for example, from 64 to 76 nanoseconds for a 12-nanosecond window. With this amount of time delay, only events that have arrival times separated by between 64 and 76 nanoseconds are accepted. No true (or scattered) prompt coincidences will be detected in the delayed window, because photons from the same decay will always arrive at the detectors within a few nanoseconds of each other. However, the rate of random coincidences will be the same in the delayed and undelayed windows because the rate at which uncorrelated photons strike the detector is the same for both windows. Thus the delayed window count provides an estimate of the number of random coincidence events. This number is subtracted from the total number of coincidence events for the detector pair. The correction occurs on-line in most PET systems and usually is transparent to the user.

The events recorded in the delayed window are not the same ones as are recorded in the undelayed window. Rather, the delayed window provides a separate and independent measure of the random event rate. Subtracting the number of random events recorded in the delayed window results in an increase in the statistical noise level for the measurement (see Chapter 9, Section C.1). Specifically, if N_{true} is the number of true coincidence events recorded, N_{scatter} the number of scatter coincidences, and N_{random} is the number of random coincidences subtracted from the total, the uncertainty in the remaining (true plus scatter) coincidences is

$$\sigma(N_{\text{true}} + N_{\text{scatter}}) = \sqrt{(N_{\text{true}} + N_{\text{scatter}}) + (2 \times N_{\text{random}})}$$

$$(18\text{-}15)$$

Thus even if accurate corrections can be made, the random coincidence rate should be minimized to avoid unduly increasing the statistical noise level of the image.

The second method for estimation of random coincidences is based on Equation 18-11. If the rate at which single (not coincidence) events occur in each detector is measured, and the coincidence timing window ΔT is known, then the rate of random coincidences for any pair of detectors can be computed. Because the rate of single events is typically at least an order of magnitude higher than the rate of coincidence events, the statistical noise level in the estimate of the number of random events is small in comparison with that in the measurement of the number of prompt coincidences, and the uncertainty in the remaining coincidences after random coincidences have been subtracted is given by

$$\sigma(N_{\text{true}} + N_{\text{scatter}}) \approx \sqrt{N_{\text{true}} + N_{\text{scatter}} + N_{\text{random}}}$$

$$(18\text{-}16)$$

This method requires that each detector module continuously monitors the rate at which it is detecting single events.

3. Correction for Scattered Radiation

Scattered radiation in PET imaging leads to a hazy background in the reconstructed images, generally more concentrated toward the center of the image. As with random coincidences, this leads to a decrease in image contrast and to errors in the quantitative relationship between image intensity and the amount of activity in the object. The fraction of scattered events in PET can be very high, especially in 3-D imaging of the abdomen, where it may be as high as 60% to 70%. This large value has three major causes. First, only one of the two annihilation photons needs to be scattered for a scatter coincidence to occur. Second, the energy resolution of PET detectors using dense scintillators such as BGO and LSO is inferior to NaI(Tl) detectors because of their lower light output. This requires the use of a wider pulse-height analyzer window to capture the photopeak events. Finally, the predominant mode of interaction in scintillators at 511 keV is Compton scattering, and many unscattered annihilation photons deposit less than 511 keV of energy in the detector. Thus to increase the detection efficiency for photons that undergo Compton scattering in the crystal, the analyzer window is widened even further to capture these events.

It is not possible to distinguish between scatter events in the body versus scatter events in the detector crystal on the basis of pulse amplitude. Therefore simple correction schemes based on dual-energy windows are far less successful in PET imaging than in SPECT imaging. Two main approaches currently are used for scatter correction in PET. The first approach uses information from the original scatter-contaminated image and transmission image (see Section D.4) to derive the correction. The emission image shows the distribution of the activity in the subject. The transmission image reflects the attenuation coefficient of the tissue. At 511 keV, virtually all attenuation is due to Compton scatter. Using these two images and computer modeling of photon interaction physics (see Chapter 6, Section C.3) with some simplifying assumptions, it is possible to derive an estimate of the underlying distribution of scattered events and their contribution to individual profiles. The estimated contribution of scattered radiation then is subtracted from the projection profiles and the reconstruction is repeated with the scatter-corrected data.

As described in Chapter 19, Section E, with the advent of hybrid PET/CT scanners, the scatter distribution also can be computed from the registered CT images, in which the CT image is used in place of a PET transmission image. This method works very well when all the sources of radioactivity that could lead to detected scatter events are contained within the FOV of the scanner. When large amounts of activity are just outside the FOV of the scanner, problems can arise. Another drawback of this approach is that it is computationally intensive.

A second method for scatter correction is based on an examination of projection profiles immediately outside the object. After correcting for random coincidences, the only events that should fall into these projection elements are those that are mispositioned because of scatter. Based on the premise that scatter is a low-frequency phenomenon with little structure, data from the tails of the projection profiles can be extrapolated using simple smoothly varying functions across the entire projection. Both gaussian and cosine functions have been used for this purpose. The extrapolated scatter distribution then is subtracted from the projections prior to image reconstruction. This method is rapid and, because it involves a direct measurement of scatter levels, it accounts for scatter from radioactivity outside the FOV. However, it can only approximate the true scatter distribution and, in situations in which the scatter distribution is complex, or when the object fills the whole FOV with no portion of the profile to examine outside the object, the technique may result in significant errors. These can range from a few percent for brain imaging to tens of percent at the heart-lung interface.

4. Attenuation Correction

Attenuation correction is by far the largest single correction in PET. Fortunately, the correction is relatively easy to derive. Consider a source located at a depth x inside an object of thickness T as shown in Figure 18-25. Both of the photons from an annihilation event in the source must be detected to record a valid event. Assuming that they are emitted in the appropriate directions, the probability that both photons will reach the detector is given by the product of their individual probabilities

$$\begin{aligned} P_{\text{det}} &= e^{-\mu x} \times e^{-\mu(T-x)} \\ &= e^{-\mu T} \end{aligned} \qquad (18\text{-}17)$$

where μ is the linear attenuation coefficient of tissue at 511 keV and is approximately 0.095 cm^{-1} for soft tissue, 0.12 to 0.14 cm^{-1} for bone, and 0.03 to 0.04 cm^{-1} for lung. Note that the probability that both photons will reach the detector is independent of the source location along the line joining the two detectors.

Equation 18-17 is similar to Equation 17-5 for the geometric mean in SPECT, except that it applies for all source distributions, whereas the geometric mean equation applies only for point or plane sources at a fixed depth, x. As was the case for SPECT, transmission measurements can be used to correct for attenuation in PET. In PET, two measurements are taken with a source located on a line joining each pair of coincidence detectors. The first

measurement, called the *blank scan,* is made without the subject in the scanner. The subject then is placed in the scanner and the measurement is repeated. This is known as the *transmission scan.* The attenuation correction factor A for a detector pair (i,j) is given by

$$A_{i,j} = \frac{Blank_{i,j}}{Trans_{i,j}} \qquad (18\text{-}18)$$

where $Blank_{i,j}$ and $Trans_{i,j}$ are the counts in the blank and transmission scans for the detector pair.

To obtain transmission data for all coincidence detector pairs, it is necessary to scan the transmission source around the scanning volume for both the blank and transmission scans. Typically, a rod source, with its length extending along the axis of the scanner, is placed in a holder near the surface of the scanner bore, and the holder rotates around the central axis so that data are acquired for all pairs. The most commonly used source material is ^{68}Ge (parent of ^{68}Ga, $T_{1/2} = 273$ days). The blank scan needs to be performed only once a day because it remains constant over a period that is short compared with the half-life of the radionuclide in the rod source. The transmission scan is performed prior to injecting the patient with the radiotracer. It is important that the patient not move between the transmission and emission scan. Otherwise serious artifacts can occur, including the appearance of areas of abnormally high or low radiotracer uptake.

Conceptually, the simplest approach is to obtain the transmission scan before injecting the radiotracer to be imaged. This eliminates any possible interference between the activity that is present for the two scans.

A second approach, called *postinjection transmission scanning,* is to perform the

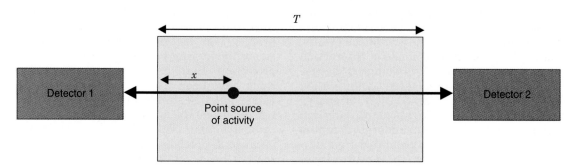

FIGURE 18-25 Parameters involved in the derivation of attenuation correction for PET (equation 18-17 in the text).

transmission scan immediately after the emission scan but while there is still radiotracer activity in the patient. This can save a significant amount of time, because the patient does not have to be on the table waiting for uptake after the transmission scan before the emission scan is performed. Another advantage of this approach is that it reduces the chances of patient motion and misalignment between the emission and transmission scans. However, it requires the ability to distinguish transmission events from emission events caused by residual radiotracer in the body. This is possible because the emission radiations generally are emitted from locations spread throughout the body, whereas the transmission radiations are emitted from a very small volume. Thus the transmission source irradiates only a small subset of detector pairs at any one time, and its counting rate in those irradiated pairs generally is much higher than the counting rate in the same detectors caused by emission radiations. To implement this method, the scanner must have the means to track the location of the transmission source and identify which detectors are being irradiated by it. A disadvantage of the postinjection approach is that the count-rate performance of the detectors must be sufficient to handle the emission and transmission activities simultaneously.

Finally, it is possible to acquire transmission and emission data at the same time. This approach, known as *simultaneous emission/transmission scanning,* is the most efficient way to use scanner time in many situations. As with postinjection transmission scanning, it is necessary to track the location of the rod source. Because the rod source irradiates only a small and known subset of detector pairs at any one time, emission data can be acquired simultaneously from the remaining nonirradiated pairs. For irradiated detector pairs, the counting rate from the transmission source is much higher than the emission counting rate, so the emission counts do not seriously affect the accuracy of the transmission data.

A disadvantage of simultaneous emission/transmission scanning is that the relatively "hot" transmission source can contribute random and scatter coincidence events to the emission data. Even with corrections, these events contribute to statistical noise and some degradation of image quality. For this reason, postinjection transmission scanning generally is the preferred approach.

Although these techniques work well, the widespread use of hybrid PET/CT scanners has significantly reduced the use of transmission scans using external radionuclide sources for attenuation correction. Instead, information from the CT scan is used to perform attenuation correction. The methods for CT-based attenuation correction are discussed in Chapter 19, Section E.

5. Dead Time Correction

Like all radiation detectors, PET detectors exhibit dead time and pile-up effects at high counting rates. The mispositioning of events caused by pile-up and possible approaches for minimizing pile-up described in the context of the gamma camera (see Chapter 14, Section A.4) apply as well to block detectors used for PET. Dead time corrections must be applied. Otherwise, the amount or concentration of radioactivity will be underestimated at high counting rates. Most PET scanners use empirical dead time models in which the observed counting rate as a function of radioactivity concentration is measured for a range of object sizes and at different energy thresholds. The resulting data are then fit with paralyzable or nonparalyzable dead time models (see Chapter 11, Section C.2). Some systems apply a global dead time correction factor for the system, whereas others apply corrections to individual pairs of detector modules.

Dead time losses are dominated by the single-channel counting rate, which are much higher than the coincidence counting rate. Corrections can be as large as a factor of 2, although generally it is desirable to keep them below this level. Situations in which the corrections can be large include first-pass cardiac studies, imaging studies near the bladder when there are high levels of excreted radioactivity, and studies with very short-lived radiotracers, such as ^{15}O, which require high starting levels of activity to maintain adequate counting statistics over the course of a study.

6. Absolute Quantification of PET Images

All of the corrections described earlier are applied to the projection or sinogram data prior to reconstruction of the image. If accurately applied, after reconstruction, the voxel intensity in the image will be directly proportional to the amount of radioactivity in that voxel. Calibration to absolute concentrations of radioactivity usually is accomplished by scanning a cylinder containing a uniform solution of a known concentration. The calibration factor CAL is defined as

$$CAL = \text{counts per pixel/radionuclide}$$

$$\text{concentration in cylinder (kBq/cm}^3)$$

$$(18\text{-}19)$$

The voxel intensity in the image of the subject is divided by the calibration factor to obtain calibrated images in kBq/cm³. To obtain the absolute amount of activity in the voxel (becquerels), one would have to multiply this result by the voxel volume.

Quantification in PET is subject to the same partial-volume effects as were discussed in Chapter 17, Section B.5, for SPECT imaging. This effect occurs for structures of size smaller than 2 × FWHM of the imaging system. Structures that have dimensions smaller than this will have their radioactivity concentrations either overestimated or underestimated depending on the regional distribution of the radioactivity. An example of where this becomes important is in the quantification of activity in thin layers of cerebral cortex.

E. PERFORMANCE CHARACTERISTICS OF PET SYSTEMS

A set of standardized methodologies have been agreed upon for measuring important parameters that describe the performance of PET scanners. Published standards exist both for clinical whole-body scanners[13] and for small-animal scanners.[14] In a similar fashion to the tests described for SPECT systems (see Chapter 17, Section C), measurements of spatial resolution and sensitivity are based on carefully designed acquisition and reconstruction protocols that enable some degree of comparison of performance across different systems. Protocols for measuring the fraction of scattered and random coincidences, losses resulting from dead time, image quality, and quantitative accuracy of attenuation and scatter corrections also are prescribed. Full details can be found in references 13 and 14.

A commonly quoted performance parameter, incorporated in these standards, is the *noise equivalent counting rate* (NECR). This parameter, which is specific to PET systems, accounts for the additional statistical noise introduced by the correction for random and

scattered coincidences. The NECR is defined as the equivalent counting rate that gives rise to the same statistical noise level as the observed counting rate after random and scattered coincidences have been corrected for.

A PET scanner measures a prompt coincidence rate (R_{prompt}) that comprises true coincidences (R_{true}), scatter coincidences ($R_{scatter}$) and random coincidences (R_{random}). The rate of true coincidence events is given by

$$R_{true} = R_{prompt} - R_{scatter} - R_{random} \quad (18\text{-}20)$$

The NECR is defined as[13]

$$NECR = \frac{R_{true}^2}{R_{true} + aR_{scatter} + bR_{random}}$$

$$(18\text{-}21)$$

It usually is plotted as a function of activity, as both the rate of random coincidences and count losses caused by dead time are activity dependent. The constant a is the fraction of the projection that is occupied by the object being imaged. The constant b is equal to 1 if the singles method is used for randoms estimation and 2 if the delayed windows method is used (see Section D.2).

The prompt and random coincidence rates are measured. The rate of scattered coincidences is estimated by imaging a phantom that contains a line source. Events in the projection data that do not intersect the known location of the line source (after random coincidences already have been subtracted) must have been scattered.

It has been demonstrated that the NECR is roughly proportional to the square of the signal-to-noise ratio of the reconstructed activity values when the object is a cylinder with a uniform activity concentration.

Figure 18-26 shows typical NECR curves for a particular PET scanner and phantom. At higher activities, the NECR actually decreases because the rate of random coincidences increases as roughly the square of the activity, and dead time losses also reduce the observed counting rate. Often the peak NECR rate, and the activity concentration at which it is achieved, is reported. However, the values strongly depend on the size of the object that is imaged, and other factors such as the energy and timing windows, and the activity distribution within the phantom. Nonetheless, assuming the phantom used to acquire the NECR data is a reasonable approximation of the object that is to be imaged in a particular clinical or research task, the NECR

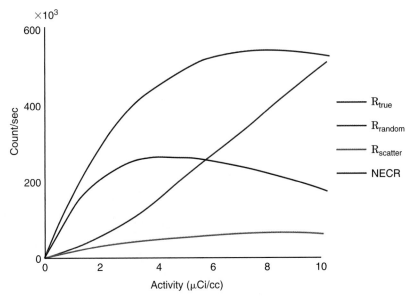

FIGURE 18-26 Example of various coincidence counting rates and noise equivalent counting rate (NECR) for a clinical whole-body scanner. These data predict that for the phantom used in this study, the best signal-to-noise (corresponding to the peak of the NECR curve) in the reconstructed image would be achieved with an activity concentration of ~4 μCi/cc (~150 kBq/cc).

provides a useful guide for estimating the activity concentrations that provide the highest signal-to-noise ratio images. In some cases, it may not be possible to reach this activity concentration because of radiation dosimetry considerations (see Chapter 22). Measurements of NECR for a well-defined phantom can be useful for comparing the performance of different scanners, and estimations of NECR also can help guide the design and development of new PET scanners.

F. CLINICAL AND RESEARCH APPLICATIONS OF PET

PET has major clinical applications in oncology, neurology, and cardiovascular disease. ^{18}F-fluorodeoxyglucose (FDG) is by far the most commonly used radiotracer for clinical studies. The uptake of FDG reflects glucose metabolism in tissues. Many pathologic conditions can cause regionally specific alterations in glucose metabolism that can be detected using FDG-PET. PET's most widespread application has been for the detection and staging of cancer, for which whole-body FDG studies have become an important tool in staging patients and for deciding patient management. A whole-body FDG study is shown in Figure 18-27.

FDG also is used diagnostically in conjunction with blood flow tracers such as

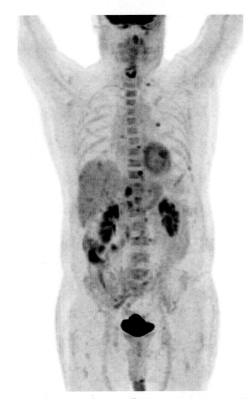

FIGURE 18-27 Whole-body ^{18}F-fluorodeoxyglucose PET scan (injected activity 370 MBq) of a patient with cancer, showing widespread metastatic disease (dark spots). The scan took 14 minutes to acquire (7 overlapping bed positions to cover thorax and abdomen, 2 minutes per bed position), with imaging commencing 60 minutes postinjection. (*Courtesy Dr. Paul Shreve, Spectrum Health, Grand Rapids, MI and Siemens, Medical Solutions USA, Inc., Knoxville, TN.*)

^{13}N-ammonia or ^{82}RbCl to evaluate myocardial viability and stratify patients with coronary artery disease with regard to bypass surgery. In the brain, PET is used diagnostically in a range of neurodegenerative diseases (Alzheimer's disease, Parkinson's disease) and dementia, for epilepsy, neurodevelopmental disorders and in psychiatric disorders. Metabolic FDG images of a patient with Alzheimer's disease compared with those of a normal control are shown in Figure 18-28.

Many other PET radiotracers are used for research studies and are being developed for future clinical use. These include radioligands that bind to specific receptors systems in the brain, and radiotracers that target cell-surface molecules specific to certain types of tumors. Positron-emitting radionuclides also have been used to radiolabel cells (e.g., stem cells) and drug delivery vehicles such as nanoparticles. Many of these radiotracers are first evaluated in animal models using a dedicated animal PET scanner. Figure 18-29 shows an image of the binding of the radioligand ^{11}C-raclopride to dopamine receptors in the rat brain.

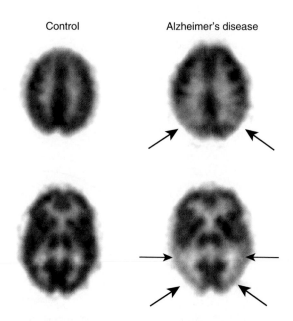

Control   Alzheimer's disease

FIGURE 18-28 Transaxial image slices showing ^{18}F-fluorodeoxyglucose uptake at two different levels of the brain in a normal volunteer (control) and in a patient at an early stage of Alzheimer's disease. Data acquisition times were ~30 minutes and injected doses were 370 MBq (10 mCi). *Arrows* indicate metabolic deficits in the patient's images. This distinct pattern of reduced metabolism is seen in all patients with Alzheimer's disease and increases in severity and extent are seen as the disease progresses.

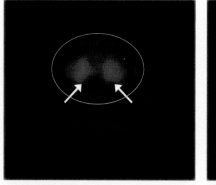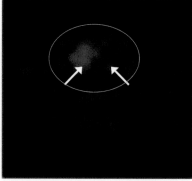

FIGURE 18-29 Coronal image of rat brain acquired on a small-animal PET scanner following injection of ^{11}C-raclopride. The location of the brain is indicated by the ellipse. This radiotracer binds to the dopamine receptors in the brain, which are located primarily in the striatum (*arrows*). Images were acquired before (*left*) and after (*right*) a pharmacologic intervention that damages the dopaminergic neurons on one side of the brain. A clear reduction in the binding of the radiotracer is observed on that side.

REFERENCES

1. Hoffman EJ, Phelps ME: Positron emission tomography: Principles and quantitation. In Phelps ME, Mazziotta JC, Schelbert HR, editors: *Positron Emission Tomography and Autoradiography: Principles and Applications for the Brain and Heart*, New York, 1986, Raven Press, pp 237-286.
2. Lewellen TK: Recent developments in PET detector technology. *Phys Med Biol* 53:R287-317, 2008.
3. Casey ME, Nutt R: A multicrystal two-dimensional BGO detector system for positron emission tomography. *IEEE Trans Nucl Sci* 33:460-463, 1986.
4. Lewellen TK, Kohlmyer SG, Miyaoka RS, et al: Investigation of the performance of the General Electric ADVANCE positron emission tomograph in 3D mode. *IEEE Trans Nucl Sci* 43:2199-2206, 1996.
5. Brambilla M, Secco C, Dominietto M, et al: Performance characteristics obtained for a new

3-dimensional lutetium oxyorthosilicate-based whole-body PET/CT scanner with the national electrical manufacturers association NU 2-2001 standard. *J Nucl Med* 46:2083-2091, 2005.

6. De Jong HWAM, van Velden FHP, Kloet RW, et al: Performance evaluation of the ECAT HRRT: an LSO-LYSO double layer high resolution, high sensitivity scanner. *Phys Med Biol* 52:1505-1524, 2007.

7. MacDonald L, Edwards J, Lewellen T, et al: Clinical imaging characteristics of the positron emission mammography camera: PEM Flex Solo II. *J Nucl Med* 50:1666-1675, 2009.

8. Constantinescu CC, Mukherjee J: Performance evaluation of an Inveon PET preclinical scanner. *Phys Med Biol* 54: 2885-2899, 2009.

9. Rowland DJ, Cherry SR: Small-animal preclinical nuclear medicine instrumentation and methodology. *Sem Nucl Med* 38:209-222, 2008.

10. Larobina M, Brunetti A, Salvatore M: Small animal PET: A review of commercially available systems. *Curr Med Imag Rev* 2:187-192, 2006.

11. Bendriem B, Townsend DW: *The Theory and Practice of 3D PET*, Netherlands, Kluwer, 1998, Dordrecht.

12. Badawi RD, Marsden PK: Developments in component-based normalization for 3D PET. *Phys Med Biol* 44: 571-594, 1999.

13. Performance Measurements of Positron Emission Tomographs: *National Electrical Manufacturers Association (NEMA) Standards Publication NU2-2007*, Rosslyn, VA, 2007, NEMA.

14. Performance Measurements of Small Animal Positron Emission Tomographs: *National Electrical Manufacturers Association (NEMA) Standards Publication NU4-2008*, Rosslyn, VA, 2007, NEMA.

BIBLIOGRAPHY

Additional general references are the texts indicated in references 1 and 11. Also pertinent are the following:

Phelps ME, editor: Molecular Imaging and its Biological Applications, New York, 2004, Springer-Verlag.

Zanzonico P: Positron emission tomography: A review of basic principles, scanner design and performance, and current systems. *Sem Nucl Med* 34:87-111, 2004.

Hybrid Imaging: SPECT/CT and PET/CT

Virtually all modern positron emission tomography (PET) scanners and an increasing number of single photon emission computed tomography (SPECT) systems are integrated with an x-ray computed tomography (CT) scanner. These *hybrid imaging systems* are capable of acquiring PET or SPECT images, along with spatially registered CT images, in quick succession. This chapter discusses the features of these hybrid systems and describes how the CT scans not only provide anatomic context to improve diagnostic interpretation of nuclear medicine studies, but also can be used as the basis for performing corrections for photon attenuation and scatter in PET and SPECT.

A. MOTIVATION FOR HYBRID SYSTEMS
....

Both SPECT and PET provide functional information, using radiotracers that are designed to measure physiologic or metabolic parameters, or that bind or interact with specific molecular targets on the cell surface, or within cells. Although for some radiotracers, the regional anatomy is obvious (e.g., cardiac perfusion studies), there are many cases in which the nuclear medicine study does not provide much in the way of anatomic information. Even in cases in which there is some anatomic information, the spatial resolution is poor compared with techniques such as CT and magnetic resonance imaging (MRI), both of which can provide images of the human body at a resolution of 1 mm or better.

Because clinical decisions may depend not only on detecting a signal (an increase or decrease in the accumulation of a radiotracer for example), but also on knowing precisely where that signal originated, it has been common practice for several decades to complement tomographic nuclear medicine scans with CT or MRI scans. Typically, these scans were acquired on separate instruments, possibly days apart. Thus correlating information in the two studies was hampered by the difficulty of spatially registering the images and the effect of any changes in the patient's condition (especially during active treatment) during the time that elapsed between the two studies.

The power of hybrid imaging systems, in which a PET or SPECT scanner is integrated with a CT scanner, is that the two scans are acquired in quick succession, and thus the data can be considered to be in fairly good spatial and temporal registration (see Sections C and D). With good spatial registration, it also becomes possible to consider using the CT scan, which provides a map of tissue attenuation values (Section B), to compute the corrections for photon attenuation and scatter for the PET and SPECT studies (see Section E). It also is more convenient for the patient and more efficient from a scheduling perspective, as both datasets are acquired in a single imaging session. Hybrid systems also are common in small-animal and preclinical imaging studies. Just as in the clinical environment, the anatomic information provided by CT can be helpful both for interpreting and quantitatively analyzing PET and SPECT studies.

To date, the focus has been largely on the integration of PET and SPECT with x-ray CT. As described in Sections C and D, from a technical perspective it has been relatively straightforward to integrate these modalities with only minor modifications to the hardware for either system. Much of the integration in hybrid PET/CT or SPECT/CT systems occurs at the software and system control level. Hybrid systems that combine PET and MRI also are under development. MRI has advantages over CT for some structural imaging applications because of the high contrast it can produce in soft tissues; however, integration is much more challenging, and the cost of a combined device could be high. Nonetheless, human scanners have been installed and a brief overview of the status of hybrid PET/MRI systems is given in Section F.

B. X-RAY COMPUTED TOMOGRAPHY

X-ray CT is a form of transmission computed tomography that is based on the reconstruction techniques described in Chapter 16. It shares many of the features of transmission scanning that already have been described for attenuation correction for SPECT (see Chapter 17, Section B.3) and PET (see Chapter 18, Section D.4) imaging. However, instead of an external radionuclide source, a CT scanner uses an x-ray tube that generates a high flux of x rays in a relatively narrow beam that are transmitted through the body.

Because the flux of x rays is so high, images with excellent counting statistics can be acquired very rapidly, on the order of seconds. This section provides a very brief overview of x-ray CT to aid in the discussion of hybrid PET/CT and SPECT/CT systems that follows. More details on x-ray CT can be found in Reference 1.

1. X-ray Tube

The x-ray tube is a vacuum tube containing a cathode, which consists of a filament (helical coil of tungsten wire) through which a current is passed. As the filament is heated, electrons are liberated, and are accelerated and focused by an applied bias voltage towards a high-density metal (typically tungsten) anode (Fig. 19-1). When the electrons interact in the anode material, they produce both continuous bremsstrahlung radiation (see Chapter 6, Section A.1) as well as discrete characteristic x rays at energies that correspond to the electron binding energies of the anode material (see Chapter 2, Section C.2).

The current passing though the filament controls the number of electrons emitted from the filament, and thus the current flowing through the x-ray tube (the *tube current*) as well as the flux of x rays produced when these electrons strike the anode. The tube current typically is approximately 10% to 20% of the current passed through the heating filament.

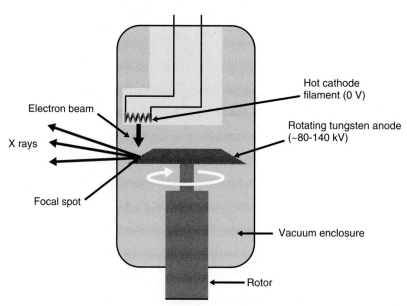

FIGURE 19-1 Schematic of an x-ray tube with a rotating anode. An alternating current is passed through the cathode filament, heating it up and releasing electrons. The electrons are accelerated by the potential difference between the cathode and anode, and strike the tungsten anode producing x rays. The anode rotates such that the location of the focal spot on the tungsten anode changes over time to avoid melting the anode at high x-ray tube currents. For further details, see text.

The voltage applied between the coil and the anode determines the energy spectrum of the x rays. The highest (peak) x-ray energy that can be produced equals the x-ray tube bias voltage setting in kV. Thus an applied voltage of 120 kV (often denoted as kVp—peak voltage in kilovolts) can produce x rays with a maximum energy of 120 keV. By varying the x-ray tube current and bias voltage, the intensity and energy spectrum of the x-ray beam transmitted through the subject can be controlled. In general, higher energy x rays are used for penetrating larger thicknesses of tissue. On the other hand, image contrast is typically better for lower energy x rays. Thus there is a trade-off in choosing the appropriate x-ray energy that also must carefully consider the radiation dose given to the tissues.

The x-ray beam is filtered by passing the x rays through a sheet of metal (typically 2.5 mm of aluminum for the x-ray energies used in CT) prior to reaching the patient to remove the lowest energy x rays. Such x rays have a very small probability of penetrating through the body, and therefore contribute mostly radiation dose to the patient, without providing any useful signal. The x-ray beam also is collimated such that x rays irradiate only the tissue slices of interest. This is one fundamental way in which x-ray imaging differs from nuclear medicine imaging. With x rays, radiation dose can be limited to the areas being imaged, whereas in nuclear medicine, the radiotracer generally distributes throughout the body and delivers a radiation dose everywhere. Radiation dosimetry is discussed in Chapter 22.

Figure 19-2 shows the typical spectrum from an x-ray tube as the high voltage is varied. Typical values for the x-ray tube settings for CT examinations are a voltage of 120 kV, and a tube current ranging between 120 and 300 mA, depending on the amount of tissue the x rays must penetrate. A lower tube current is used for CT imaging of the head and neck, and a larger current for the abdomen. The number of x rays that are used to form an image is proportional to the product of the x-ray tube current and the exposure time for a given tissue slice and is given in units of milliamps × seconds, commonly abbreviated to mAs for short. The time for which a slice of tissue is exposed to x rays in a CT examination typically is on the order of 0.5 to 2 seconds.

One factor that ultimately limits the spatial resolution of x-ray techniques is the spot size of the electron beam on the anode target (also called the *focal spot*). For clinical CT

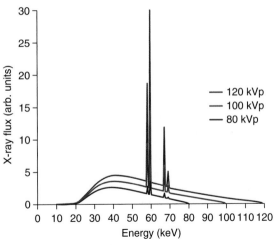

FIGURE 19-2   Representative x-ray spectra from a tungsten anode tube operated at three different bias voltages. Low-energy x rays are removed by the use of an aluminum filter. The sharp spikes in the spectrum correspond to the production of characteristic x rays from the tungsten anode. The continuous spectrum arises from bremsstrahlung radiation. Changes in the x-ray tube current, for a given bias voltage, do not change the shape of the spectrum, just the magnitude of the x-ray flux. Thus doubling the tube current doubles the x-ray flux at all energies.

scanners, the typical focal-spot size is on the order of 0.5 to 1 mm. To avoid overheating of the anode by the intense electron beam, the anode is rotated continuously such that the beam location on the anode material itself is constantly changing.

2. X-ray Detectors

The detectors used in a CT scanner typically are scintillator materials [e.g. CsI(Tl) or ceramic scintillators] read out by silicon photodiodes (see Chapter 7, Section C). Some older or low-cost CT systems employ detectors consisting of thin ionization chambers containing high-pressure xenon gas (Chapter 7, Section A.2). Because the x-ray photons arrive at the detectors at very high rates, individual pulses cannot be separated for each interacting x-ray photon. Therefore, unlike nuclear medicine detectors that operate in pulse mode, x-ray detectors operate in an integrating current mode. That is, they produce an output current proportional to the x-ray flux impinging on them.

Older single-slice CT scanners consisted of a single linear detector array, with elements measuring, for example, 1 mm in the transaxial direction and 15 mm in the axial direction. Slice thickness was controlled by collimating the x-ray beam and therefore could be considerably smaller than the physical height of the detector in the axial direction. Slice thicknesses of 1 to 5 mm were typical. To acquire multiple image slices, the patient bed was stepped incrementally through the system.

Two-dimensional (2-D) detector arrays are used in most modern scanners, allowing multiple slices through the body to be acquired simultaneously. To ensure complete sampling for tomographic reconstruction, the patient bed usually is translated while the detectors rotate about the patient, creating a helical trajectory. Individual detector sizes are small, typically in the range of 0.25 to 1.25 mm, in order to achieve high spatial resolution. Systems typically come with 16, 32, or 64 rows of detectors, allowing this many slices to be obtained in a single rotation of the system. Each row may consist of approximately 1000 detector elements; thus modern multi-slice CT scanners can have a total of approximately 10,000 to 100,000 individual detector elements. Signals from smaller detectors sometimes may be binned together to improve signal-to-noise ratio or reduce dose at the expense of spatial resolution. For example, signals from 2×2 detectors of size 0.5×0.5 mm^2 may be combined to create a "virtual" detector measuring 1×1 mm^2.

CT detectors have very high efficiency for the x-ray energies of interest (40-140 keV). Virtually all x-rays incident on a detector element interact and produce a signal. However, small gaps between the individual detector elements cause the overall efficiency to be less than 100%.

3. X-ray CT Scanner

The CT scanner consists of the x-ray tube and a step-up high-voltage power supply, along with the detector, mounted on a rotating gantry (Fig. 19-3). Low-voltage power is supplied to the gantry via slip-ring technology rather than by cables. The slip rings consist of conductive brushes that slide over the surface connections providing continuous electrical contact and allowing the gantry to rotate continuously. Data can be taken off the gantry in the same way or, more commonly, by wireless transmission. This enables very fast data acquisition as the patient moves through the gantry while it rotates at speeds of 1 to 3 revolutions per second. A large axial stack of images can be acquired in just a few seconds.

4. CT Reconstruction

The projection data for CT typically are collected as the scanner performs a helical trajectory, where the patient bed is translated through the scanner while the x-ray tube and detector rotate. They are readily sorted into stacks of sinograms with each sinogram containing the data for a single transaxial section through the patient. Images commonly are reconstructed using the same types of filtered backprojection methods discussed in Chapter 16, Section B.

For CT, the measured projection data correspond to the transmission of x rays through the subject. For a given line of response (a ray connecting the x-ray focal spot and a detector element), the measured intensity, I, is

$$I = I_0 e^{-\sum_i \mu_i \Delta x_i} \qquad (19\text{-}1)$$

where I_0 is the initial x-ray intensity directed along the line of response, μ_i is the linear attenuation coefficient (Chapter 6, Section D) for the ith pixel and Δx_i is the pathlength of the line of response through the ith pixel. X-ray CT seeks to reconstruct the linear attenuation coefficient μ; therefore the measured data must be transformed prior to reconstruction

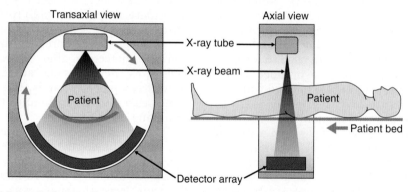

FIGURE 19-3 Schematic showing the components and geometry of a CT scanner. The x-ray tube and detector rotate in the transaxial plane while the patient bed translates along the axial direction, resulting in a helical trajectory of the x-ray beam around the patient. Multiple image slices can be rapidly acquired.

by taking the logarithm of the ratio of the incident x-ray intensity with respect to the measured intensity. The transformed projection, p, used in the reconstruction algorithm is therefore given by:

$$p = \log \frac{I_0}{I} = \sum_i \mu_i \Delta x_i \qquad (19\text{-}2)$$

The reconstruction of these projections yields images in which the pixel values correspond to μ, the linear attenuation coefficient of the tissue in that pixel. Thus a CT scan provides an image of the linear attenuation coefficients inside the body.

Given the x-ray energies used and the relatively low atomic number of most tissue constituents, the majority of x-ray interactions in the body are by Compton scatter (see Chapter 6, Section C.3). Because Compton scatter depends primarily on tissue density (see Chapter 6, Section D.1), the pixel value in CT images is roughly proportional to tissue density. Higher-density materials such as bone show up with high pixel values, whereas lower-density tissues such as lung show up with low pixel values.

Note that the reconstructed values represent some average of μ across the range of energies present in the x-ray beam (see Fig. 19-2) and therefore depend strongly on the x-ray tube voltage and filtration. For this reason, pixel values in CT images usually are expressed on a normalized scale, with respect to the values for water, as follows:

$$CT\ number(x, y) = 1000 \times \frac{\mu(x, y) - \mu_{water}}{\mu_{water}}$$

$$(19\text{-}3)$$

For soft tissues (which are fairly similar to water in density and attenuation coefficient),

this normalization results in roughly equivalent pixel intensities in the reconstructed CT image, independent of x-ray tube and filtration settings.

CT numbers calculated from Equation 19-3 are called Hounsfield units (HU), named after the inventor of medical CT scanning. By definition, water has a value of 0 HU, air spaces in the lungs have values that can approach −1000 HU, and bone can have values as high as 3000 HU.

CT projection data are subject to the same sampling requirements as nuclear medicine data (see Chapter 16, Section C.1). Angular projection data are captured as the system rotates and acquired at ~1000 to 2500 projection angles. The linear sampling distance is governed by the detector element spacing and typically is in the range of 0.25 to 1 mm. Thus the sampling of projection data in CT datasets is sufficient to support much higher spatial resolution than is obtained in nuclear medicine studies; however, the size of the projection datasets also is much larger. In practice, the spatial resolution obtained in CT is limited by the focal-spot size of the x-ray tube and also by radiation dose considerations. A reconstructed spatial resolution of approximately 1 mm is typical in many clinical scenarios.

Figure 19-4 shows examples of reconstructed CT images of the human body. In this study, a large number of transverse slices were obtained to produce a volumetric dataset covering the entire abdominal region. A transverse and coronal view are shown. In many clinical studies, native tissue contrast is augmented by the injection or ingestion of a *contrast agent*. These agents, typically containing high-Z elements such as iodine (injected agents) or barium (ingested agents), serve to increase x-ray attenuation in tissues

FIGURE 19-4 Examples of volumetric clinical CT images. *A*, Transaxial section at level of liver. *B*, Coronal section, obtained by slicing through a contiguous stack of transaxial slices, showing liver, descending aorta, intestines, and other structures. (*Courtesy GE Healthcare, Waukesha, WI.*)

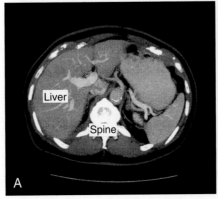
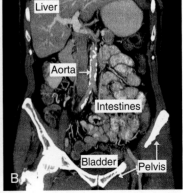

in which they are present. They are used to increase contrast in the vasculature, in highly perfused organs and tissues with leaky blood vessels (e.g., tumors), to detect internal bleeding, or to highlight the digestive and excretory tracts of the body.

C. SPECT/CT SYSTEMS

1. Clinical SPECT/CT Scanners

Early attempts to develop hybrid SPECT/CT systems were pioneered by Dr. Bruce Hasegawa and co-workers in the early 1990s.[2] One concept was to develop a single detector, based on high-purity germanium, that could be used both to efficiently detect the gamma-ray emissions from radionuclides as well as x rays. Although a prototype system was developed, the challenge of devising a single detector technology that could span the quite differing requirements of SPECT (e.g., good energy resolution) and CT (e.g., high x-ray flux) was considerable. Therefore approaches that integrated existing SPECT and CT systems were pursued. Figure 19-5 shows the first such system, developed at the University of California–San Francisco, by combining a single-headed SPECT system with a single-slice CT scanner.[3]

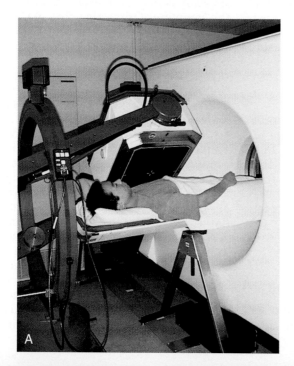

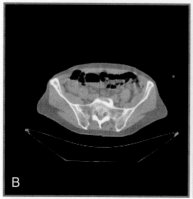 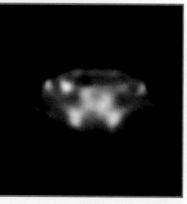

CT   SPECT

FIGURE 19-5 *A,* First prototype SPECT/CT scanner, developed at the University of California–San Francisco. On the *left* is a single-headed rotating gamma camera, which is placed immediately adjacent to a single-slice x-ray CT scanner. *B,* Images obtained with this system show the distribution of the radiopharmaceutical ^{111}In-prostascint (SPECT) in a patient with prostate cancer and the corresponding anatomic image slice (CT). (*Courtesy of the late Dr. Bruce Hasegawa, University of California–San Francisco, San Francisco, CA.*)

Modern hybrid SPECT/CT scanners typically consist of a dual-headed SPECT system combined with a multi-slice CT system. Compared with earlier systems, the detectors and electronics now are integrated in a single gantry with a single computer controlling data acquisition. Typically, the imaging fields-of-view of the two modalities are offset from each other in the axial direction; thus scans are acquired sequentially, with a displacement of the patient bed between the CT scan and the SPECT scan to place the organ or region of interest in the appropriate scanner field-of-view. The SPECT detector heads can be placed in different orientations (e.g., 90° and 180° configurations; see Fig. 13-12) for different types of studies. Integration of the SPECT system with a CT system typically does not result in any compromise in SPECT performance. The major practical issue is the increase in physical size of the system that may cause siting challenges in departments designed for more compact SPECT-only systems. High-end CT scanners in tandem with a dual-headed SPECT camera may require rooms as large as 7.5 m × 5 m.

A wide range of different CT systems are employed in hybrid SPECT/CT scanners. Some manufacturers use relatively simple CT scanners with a single or small number of slices, and with the detectors mounted on a slowly rotating gantry that does not employ slip-ring technology. In other cases, state-of-the-art 64-slice CT scanners with fast helical acquisition have been integrated with SPECT. The choice of system at any given institution depends on a number of factors, including intended clinical use, whether the purpose of the CT is to provide diagnostic-quality scans or primarily for attenuation correction (see Section E), cost, throughput, and any site limitations.

Two commercially available SPECT/CT systems are shown in Figure 19-6. The system in Figure 19-6A combines a dual-headed SPECT system in tandem with a 16-slice CT scanner (1-, 2-, and 6-slice versions also are available). The axial separation between the SPECT and CT field-of-view is 136 cm. The SPECT system consists of two gamma camera detectors with a 9.5-mm-thick 59-cm × 44-cm NaI(Tl) scintillator crystal read out by 59 50-mm-diameter hexagonal photomultiplier (PM) tubes. This provides a detector head with an active detector area 53 cm transaxially by 39 cm axially. The energy resolution (full width at half maximum) of the detectors is approximately 9.5% at 140 keV. When used with a low-energy, high-resolution collimator to image a source at a distance of 10 cm from the collimator face, the sensitivity is reported to be ~200 cpm/μCi. The reconstructed spatial resolution using the same collimator is approximately 10 mm. The x-ray tube has a focal spot measuring 0.8 mm × 0.7 mm. The tube voltage can be selected as 80, 110, or 130 kVp and the tube current varied between 20 and 345 mA. The detectors consist of a ceramic scintillator with 16 rows of detector elements by 1472 detectors per row. A complete rotation can be performed in as little as 0.5 second, collecting data from 16 slices simultaneously with a slice thickness of either 0.6 mm or 1.2 mm.

A second example is shown in Figure 19-6B. This system employs two gamma camera heads, each with an active area of 54 cm in the transaxial direction and 40 cm in the axial direction. The two heads can be configured in a variety of orientations with respect to each other. In this system, the x-ray CT is coplanar with the SPECT detector heads; thus the SPECT and CT imaging fields-of-view coincide with each other. The x-ray tube operates with a tube voltage of up to 120 kVp and a tube current of 5-80 mA. The focal spot size is 0.4 mm.

The flat panel x-ray detector uses columnar CsI(Tl) scintillator coupled onto an amorphous silicon photodetector with detector pixels measuring 0.2 × 0.2 mm. The x-ray detector measures 40 cm (transaxial direction) by 30 cm (axial direction) and can acquire images as fast as 60 frames per second. The x-ray system rotates through 360° in 12 seconds and provides CT images over a 14-cm axial field-of-view with a slice thickness that can be varied from 0.3 to 2 mm. Because of the large axial field-of-view of this detector, the data collected are no longer approximated well by 2-D parallel slices, and 3-D cone-beam reconstruction methods are used. These algorithms are entirely analogous to the 3-D reconstruction methods described for cone-beam and pinhole collimators in SPECT in Chapter 16, Section E.2.

SPECT/CT systems are being employed for a range of clinical applications. Studies in the literature have shown that SPECT/CT can improve both the sensitivity and specificity of nuclear medicine studies, and also reduce the incidence of indeterminate scans.[4] The advantages of SPECT/CT versus SPECT alone have been shown for a range of oncologic applications (e.g., thyroid cancer, skeletal cancers and neuroendocrine cancers arising in organs

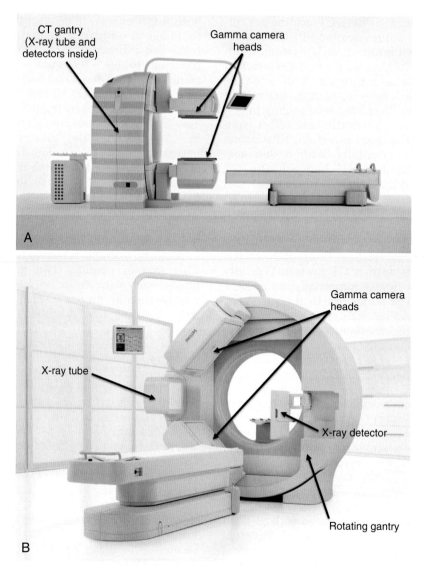

FIGURE 19-6   Examples of two clinical SPECT/CT scanners. *A*, The two gamma camera heads are mounted on a rotating gantry that is separate from the high-speed slip-ring CT system. *B*, Two gamma camera heads and the x-ray tube and detector are mounted on a common rotating gantry. See text for further details on these two systems. (*A, Courtesy Siemens Medical Solutions, Malvern, PA. B, Courtesy Philips Healthcare, Andover, MA.*)

such as the pituitary), and initial data on the use of SPECT/CT in the heart and brain has appeared. An example of a SPECT/CT imaging application is shown in Figure 19-7. SPECT/CT also has potential as a useful tool for computing individualized radiation dosimetry for radionuclide therapies and is increasingly used in the research setting for evaluating the biodistribution of new radiotracers.

2. Small-Animal SPECT/CT Scanners

SPECT/CT systems also have been developed for small-animal imaging. An example of such a system is shown in Figure 19-8A. The CT

components of these systems typically use lower-voltage (up to 80 kVp) and lower-flux x-ray tubes compared with clinical systems, because of the smaller size of the objects being imaged. Generally, they also use x-ray tubes with a smaller focal-spot size (typically 50-100 microns), and higher-resolution detectors to improve spatial resolution. Resolution also can be improved by increasing the object-to-detector distance and magnifying the projections on the detector when the subject is significantly smaller in dimensions than the detector area.

Most systems employ a 2-D x-ray detector that co-rotates with the x-ray tube around the

SPECT   Fused SPECT/CT   CT

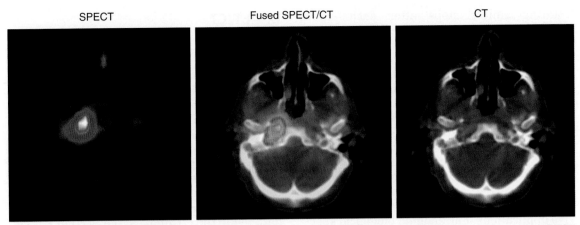

FIGURE 19-7 SPECT/CT image of a patient with osteomyelitis (infection of the bone) behind the right eye. The SPECT scan was performed following the injection of 22.2 MBq (600 μCi) of ^{111}In-labeled white blood cells. The helical CT scan used 140 kVp x rays with an exposure of 30 mAs per slice. A single transaxial slice is shown. The SPECT scan shows accumulation of the radiolabeled white blood cells at the site of infection, and the fused SPECT/CT scan (SPECT in *pseudocolor*, CT in *grayscale*) allows the precise anatomic location of the infection to be determined. Total acquisition time for the two scans was approximately 20 minutes. (*Courtesy GE Healthcare, Waukesha, WI.*)

subject. The resulting projection data correspond to a cone-beam geometry (similar to that obtained from a cone-beam or pinhole collimator in SPECT; see Chapter 16, Section E.2) and must be reconstructed using appropriate 3-D algorithms. Some systems employ a helical scanning mode similar to those used in clinical systems.

Currently, small-animal SPECT/CT systems do not use slip-ring technology and thus rotate more slowly than clinical CT systems. They can undergo only one rotation before the gantry must be reversed because of the attached cables that supply the high voltage and read out the detector. Data are acquired either in "step-and-shoot mode"

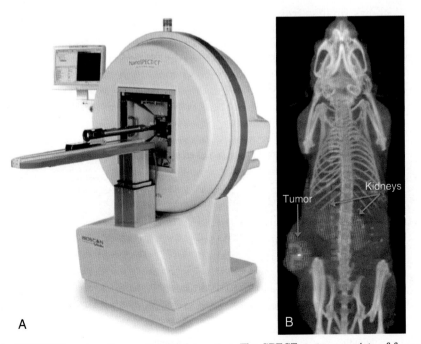

FIGURE 19-8 *A,* SPECT/CT system for small-animal imaging. The SPECT system consists of four gamma camera detector heads with multi-pinhole collimators. *B,* SPECT/CT images of an ^{111}In-labeled radiotracer targeted to a tumor in a mouse model of cancer. The SPECT image (*pseudocolor*) shows accumulation of the radiotracer in the tumor, as well as excretion via the kidneys. The CT image (*grayscale*) shows the underlying mouse anatomy (*A, Courtesy Professor Stephen Mather, Barts Cancer Institute, London.*)

(rotating gantry stops after each small angular increment and data are then acquired) or in "continuous mode" (sequential projection angle images acquired continuously while gantry slowly rotates around subject). The former produces higher resolution (no blurring caused by rotation during acquisition), whereas the latter is faster (data is acquired continuously and there is no need to start and stop rotation). Typical scan times range from tens of seconds to several minutes for a complete 180° or 360° set of projection angles.

Reconstructed spatial resolution in the range of 0.1 to 0.5 mm is quite typical for in vivo applications. For imaging biological specimens (e.g., an excised piece of bone from an animal model), for which radiation dose is of no concern and there is no motion, resolution on these systems can approach tens of microns. Reference 5 provides a useful overview of small-animal CT systems and applications.

The SPECT components of these small-animal systems are the same as those already described in Chapter 17, Section A.4, and will therefore not be repeated here. Most systems use two or more gamma camera heads with multi-pinhole collimators. Virtually all commercial small-animal SPECT systems now come with the option of a CT component for hybrid imaging.

Figure 19-8B shows images from a small-animal SPECT/CT system. Such systems are used for a variety of research applications, including imaging of the biodistribution of novel radiotracers. SPECT/CT systems also are used in the pharmaceutical industry to evaluate new therapeutic agents.

D. PET/CT

1. Clinical PET/CT Scanners

The first prototype hybrid PET/CT scanner was developed in the late 1990s by David Townsend, Ron Nutt, and colleagues.[6] The system consisted of a partial-ring PET scanner and a single-slice CT scanner, with all components mounted on a single rotating gantry (Fig. 19-9) to provide the necessary angular projection views for tomographic reconstruction. Methods also were developed for rapid, low-noise attenuation correction of the PET data using the CT images (see Section E). The ability of hybrid PET/CT systems to accurately identify the anatomic location of "hot spots" seen in ^{18}F-fluorodeoxyglucose (FDG) whole-body cancer studies (such as the example shown in Fig. 18-27) and to provide attenuation-corrected images led to the rapid introduction of commercial systems by the major medical imaging companies at the start of the 21st century. At present, hybrid PET/CT technology has been more widely adopted than SPECT/CT technology. Although many hospitals still use stand-alone SPECT systems, since 2006 essentially all clinical PET scanners sold have come combined with CT.

Modern clinical PET/CT scanners are configured in a tandem ("back-to-back") arrangement, with the center of the field-of-view of the PET and CT components separated in the axial direction by a distance on the order of 60 to 120 cm. These hybrid systems consist of separate PET and CT components, with some mechanical integration related to the motion

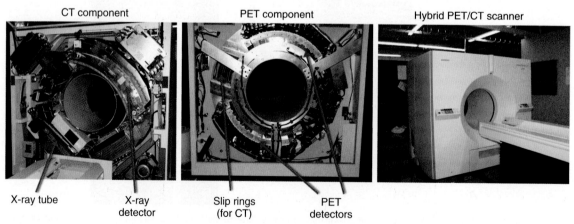

FIGURE 19-9 The first hybrid PET/CT scanner. The CT component (*left*) consists of a x-ray tube and single row of x-ray detectors mounted on a rotating gantry. The PET detector system (*center*) is mounted on the rear of the same rotating gantry and consists of a partial ring of block detectors made from the scintillator bismuth germanate. The photograph on the *right* shows the completed prototype system (*Courtesy Dr. David Townsend, University of Singapore.*)

of the patient bed through the two systems and a large degree of software integration that controls the position of the patient bed and sequential acquisition of PET and CT data, as well as image reconstruction, attenuation correction, and visualization and analysis of the reconstructed PET/CT datasets. One important practical consideration, resulting from the long travel length of the patient bed, is to ensure that the bed deflection does not change as the bed is cantilevered through the two imaging systems, which would result in misalignment of PET and CT images. Mechanical designs typically incorporate a rail system or additional supports for the bed to prevent this.

The description of PET scanner designs and configurations presented in Chapter 18, Section B, applies also to the PET component of current hybrid PET/CT systems. These PET systems all consist of multiple circumferential rings of scintillation detectors arranged around the patient to provide a complete set of projection angles without any rotation of the detectors.

The CT component may range from a relatively simple single-slice system to state-of-the-art 64-slice CT scanners. These systems use slip-ring technology (see Section B.3) to permit rapid rotation of the x-ray tube and detectors, and are capable of acquiring complete angular projections for a large number of contiguous slices in times ranging from a few seconds to a few minutes, depending on how many slices can be acquired simultaneously (which depends on the number of rows of x-ray detector elements). Most manufacturers offer a range of options for the number of CT slices available in their PET/CT scanners.

Figure 19-10 shows an example of a modern clinical PET/CT scanner.[7] The PET component of this system employs 28,336 LYSO scintillator elements, each with dimensions of 4×4 mm in cross-section, by 22 mm in thickness. These elements are arranged in 28 flat-panel modules, each consisting of a 23×44 array of elements, and read out by 15 PM tubes. Anger logic (see Chapter 18, Section B.1 and B.2) is used to determine the element in which an interaction took place. The scanner has an axial field-of-view of 18 cm, a transverse field-of-view of 57.6 cm, and a patient port diameter of 71.7 cm.

The scanner also incorporates time-of-flight information (see Chapter 18, Section A.2), with a timing resolution on the order of 600 psecs. The CT component most commonly has a 16- or 64-slice detector. The 64-slice detector consists of 64 rows of 690 detectors in each row, with each detector measuring 0.625 mm in the axial direction, giving an axial coverage of 64×0.625 mm = 400 mm. The x-ray tube used with this system has a maximum voltage of 140 kV. At this voltage, the maximum x-ray tube current is 430 mA. At a voltage of 120 kV, the x-ray tube can be operated at a current of up to 500 mA. Rotation times can be as fast as 0.4 seconds.

The primary application of PET/CT is for whole-body imaging of cancer patients with the radiotracer FDG. The CT component helps to identify the precise anatomic location of foci of increased FDG uptake that may be due to primary tumors or their metastases. An example of such a study is shown in Figure

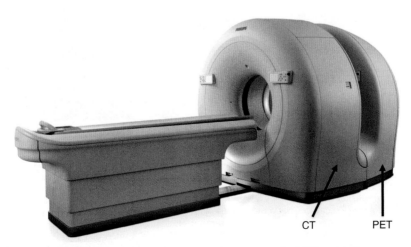

CT   PET

FIGURE 19-10 Hybrid PET/CT scanner. The axial separation between the PET and CT components is adjustable. (*Courtesy Philips Healthcare, Andover, MA.*)

19-11. Images of tissue attenuation from the CT scan usually are used to derive the PET attenuation correction factors (see Section E). However, some PET/CT systems also still have an option to use radionuclide transmission sources for attenuation correction as described in Chapter 18, Section D.4.

2. Small-Animal PET/CT Scanners

CT scanners, identical or very similar to those offered in small-animal hybrid SPECT/CT systems (see Section C.2), also are available with the small-animal PET scanners described in Chapter 18, Section B.5. Figure 19-12 shows a PET/CT scanner that consists of independent PET and CT components that can be docked together to form a hybrid

scanner. This design allows each modality to be used separately if desired. Like small-animal SPECT/CT, applications of PET/CT relate to imaging the biodistribution of new radiopharmaceuticals, studying the biodistribution of new therapeutics, and monitoring the effect of new therapeutics in animal models of human disease.

E. ATTENUATION AND SCATTER CORRECTION USING CT

One of the key advantages of hybrid SPECT/CT or PET/CT scanners is the ability of CT to rapidly acquire transmission images of high statistical quality that can be used for

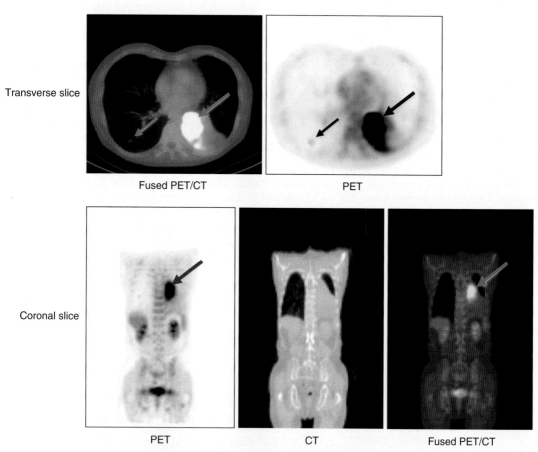

Transverse slice

Fused PET/CT   PET

Coronal slice

PET   CT   Fused PET/CT

FIGURE 19-11 Whole-body ^{18}F-fluorodeoxyglucose (FDG) study performed on a hybrid PET/CT scanner. The *top row* shows images of a single transverse section through the body at the level of the thorax. In the fused PET/CT image, the PET image is displayed using a hotwire color scale (*red, yellow, white* in order of increasing pixel values) and is superimposed on the CT scan, which is shown in *grayscale*. A primary lung tumor shows up in the left lung as a site of increased FDG accumulation (*large arrows*). There also is a small region of suspicious uptake in the posterior part of the right lung (*small arrows*) that may be a metastasis. The *bottom row* shows images of a coronal slice to illustrate the large fraction of the body covered in this PET/CT examination. The primary tumor is indicated with an arrow. The PET scan was obtained 60 minutes after the injection of 455 MBq (12.3 mCi) of FDG. The scan consisted of nine overlapping bed positions and took approximately 15 minutes to complete. The helical CT scan was acquired with a tube voltage of 140 kV and an exposure of 44 mAs/slice with 5-mm slices. (*Courtesy Philips Healthcare, Andover, MA, and University Hospitals Case Medical Center, Cleveland, OH.*)

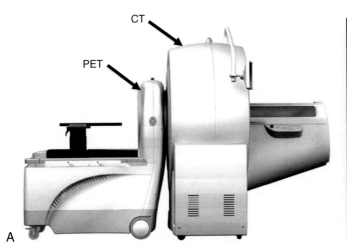
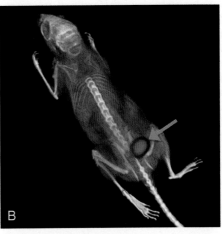

FIGURE 19-12 *A,* Dockable PET/CT scanner for small-animal imaging. *B,* Example image from such a system in which the PET (*pseudocolor*) and CT (*grayscale*) data have been fused and rendered in 3D. The PET scan shows accumulation of the radiopharmaceutical ^{64}Cu-Oxo-DO3A-trastuzumab in a tumor growing on the flank of a mouse (*arrow*). (*A, Courtesy Siemens Preclinical Solutions, Knoxville, TN. B, Courtesy D. Yapp, C. Ferreira, and F. Benard, The British Columbia Cancer Agency and Nordion.*)

attenuation correction and for scatter correction of PET and SPECT studies. The CT scan replaces the need for a lengthy transmission scan using external radionuclide sources (see Chapter 17, Section B.3, and Chapter 18, Section D.4) and thus decreases total imaging time for attenuation-corrected PET and SPECT studies. As was shown in Chapters 17 and 18, attenuation correction is critical for reconstructing quantitatively accurate SPECT or PET images. Even for more qualitative clinical evaluation, attenuation correction often is necessary to avoid distracting artifacts and nonuniformities, and to enable accurate diagnostic interpretation. Reference 8 discusses the methods by which CT scans are used for attenuation correction and compares them with attenuation correction based on the use of external radionuclide transmission sources.

1. Computing Attenuation Correction Factors from CT Scans

Use of the CT scan to correct for photon attenuation in PET or SPECT requires that the pixel intensity in the CT image (expressed as CT numbers in HU; see Section B.4) be converted into the linear attenuation coefficient, $\mu(E)$, at the photon emission energy E of the radionuclide used in the nuclear medicine study. For PET, this energy always is 511 keV. For SPECT, the energy depends on the radionuclide used.

A simple approach is to use image segmentation (see Chapter 20, Section B.5) to separate tissues based on their pixel values in the CT scan into three broad classes: air, soft tissue, and bone. Then the known linear attenuation coefficients for each of these materials at the emission energy of the radionuclide can be substituted for the CT number to yield an attenuation map appropriate for attenuation correction at that energy. This attenuation map then is used in iterative reconstruction algorithms or forward-projected into sinogram space, to compute the attenuation correction factors to apply to the emission study in the same way that transmission scans acquired using external radionuclide sources were applied in Chapter 17, Section B.3. This approach suffers from significant limitations, because pixels at tissue boundaries can contain a mixture of materials (e.g., bone and soft tissue) and do not fall into any of these three categories. Also, many tissues, most notably the lung, have a range of densities, which causes the linear attenuation coefficient to vary from location to location.

Therefore more sophisticated techniques, which provide a continuous mapping from CT number to μ at the radionuclide emission energy, typically are used to create the attenuation map. Two factors, however, complicate this mapping. First, the x-ray beam consists of photons with a range of energies with varying attenuation coefficients. Fortunately, a relatively straightforward solution is to use the effective energy of a known x-ray spectrum. The effective energy is defined as the energy of a monochromatic (single-energy) x-ray beam having the same attenuation coefficient as the polychromatic x-ray spectrum used in the measurement. For example, for a filtered 130-kVp x-ray spectrum suitable for CT, the effective energy is approximately 80 keV.

The second problem is that the energy scaling for pixels containing significant amounts of bone is quite different than that for pixels containing primarily soft tissue, blood, or air (Fig. 19-13). This is because bone, with its relatively high atomic number (caused by its calcium content) has a significant probability of causing photoelectric interactions at x-ray energies. However, at typical nuclear medicine energies (140 keV and above), the strong drop in photoelectric absorption probability with increasing energy (see Chapter 6, Section D) means that almost all interactions are by Compton scatter. Interactions in the lower-Z soft tissues, on the other hand, are almost exclusively by Compton scatter, at both x-ray and nuclear medicine energies. For this reason, no single energy-scaling relationship holds across all tissue types.

Therefore a bilinear fitting approach often is taken, as illustrated in Figure 19-13. In this specific example, the mapping from an effective x-ray energy of 80 keV to the 511-keV energy appropriate for PET is shown, although conceptually the same approach can be taken for radionuclides such as ^{99m}Tc (140-keV emission) used in SPECT. Pixels in the CT scan with CT numbers ≤ 60 HU are scaled to $\mu(E)$ using a linear fit to the data points for air, water and soft tissue. Pixels with CT numbers ≥ 60 HU are scaled with a second linear fit that passes through data points for bone. This plot provides a continuous mapping from reconstructed CT number to the energy-specific attenuation coefficient $\mu(E)$. Note that the slopes of the fitted lines are energy-dependent and must be computed for each radionuclide energy. Furthermore, the slope of the line for bone varies somewhat with x-ray tube voltage and filtration and, strictly speaking, should be computed for each different kVp setting used.

2. Possible Sources of Artifacts for CT-Based Attenuation Correction

There are a number of possible sources for introducing artifacts or biases when using CT-based attenuation correction in nuclear medicine:

- The CT image typically is acquired at much higher spatial resolution than the PET or SPECT study. If the resolutions are not matched, then artifacts can occur

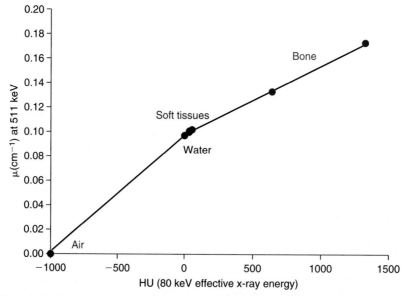

FIGURE 19-13 Plot of the linear attenuation coefficient at 511 keV versus CT number in Hounsfield units for an effective x-ray energy of 80 keV for soft tissues, air, water, and bone. A bilinear fit to these points commonly is used to scale the CT numbers for attenuation correction of the PET data. See text for further details. Data obtained from reference 9.

at tissue boundaries. Therefore the attenuation map derived from the CT images typically is rebinned into large pixels or blurred to match the resolution of the PET or SPECT images to which it will be applied.

- A fundamental assumption in applying CT-based attenuation correction is that the patient is positioned identically in the two systems and that the images are therefore in perfect spatial registration. Any movement of the patient between the sequential CT and PET/SPECT studies can result in artifacts. Even if the patient does not move, respiratory motion is a problem. Fast, helical CT scans typically take several seconds and are acquired under "breath-hold" conditions to reduce artifacts that would occur if different projection angles were acquired at different times during the respiratory cycle. Nonetheless, respiratory artifacts are not uncommon, as the "breath-hold" state is hard for some patients to maintain for the duration of the CT scan. Nuclear medicine studies typically take several minutes, and the images thus represent an average over many respiratory cycles, which results in motion blurring. The use of breath-hold CT images to reconstruct motion-blurred nuclear medicine images can lead to noticeable artifacts, especially around the diaphragm and the top of the liver. An example of a PET/

CT scan showing these types of artifacts is provided in Figure 19-14A. One solution for reducing such artifacts is to use respiratory gating (see Chapter 20, Section A.4) for the nuclear medicine study. This, however, reduces the statistical quality of the data for a fixed acquisition time.

- In some instances, especially in the region of the shoulders, the CT field-of-view may be smaller than the patient, leading to truncation of the images (see Fig. 19-14B). Tissue outside the CT field-of-view that attenuates the PET signal is not visualized in the CT scan; therefore attenuation correction factors are underestimated. Artifacts may also occur because of artifacts in the reconstructed CT image at the edges of the field-of-view caused by missing data in some projection angles.

- Many diagnostic CT examinations are performed with the administration of a contrast agent. Intravenous contrast agents typically contain iodinated compounds, whereas oral contrast agents usually are based on barium. Both these agents increase x-ray attenuation (and therefore CT numbers) where present, because of their high-Z values relative to soft tissues. In turn, this leads to an increase in the probability of photoelectric absorption. However, their presence has very little effect on the attenuation coefficient at nuclear medicine energies;

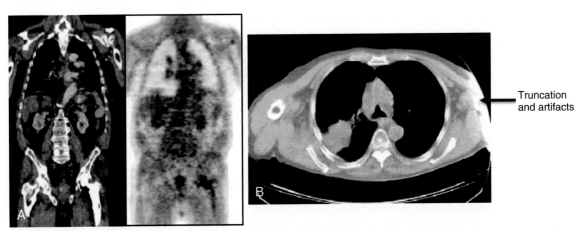

Truncation and artifacts

FIGURE 19-14 Two sources of artifacts in CT-based attenuation correction of PET data. *A,* Respiratory motion artifacts in the CT scan (*arrow*) leads to artifacts in the activity distribution in the reconstructed PET images. *B,* Truncation of the body in a CT scan leads to underestimation of attenuation from the missing tissues and artifacts at the edge of the field-of-view. Both of these factors can introduce substantial quantitative errors and artifacts in PET images reconstructed with attenuation correction based on the CT images. (*Images courtesy Dr. Paul Kinahan, University of Washington, Seattle, WA.*)

thus they change the relationship between CT number and μ(E) shown in Figure 19-13. Although this is theoretically a problem, empirically, it is rare that CT-based attenuation correction leads to artifacts from exogenous contrast administration. Furthermore, if the CT scan is acquired solely for attenuation correction and anatomic localization, it will often be performed as a low-dose study without administration of contrast agent.

• For similar reasons, care must be exercised when imaging patients that have surgical or dental implants made from high-Z materials. These have the potential to cause local artifacts because their appearance on the CT image (high CT number) will not lead to the correct prediction of their effect on the attenuation of photons at nuclear medicine energies. In addition to leading to the wrong scaling, large metal implants can also cause significant artifacts in the CT images that can propagate into the attenuation maps used for attenuation correction. Modern CT scanners, however, use algorithms that can significantly reduce these artifacts and their effects on CT-based attenuation correction.

• At the time of the CT scan, the patient often already has been injected with the radiopharmaceutical for the nuclear medicine scan. However, the flux of photons from the radiopharmaceutical is so low compared with the x-ray flux that there is no detectable bias to the x-ray measurements and this potential problem can be ignored.

Although these factors need to be considered, the advantages of the speed and high statistical quality of CT-based attenuation correction almost always outweigh such concerns.

3. Scatter Correction

In addition to being used for attenuation correction, the scaled attenuation map also can be used for calculation of a correction for scattered radiation in the tomographic nuclear medicine study. At nuclear medicine energies, virtually all interactions in biological tissues are by Compton scatter, and thus the attenuation map of μ(E) is equivalent to a map of the coefficients for Compton scatter interactions. Chapter 17, Section B.4, and Chapter 18, Section D.3, describe how these

transmission maps can be used for scatter correction in SPECT and PET respectively.

F. HYBRID PET/MRI AND SPECT/MRI

There are many clinical situations in which MRI,[10] rather than CT, is the anatomic imaging technique of choice. Furthermore, MRI can image functional and physiologic processes and, via magnetic resonance spectroscopic imaging, can also study metabolic pathways. Thus an argument for the integration of nuclear medicine tomography with MRI systems appears to be compelling. However, a number of considerations led to CT-based hybrid systems emerging first, and only in 2005 were the first practical hybrid PET/MRI systems produced.[11] SPECT/MRI systems are even less well developed.

The integration of PET and SPECT with MR is challenging for a number of reasons. First, clinical MRI systems typically have magnetic field strengths of 1 to 3 T,[*] whereas the field strengths of research and small-animal systems may be over 10T. Strong magnetic fields exist not only inside but often around the edges of a MR system as well. The PM tube detector technology used in most nuclear medicine systems is highly sensitive to even small magnetic fields. For example, in SPECT, the PM tubes are shielded to guard against magnetic field effects when rotating the gamma camera detector heads though the earth's magnetic field, which measures a mere 30 to 60 μT. Most PM tubes cannot operate at all in fields higher than approximately 0.01 T. Thus PM tube-based detectors must be kept far away from the magnet; however, they can still be used in hybrid PET/MRI systems in which the two scanners are physically separate, and a bed on rails moves the patient from one scanner into the other.

Second, unlike CT, magnetic resonance is a relatively slow modality for acquiring high-resolution, diagnostic-quality images over large tissue volumes. Typical clinical examinations require several minutes, much like PET and SPECT. This makes sequential imaging (which is the standard approach in PET/CT and SPECT/CT) less attractive, because the total time of the examination can be rather long. Simultaneous imaging to improve throughput is possible, but only by

---

[*]T, Tesla: the standard unit of magnetic field strength.

creating a very compact PET or SPECT system that can fit inside the bore of a magnet, and operate in very strong magnetic fields. Early approaches used long light guides or optical fibers to transport scintillation light from PET detectors residing inside the bore of the magnet, to PM tubes placed several meters outside the MRI scanner where the fringe magnetic field is low enough for their operation. More recently, practical hybrid PET/MRI systems have been developed for both small-animal and human applications using detectors based on scintillator elements read out by avalanche photodiodes (see Chapter 7, Section C.3), which are relatively immune to magnetic field effects.

Third, cost is an important consideration in the clinical adoption of hybrid PET/MRI scanners. Integration of PET and MRI typically requires at least some modification of

stand-alone systems to reduce magnetic field and radiofrequency interference to acceptable levels. As well, it combines two expensive diagnostic imaging modalities. It has yet to be demonstrated whether clinical applications will justify the cost of these systems. Nonetheless, PET/MRI is likely to emerge as a powerful tool for biomedical research. Figure 19-15 shows an early hybrid PET/MRI system designed for imaging the human brain, and examples of the images it produces.

REFERENCES

1. Bushberg JT, Seibert JA, Leidholdt EM, Boone JM: *The Essential Physics of Medical Imaging*, ed 3, Chapter 10, Philadelphia, PA, 2012, Lippincott Williams & Wilkins.
2. Lang TF, Hasegawa BH, Liew SC, et al: Description of a prototype emission-transmission computed tomography imaging system. *J Nucl Med* 33:1881-1887, 1992.
3. Tang HR, et al: Implementation of a combined x-ray CT-scintillation camera imaging system for localizing and measuring radionuclide uptake: experiments in phantoms and patients. *IEEE Trans Nucl Sci* 46:551-557, 1999.
4. Buck AK, Nekolla S, Ziegler S, et al: SPECT/CT. *J Nucl Med* 49:1305-1319, 2008.
5. Badea CT, Drangova M, Holdsworth DW, Johnson GA: In vivo small animal imaging using microCT and digital subtraction angiography. *Phys Med Biol* 53:R319-350, 2008.
6. Beyer T, Townsend DW, Brun T, et al: A combined PET/CT scanner for clinical oncology. *J Nucl Med* 41:1369-1379, 2000.
7. Surti S, Kuhn A, Werner ME, et al: Performance of Philips Gemini TF PET/CT scanner with special consideration for its time-of-flight imaging capabilities. *J Nucl Med* 48:471-480, 2007.
8. Kinahan PE, Hasegawa BH, Beyer T: X-ray based attenuation correction for positron emission tomography/computed tomography scanners. *Semin Nucl Med* 33:166-179, 2003.
9. Hubbell JH, Seltzer SM: Tables of x-ray mass attenuation coefficients and mass-energy absorption coefficients from 1 keV to 20 MeV for elements Z = 1 to 92 and for 48 additional substances of dosimetric interest. Available at http://physics.nist.gov/PhysRefData/XrayMassCoef/cover.html [accessed August 26th, 2011].
10. Bushberg JT, Seibert JA, Leidholdt EM, Boone JM: *The Essential Physics of Medical Imaging*, ed 3, Chapters 12 and 13,, Philadelphia, PA, 2012, Lippincott Williams & Wilkins.
11. Cherry SR, Louie AY, Jacobs RE: The integration of positron emission tomography with magnetic resonance imaging. *Proc IEEE* 96:416-438, 2008.

BIBLIOGRAPHY

The following is an informative general review on hybrid imaging:

Townsend DW: Multimodality imaging of structure and function. *Phys Med Biol* 53:R1-R39, 2008.

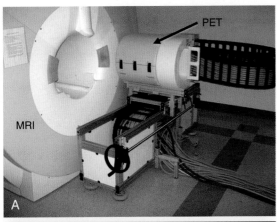
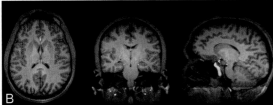

FIGURE 19-15 *A,* Hybrid PET/MRI scanner. The PET scanner is inserted inside the bore of the magnet to allow simultaneous acquisition of PET and MRI images of the brain. The field strength of the MRI scanner is 3 T. *B,* Simultaneously acquired ¹⁸F-fluorodeoxyglucose PET images (*red color scale*) superimposed on MRI images (*grayscale*) in a normal subject. A single transverse, coronal, and sagittal slice from the three-dimensional PET/MRI dataset is shown. (*Courtesy Dr. Ciprian Catana and Dr. A. Gregory Sorenson, MGH/HST, Athinoula A. Martinos Center for Biomedical Imaging, Charlestown, MA.*)

Digital Image Processing in Nuclear Medicine

Image processing refers to a variety of techniques that are used to maximize the information yield from a picture. In nuclear medicine, computer-based image-processing techniques are especially flexible and powerful. In addition to performing basic image manipulations for edge sharpening, contrast enhancement, and so forth, computer-based techniques have a variety of other uses that are essential for modern nuclear medicine. Examples are the processing of raw data for tomographic image reconstruction in single photon emission computed tomography (SPECT) and positron emission tomography (PET) (see Chapters 16 to 18), and correcting for imaging system artifacts (e.g., Chapter 14, Section B, and Chapter 18, Section D). Another important example is time analysis of sequentially acquired images, such as is done for extracting kinetic data for tracer kinetic models (see Chapter 21). Computer-based image displays also allow three-dimensional (3-D) images acquired in SPECT and PET to be viewed from different angles and permit one to fuse nuclear medicine images with images acquired with other modalities, such as computed tomography (CT) and magnetic resonance imaging (MRI) (see Chapter 19). Computer-based acquisition and processing also permit the raw data and processed image data to be stored digitally (e.g., on computer disks) for later analysis and display.

All of these tasks are performed on silicon-based processor chips, generically called *microprocessors*. The *central processing unit* (CPU) of a general purpose computer, such as a personal computer, is called a *general purpose microprocessor*. Such devices can be programmed to perform a wide variety of tasks, but they are relatively large and not very energy efficient. For very specific tasks, an *application-specific integrated circuit* often is used. ASICs are compact and energy efficient, but their functionality is hardwired into their design and cannot be changed. Examples of their uses include digitizing signals (analog-to-digital converters) and comparing signal amplitudes (pulse-height analyzers and multichannel analyzers). Other categories of microprocessors include *digital signal processors* (DSPs) and *graphics processing units*. These devices have limited programmability, but they are capable of very fast real-time signal and image processing, such as 3-D image rotation and similar types of image manipulations.

The technology of microprocessors and computers is undergoing continuous and rapid evolution and improvement, such that a "state-of-the-art" description rarely is valid for more than a year or, in some cases, even a few months. However, the end result is that the usage of computers and microprocessors in nuclear medicine is ubiquitous. They are used not only for acquisition, reconstruction, processing, and display of image data but also for administrative applications such as scheduling, report generation, and monitoring of quality control protocols.

In this chapter, we describe general concepts of digital image processing for nuclear medicine imaging. Additional discussions of specific applications are found in Chapters 13 to 19 and Chapter 21.

A. DIGITAL IMAGES

1. Basic Characteristics and Terminology

For many years, nuclear medicine images were produced directly on film, by exposing the film to a light source that produced flashes of light when radiations were detected by the imaging instrument. As with ordinary photographs, the image was recorded with a virtually continuous range of brightness levels and x-y locations on the film. Such images sometimes are referred to as *analog images*. Very little could be done in the way of "image processing" after the image was recorded.

Virtually all modern nuclear medicine images are recorded as *digital images*. This is required for computerized image processing. A digital image is one in which events are localized (or "binned") within a grid comprising a finite number of discrete (usually square) picture elements, or *pixels* (Fig. 20-1). Each pixel has a digital (nonfractional) location or *address,* for example, "x = 5, y = 6." For a gamma camera image, the area of the detector is divided into the desired number of pixels (Fig. 20-2). For example, a camera with a field-of-view of 40 cm × 40 cm might be divided into a 128 × 128 grid of pixels, with each pixel therefore measuring 0.3125 mm × 0.3125 mm. Each pixel corresponds to a range of possible physical locations within the image. If an event were determined to have interacted at

a location x = 4.8 cm, y = 12.4 cm, the appropriate pixel location for this event would be

$$x\text{-pixel location} = 4.8 \text{ cm}/0.3125 \text{ cm/pixel}$$
$$= \text{int}(15.36) = 15$$
$$y\text{-pixel location} = 12.4 \text{ cm}/0.3125 \text{ cm/pixel}$$
$$= \text{int}(39.68) = 40$$

where int(x) denotes the nearest integer of x, and the pixels are labeled from 0-127 with the coordinate system defined as shown in Figure 20-2.

A similar format is used for digital multi-slice tomographic images, except that the discrete elements of the image would correspond to discrete 3-D volumes of tissue within a cross-sectional image. The volume is given by the product of the x- and y-pixel dimensions multiplied by the slice thickness. Thus they are more appropriately called volume elements, or *voxels*. However, when discussing an individual tomographic slice, the term *pixel* still is commonly used. In tomographic images, the "intensity" of each voxel may or may not have a discrete integer value. For example, voxel values for a reconstructed image will generally have noninteger values corresponding to the calculated concentration of radionuclide within the voxel.

Depending on the mode of acquisition (discussed in Section A.4), either the x-y address of the pixel in which each event occurs, or the pixel value, $p(x,y)$, is stored in computer memory. For 3-D imaging modes, such as 3-D SPECT or PET, individual events are localized within a 3-D matrix of voxels, and the reconstructed value in a voxel is denoted as $v(x,y,z)$. Depending on how data are acquired and processed by the imaging system, the pixel or voxel value may correspond to the number of counts, counts per unit time, the reconstructed pixel or voxel value, or absolute radionuclide concentrations (kBq/cc or μCi/cc).

Although most interactions between the user and a computer system involve conventional decimal numbers, the internal operations of the computer usually are performed using *binary numbers*. Binary number representation uses powers of 2, whereas the commonly used decimal number system uses powers of 10. For example, in decimal representation, the number 13 means $[(1 \times 10^1) + (3 \times 10^0)]$. In the binary number system, the same number is represented as 1101, meaning $[(1 \times 2^3) + (1 \times 2^2) + (0 \times 2^1) + (1 \times 2^0)]$, or $(8 + 4 + 0 + 1) = 13$. Each digit in the binary number representation is called a *bit* (an abbreviation

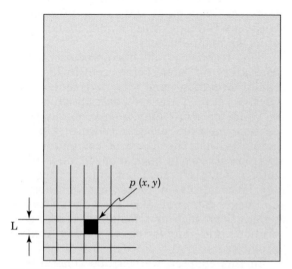

FIGURE 20-1 A digital image consists of a grid or matrix of pixels, each of size L × L units. Each pixel has an x-y address location, with pixel value, *p(x,y)*, corresponding to the number of counts or other quantity associated with that pixel.

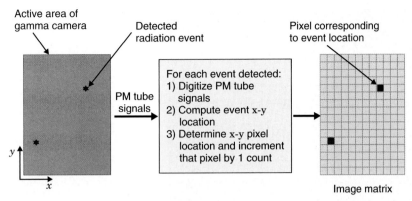

FIGURE 20-2 Subdivision of the gamma camera detector area for generating a digital image. The photomultiplier tube signals are analyzed using analog-to-digital converters to assign the digital matrix location for each detected event.

for "*bi*nary digi*t*"). In general, an *n*-bit binary number can represent decimal numbers with values between zero and $(2^n - 1)$.

Binary numbers are employed in computer systems because they can be represented conveniently by electronic components that can exist only in an "on" or "off" state. Thus an *n*-bit binary number can be represented by the "on" or "off" state of a sequence of *n* such components. To communicate sensibly with the outside world, the binary numbers used within the computer must be converted into decimal integers or into decimal numbers and fractions. The latter are called *floating point* numbers. The methods by which binary numbers are converted to decimal format are beyond the scope of this presentation and can be found in more advanced texts on computer systems.

Digital images are characterized by matrix size and pixel depth. *Matrix size* refers to the number of discrete picture elements in the matrix. This in turn affects the degree of spatial detail that can be presented, with larger matrices generally providing more detail. Matrix sizes used for nuclear medicine images typically range from (64×64) to (512×512) pixels. Matrix size virtually always involves a power of 2 (2^6 and 2^9 in the previous examples) because of the underlying binary number system used in the computer.

Pixel depth refers to the maximum number of events that can be recorded per pixel. Most systems have pixel depths ranging from 8 bits ($2^8 = 256$; counts range from 0 to 255) to 16 bits ($2^{16} = 65,536$; counts range from 0 to 65,535). Note again that these values are related to the underlying binary number system used in the computer. When the number of events recorded

in a pixel exceeds the allowed pixel depth, the count for that pixel is reset to 0 and starts over, which can lead to erroneous results and image artifacts.

Pixel depth also affects the number of gray shades (or color levels) that can be represented within the displayed image. In most computer systems in use in nuclear medicine, 8 bits equals a byte of memory and 16 bits equals a word of memory. The pixel depth, therefore, frequently is described as "byte" mode or "word" mode.*

2. Spatial Resolution and Matrix Size

The spatial resolution of a digital image is governed by two factors: (1) the resolution of the imaging device itself (such as detector or collimator resolution) and (2) the size of the pixels used to represent the digitized image. For a fixed field-of-view, the larger the number of pixels, that is, the larger the matrix size, the smaller the pixel size (Fig. 20-3). Clearly, a smaller pixel size can display more image detail, but beyond a certain point there is no further improvement because of resolution limitations of the imaging device itself. A question of practical importance is, At what point does this occur? That is, how many pixels are needed to ensure that significant detail is not lost in the digitization process?

The situation is entirely analogous to that presented in Chapter 16 for sampling requirements in reconstruction tomography. In particular, Equation 16-13 applies—that is, the

---

*Most modern computer CPUs have 32-bit or 64-bit processors. This means they can process data 32 or 64 bits at a time; however, this is largely independent of image display and how pixel values are stored.

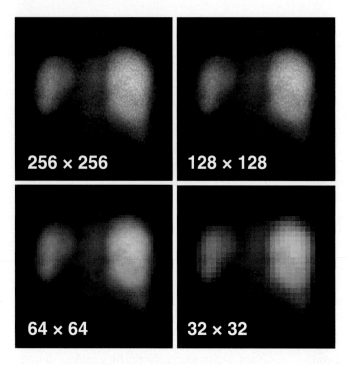

FIGURE 20-3 Digital images of the liver and spleen (posterior view) displayed with different matrix sizes. The larger the matrix size, the smaller the pixels and the more detail that is visible in the image. (*Original image courtesy GE Medical Systems, Milwaukee, WI.*)

linear sampling distance, d, or pixel size, must be smaller than or equal to the inverse of twice the maximum spatial frequency, k_{max}, that is present in the image:

$$d = 1/(2 \times k_{max}) \tag{20-1}$$

This requirement derives directly from the sampling theorem discussed in Appendix F, Section C.

Once this sampling requirement is met, increasing the matrix size does not improve spatial resolution, although it may produce a cosmetically more appealing image with less evident grid structure. If the sampling requirements are not met (too coarse a grid), spatial resolution is lost. The maximum spatial frequency that is present in an image depends primarily on the spatial resolution of the imaging device. If the resolution of the device is specified in terms of the full width at half maximum (FWHM) of its line-spread function (Chapter 15, Section B.2), then *the sampling distance (pixel size) should not exceed about one third of this value to avoid significant loss of spatial resolution,* that is,

$$d \lesssim \frac{\text{FWHM}}{3} \tag{20-2}$$

This applies for noise-free image data. With added noise it may be preferable to relax the sampling requirement somewhat (i.e., use larger pixels) to diminish the visibility of noise in the final digitized image.

EXAMPLE 20-1

What is the approximate spatial resolution that can be supported for a 30-cm diameter field-of-view using a 64×64 matrix? A 128×128 matrix? Assume that the original data are noise free.

Answer

64 × 64 matrix

A 64×64 image matrix results in a pixel size of 300 mm/64 = 4.69 mm. From Equation 20-2, this would be suitable for image resolution given by

$$\text{FWHM} \gtrsim 3 \times \text{pixel size} = 14.06 \text{ mm}$$

128 × 128 matrix

$$\text{FWHM} \gtrsim 3 \times 300 \text{ mm}/128 = 7.03 \text{ mm}$$

The values calculated in Example 20-1 represent the approximate levels of imaging system resolution that could be supported without loss of imaging resolution for the specified image and matrix sizes. The practical effects of undersampling depend as well on the information contained in the image and whether it has a significant amount of actual spatial frequency content near the resolution limit of the imaging device. Practical experimentation sometimes is required to determine this for a particular type of imaging procedure.

3. Image Display

Digital images in nuclear medicine are displayed on cathode ray tubes (CRTs) or flat-panel displays such as liquid crystal displays (LCDs). In addition to their use at the site of the imaging device, displays are an essential component of picture archival communications systems (PACS) networks, for remote viewing of images (see Section C). The spatial resolution of the display device should exceed that of the underlying images so as not to sacrifice image detail. In general, the display devices used in nuclear medicine computer systems and in radiology-based PACS networks comfortably exceed this requirement. Typical high-resolution CRTs have 1000 or more lines and a typical LCD might have 1536×2048 elements.

Individual pixels in a digital image are displayed with different brightness levels, depending on the pixel value (number of counts or reconstructed activity in the pixel) or voxel value. On *grayscale* displays, the human eye is capable of distinguishing approximately 40 brightness levels when they are presented in isolation and an even larger number when they are presented in a sequence of steps separated by sharp borders. Image displays are characterized by the potential number of brightness levels that they can display. For example, an 8-bit grayscale display can potentially display $2^8 = 256$ different brightness levels. Such a range is more than adequate in comparison with the capabilities of human vision. In practice, the effective brightness scale often is considerably less than the physical limits of the display device because of image noise. For example, if an image has a root mean square noise level of 1%, then there are not more than 100 significant brightness levels in the image, regardless of the capabilities of the display device.

Digital images also can be displayed in color by assigning color hues to represent different pixel values. The human eye can distinguish millions of different colors, and color displays are capable of producing a broader dynamic range (i.e., number of distinguishably different levels) than can be achieved in black-and-white displays. For example, a *true-color* display with 24-bit graphics can generate nearly 16.8 million different colors $[2^{24} = (2^8)^3$, in which the 3 represents the independently generated red, green, and blue color channels].

One commonly used color scale, the *pseudo-color* scale (sometime known as the *rainbow* or *spectrum* color scale), assigns different colors from the visible spectrum, ranging from blue at the low ("cool") end, through green, yellow, and red ("hot"), for progressively increasing pixel values. This is an intrinsically nonlinear scale, because the viewer does not perceive equal significance for successive color steps. A somewhat more natural scale, the so-called *heat* or *hot-body* scale, assigns different shades of similar colors, such as red, yellow, and white, to progressively increasing pixel values, corresponding to the colors of an object heated to progressively higher temperature. In both examples, the colors are blended to produce a gradual change over the full range of the scale. Figure 20-4 shows the same image displayed with different color scales.

The major problem with the use of color scales to represent pixel count levels is that they are somewhat unnatural and also can produce contours, such as apparently sharp changes in pixel values, where none actually exist. A more practical use of color displays is for color coding a second level of information on an image. For example, in combined-modality imaging of PET or SPECT with CT (see Chapter 19), the anatomic (CT) image often is displayed using a standard gray scale, whereas the functional (PET) image is shown using a color scale. Such a display clearly differentiates between the two types of images, whereas a simple overlay of two grayscale images would be confusing.

Hard-copy images can be produced on black-and-white transparency film from a CRT display. Single-emulsion films are used to minimize blurring of the recorded image, especially when images are minified for compact display on a single sheet of film. The CRT display intensity must be calibrated to compensate for the sensitometric properties of the recording film to match the monitor display. Computer printers are now commonly used to record hard-copy images and a range of different technologies and media are available depending on requirements such as quality (resolution and gray-scale range), cost and printing speed.

4. Acquisition Modes

Digital images are acquired either in frame mode or in list mode. In *frame-mode* acquisition, individual events are sorted into their

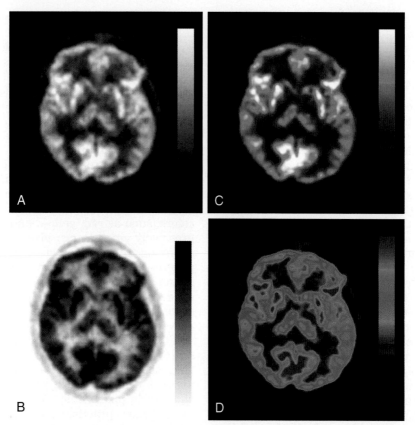

FIGURE 20-4 The same reconstructed transaxial image slice rendered in different color scales. *A,* Grayscale, high-intensity white; *B,* inverted grayscale, high-intensity black; *C,* hot-wire or hot-body scale; *D,* pseudocolor spectral scale. The slice is from a PET scan of the brain using the radiotracer ^{18}F-fluorodeoxyglucose. (*Original image courtesy Siemens Molecular Imaging, Knoxville, TN*).

appropriate x-y locations within the digital image matrix immediately after their position signals are digitized. After a preset amount of time has elapsed or after a preset number of counts have been recorded, the acquisition of data for the image is stopped and the pixel values [$p(x,y)$ = number of counts per pixel] are stored in computer memory.

When a series of such images is obtained sequentially, individual images in the sequence are referred to as "frames." Clearly, the image matrix size (e.g., 64×64, 128×128, and so forth) must be specified before the acquisition begins. Additionally, the time duration of the frame sets a limit on the temporal accuracy of the data. For example, if the frame is acquired during a 1-minute period, the number of counts recorded in each pixel represents the integrated number of counts during the 1-minute acquisition period and cannot be subdivided retrospectively into shorter time intervals. When faster framing rates are used, such as for cardiac blood-pool imaging, temporal sampling accuracy is improved, but the total counts per frame and per pixel are reduced compared with slower frame rates.

In *list-mode* acquisition, the incoming x and y position signals from the camera are digitized, but they are not sorted immediately into an image grid. Instead, the x and y position coordinates for individual events are stored, along with periodic clock markers (e.g., at millisecond intervals). This permits retrospective framing with frame duration chosen after the data are acquired.

List-mode acquisition permits greater flexibility for data analysis. However, it is not an efficient method for using memory space during acquisition for conventional imaging, especially for high-count images, because every recorded event occupies a memory location. Thus a 1-million count 128×128 image recorded in list mode would require 1 million memory locations, whereas in frame mode the same image would require only approximately 16,000 memory locations. However, list mode can actually be more efficient in some situations. This would apply, for example, if the average number of counts is less than 1 per pixel in an image frame. In this case, list mode would require fewer memory locations to record the image than frame mode. Such situations can arise (e.g., in fast dynamic studies).

Another commonly used acquisition mode is called *gated imaging*. In this mode, data are acquired in synchrony with the heart beat or with the breathing cycle, so that all images are acquired at the same time during the motion cycle. This helps reduce blurring and other possible image artifacts induced by body motion (see Fig. 15-1). To perform gated imaging, it is necessary to have some sort of monitoring system in place, such as electrocardiogram leads for the heart beat or a pneumatically operated "belly band" that produces an electrical signal when it expands during the breathing cycle.

In frame-mode gated imaging, the signals from the motion-monitoring device are used to initiate an image-acquisition cycle. The "cycle" may consist of several images, each of which represents the object at the same location, assuming the motion is repeated reproducibly during the cycle. With list-mode acquisition, retrospective synchronization is possible by recording the motion-monitoring signal along with the signals from the detector. In either case, data usually are acquired over a large number of cycles and the images are added together, until the total number of counts is sufficient to provide adequate counting statistics.

B. DIGITAL IMAGE-PROCESSING TECHNIQUES
....

All modern nuclear medicine systems are provided with fairly sophisticated software for displaying and processing the images that are acquired on those systems. There also are many "third-party" software packages available that incorporate extensive tools for image processing and allow images from different manufacturers and different modalities to be analyzed. A variety of digital image-processing techniques are used in nuclear medicine, some of which are fully automatic (i.e., performed entirely by the computer), whereas others are interactive (i.e., require some user input). In this section, we describe briefly a subset of the major image-processing tools that are commonly used in nuclear medicine.

1. Image Visualization

Commonly, a single projection image, or in the case of tomographic data, a set of contiguous image slices, are displayed on the screen. The display of the images can be manipulated in a number of ways to aid in interpretation. This includes changing from a linear gray scale to a color scale or to a nonlinear (e.g., logarithmic) gray scale, or limiting the range of pixel values displayed. The latter is known as *windowing*. For example, if greater contrast is desired in one region of an image, the full brightness range of the display device can be used to display only the range of pixel values found within that region (Fig. 20-5). This increases the displayed contrast in the selected area, but other parts of the image may have diminished contrast as a result (i.e., the counts per pixel may be beyond the upper or lower range of the selected grayscale window). Whenever image data are displayed or reproduced, it is desirable to show a grayscale or color-scale bar that has undergone the same manipulations as the images, so that the viewer can interpret the image in the context of how the display has been modified.

Tomographic nuclear medicine data consists of a 3-D volume that can be displayed in conventional transaxial views or in coronal or sagittal views (Fig. 20-6). Often it is useful to display all three views simultaneously on the screen. Typically, a point within the object is chosen using the cursor, and the three orthogonal images that pass through that point are displayed. As the cursor is moved, the transaxial, coronal, and sagittal images are updated. This is an efficient way of navigating through a large 3-D dataset. The dataset also can be resliced at an arbitrary orientation to provide oblique views. This is useful for objects whose line of symmetry does not fall naturally along one of the perpendicular axes of the 3-D volume. An example is reslicing the 3-D dataset to provide short axis views of the heart.

Another useful visualization tool for 3-D tomographic datasets is the *projection tool*. This collapses the 3-D dataset into a single 2-D image for a specified viewing angle and allows all the data to be seen at once. An example is shown in Figure 20-7. A number of different algorithms are available to "render" the projection images. The simplest approach is to simply sum the intensities along the projection direction. This is essentially equivalent to what would be obtained by acquiring

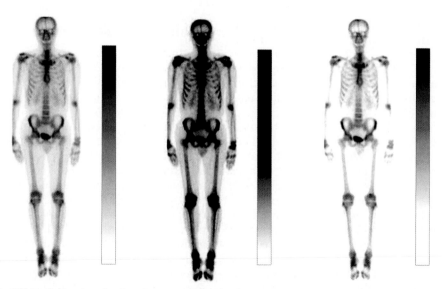

FIGURE 20-5 Effect of changing the distribution of gray levels on image contrast. *Left,* Original image with uniform distribution of gray levels. *Center,* Gray scale compressed (fewer levels) in high-count (*dark*) regions to improve the visualization of soft tissues. *Right,* Gray scale compressed in low-count (*light*) regions to suppress soft tissues and visualize only bone. (*Image courtesy Siemens Medical Solutions, USA, Inc.*)

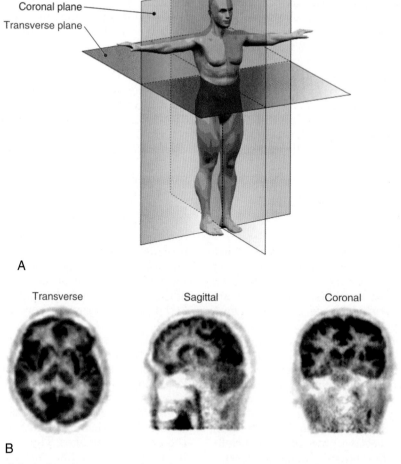

FIGURE 20-6 *A,* Orientation of transverse (also known as transaxial), coronal and sagittal sections. *B,* Orthogonal views (transverse, coronal, sagittal) of an ^{18}F-fluorodeoxyglucose PET brain study in which the imaging field-of-view covers the entire head. (*A, Reproduced from* http://en.wikipedia.org/wiki/File:Human_anatomy_planes.svg#file. *B, Images courtesy CTI PET Systems, Inc., Knoxville, TN.*)

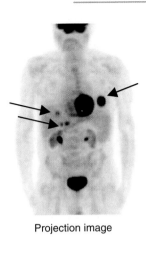

Projection image

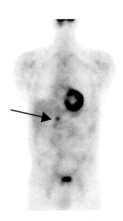
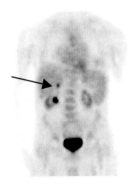
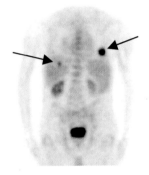

Individual coronal slices

FIGURE 20-7   Whole-body ^{18}F-fluorodeoxyglucose PET study displayed as a single projection image (*top*) and as a series of coronal image slices (*bottom row*). The projection image shows the distribution of radiotracer in all slices (*arrows* correspond to metastatic disease in this patient with cancer), but image resolution and contrast are lost relative to the individual tomographic slices. Projection images often are a convenient way to initially view large tomographic datasets, after which individual tomographic slices can be used for more detailed examination or quantitative analysis. (*Courtesy Dr. Magnus Dahlbom, University of California–Los Angeles.*)

2-D views of the object from many projection directions. Another approach is to display only the surface pixel values (*surface rendering*). An approach that highlights internal features is to display only the pixel with the maximum value along the projection direction (*maximum intensity projection*).

By computing projection views at a set of angles around the object and presenting them in a continuous loop, one can create movies in which it appears that the object is rotating in space. This sometimes is called *cine mode.* These and other rendering and display algorithms are discussed in some of the suggested readings at the end of the chapter.

Another important application of image processing is *image arithmetic.* There are a number of applications in which one wishes to see differences between images or to combine images acquired with different radionuclides or acquired with different modalities. Most image-processing software allows one to add, subtract, multiply, and divide single images or 3-D image volumes. These operations typically are applied on a pixel-by-pixel basis. Figure 20-8 is an example of a simple frame arithmetic operation: subtraction. The study illustrated is a visual stimulation using ^{15}O-labeled water as a flow tracer. Visual stimulation, created by having the subject view a strobe light, caused an increase in blood flow to the occipital (visual) cortex, while the remainder of the brain remained largely unaffected. Subtraction of an image taken from a resting control study from the image obtained in the stimulation study provides a display of the blood flow changes occurring as a result of stimulation.

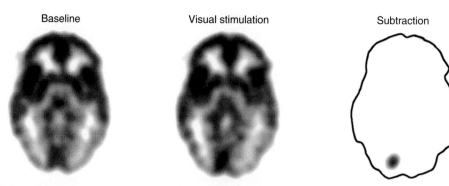

Baseline   Visual stimulation   Subtraction

FIGURE 20-8   Cerebral blood-flow images ($H_2^{15}O$ PET) acquired in the resting state (*left*) and during visual stimulation using a flashing light (*center*). The stimulus causes a small increase in blood flow in the visual cortex that is virtually invisible on the image acquired during visual stimulation; however, the increase is clearly visible when the resting-state image is subtracted from it (*right*).

Most digital images in nuclear medicine are, in essence, pictures of the count density in the organ or tissue of interest. Instead of presenting the data in this format, one may desire to first process the image data on a pixel-by-pixel basis using a model that represents the functional process and display the calculated result. Such an image, in which the pixel values represent a calculated parameter, sometimes is called a *parametric image*. For example, a digital ventilation image can be divided by a perfusion image to produce a parametric image that shows the ventilation/perfusion ratio. Other examples of calculated functional parameters are discussed in Chapter 21.

2. Regions and Volumes of Interest

Both PET and SPECT can provide semiquantitative, or when all appropriate corrections are applied, quantitative images of the radiotracer distribution in the body. Conventional 2-D images also can provide information about the relative concentration of radiotracer in different areas. *Regions of interest* (ROIs) are used to extract numerical data from these images. The size, shape, and position of ROIs can be defined and positioned by the user, using a selection of predefined geometric shapes (e.g., rectangles, circles). Alternatively, irregular ROIs can be created using a cursor on the image display. The computer then reports ROI statistics such as the mean pixel value, the standard deviation of the pixel values, and the number of pixels in the ROI (Fig. 20-9). Software tools that use edge-detection algorithms also are available for automated definition of ROIs (see Section B.5).

Care must be taken in the use of ROIs to accurately place them on the tissues of interest, especially for applications in which radiotracer uptake or concentration are

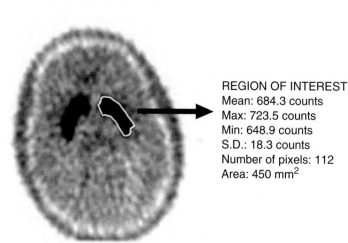

REGION OF INTEREST
Mean: 684.3 counts
Max: 723.5 counts
Min: 648.9 counts
S.D.: 18.3 counts
Number of pixels: 112
Area: 450 mm²

FIGURE 20-9   Manually drawn region of interest (ROI) placed over the right striatum (a small, gray matter structure deep in the brain) in an ^{18}F-fluoroDOPA PET brain study. This tracer reflects dopamine synthesis and accumulates in the dopaminergic (dopamine-producing) neurons of the striatum. Typically, ROI software programs provide a set of statistics on the number of counts per pixel per second (proportional to the radioactivity concentration) and the area of the ROI.

monitored during a longer period. Such measurements sometimes are made, for example, to observe and monitor the metabolic status of a tumor. Automated methods are provided on some computer systems to assist with this task. However, even with an automated method, one must be aware of and careful to avoid errors caused by the partial-volume effects discussed in Chapter 17, Section B.5. For example, an important question for staging in tumor imaging is the concentration of radiotracer in the tumor. A "small" tumor with high uptake (Bq/cm^3) is different from a somewhat larger tumor with low uptake. An apparent increase in radiotracer concentration in a "small" tumor (one that is near the resolution limits of the imaging device) during a longer period actually could be due to tumor growth in size and an associated reduction in partial-volume effects. ROI analysis is therefore particularly difficult when the volume of tissue that is of interest is changing with time.

Most structures of interest extend in all three dimensions and cover multiple image planes in a tomographic dataset. To obtain the best signal-to-noise ratio (SNR), and to ensure that the ROI values are representative of the entire structure, it is necessary to draw ROIs on multiple contiguous image planes that contain the object. In some software these ROIs can be connected to form a *volume of interest* (VOI). Automated tools for defining VOIs also have been developed.

3. Time-Activity Curves

As discussed in Chapter 21, the rate of change of radiotracer uptake in a specific organ or tissue often is of interest. To determine this, the data are acquired as a series of frames over time (see Section A.4). The data typically are analyzed by defining an ROI on one frame, or on the sum of all frames, and then copying the ROI across all of the frames. This is accurate provided that the patient has not moved between frames. The process is illustrated in Figure 20-10 for a time series of images of the brain following the injection of ^{18}F-fluoro DOPA, a compound that localizes in the striatum whose uptake is related to the rate at which dopamine is synthesized.

The ROI data from a series of frames can be used to create a time-activity curve (TAC), showing the radiotracer concentration as a function of time in the tissue defined by the ROI. Figure 20-10 also shows the TAC extracted from the ROI data in the time series of images.

4. Image Smoothing

All nuclear medicine computer systems provide image-smoothing algorithms. Figure 20-11 illustrates the effect of smoothing on an image that has a poor SNR in the unprocessed state. Smoothing operations are, in essence, techniques that average the local pixel values to reduce the effect of pixel-to-pixel variation. Two simple algorithms for 2-D images are 5-point and 9-point smoothing, in which a pixel value is averaged with its nearest 4 or 8 neighbors (Fig. 20-12).

In the previous examples, all four or eight neighboring pixels, as well as the center pixel, are given equal weight in the smoothing process. Image smoothing also can be performed using filters that are weighted according to the distance from the pixel that is being smoothed. One such example is a gaussian smoothing filter. In general, one can write the following:

smoothed image

$$= \text{original image} * \text{smoothing filter} \qquad (20\text{-}3)$$

where * represents the operation of convolution (see Appendix G). Although smoothing frequently produces a more appealing image by reducing noise (and improving the SNR), it also results in blurring and potential loss of image detail. Sometimes it is convenient to perform analytic studies (e.g., integrating pixel count values over an area) on an unsmoothed image after identifying ROIs on a smoothed image. In such applications, a practical compromise between resolution and visual appeal must be reached.

5. Edge Detection and Segmentation

Edge detection and segmentation are two image-processing tools that can be used to assist in automatically defining ROIs. They also are used for classifying different types of tissue based on their radiotracer uptake and for defining the body and lung contours for attenuation correction (see Chapter 17, Section B.2).

Edge-detection algorithms work best with edges that are very clearly defined as a result of a sharp boundary in radiotracer uptake. One of the most common is the *Laplacian technique*. The 2-D Laplacian is defined by

$$\text{Laplacian} = \frac{\partial^2}{\partial x^2} + \frac{\partial^2}{\partial y^2} \qquad (20\text{-}4)$$

where $\partial^2/\partial x^2$ represents the second partial derivative of the function [i.e., the pixel

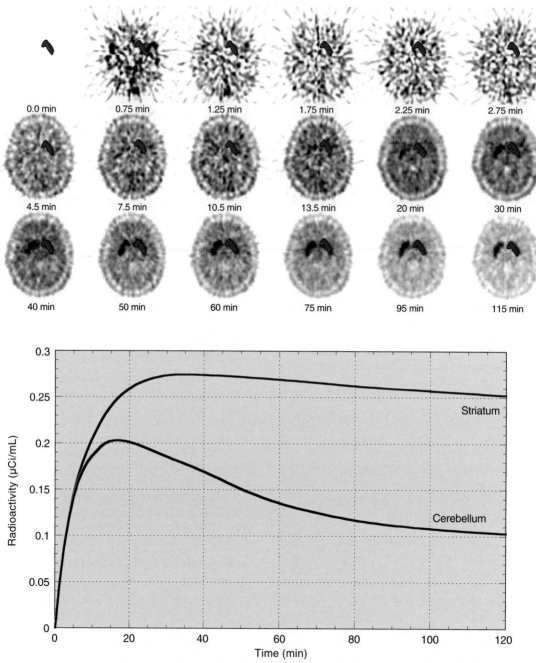

FIGURE 20-10   *Top,* PET images of the same two-dimensional (2-D) slice through the brain at different times after administration of a bolus injection of ^{18}F-fluoroDOPA. A region of interest (ROI) is drawn over the right striatum on the last image and then copied to all other time points. *Bottom,* Time-activity curve (TAC) showing the mean value in the ROI, converted with a calibration factor from counts per second per pixel to absolute concentration of radioactivity, versus time for the striatum. Also shown is a TAC for the cerebellum, taken from a different 2-D image slice, demonstrating how different brain regions can have different kinetics. Analysis of such TACs is discussed in Chapter 21. *(Adapted from Cherry SR, Phelps ME: Positron emission tomography. In Sandler MP, Coleman RE, Patton JA, et al [eds]:* Diagnostic Nuclear Medicine, *4th ed. Baltimore, Williams & Wilkins, 2002, p. 79.)*

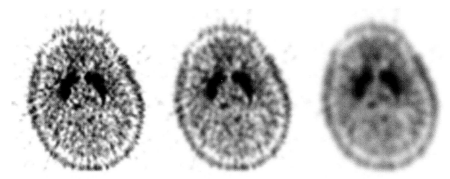

FIGURE 20-11 Effect of image smoothing using a gaussian filter with a full width at half maximum of 4 mm (*center*) and 8 mm (*right*). Smoothing improves the signal-to-noise ratio in the images but at the expense of spatial resolution.

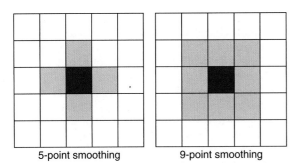

FIGURE 20-12 Illustration of pixels used in 5-point and 9-point smoothing. The value in the pixel of interest (*dark blue*) is averaged with the values in the surrounding pixels (*light blue*).

values, $p(x,y)$], with respect to spatial coordinate x, and similarly for the spatial coordinate y. An analogous definition can be made for the 3-D Laplacian, which would be applied to 3-D datasets. As illustrated by the 1-D example presented in Figure 20-13, the point at which the Laplacian crosses zero represents a region with a high rate of change between neighboring pixel values and therefore reflects an edge. In practice, the operator often specifies a starting point for the algorithm, which then searches in all possible directions and constructs a line (edge) where the Laplacian crosses zero. Eliminating small changes (i.e., setting limits) in the Laplacian reduces the effect of noise. Figure 20-14 shows an image in which the lung contours have been defined automatically by the Laplacian algorithm.

The goal of image segmentation is to group all pixels that have certain defined characteristics. In nuclear medicine, this usually refers to pixels that have a certain range of pixel intensities and thus a certain level of radiotracer uptake. The simplest method of segmentation is just to select pixels having values within a specified range: $A < p(x,y) < B$. Because of image noise, this simple method rarely is sufficient for accurate segmentation. More sophisticated algorithms that consider the underlying resolution and noise properties of the images and that also seek clusters of contiguous pixels usually are employed. Edge-detection algorithms also can be used in segmentation. For example, an edge-detection algorithm can be used to define the contours of the lungs and all pixels within that contour defined as lung tissue. Other examples of edge-detection and segmentation algorithms are discussed in the suggested readings at the end of the chapter.

6. Co-Registration of Images

It is quite common to perform multiple nuclear medicine imaging studies at different times on the same subject. This is useful, for example, to monitor the progression of Alzheimer's or Parkinson's disease, to measure the change in tumor blood flow or metabolism following cancer treatment, or to study the response of brain blood flow to various stimuli. To accurately compare nuclear medicine studies on the same subject performed at different times (e.g., by subtraction of the images), it is necessary that the images be accurately aligned. This is known as *intrasubject intramodality image co-registration*.

Many algorithms have been developed to co-register nuclear medicine studies. They have been particularly successful in the brain, because the brain is rigidly held within the skull and the transformations required to co-register the images are limited to simple translations and rotations. Co-registration accuracy can be as high as 1 to 2 mm. Figure

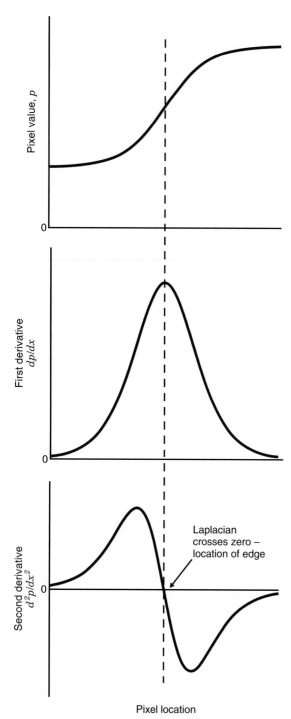

FIGURE 20-13 A one-dimensional example showing a count profile of pixel value p versus pixel location x across a boundary with two different activity levels. The first and second (Laplacian) derivatives of the count profile with respect to x are shown. The first derivative is the difference between neighboring pixel values, or equivalently the local slope of the count profile. At an edge, pixel values are changing rapidly with location and the slope has a high absolute value. The second derivative, the Laplacian, is the difference between adjacent first derivative values. The x-location where this second derivative crosses zero defines the location of the edge.

20-15 shows co-registered PET images of the same subject acquired at different times. Outside the brain, image co-registration becomes much more difficult, because organs can shift relative to each other depending on the exact positioning of the patient on the bed. In general, this requires nonlinear co-registration algorithms that attempt to "warp" one image to fit with the other.

Intrasubject, cross-modality co-registration involves registering a nuclear medicine study with a study of the same subject performed with a different imaging modality (e.g., MRI or CT). This requires more sophisticated algorithms because the information content of the images as well as their spatial resolution and SNR characteristics are different. Nonetheless, algorithms have been successful for co-registering PET and SPECT studies onto MRI or CT images, thereby providing co-registered volumetric datasets that reflect both biologic function (PET or SPECT) and anatomy (CT or MRI). Again, because of its precise geometric relationship to the skull, the most successful applications of these algorithms have been in brain imaging.

There also is interest in intersubject co-registration of nuclear medicine images for comparisons between multiple subjects. For example, this can be used to create images based on a large database of subjects, showing the distribution of a radiotracer across a specific population of subjects. If such a database of images of normal subjects is available for a particular radiotracer, a patient then can be compared with the normal database to see if the distribution is significantly different from normal controls. A summary of modern image co-registration techniques is provided in reference 1.

C. PROCESSING ENVIRONMENT

The digital image-processing environment may be limited to a single gamma camera, or, in a large department, it may involve a collection of gamma cameras and tomographic imaging devices. For all modes of imaging, digital image processing involves several steps: (1) acquisition, (2) processing, (3) display, (4) archiving (storing the raw or processed data or images), and (5) retrieval. Acquisition obviously takes place on the imaging device itself; however, it is useful if the other steps in the chain can be performed not only on the imaging system console but also on any other computers in the hospital or laboratory. In

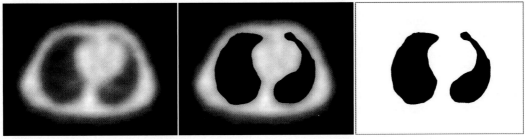

FIGURE 20-14 Series of images illustrating the segmentation of the lungs on a transmission scan acquired on a single-photon emission computed tomography system. (*Original image courtesy Dr. Freek Beekman, University Medical Delft University of Technology, The Netherlands.*)

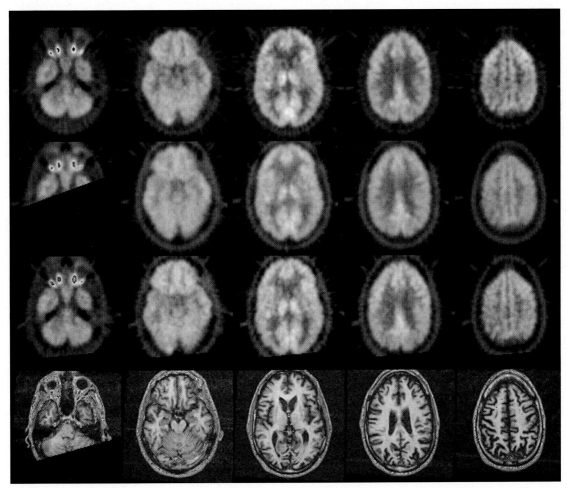

FIGURE 20-15 *Top three rows,* Co-registered slices from three [18]fluorodeoxyglucose PET scans acquired at 1-year intervals on the same subject. Images in each column represent the same anatomic slice, after co-registration. *Bottom row,* Corresponding co-registered slices from a magnetic resonance imaging scan acquired at the time of the third PET scan (intermodality co-registration). Images were co-registered using the Automated Image Registration software developed by Roger Woods of the University of California-Los Angeles. Note the excellent agreement in structures included and their locations in each slice. Some images were truncated (particularly in the *left column*) because parts of the brain were outside the field-of-view in some scans. (*From Woods RP, Mazziotta JC, Cherry SR: Optimizing activation methods: Tomographic mapping of functional cerebral activity. In Thatcher RW, Hallett M, Zeffiro T, et al [eds]:* Functional Neuroimaging: Technical Foundations. *San Diego, Academic Press, 1994, p. 54.*)

addition to freeing up the image acquisition computer for additional studies, it allows a variety of other activities to proceed simultaneously. For example, it allows a medical physicist to reprocess a study from his or her office while physicians are viewing the same images in the reading room or even at a different hospital and researchers are downloading the studies onto a computer in the research laboratory.

Many nuclear medicine departments therefore employ high-speed networks to connect their imaging systems together with other computer systems in the institution and, via the Internet, to the outside world. These departments use PACS to store and move images from acquisition sites to more convenient viewing stations and to provide a common basis for handling nuclear medicine and all other diagnostic imaging modalities.[2] PACS systems in hospitals with large radiology and nuclear medicine departments must be capable of handling huge amounts of data, typically several gigabytes per day (1 Gb = 10^{12} bytes).

To facilitate the exchange and handling of images from multiple different imaging modalities and from different vendors each with their own custom software, image file format standards have been developed. The central standard in radiology is the Digital Imaging and Communications in Medicine (DICOM) standard described in reference 3, and all manufacturers producing diagnostic imaging equipment support this standard. The objective of DICOM is to enable vendor-independent communication not only of images but also of associated diagnostic and therapeutic data and reports.

A true archival system not only should store the image or processed image data but also should be organized around a logical retrieval system that permits correlation of images with other types of data (e.g., reports) for a given patient study. That is, it should have the capacity of a computer database system and interface seamlessly with radiology, nuclear medicine, and hospital information systems. In addition, the system must be capable of protecting patient information by providing access only to authorized users.

REFERENCES

1. Hill DLG, Batchelor PG, Holden M, Hawkes DJ: Medical image registration. *Phys Med Biol* 46:R1-R45, 2001.
2. Bick U, Lenzen H: PACS: The silent revolution. *Eur Radiol* 9:1152-1160, 1999.
3. Mildenberger P, Eichelberg M, Martin E: Introduction to the DICOM standard. *Eur Radiol* 12:920-927, 2002.

BIBLIOGRAPHY

General references on image processing
Gonzalez RC, Woods RE: *Digital Image Processing*, ed 3, Upper Saddle River, NJ, 2008, Pearson Prentice Hall. (Chapters 2, 3, 6, and 10 are especially relevant.)
Robb RA: *Biomedical Imaging, Visualization, and Analysis*, ed 2, New York, 1999, Wiley-Liss.

Tracer Kinetic Modeling

The spatial distribution of a radiotracer in the body is time varying and depends on a number of components such as tracer delivery and extraction from the vasculature, binding to cell surface receptors, diffusion or transport into cells, metabolism, washout from the tissue, and excretion from the body. Thus the temporal component often is very important in nuclear medicine studies, and the timing of the imaging relative to the administration of the radiopharmaceutical must be carefully chosen such that the images reflect the biologic process of interest. Furthermore, the rate of change of radiotracer concentration often provides direct information on the rate of a specific biologic process. This chapter discusses how the temporal information that can be obtained from nuclear medicine studies is incorporated to provide quantitative measures of physiologic parameters, biochemical rates, or specific biologic events. Further examples are provided in reference 1.

A. BASIC CONCEPTS

Dynamic nuclear medicine studies enable the radiotracer concentration to be measured as a function of time, as shown in Figure 20-10. With an understanding of the biologic fate of the radiotracer in the body, it is possible to construct mathematical models with a set of one or more parameters that can be fit to explain the observed time-activity curves. In some cases the model parameters can be related directly to physiologic or biologic quantities. Examples include tissue perfusion (measured in mL/min/g) and the rate of glucose use (measured in mol/min/g). The mathematical models that describe the time-varying distribution of radiopharmaceuticals in the body are known as *tracer kinetic models*.

Tracer kinetic models may be very simple. For example, one method for evaluating renal function is to measure the uptake of ^{99m}Tc-labeled dimercaptosuccinic acid (DMSA) using a single region of interest (ROI) positioned over each kidney at one instant in time. "Function" in this case is determined in relative rather than absolute physiologic units. A more rigorous approach for evaluating kidney function is to measure glomerular filtration rates (GFRs), in mL/min, using a tracer that is filtered by the kidneys, such as ^{99m}Tc-labeled diethylene triamine pentaacetic acid (DTPA). In this case, it is necessary to obtain serial images of the kidneys and also to collect blood samples to measure tracer concentration in the blood as a function of time. Using these data and applying an appropriate mathematical model, one can then calculate the GFR.

Each of these approaches permits an assessment of "renal function" that is based on a different model for the behavior of the kidneys. The approach of choice depends on the medical or biologic information desired, as well as on the equipment available and acceptable level of technical complexity. Developing a model requires the investigator to synthesize a large amount of biologic information into a comprehensive description of the process of interest. This chapter summarizes some of the principles and techniques in developing these models and presents some examples of tracer kinetic models currently used in nuclear medicine.

The following example illustrates the principle of tracer kinetic techniques. Figure 21-1 shows a hollow tube with a substance flowing through it. If a small amount of tracer is injected instantaneously at time t and at point A and the measured activity at point B is plotted as a function of time, the resultant time-activity curve represents a histogram of the transit times for the tracer molecules

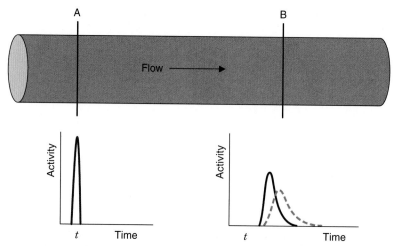

FIGURE 21-1 Illustration of use of tracer kinetics for measurement of flow. The system consists of a hollow tube characterized by flow in the direction indicated. A bolus of tracer introduced at time t and point A produces a time-activity curve at location B that depends on flow. Relatively higher flow (*solid line*) results in less dispersion and a shorter average transit time, whereas a lower flow rate (*dashed line*) produces a longer average transit time. (*Adapted from Phelps ME, Mazziotta JC, Schelbert HR:* Positron Emission Tomography and Autoradiography: Principles and Applications for the Brain and Heart. *New York, 1986, Raven Press.*)

from point A to point B. If the flow rate through the tube is decreased (*dashed curve* in Fig. 21-1), then the tracer molecules will on average take longer to get from point A to point B and the shape of the measured time-activity curve will change accordingly. This simple example illustrates conceptually how the kinetic information (i.e., the time-activity curve) varies in response to a change in a parameter in the system (flow rate). The flow rate, F, through the tube can be calculated as

$$F(mL/min) = V(mL)/\tau(min) \quad (21\text{-}1)$$

where V is the volume of the tube and τ is the mean transit time of the tracer molecules between points A and B. This is known as the *central volume principle*.

B. TRACERS AND COMPARTMENTS

Most applications of tracer kinetic principles in nuclear medicine are based on *compartmental models*. In this section, we review the basic principles of compartmental modeling.

1. Definition of a Tracer

A tracer is a substance that follows ("traces") a physiologic or biochemical process. In this chapter, tracers are assumed to be radio-nuclides or, more commonly, small molecules or larger biomolecules (e.g., antibodies and peptides) that are labeled with radionuclides. These labeled molecules are also known as *radiotracers* or *radiopharmaceuticals*. For simplicity, we refer to them as *tracers* in the remainder of this discussion. Tracers can be naturally occurring substances, analogs of natural substances (i.e., substances that mimic the natural substance), or compounds that interact with specific physiologic or biochemical processes in the body. Examples include diffusible tracers for blood flow, tracers that follow important metabolic pathways in cells, and tracers that bind to specific receptors on cell surfaces. Table 21-1 lists some examples of tracers that are used in nuclear medicine and their applications.

Some specific requirements for an ideal tracer include the following:

1. The behavior of the tracer should be identical or related in a known and predictable manner to that of the natural substance.
2. The mass of tracer used should not alter the underlying physiologic process being studied or should be small compared with the mass of endogenous compound being traced (a typical "rule of thumb" is that the mass of tracer should be <1% of the endogenous compound).
3. The specific activity of the tracer should be sufficiently high to permit imaging and blood or plasma activity assays without violating the first two requirements.
4. Any isotope effect (see Chapter 3, Section B) should be negligible or at least quantitatively predictable.

TABLE 21-1
SELECTED EXAMPLES OF TRACERS USED IN NUCLEAR MEDICINE

Process	Tracer
Blood flow/perfusion:	
Diffusible (not trapped)	$H_2^{15}O$, ^{133}Xe, ^{99m}Tc-teboroxime (heart)
Diffusible (trapped)	$^{201}TlCl$ (heart), ^{99m}Tc-sestamibi (heart),
	$^{13}NH_3$ (heart), $^{82}RbCl$, ^{99m}Tc-ECD (brain),
	^{99m}Tc-tetrofosmin (heart), ^{62}Cu-PTSM, ^{99m}Tc-HMPAO (brain)
Nondiffusible (trapped)	^{99m}Tc-macroaggragated albumin (lung)
Blood volume	^{11}CO, ^{51}Cr-RBC, ^{99m}Tc-RBC
Ventricular function	^{99m}Tc-pertechnetate, ^{99m}Tc-DTPA
Esophageal transit time/reflux	^{99m}Tc-sulphur colloid
Gastric emptying	^{99m}Tc-sulphur colloid, ^{111}In-DTPA
Gallbladder dynamics	^{99m}Tc-disofenin, ^{99m}Tc-mebrofenin
Infection	^{111}In-WBC, ^{67}Ga-citrate, ^{99m}Tc-WBC
Lung ventilation	^{133}Xe, ^{81}Kr, ^{99m}Tc-technegas™
Metabolism:	
Oxygen	$^{15}O_2$
Oxidative	^{11}C-acetate
Glucose	^{18}F-fluorodeoxyglucose
Free fatty acids	^{11}C-palmitic acid, ^{123}I-hexadecanoic acid
Osteoblastic activity	^{99m}Tc-MDP, $^{18}F^-$
Hypoxia	^{18}F-fluoromisonidazole, ^{62}Cu-ATSM
Proliferation	^{18}F-fluorothymidine
Protein synthesis	^{11}C-leucine, ^{11}C-methionine
Receptor systems:	
Dopaminergic	^{18}F-fluoro-L-dopa, ^{11}C-raclopride, ^{18}F-fluoroethylspiperone, ^{11}C-CFT
Benzodiazepine	^{18}F-flumazenil
Opiate	^{11}C-carfentanil
Serotonergic	^{11}C-altanserin
Adrenergic	^{123}I-mIBG
Somatostatin	^{111}In-octreotide
Estrogen	^{18}F-fluoroestradiol

ATSM, diacetyl-bis (N^4-methylthiosemicarbazone); CFT, [N-methyl-^{11}C]-2-β-carbomethoxy-3-β-(4-fluorophenyl)-tropane; DOPA, 3,4-dihydroxyphenylalanine; DTPA, diethylenetriamine penta-acetic acid; ECD, ethyl cysteinate dimer; HMPAO, hexamethyl propylene amine oxime; MDP, methylene diphosphonate; mIBG, metaiodobenzylguanidine; PTSM, pyruvaldehyde bis(N^4-methylthiosemithiocarbazone); RBC, red blood cell; WBC, white blood cell.

If a tracer is labeled with an element not originally present in the compound (this is often the case with radionuclides such as ^{99m}Tc, ^{123}I, and ^{18}F), it should behave similarly to the natural substance or in a way that differs in a known manner. The strictness of this requirement depends on the process under investigation. One common use of tracers in clinical nuclear medicine is to examine gross function and distribution, including blood flow, filtration, and ventilation. Although the elements represented by radionuclides such as ^{99m}Tc, ^{67}Ga, ^{111}In, and ^{123}I are not normally present in biologic molecules, it is possible to incorporate these radionuclides in physiologically relevant tracers that can measure simple parameters that are related to distribution, transport, and excretion.

However, these same elements are not normally present in human biochemistry (iodine is an exception when used to study thyroid metabolism). It is therefore much more difficult to mimic a biochemical reaction sequence with these radionuclides. The biochemical systems of the body are more specific than the transport processes that move or filter fluids or gases. Biochemical systems can selectively require that compounds be of one optical polarity versus the other; that compounds fit within angstroms in the cleft of an enzyme; that chemical bond angles, lengths, and strengths are appropriate; and so forth. When a compound is labeled with a foreign species, such as ^{99m}Tc, one cannot be sure that it will retain its natural properties and a careful examination and characterization of the compound must be undertaken. One of the advantages of radionuclides that represent elements normally involved in biochemical processes, such as ^{11}C, ^{13}N, and ^{15}O, is that they generally do not alter the behavior of the labeled compound.

Analog tracers are compounds that possess many of the properties of natural compounds but with differences that change the way the analog interacts with biologic systems. In many cases, analog tracers are deliberately created to simplify the analysis of a biologic system. For example, analogs that participate through only a limited number of steps in a sequence of biologic reactions have been developed in biochemistry and pharmacology. Analogs are used to decrease the number of variables that must be measured, to increase the specificity and accuracy of the measurement, or to selectively investigate a particular step in a biochemical sequence. In other cases analog tracers are used because of the need to label the tracer with an element that is not normally present in the molecule of interest. As discussed earlier, this can lead to very significant deviations in the biologic properties (particularly in small molecules) compared with the natural compound. Correction factors based on the principles of competitive substrate or enzyme kinetics are employed in studies using analog tracers to account for differences between the analog and the natural compound. A well-known and widely used example of an analog tracer in nuclear medicine is 2-deoxy-2[^{18}F]fluoro-D-glucose (FDG) to measure glucose metabolism (see Section E.5).

2. Definition of a Compartment

A *compartment* is a volume or space within which the tracer rapidly becomes uniformly distributed; that is, it contains no significant concentration gradients. In some cases, a compartment has an obvious physical interpretation, such as the intravascular blood pool, reactants and products in a chemical reaction, substances that are separated by membranes, and so forth. For other compartments, the physical interpretation may be less obvious, such as a tracer that may be metabolized or trapped by one of two different cell populations in an organ, thus defining the two populations of cells as separate compartments. Additionally, although the definition of a particular compartment may be appropriate for one tracer [e.g., the distribution of labeled red blood cells (RBCs) in the intravascular blood pool], it might not apply for a different tracer (e.g., the distribution of thallium or rubidium, which has both an intravascular and an extravascular distribution). Thus the number, interrelationship, organization, and definition of compartments in a compartmental model must be developed from knowledge of physiologic and biochemical principles.

3. Distribution Volume and Partition Coefficient

A compartment may be *closed* or *open* to a tracer. A closed compartment is one from which the tracer cannot escape, whereas an open compartment is one from which it can escape to other compartments. Whether a compartment is closed or open depends on both the compartment and the tracer. Indeed, a compartment may be open to one tracer and closed to another. If a tracer is injected into a closed compartment, such as a nondiffusible tracer in the vascular system, conservation of mass requires that after the distribution of the tracer reaches equilibrium or steady-state conditions, the amount of tracer injected, A (in becquerels or other units of activity), must equal the concentration of the tracer in the compartment, C (in Bq/mL), multiplied by the *distribution volume*, V_d, of the compartment. Thus,

$$V_d = A/C \text{ (at equilibrium)} \qquad (21\text{-}2)$$

This equation is the basis for the *dilution principle,* which provides a convenient method for determining the distribution volume of a closed compartment, as shown by the following example.

EXAMPLE 21-1

What is the distribution volume of the RBCs if 1 MBq ^{51}Cr-labeled RBCs is injected into the blood stream and an aliquot of blood taken after an equilibration period (10 minutes) contains 0.2 kBq/mL? Assume the hematocrit, H (fraction of the total blood volume occupied by RBCs), is 0.4.

Answer
From Equation 21-2:

$$V_d = (1000 \text{ kBq})/(0.2 \text{ kBq/mL})$$
$$= 5000 \text{ mL}$$

This result gives the total distribution volume, that is, total blood volume. The RBC volume is given by

$$V_{RBC} = H \times V_d$$
$$= 0.4 \times 5000 \text{ mL}$$
$$= 2000 \text{ mL}$$

More commonly, a compartment will be open; that is, the tracer will be able to escape from it. This applies, for example, to tracers that are distributed and exchanged between blood and tissue. In this case, after the tracer reaches its equilibrium distribution,* the concentration in blood will typically be different from that in the tissue (Fig. 21-2A). The ratio of tissue concentration C_t (Bq/g) to blood concentration C_b (Bq/mL) at equilibrium, is called the *partition coefficient*, λ, defined by

$$\lambda (\text{mL}/\text{g}) = C_t(\text{Bq/g})/C_b(\text{Bq/mL}) \quad (21\text{-}3)$$

The equilibrium blood concentration, C_b, can be directly measured by taking blood samples. If one assumes that the concentration of tracer in tissue is the same as the concentration in blood (Fig. 21-2B), and applies Equation 21-2, this leads to an apparent distribution volume in tissue given by $V_1 = A_t/C_b$, in which A_t is the activity in the tissue. One also knows that $A_t = C_t \times V_t$, in which V_t is the volume (or

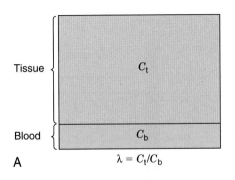

A

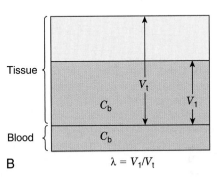

B

FIGURE 21-2 *A,* Partition coefficient, λ, for tracers that can diffuse or be transported into tissue from blood. The value of λ is given by the ratio of tissue-to-blood concentrations of the tracer when it has reached an equilibrium or steady-state condition. *B,* Partition coefficient also equals the ratio of the apparent distribution volume in tissue, V_1, assuming the same tracer concentration as blood-to-tissue volume (or mass), to V_t. *(From Phelps ME, Mazziotta JC, Schelbert HR: Positron Emission Tomography and Autoradiography: Principles and Applications for the Brain and Heart. New York, 1986, Raven Press.)*

mass) of tissue; therefore combining these relationships and Equation 21-3 yields

$$\lambda = V_1/V_t \quad (21\text{-}4)$$

Thus another interpretation of the partition coefficient is that it is the distribution volume per unit mass of tissue for a diffusible substance or tracer. This interpretation is employed in some models for estimating blood flow and perfusion, as discussed in Section E.

4. Flux

Flux refers to the amount of substance that crosses a boundary or surface per unit time (e.g., mg/min or mol/min) (Fig. 21-3). It also can refer to the transport of a substance between different compartments in terms of flux per unit volume or mass of tissue (e.g., mol/min/mL or mg/min/g).

Flux is a general term that can refer to a variety of processes. For example, the total mass of RBCs moving through a blood vessel

---

*Note that "equilibrium" in this case means that the concentration of the tracer in the compartments has reached a constant value with time. It does not imply equilibrium in the thermodynamic sense, that is, that there is no further transport of tracer between tissue and blood. Thus tracer equilibrium is synonymous with the term *steady state* (see Section B.6).

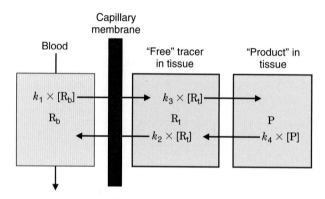

FIGURE 21-3 Three-compartment system consisting of reactants in blood (R_b) and tissue (R_t) and product in tissue (P). Fluxes between the compartments, indicated by *arrows,* are products of the first-order rate constants and the respective compartmental concentrations. (*Adapted from Phelps ME, Mazziotta JC, Schelbert HR:* Positron Emission Tomography and Autoradiography: Principles and Applications for the Brain and Heart. *New York, 1986, Raven Press.*)

per unit time is a flux. The "boundary" or "surface" in this case could be any transverse plane through the vessel. The amount of glucose moving across a cell membrane per unit time also is a flux. Fluxes therefore may either be closely related or unrelated to blood flow.

5. Rate Constants

Rate constants describe the relationships between the concentrations and fluxes of a substance between two compartments. For simple *first-order processes,* the rate constant, *k,* multiplied by the amount (or concentration) of a substance in a compartment determines the flux:

$$\text{flux} = k \times \text{amount of substance in compartment} \tag{21-5}$$

For first-order processes, the units of k are $(\text{time})^{-1}$. If "amount" refers to the mass of tracer in the compartment, the units of flux are mass/time (e.g., mg/min). If "amount" refers to concentration of tracer in the compartment, the units of flux are mass/time per unit of compartment volume (e.g., mg/min/mL), or mass/time per unit of compartment mass (e.g., mg/min/g). Note that, as illustrated by Figure 21-3, different directions of transport between two compartments can be characterized by different rate constants.

A first-order rate constant also may represent the fractional rate of transport of a substance from a compartment per unit time. For example, a rate constant of $0.1\ \text{min}^{-1}$ corresponds to a transport of 10% of the substance from the compartment per minute. The inverse of the rate constant, $1/k$, is sometimes referred to as the *turnover time,* or *mean transit time,* τ, of the tracer in the compartment (in this example, 10 minutes). Similarly,

the *half-time of turnover,* $t_{1/2}$, that is, the time required for the original amount of tracer in the compartment to decrease by 50% (assuming no back transfer into the compartment), is given by

$$t_{1/2} = 0.693/k \tag{21-6}$$

Thus the fractional rate constant k is analogous to the decay constant λ for radioactive decay, whereas the mean transit time is analogous to the average lifetime of a radionuclide (see Chapter 4, Section B.3). In first-order models, transport out of a compartment through a single pathway (without back-transport) is described by a single exponential function, e^{-kt}, analogous to the radioactive decay factor $e^{-\lambda t}$.

If there is more than one potential pathway for a tracer to leave a compartment, each characterized by a separate rate constant, k_i, then the turnover time of the tracer in the compartment is the inverse of the sum of all these rate constants and the half-time of turnover is

$$t_{1/2} = 0.693/(k_1 + k_2 + \ldots + k_m) \tag{21-7}$$

where m is the number of pathways by which the tracer can leave the compartment.

Most compartment models used in nuclear medicine are based on the assumption that first-order kinetics describe the dynamics of the system of interest. The tracer kinetics of such systems are linear. That is, doubling the input (amount or concentration) doubles the output (flux) of the system. As shown in Section E, linear first-order tracer kinetic models adequately describe many systems even when the dynamics of the natural substances are nonlinear.

A more general expression for the relationship among rate constants, fluxes, and concentrations (or masses) is

$$\text{flux} = k \times (\text{mass or concentration of substance})^n \tag{21-8}$$

where n refers to the *order* of the reaction. The units of rate constants for n^{th} order reactions (in terms of concentration) are [concentrations$^{(1-n)} \cdot \text{time}^{-1}$]. Thus only first-order rate constants represent a constant fractional turnover and Equations 21-6 and 21-7 apply only to first-order processes.

Figure 21-3 illustrates a three-compartment system consisting of a blood compartment separated by a membrane barrier (e.g., capillary wall) from two sequential tissue compartments. R and P refer to chemical reactant and product, whereas the subscripts b and t refer to reactant in blood and tissue compartments, respectively. $[R_b]$,* $[R_t]$, and $[P]$ are the blood and tissue concentrations of reactant and product, whereas the fluxes between the compartments are the first-order rate constants, k_1, k_2, k_3, and k_4, multiplied by corresponding concentrations. The thicknesses of the arrows in Figure 21-3 are proportional to the magnitude of the corresponding rate constant. In this example, the rate constants into and out of tissue are larger than the corresponding rate constants between the reactant and product compartments in tissue. Thus the majority of the reactant initially transported into the tissue space is transported

---

*The notation $[X]$ is used to denote the concentration (usually in units of g/mL or mol/mL) of substance X.

back into blood without undergoing any biochemical reactions. This is a common occurrence in actual biochemical systems and introduces a reserve capacity into the system that can accommodate changes in metabolic supply and demand (e.g., by changing k_3).

Figure 21-4 illustrates the relationship between first-order rate constants and the relative concentrations of the substrates in a biochemical sequence. If a substrate (S) and enzyme (E) combine to form a substrate-enzyme complex (SE), which then dissociates into a product (P) with release of the enzyme, the fluxes of the first-order reaction steps are concentrations multiplied by the corresponding rate constants. If a small amount of labeled substrate is introduced into the system at time zero, the tracer will go through the reaction steps, producing concentrations of labeled S, SE, and P as shown in the graphs in Figure 21-4. If k_3 (the forward rate constant for the reaction converting SE to E and P) is reduced by 50% with all the other rate constants remaining unchanged, the concentrations of labeled S, SE, and P are then represented by the dotted orange lines in Figure 21-4. Decreasing k_3 causes a slower production of P and causes a compensatory increase in labeled S and SE.

6. Steady State

The term *steady state* refers to a condition in which a process, parameter, or variable is not changing with time. For example, a flux through a biochemical pathway is said to be in a steady state when the concentration of reactants and products are not changing with time. In all tracer kinetic models, it is assumed that the underlying process that is being measured

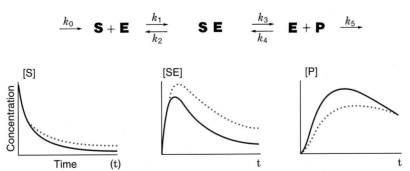

FIGURE 21-4 Illustration of tracer kinetics of a chemical reaction sequence. First-order rate constants (k_0, k_1, k_2, k_3, k_4, k_5) characterize the various reaction steps, whereas S refers to substrate, E refers to enzyme, P is product, and SE is the substrate-enzyme complex. The time-activity relationships for concentrations of labeled S, SE, and P are shown for a particular value of k_3 (*solid purple line*) and with k_3 reduced by 50% (*dotted orange line*). (*From Phelps ME, Mazziotta JC, Schelbert HR: Positron Emission Tomography and Autoradiography: Principles and Applications for the Brain and Heart. New York, 1986, Raven Press.*)

by the tracer is in a steady state. Because of biorhythms, steady states almost never exist in the body; however, if the magnitude or temporal period of change is small compared with the process being measured, then the steady-state assumption is reasonable. In many cases, the experimental sampling rate is slow compared with the biorhythm (e.g., blood sampling rate vs. pulsatile nature of blood flow) and it is not perceived in the measured data. In these cases, the measured parameters represent average values of the function measured. However, if the experimental sampling rate is fast compared with the biorhythm, significant errors can be introduced in the model calculations. In this case, the calculated parameters typically do not represent a simple average of the non-steady-state values.

Steady state of a process should not be confused with steady state of the tracer. Measurements of the tracer commonly are made when the tracer itself is not in steady state but rather while it is distributing through the process under study. Some tracer kinetic models are used in which measurements are made when both the tracer and process studied are in a steady state. These methods usually are referred to as "equilibrium" models.

An important and useful property of a steady-state condition is that the rates (fluxes) of all steps in a nonbranching transport or reaction sequence are equal. Thus if a tracer technique is used to measure one step in a sequence, the rate for each step in the entire sequence can be determined. If the reaction branches into two or more separate pathways then the sum of each pathway must equal the rate of the preceding step. In this case, if one determines the rate of any of the preceding steps and also knows the branching fractions, then the rate of each branch can be determined by multiplying the rate of the preceding step by the branching fraction. For example, if the reaction sequence in Figure 21-5 is in a steady state and the rate of disappearance of A is R_A, the rates of formation of B, C, D, and E are R_B, R_C, R_D, and R_E, respectively, and f_d and f_e are the branching fraction down the corresponding pathways, then

$$R_A = R_B = R_C = (R_D + R_E) \qquad (21\text{-}9)$$

and

$$R_D = f_D \times R_C \qquad (21\text{-}10)$$

$$R_E = f_E \times R_C \qquad (21\text{-}11)$$

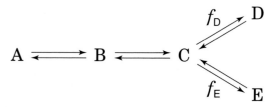

FIGURE 21-5   Example of a multistep reaction sequence that branches into two pathways. The terms f_D and f_E are the branching fractions for the corresponding pathways ($f_D + f_E = 1$).

where

$$f_D + f_E = 1 \qquad (21\text{-}12)$$

C. TRACER DELIVERY AND TRANSPORT

A tracer that is injected into the body must follow several steps in sequence before it can enter a biochemical pathway: delivery to the capillary via blood flow, extraction across the capillary wall into the tissue space, and finally, incorporation into a biochemical reaction sequence. Although only one of the steps in a process may be of interest in a particular application, it may be influenced by other steps in the process of tracer delivery. In this section we examine tracer techniques for describing these processes.

1. Blood Flow, Extraction, and Clearance

Blood flow through vessels is described in units of volume per unit time (usually in units of mL/min). For regional tissue measurements it is blood flow per mass of tissue that is determined (mL/min/g). Blood flow per mass of tissue is more properly referred to as *perfusion*; however, in the literature the term *blood flow* is used to indicate both blood flow and blood flow per mass of tissue. In both cases the basic phenomenon is still blood flow. Thus relationships involving blood flow apply equally to blood flow and perfusion, provided that care is taken to ensure that the units are consistent. For example, in the relationship between blood flow and blood volume (see Section E.4), if blood flow is in units of mL/min, then blood volume must be in units of mL. If blood flow is in units of mL/min/g, then blood volume must be in units of volume per mass of tissue (mL/g). In this text, the term *blood flow,* symbolized by F, is used to denote either blood flow or blood flow per mass of tissue. The units indicate which quantity is being discussed.

In addition to its dependence on blood flow, the uptake of a tracer by tissue depends on tissue extraction and clearance. Extraction is defined in two different contexts: net and unidirectional. *Net extraction* refers to the difference in steady-state tracer concentrations between the input and output blood flow of an organ. If the input (arterial) concentration is C_A, and the output (venous) concentration is C_V, the net extraction fraction, E_n, is defined as

$$E_n = (C_A - C_V)/C_A \qquad (21\text{-}13)$$

If there is no metabolism of the tracer, that is, if all the tracer delivered to the tissue eventually is returned to the blood, the net extraction is zero. This situation applies, for example, to inert diffusible blood flow tracers when steady-state conditions for the tracer are reached.

Unidirectional extraction refers to the amount of tracer extracted only from blood to tissue. It does not include the amount transferred back from tissue to blood. Thus the unidirectional extraction fraction, E_u, generally is larger than the net extraction fraction. An exception to this general rule occurs with O_2. Virtually all oxygen extracted by tissue is metabolized; thus the net and unidirectional extraction fractions are the same. For essentially all other substances, a major portion of what is extracted by the tissue is transported back to blood. This is the situation represented by the bidirectional transport in the model shown in Figure 21-3.

Extraction fractions are expressed as fractions or as percentages and can be measured using tracer kinetic techniques. To determine net extraction, it is necessary to measure the input and output concentrations of the tracer in the blood under steady-state conditions, that is, after the concentrations have reached constant values. The route of administration of the tracer is unimportant in this case. For example, the tracer can be administered by constant infusion or as a bolus into a peripheral vein.

Unidirectional extraction can be measured by observing the rate of uptake by the tissue or organ immediately after injection of the tracer, that is, when the blood concentration is maximum and the tissue concentration is zero. Measurements of unidirectional

extraction are useful for studying the transport properties of substrates and drugs.

An important concept relating the processes of blood flow, flux, and extraction is the *Fick principle.* This principle is based on the conservation of mass and states that, under steady-state conditions, the net uptake of a tracer (or other substance) is simply the difference between the input to and output from the organ or tissue. If the input (arterial) concentration of the tracer is C_A (mg/mL) and the output (venous) concentration is C_V (mg/mL), and the blood flow to the organ is F (mL/min), then the net uptake rate, U (mg/min), is given by

$$U = F \times (C_A - C_V) \qquad (21\text{-}14)$$

As an example, if the arterial and venous concentrations of oxygen and the blood flow to an organ are measured, Equation 21-14 can be used to determine the oxygen utilization rate for that organ. If blood flow F in Equation 21-14 is replaced by blood flow per mass of tissue (perfusion), then the uptake or utilization is given in units of utilization per mass of tissue (mg/min/g).

The Fick principle can be employed only under steady-state conditions. An alternative approach that is applicable to non–steady-state conditions is the Kety-Schmidt method, which is discussed in Section E.4.

The extraction of tracers generally occurs across membranes or through the fenestrations* found in capillaries. The extraction fraction of a tracer depends on the capillary surface area, S, the capillary permeability for the tracer, P, and blood flow through the capillaries, F. A simple model relating these quantities was developed by Renkin[2] and Crone.[3] Figure 21-6 illustrates an idealized capillary

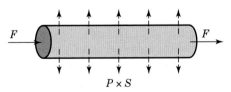

FIGURE 21-6 Renkin-Crone capillary model. The capillary is assumed to be a rigid tube, and extraction from blood to tissue is characterized by the product of the permeability, P, and surface area, S. Blood flow through the capillary is F.

---

*Fenestrations are small gaps found between the junctions of cells in the capillary wall.

(i.e., rigid tube) through which is passing a tracer with flow F. It is assumed that the concentration of tracer across the cross-section of the capillary at any point along its length is constant and that the extraction of tracer from the capillary to the tissue at any point is proportional to the concentration of the tracer in the blood. It is further assumed that extraction is unidirectional; that is, there is no back-transfer of the tracer from tissue to blood.

For this simple model, it can be shown that the unidirectional extraction fraction E_u for the capillary is given by

$$E_u = 1 - e^{-(P \times S/F)} \qquad (21\text{-}15)$$

Thus the extraction fraction depends only on the permeability-surface area product, $P \times S$ (mL/min), and on flow, F. This equation also can be stated in terms of perfusion, by replacing blood flow with perfusion (mL/min/g) and the permeability-surface area product with $P \times S$ for a capillary network per mass of tissue (mL/min/g).

EXAMPLE 21-2

What is the unidirectional extraction fraction for the diffusible tracer $^{13}NH_3$ in the brain if $P \times S = 0.25$ mL/min/g and blood flow to the brain (perfusion) is 0.50 mL/min/g?

Answer
From Equation 21-15:

$$E_u = 1 - e^{-(0.25/0.50)} = 0.39$$

The Renkin-Crone model is not completely realistic, because it assumes no back-transfer of tracer from tissue to blood, that the permeability-surface area product, $P \times S$, does not depend on blood flow, and that the capillary is a rigid tube. Nevertheless, it is instructive for illustrating the relationships between extraction fraction, blood flow, and the permeability-surface area product.

The ratio $[(P \times S)/F]$ sometimes is referred to as the *extraction coefficient*. One interpretation of the product, $P \times S$, is that it represents the flow of the tracer from blood to tissue through the capillary wall, whereas F represents flow through the capillary itself. These two "flows" represent competing processes for removal of the tracer from the capillary. Thus the fraction of a substance that is extracted by tissue can be increased either by increasing the flow through the capillary

wall, $P \times S$, or decreasing the blood flow through the capillary. For a given value of $P \times S$, the greater the flow, F, the shorter the residence time of the tracer in the capillary, and, thus, the less the chance that the tracer will escape through the capillary wall. This is illustrated graphically in Figure 21-7.

On the other hand, the *amount* of material entering the tissue depends on the product of blood flow times extraction fraction, $F \times E$. This product is sometimes referred to as *clearance* and has the same units as flow (i.e., mL/min or mL/min/g). In essence, it represents "virtual flow" from the capillary into the tissue. Typically the increased amount of tracer that is delivered to the capillary with increasing blood flow more than offsets the decrease in the extraction fraction, with the net result that clearance increases with flow. This is illustrated graphically in Figure 21-8. Note that for small values of $P \times S$, that is, low flow across the capillary membrane, clearance is low and reaches a plateau value for relatively low values of capillary blood flow, F, whereas for large values of $P \times S$, clearance continues to increase with increasing flow through the capillaries. This indicates that the clearance or deposition of a tracer into tissue will have a high or low dependence on blood flow, depending on whether it has a high or low value of $P \times S$.

Clearance of tracers from tissue to blood also is used in tracer kinetic modeling. The most common use of tissue-to-blood clearance is in the measurement of bidirectional transport and blood flow.

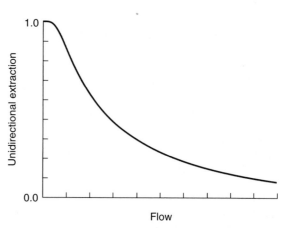

FIGURE 21-7 Unidirectional extraction fraction versus flow for the Renkin-Crone model. The extraction fraction is 1 when flow is near zero and decreases toward zero as flow increases.

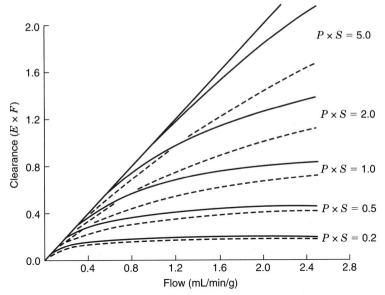

FIGURE 21-8 Clearance (flow × unidirectional extraction fraction) versus flow for various values of the permeability-surface area product, $P \times S$, for the Renkin-Crone model (*solid red lines,* Equation 21-15) and for a compartmental model description (*dashed red lines,* Equation 21-20). $P \times S$ is in units of mL/min/g. The *solid blue line* corresponds to the situation where $E=1$. (*From Phelps ME, Mazziotta JC, Schelbert HR:* Positron Emission Tomography and Autoradiography: Principles and Applications for the Brain and Heart. *New York, 1986, Raven Press.*)

2. Transport

Three different mechanisms exist for the transport of substances across a membrane or capillary wall. *Active* transport mechanisms require energy and can move substances against concentration gradients. Usually the energy source is adenosine triphosphate. Examples of active transport are the sodium-potassium "pump" that maintains the difference between intracellular and extracellular concentrations of these ions and the renal tubular reabsorption of glucose.

Passive transport mechanisms do not require energy and move substances in the same direction as the concentration gradient. The passive mechanisms include *carrier-mediated* diffusion, such as glucose and amino acid transport from blood to brain, and *passive* diffusion, which depends only on the existence of a concentration gradient, such as the diffusion of ^{99m}Tc-pertechnetate from blood to brain through a disrupted blood-brain barrier. Bulk flow across fenestrations in capillaries also accounts for a fraction of passive transport but varies from tissue to tissue; for example, it is insignificant in the brain when the blood-brain barrier is intact, but it is a major source of capillary/tissue transport in the heart.

Simple passive diffusion for a given membrane and molecule is characterized by a *diffusion constant, D.* Membrane permeability is related to the diffusion constant by

$$P \text{ (cm/min)} = D \text{ (cm}^2\text{/min)}/x \text{ (cm)} \quad (21\text{-}16)$$

where x is the thickness of the membrane or the diffusion path length. One usually deals with the $P \times S$ product rather than P itself because in most applications the regional capillary surface area S is not known accurately. The larger the value of D (or $P \times S$), the more rapid the passive diffusion process and the greater the clearance of the substance from blood to tissue or from tissue to blood (Fig. 21-8). Many substances and in vivo processes depend on passive diffusion mechanisms, such as water, oxygen, ammonia, and carbon dioxide.

Carrier-mediated diffusion is somewhat more complicated. It also transports substances in the same direction of a concentration gradient and is characterized by the following reaction process:

$$S + C \underset{k_2}{\overset{k_1}{\rightleftharpoons}} SC \underset{k_4}{\overset{k_3}{\rightleftharpoons}} C + S \quad (21\text{-}17)$$

where S is the substrate and C is the carrier molecule. SC is a carrier/substrate complex that physically moves across the membrane and then dissociates into C and S, and k_1, k_2, k_3, and k_4 are first-order rate constants that characterize the respective steps of the process. Because only a finite number of carrier molecules are available, this type of transport process can be saturated. The rate of the process increases as the substrate concentration S increases, but only to the point of saturating the number of available carrier sites. Generally, the carrier C is a protein enzyme that is neither created nor destroyed in the reaction but only enhances its rate. The kinetics of carrier-mediated diffusion are described further in Section E.5.

D. FORMULATION OF A COMPARTMENTAL MODEL

Most models commonly in use in nuclear medicine are compartmental models with first-order rate constants describing the flux of material between compartments. Consider the simple two-compartment model illustrated in Figure 21-9. If a tracer is present in compartment A with concentration defined by $C_A(t)$ (e.g., a tracer administered as a bolus injection so that the concentration in A is time dependent), then the rate of change in concentration of compartment B [i.e., $dC_B(t)/dt$] is described by

$$dC_B(t)/dt = \text{flux into B} - \text{flux out of B} \quad (21\text{-}18)$$

Because first-order kinetics apply, the flux into B is simply $k_1C_A(t)$ and the flux out of B is $k_2C_B(t)$; therefore

$$dC_B(t)/dt = k_1C_A(t) - k_2C_B(t) \quad (21\text{-}19)$$

This first-order ordinary differential equation with constant coefficients is a typical, although simple, example of the equations necessary to mathematically define a tracer kinetic model. The time course of the tracer in the delivery compartment (usually the blood) $[C_A(t)]$ is known as the *input function,* whereas the rate constants k_1 and k_2 are model parameters.

Typically, compartment A would represent the vascular compartment and compartment B might represent the amount of tracer present in the tissue. The parameters k_1 and k_2 can be estimated numerically with a technique known as *regression analysis.* The input function $C_A(t)$ is directly measured from blood samples or images of the blood (typically the left ventricular blood pool is chosen because of its large size), and the time course of activity in tissue $C_B(t)$ is determined from a time sequence of nuclear medicine images (see Fig. 20-10) or other counting measurements. A detailed discussion of regression analysis and other numerical methods related to tracer kinetic models is beyond the scope of this chapter, and the interested reader is referred to reference 1 for further details (see also Chapter 9, Section E.4).

The simple compartmental model illustrated in Figure 21-10 also can be used to describe the relationship between extraction and blood flow discussed in Section C.1 in reference to the Renkin-Crone model (Equation 21-15). In the compartmental model, the vascular space in the capillary is assumed to be a compartment of uniform concentration. The extraction and venous blood flow compete through the common vascular pool for removal of tracers. According to this compartmental model, the extraction E is related to flow F and the $P \times S$ product as:

$$E = (P \times S)/[(P \times S) + F] \quad (21\text{-}20)$$

Tissue extraction as a function of blood flow for this compartmental model is illustrated by the *dashed red lines* in Figure 21-8. The amount of extraction predicted with the compartmental model is lower than that predicted with the Renkin-Crone model for a given $P \times S$ value. In addition, the change in extraction with blood flow is somewhat different. In most

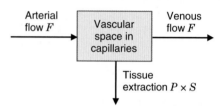

FIGURE 21-10 Compartmental model of capillary in which tissue extraction is competing with the venous flow for the tracer that is delivered by the arterial blood flow. *(From Phelps ME, Mazziotta JC, Schelbert HR:* Positron Emission Tomography and Autoradiography: Principles and Applications for the Brain and Heart. *New York, 1986, Raven Press.)*

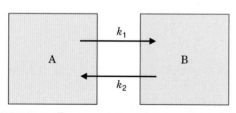

FIGURE 21-9 Transport between two compartments, A and B, is described by rate constants k_1 and k_2.

modeling applications of interest in nuclear medicine, the Renkin-Crone model (Equation 21-15) and the compartmental model (Equation 21-20) of capillary extraction yield substantially equivalent results.

In nuclear medicine studies, there generally are two sources of data that can serve as input to a tracer kinetic model: the time course of the injected radiotracer in whole blood or plasma (the *input function*), and the measured amount of activity in tissue (the *tissue response*), usually obtained from ROI analysis of a dynamic series of images. The ROI count values are equivalent to local tissue concentrations of the tracer of interest if appropriate corrections are made for attenuation and other causes of counting inaccuracy. Additionally, the devices used to measure radioactivity in blood and tissue (e.g., a well counter and a gamma camera) must be calibrated so corrections can be made to account for their differences in counting efficiencies.

Figure 21-11 illustrates the relationships between input function and tissue response curves for two different models and two different input functions. The curves for the first model (a three-compartment model with three rate constants) illustrate how the shape of the tissue response curve changes as the shape of the input function changes. In both cases, the kinetics of the tracer in tissue are identical, but the temporal response in tissue is influenced by the input function (e.g., changing the rate of injection could produce this pattern). The parameters of the model (the rate constants) are unchanged in these two situations. If one did not measure the input function in this example and measured only the tissue response function, the different tissue response functions might be interpreted incorrectly as a change in the rate of the physiologic or biochemical process in the tissue.

The second example (bottom row of Figure 21-11), shows a four-compartment model representing a more-complicated biological process. This model produces a different tissue response when presented with the same input function as the first example.

Thus Figure 21-11 shows that the shape of the tissue response is a function of both the input function and a characteristic of the tissue "system" called the *impulse response*.

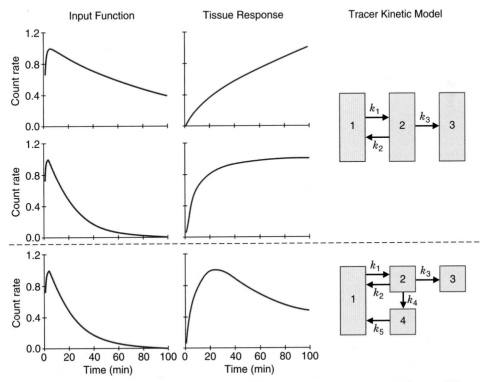

FIGURE 21-11 Input function and tissue response (sum of the tracer concentration in tissue) for two different models and for two different shapes of input functions. Different input functions produce different tissue responses for a given model configuration (*top two rows*), whereas the same input function produces different tissue responses for two different model configurations (*bottom two rows*). (*From Phelps ME, Mazziotta JC, Schelbert HR:* Positron Emission Tomography and Autoradiography: Principles and Applications for the Brain and Heart. *New York, 1986, Raven Press.*)

The impulse response of a linear system is the system's response when presented with an impulse as an input function. An impulse is in essence a function of infinitely short duration with an undefined magnitude at the origin and that has an integrated value of one if summed over all time. It can be thought of as an "idealized bolus" input given instantaneously (i.e., beyond zero time it has a zero value). In reality, a practical "impulse" delivery of tracer to a system has a duration shorter than the shortest vascular transit time through the organ. In this case there is no significant clearance of the tracer from the organ until after all the tracer has been delivered.

E. EXAMPLES OF DYNAMIC IMAGING AND TRACER KINETIC MODELS

In this section we show examples of the use of dynamic imaging and tracer kinetic models that are relatively simple and instructive. In research applications of nuclear medicine, a wide variety of more complex models are employed that are beyond the scope of this text. In many cases, there is considerable debate over the exact formulation of these complex models and the interpretation of the model parameters. Research in this area is of continued importance, particularly with the ongoing development of new radiotracers.

1. Cardiac Function and Ejection Fraction

^{99m}Tc-labeled RBCs are used for dynamic (first-pass) and equilibrium-gated studies of the heart. If a bolus of ^{99m}Tc-labeled RBCs is injected intravenously and their distribution through the thorax as a function of time is imaged dynamically with a gamma camera, the images will show the flow of blood through the venous system into the right atrium and right ventricle, to the lungs, and then to the left atrium, left ventricle, and out through the aorta to the systemic circulation. Abnormalities in this flow pattern (such as those produced by structural defects in the heart that cause intracardiac shunts) result in an abnormal distribution of the bolus of activity. For example, a ventricular septal defect with shunting of blood from the right to the left ventricle will result in an early appearance of the bolus of activity in the left ventricle as some of the blood (and labeled RBCs) bypasses the lungs and travels directly to the left ventricle. The amount of shunting can be estimated if time-activity curves of

the radiotracer distribution are generated and an appropriate model is applied.

Once the RBCs have distributed uniformly throughout the vascular system (i.e., equilibrium is reached), the gamma camera images will reflect regional blood volume. If the distribution of activity over the left ventricle is plotted as a function of time by gating the acquisition to the electrocardiogram signal, changes in ventricular volume caused by cardiac contractions can be seen (Fig. 21-12). This permits calculation of the ventricular ejection fraction (EF) because the relative number of counts within the ventricle is proportional to the blood volume of the ventricle (see Section B.3).

EXAMPLE 21-3

^{99m}Tc-labeled RBCs are injected intravenously and the number of counts within the left ventricle are plotted as a function of time (Fig. 21-12) after the labeled RBCs have equilibrated within the blood pool. Calculate the left ventricular EF using the data shown.

Answer
The EF is defined as the difference between the end-diastolic volume (EDV) and the end-systolic volume (ESV) divided by EDV. Therefore,

$$EF = (EDV - ESV)/EDV$$
$$= (13 - 7)/13$$
$$= 46\%$$

2. Blood Flow Models

Many tracer kinetic methods and models to measure blood flow exist. Virtually all such methods are included in one of three

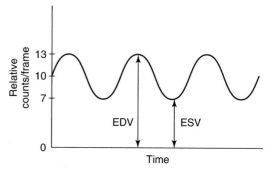

FIGURE 21-12 Sinusoidal curve illustrating relative counts per frame in a gated blood-pool image. The counts measured at the peaks and troughs of the curve are proportional to end-diastolic volume (EDV) and end-systolic volume (ESV), respectively.

categories: trapping, clearance, and equilibrium techniques. These techniques can be implemented by administering compounds labeled with either gamma-emitting or positron-emitting isotopes and imaging their distribution with gamma cameras, single-photon emission computed tomography (SPECT), or positron emission tomography (PET) scanners (see Table 21-1). All three techniques require measuring the concentration of tracer in arterial blood if quantitative estimates of blood flow in units of mL/min or mL/min/g (perfusion) are desired.

In *trapping* methods, tracers are used that are distributed to organs in proportion to blood flow but that are then trapped, either physically in the circulation (e.g., trapping of macroaggregated albumin labeled with ^{99m}Tc in the pulmonary capillaries in lung perfusion studies), or metabolically in tissue (e.g., $^{13}NH_3$ and ^{99m}Tc-sestamibi).

Clearance methods require injecting metabolically inert labeled tracers that also are distributed in proportion to flow but that do not remain trapped in the vascular or tissue spaces. The rate of washout, or clearance, of these tracers depends on blood flow. Either nondiffusible tracers that remain within the vascular compartment or diffusible tracers that distribute in both the tissue and vascular compartments can be used. As discussed later, with both classes of tracers, the mathematical approach to measuring blood flow is similar.

Equilibrium techniques require administering a continuous supply of a diffusible blood flow tracer, waiting until a tracer steady state has been reached, and then imaging the distribution. This approach uses very short half-life radionuclides such as ^{15}O ($T_{1/2} = 122$ seconds) in which equilibrium is established by removal of the tracer from tissue by the rate of blood flow and the rate of radioactive decay. This method is not commonly used and is not discussed further here.

3. Blood Flow: Trapped Radiotracers

A simple trapping method for measuring blood flow employing nondiffusible tracers is the *labeled microsphere* technique. Microspheres are spherical or irregularly shaped small particles that are larger than capillaries and embolize (become lodged in) the first capillary bed they encounter. They remain within the capillary system for a time that is sufficiently long so that particle breakdown and excretion are insignificant during the measurement period.

If labeled microspheres are injected into the left atrium or ventricle, they are distributed to individual organs in proportion to the blood flow to the organ. If the total activity of microspheres injected is A_t, blood flow to an organ can be calculated from

$$F(mL/min) = C.O.(mL/min) \times (A_o/A_t)$$

(21-21)

where A_o is the activity of microspheres accumulated in the organ (assuming 100% are trapped) and $C.O.$ is the cardiac output (mL of blood per minute). A_o is determined by tissue sampling or from quantitative imaging measurements. Cardiac output must be measured independently as described later. Organ perfusion, that is, blood flow per gram of tissue, is then given by

$$F(mL/min/g) = C.O. \times [(A_o/m_o)A_t]$$

(21-22)

where A_o/m_o is the concentration of activity in the organ or the measured tissue sample.

If the labeled microspheres are not 100% trapped, A_o will be reduced and Equations 21-21 and 21-22 will lead to underestimations of flow or organ perfusion. This effect is illustrated graphically in Figure 21-8 for tracers with a $P \times S$ product below the value required for 100% clearance over the blood flow range studied.

Frequently, cardiac output is difficult to measure. In such cases, flow to a single organ can be determined by a modification of the method just described, known as the *reference sample technique*. In this technique, arterial blood is withdrawn at a rate S (mL/min) during the time when the microspheres are flowing to the organ, and the total activity of the blood sample withdrawn, A_s, is determined. Blood flow to the organ then is calculated from

$$F = S(mL/min) \times (A_o/A_s) \quad (21-23)$$

where A_o is the activity of microspheres trapped in the organ. *Perfusion* of blood into the organ is calculated from

$$F = S \times [(A_o/m_o)/A_s] \quad (21-24)$$

Although the microsphere technique is conceptually simple, it requires injection of the tracer into the left atrium (rather than intravenously) to avoid extraction of particles by the lungs. This technique also causes microembolization of capillaries exposed to the particles, blocking blood flow through those capillaries. To minimize perturbation of the system being studied, only a small number of microspheres are injected. An example of the labeled microsphere technique is the use of labeled albumin (which dissolves with time) to determine lung perfusion or to detect right-to-left cardiac shunts. The microsphere technique is used infrequently in humans but is quite commonly used in research studies involving animals.

More commonly, diffusible tracers that do not remain within the vascular space and that are trapped in tissue are used to trace and quantitate blood flow in human studies. Indeed, all tracers injected into the vascular system initially behave as "blood flow tracers" to some degree, but not all tracers permit convenient quantitation of blood flow independent of other processes such as extraction, metabolism, and so forth. A good diffusible tracer used to estimate blood flow by the trapping method must have high first-pass extraction into the tissue (Equation 21-15) so that tissue uptake reflects blood flow. It must also be rapidly trapped in tissue, minimizing the amount of tracer that can be transported back out of the tissue and into the blood. Examples of diffusible tracers that are trapped in tissue in proportion to blood flow include ^{99m}Tc-sestamibi, ^{13}NH$_3$, ^{62}Cu-PTSM, ^{201}TlCl, and ^{82}RbCl. Blood flow can be estimated using Equation 21-24 as long as a reference arterial blood sample is available.

EXAMPLE 21-4

Suppose that microspheres are labeled with a radionuclide that permits quantitative measurement of concentrations in tissues (e.g., ^{18}F measured by PET). For simplicity, assume that these measurements can be related to the actual number of microspheres present. Calculate blood flow per gram of tissue (i.e., perfusion) to the brain if 500,000 microspheres are injected into the left atrium, and the average concentration of microspheres measured in the brain is 50 microspheres/g. Assume the cardiac output is 5 L/min.

Answer

From Equation 21-22, blood flow per gram of tissue is given by

$$F = \frac{5000(\text{mL}/\text{min}) \times 50(\text{microspheres/g})}{500,000 \text{ microspheres}}$$

$$= 0.50 \text{ mL}/\text{min}/\text{g}$$

EXAMPLE 21-5

Assume in the situation of Example 21-4 that the cardiac output is unknown. After injecting 500,000 microspheres into the left atrium, arterial blood is sampled at a rate of 10 mL/min. The average microsphere concentration in the whole brain is again 50 microspheres per gram. Assume the microspheres are 100% trapped in capillaries in one pass through the circulation (e.g., in 1 minute). A total of 1000 microspheres are counted in the radial arterial sample obtained in a 1-minute period. What is the average brain blood flow per gram of tissue?

Answer

From Equation 21-24:

$$F = \frac{10 \text{ (mL}/\text{min}) \times 50 \text{ (microspheres/g)}}{1,000 \text{ microspheres}}$$

$$= 0.50 \text{ mL}/\text{min}/\text{g}$$

4. Blood Flow: Clearance Techniques

Most clearance techniques for measuring blood flow are based on the *central volume principle,* which was defined in Equation 21-1. In the case of blood flow in units of mL/min, V is in units of mL. For blood flow in units of mL/min/g, V is in units of mL/g (volume per mass of tissue).

For a nondiffusible tracer, that is, one that stays within the vascular space, V is the blood volume in mL or mL/g, depending upon the units of blood flow. The determination of F with a nondiffusible tracer requires the measurement of both the transit time, τ, and V. Normally, with nondiffusible tracers only τ is measured, which thus provides a measure of F/V. Because F changes with both τ (τ decreases as F increases) and V (V increases as F increases), there is a nonlinear relationship between F/V and τ, with the magnitude of changes in τ being less than the magnitude of changes in F.

It is more common and easier to measure blood flow with freely diffusible tracers. Freely diffusible tracers are those that have 100% extraction from blood to tissue during a single transit through the vascular bed of the tissue. Extraction occurs almost exclusively at the level of capillaries because of their large vascular surface area. If the tracer is chemically inert, it will be cleared from tissue in proportion to blood flow and F can be determined by

Equation 21-1. For diffusible tracers, V is the volume into which the tracer is distributed. The fraction of the total tissue that a diffusible tracer will occupy depends on the specific diffusion and solubility properties of the tracer. This volume is denoted by V_1, as described in Section B.3. Thus, by substituting V_1 for V in Equation 21-1 and rearranging,

$$F/V_1 = 1/\tau \qquad (21\text{-}25)$$

Multiplying each side of Equation 21-25 by V_1/V_t, in which V_t is the total tissue volume, gives

$$(F/V_1)(V_1/V_t) = (1/\tau)(V_1/V_t)$$
$$F/V_t = \lambda / \tau \qquad (21\text{-}26)$$

where V_1/V_t is the partition coefficient (λ) for the tracer as given by Equation 21-4. As discussed earlier, λ typically is measured using Equation 21-3. F/V_t is blood flow per volume or mass of tissue (perfusion). As discussed in Section B.1, by convention in the literature F/V_t is also referred to as blood flow, F. Thus Equation 21-26 also may be written as $F = \lambda/\tau$.

The problem of V varying with flow, discussed earlier for nondiffusible tracers, does not exist to any significant degree with diffusible tracers. The term V with diffusible tracers is the tissue volume that does not change appreciably with changes in blood flow. Thus changes in $1/\tau$ are directly proportional to flow. In addition, τ is longer for diffusible tracers (i.e., 30 to 100 seconds for brain) than for nondiffusible tracers (3 to 6 seconds for brain). Because of the longer values of τ and the linear relationship between τ and blood flow with diffusible tracers, they usually provide more accurate and convenient measurements of blood flow than those obtained with nondiffusible tracers. However, λ usually is measured in normal tissue, and it may vary from this value in pathologic tissues. For example, ^{15}O-labeled water has little variability,[1] whereas ^{133}Xe has considerable variability[4] between normal and diseased tissue.

An example of the application of the central volume principle is the *Kety-Schmidt method* for measuring cerebral blood flow with inhaled, diffusible, inert gasses (e.g., nitrous oxide or krypton).[5] This approach is based on the assumption of a constant partition coefficient of 1 mL/g in the case of nitrous oxide and krypton in the brain. Blood flow is given by (Equation 21-26)

$$F = \lambda / \tau = 1.0 \text{ mL} / \text{g} / \tau \text{ (min)}$$
$$= \tau^{-1} \text{ (mL} / \text{min} / \text{g)} \qquad (21\text{-}27)$$

The technique thus requires measuring the mean transit time of the gas through the brain. The original technique involved giving a subject a continuous inhalation of the gas and sampling the gas concentrations in the arterial supply to the brain (i.e., any convenient arterial source) and the venous drainage (internal jugular bulb). Gas was breathed until an equilibrium (or near equilibrium) concentration was reached in the brain and blood. The law of conservation of mass (Fick principle, Equation 21-14) states that the total amount of tracer in the brain at equilibrium must be the difference between the cumulative input and output. Therefore

$$V \times C_E = F \int_0^\infty C_A(t)dt - F \int_0^\infty C_V(t)dt \qquad (21\text{-}28)$$

But $\lambda/F = \tau$ (Equation 21-26); therefore rearranging Equation 21-28 and writing it in terms of blood flow per volume of tissue

$$\frac{\lambda}{F} = \frac{\int_0^\infty (C_A(t) - C_V(t))dt}{C_E} = \tau \qquad (21\text{-}29)$$

where $C_A(t)$ and $C_V(t)$ are the arterial and venous concentrations as a function of time and C_E is the equilibrium concentration in the blood, usually approximated by a venous value at a selected time (e.g., 10 minutes) after the initiation of the inhalation. Typical values for this approach are $\tau = 2$ minutes, which, from Equation 21-27 yields a flow of approximately 0.5 mL/min/g of brain tissue. A completely analogous approach with nuclear medicine imaging devices (e.g., PET and SPECT) permits one to perform these calculations regionally in the brain or other tissues and also produces estimates of blood flow using compartmental modeling approaches with a measured arterial input function and serial images of the tissue (i.e., venous efflux concentration measurements are not necessary).[6] Freely diffusible tracers used for studies of cerebral blood flow include ^{15}O-water and ^{15}O-butanol, and these are typically introduced into the body by intravenous bolus injection. Equilibrium approaches to measuring blood flow using continuous infusion of short half-life (usually ^{15}O-labeled) diffusible tracers also have been developed.[7]

One very simple method for determining the mean transit time in nuclear medicine studies (and if the partition coefficient is known, blood flow) is the "area/height" method. A very short bolus of a freely diffusible tracer such as ^{133}Xe is injected. A gamma camera then is used to image the washout of the tracer from the organ of interest over time. It can be shown that the mean transit time is[8]

$$\frac{\lambda}{F} = \tau = \frac{\int_0^\infty A(t)dt}{A(0)} \qquad (21\text{-}30)$$

in which $A(t)$ is the amount of the tracer in tissue as a function of time after the injection and $A(0)$ is the amount of tracer in the tissue just after the bolus has been delivered. Figure 21-13 illustrates this approach graphically.

5. Enzyme Kinetics: Glucose Metabolism

Enzymes catalyze many biochemical reactions that are of interest from the modeling standpoint. An example is the hexokinase-catalyzed phosphorylation of glucose that is the step initiating glycolysis and also is the focal reaction for the deoxyglucose and 2-deoxy-2-[^{18}F]fluoro-D-glucose (FDG) model for determining rates of glucose metabolism (references 1 and 9). FDG is by far the most commonly used PET radiopharmaceutical in clinical studies, because many different diseases can result in changes in glucose metabolism (see Chapter 18, Section F). The model for determining the rate of glucose use from a PET study therefore is one of the most studied tracer kinetic models in nuclear medicine.

The *Michaelis-Menten* hypothesis states that an intermediate complex is formed between a reactant (also known as the *substrate*) and an enzyme. This complex then is converted to the chemical product with release of the enzyme. The reaction can be written in form similar to that for carrier-mediated transport (Equation 21-17):

$$S + E \underset{k_2}{\overset{k_1}{\rightleftharpoons}} SE \overset{k_3}{\underset{}{\rightleftharpoons}} P + E \quad (21\text{-}31)$$

where S is the substrate, P is the product, E is the enzyme, and k_1, k_2, and k_3 are the rate constants for the steps of the reaction process. Note that it is generally assumed that there is no reverse association of P and E; therefore no k_4 rate constant is present. However, in biologic systems there usually are separate

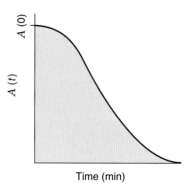

FIGURE 21-13 Tissue time-activity curve following bolus injection of a blood flow tracer. The total amount of activity detected in a given tissue region is plotted as a function of time. The maximum activity at time zero represents $A(0)$. From Equation 21-30, the mean transit time, τ, is given by the area under the curve divided by the height of the curve, $A(0)$. This method often is referred to as the "area/height" method.

enzymes for conversion of P back to S for additional regulation of reaction sequences.

The reaction rate R (conversion of S to P) is

$$R = \frac{V_m \times [S]}{[S] + K_m} \qquad (21\text{-}32)$$

and is known as the *Michaelis-Menten equation*. The term V_m (mg/min) is the maximum rate of the reaction, whereas K_m is the concentration of S that produces a reaction rate of one half the maximum value, as shown in Figure 21-14.

Equation 21-32 predicts that the reaction rate approaches the maximum value V_m as [S] approaches infinity. It is clear from Equation 21-32 and Figure 21-14 that the reaction rate is not a linear function of [S]; however, it still is possible to model such processes with linear tracer compartmental models because of the following relationship. If more than one substrate is competing for the enzyme E (i.e., S and S′), it can be shown[1] that the reaction rates R and R' for the competing processes are

$$R = \frac{V_m \times [S]/K_m}{([S]/K_m) + ([S']/K'_m) + 1} \qquad (21\text{-}33)$$

$$R' = \frac{V'_m \times [S']/K'_m}{([S]/K_m) + ([S']/K'_m) + 1} \qquad (21\text{-}34)$$

where V'_m and K'_m are the Michaelis-Menten constants for the reaction with substrate S′. If S′ represents a tracer of the original

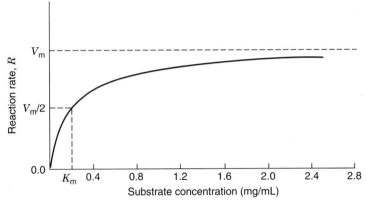

FIGURE 21-14 Graphical illustration of Michaelis-Menten enzyme kinetics. The rate of a reaction (i.e., the reaction flux R) is plotted against substrate concentration [S] using Equation 21-32. Note that when the substrate concentration equals K_m, the reaction rate is $V_m/2$. As the substrate concentration increases, the reaction rate gradually approaches V_m. (*From Phelps ME, Mazziotta JC, Schelbert HR:* Positron Emission Tomography and Autoradiography: Principles and Applications for the Brain and Heart. *New York, 1986, Raven Press.*)

substrate S, then [S′] is of a much lower value than [S]. Therefore

$$R \cong \frac{V_m \times [S]}{[S] + K_m} \qquad (21\text{-}35)$$

$$R' \cong \frac{V'_m \times K_m / K'_m}{[S] + K_m}[S'] \qquad (21\text{-}36)$$

R is simply the original rate of the process (Equation 21-32) unaffected by the presence of the tracer S′, whereas R' is a linear function of [S′] as long as [S′] remains much less than [S] (i.e., as long as the tracer condition holds). Therefore the tracer concentration [S′] is linearly related to the reaction rate R' and linear modeling techniques are appropriate for describing this process. Dividing Equation 21-36 by Equation 21-35 yields

$$\frac{R'}{R} = \frac{V'_m \times K_m [S']}{V_m \times K'_m [S]} \qquad (21\text{-}37)$$

The reaction rate of the measured or "traced" process, R', is therefore directly related to the "natural" or unknown rate R by a ratio of the Michaelis-Menten constants and the relative concentrations of S and S′. In some cases (e.g., direct isotopic substitution labeling of biologic compounds such as ^{11}C for ^{12}C), the Michaelis-Menten constants for the tracer and natural substance are essentially the same and Equation 21-37 would reduce to a simple ratio of [S′] to [S]. With analog tracers the Michaelis-Menten constants are different than the natural substance but Equation 21-37 still applies.

As an example of an analog tracer approach to modeling an enzymatic process, consider the Sokoloff deoxyglucose method[9] or the analogous approach with PET using ^{18}F-labeled FDG for measuring glucose metabolism in the brain.[10] Glucose supplies 95% to 99% of the brain's energy in normal physiologic states, and the rate of glucose use is an excellent indicator of energy-requiring functions of the brain.

FDG is an analog of glucose that is similar to glucose in several respects. Like glucose, it is transported from the blood to the brain by a carrier-mediated diffusion mechanism. Hexokinase catalyzes the phosphorylation of glucose to glucose-6-PO$_4$ and FDG to FDG-6-PO$_4$. In both the transport and phosphorylation steps, FDG is a competitive substrate with glucose. FDG-6-PO$_4$, however, is not a significant substrate for further metabolism. As shown in Figure 21-15, it is not converted into glycogen to any significant extent and is not further metabolized in the glycolytic pathway. The FDG-6-PO$_4$ also does not diffuse across cell membranes and is therefore metabolically trapped in tissues, which is convenient both from an imaging and modeling viewpoint.

If glucose metabolism in the tissue of interest is assumed to be in a steady state (i.e., it has a constant rate), then the rate of the hexokinase reaction will be the rate of the entire process of glycolysis (see Section B.6). A compartmental configuration for the FDG model is shown in Figure 21-16. The three-compartment model consists of FDG in plasma, FDG in tissue, and FDG-6-PO$_4$ in

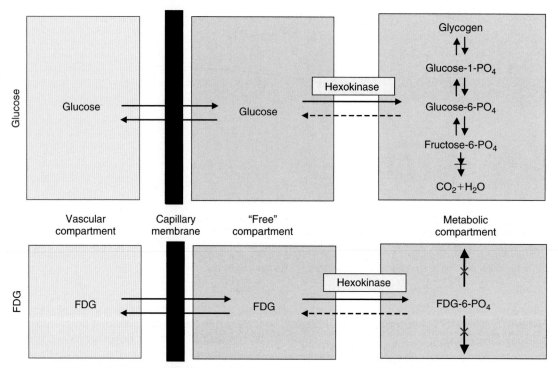

FIGURE 21-15 Transport and metabolic pathways for glucose and 2-deoxy-2[^{18}F]fluoro-D-glucose (FDG). Physically, the "free" compartment represents a combination of the interstitial space and the cytosol (the fluid component inside the cell, excluding the nucleus) in both of which unphosphorylated glucose and FDG are uniformly distributed. The red crosses indicate that FDG-6-PO$_4$, unlike glucose-6-PO$_4$, is not further metabolized.

tissue corresponding to comparable distributions of glucose, although glucose continues on through metabolism. The first-order rate constants k_1^* and k_2^* describe the transport of FDG from the blood to brain and brain to blood, respectively, whereas the first-order rate constants k_3^* and k_4^* describe the phosphorylation of FDG and dephosphorylation of FDG-6-PO$_4$. The asterisk refers to FDG

indices, whereas the corresponding terms for glucose do not have an asterisk.

Let *MRGlc** (μmol/min/g) refer to the metabolic rate of FDG and *MRGlc* (μmol/min/g) be the metabolic rate for glucose. If C_t is the concentration of free (unphosphorylated) glucose in the tissue space and C_t^* is the corresponding concentration for FDG (second compartment, top row, Fig. 21-16), then, from

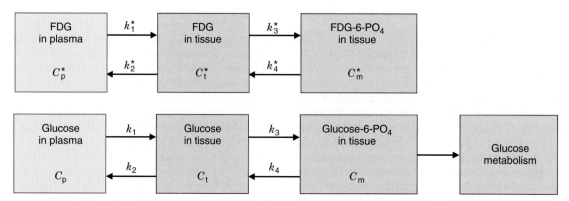

FIGURE 21-16 Three-compartment FDG model with the four first-order rate constants describing transport between the compartments. C_p, C_t, and C_m are the concentrations of glucose in plasma, tissue, and metabolized glucose (glucose-6-PO$_4$) in tissue, respectively. C_p^*, C_t^*, and C_m^* are the corresponding concentrations for FDG.

Equation 21-37 and the fact that the rate of phosphorylation (i.e., flux as described in Sections B.4 and B.5) equals $k_3 \times C_t$ for glucose and $k_3^* \times C_t^*$ for FDG, it follows that

$$\frac{MRGlc^*}{MRGlc} = \frac{V_m^* \times K_m \times C_t^*}{V_m \times K_m^* \times C_t} = \frac{k_3^* \times C_t^*}{k_3 \times C_t} \quad (21\text{-}38)$$

Equation 21-38 assumes that k_4^* and k_4 are very small and the forward rate of phosphorylation of glucose and FDG approximate the net metabolic rates. V_m and K_m are the Michaelis-Menten constants for the hexokinase-mediated phosphorylation of FDG (*) and glucose (no *).

If the ratio of terms defined by Equation 21-38 is a constant, then

$$MRGlc = MRGlc^*/\text{constant} \quad (21\text{-}39)$$

If plasma glucose (C_p) and plasma FDG (C_p^*) concentrations are constant and both sides of Equation 21-38 are multiplied by C_p/C_p^*, then

$$\frac{MRGlc^*/C_p^*}{MRGlc/C_p} = \frac{V_m^* \times K_m \times C_t^*/C_p^*}{V_m \times K_m^* \times C_t/C_p} \quad (21\text{-}40)$$

But, from Equation 21-3, the ratios C_t^*/C_p^* and C_t/C_p are the partition coefficients (i.e., tissue-to-blood concentration ratios) of FDG (λ^*) and glucose (λ). If the numerator and denominator of the left side of Equation 21-40 are divided by blood flow (F), the numerical value of the equation does not change; thus, with these substitutions,

$$\frac{MRGlc^*/(C_p^* \times F)}{MRGlc/(C_p \times F)} = \frac{V_m^* \times K_m \times \lambda^*}{V_m \times K_m^* \times \lambda} \quad (21\text{-}41)$$

This ratio of terms is defined as the *lumped constant* (*LC*) of the FDG model. The left side of Equation 21-41 is simply the net extraction of FDG (E_{net}^*) divided by the net (i.e., steady-state) extraction of glucose (E_{net}). Therefore,

$$LC = (E_{net}^*)/(E_{net}) \quad (21\text{-}42)$$

That is, the lumped constant of the FDG model is just the steady-state ratio of the net extraction of FDG to that of glucose at constant plasma levels of FDG and glucose. The lumped constant is a direct consequence of Equation 21-37 and illustrates the principle of competitive enzyme kinetics applied to tracer kinetic modeling. Intuitively, the lumped constant is simply a correction term that measures the net difference in the way

a tissue uses FDG and glucose. The full expression of *LC* also includes an additional term including the influence of dephosphorylation of glucose on the net metabolic rates of FDG and glucose. This latter term is normally quite close to 1 but is intrinsically included in the measured values of *LC* by Equation 21-42.

Although the lumped constant describes the differences between FDG and glucose metabolism, it is actually glucose metabolism itself that is of interest physiologically. It can be shown that *MRGlc* is given by[1]

$$MRGlc = \frac{C_p}{LC}\left(\frac{k_1^* \, k_3^*}{k_2^* + k_3^*}\right) \quad (21\text{-}43)$$

The term $k_1^*/(k_2^* + k_3^*)$ is the partition coefficient (λ^*) for FDG. To understand this, consider the tracer steady state when the flux of FDG into tissue is balanced by the flux out. Thus,

$$\text{flux in} = \text{flux out} \quad (21\text{-}44)$$

$$k_1^* C_p^* = k_2^* C_t^* + k_3^* C_t^* \quad (21\text{-}45)$$

$$\frac{k_1^*}{k_2^* + k_3^*} = \frac{C_t^*}{C_p^*} \quad (21\text{-}46)$$

It is apparent from Equation 21-46 that the partition coefficient, $k_1^*/(k_2^* + k_3^*)$, is simply the tissue-to-plasma concentration ratio. Because *LC* is the correction factor that converts the transport and phosphorylation steps measured with FDG to those for glucose, one can perform the following transformation from values for FDG to those for glucose:

$$\frac{[k_1^*/(k_2^* + k_3^*)] \times k_3^*}{LC} = [k_1/(k_2 + k_3)] \times k_3 \quad (21\text{-}47)$$

Since $k_1/(k_2 + k_3) = C_t/C_p$, the tissue concentration of glucose, C_t, can be obtained from the plasma concentration, C_p, by

$$C_p \times k_1/(k_2 + k_3) = (C_t/C_p) \times C_p = C_t \quad (21\text{-}48)$$

Thus, from the general equation for calculating fluxes (Equation 21-5),

$$MRGlc \ (\mu mol/g/min) = k_3 \times C_t \quad (21\text{-}49)$$

Equations 21-43 and 21-49 are valid only if the dephosphorylation rate is negligible compared with the rate of phosphorylation.

Models that include dephosphorylation are described in reference 1.

Equation 21-43 produces a local estimate of *MRGlc* if C_p (the steady-state plasma glucose value) is measured, if *LC* is known and if the rate constants for FDG (k_1^*, k_2^*, and k_3^*) are determined for each region of organ of interest. The equations and procedures for measuring these rate constants are given in references 1 and 10. Although this technique generates local estimates of *MRGlc*, it does require imaging the organ over time and iteratively solving for the rate constants in Equation 21-43. In actual practice, an operational equation for the FDG model usually is employed that uses predetermined population values for the FDG rate constants and requires only a single tissue measurement of the total tissue concentration of ^{18}F (^{18}FDG + ^{18}FDG-6-PO$_4$) obtained from ROI analysis of the image data. This operational equation originally was developed by Sokoloff and coworkers[9] for autoradiographically determining *MRGlc* with ^{14}C-deoxyglucose. This technique still requires knowledge of the time course of FDG input function [$C_p^*(t)$], the plasma glucose concentration C_p, the lumped constant value (*LC*), and average values of the rate constants. The Sokoloff operational equation of the deoxyglucose model, which does not include the dephosphorylation of DG-6-PO$_4$, is given by

MRGlc =

$$\frac{C_i^*(T) - [k_1^* \times e^{-(k_2^*+k_3^*)T} \times \int_0^T C_p^*(t) \times e^{(k_2^*+k_3^*)t} dt]}{LC\left[\int_0^T \frac{C_p^*(t)}{C_p} dt - e^{-(k_2^*+k_3^*)T} \times \int_0^T \frac{C_p^*(t)}{C_p} \times e^{(k_2^*+k_3^*)t} dt\right]}$$

(21-50)

where $C_i^*(T)$ is the total ^{18}F tissue concentration at time *T*.

The quantities highlighted in blue in Equation 21-50 are measured in each study. Average estimates of the rate constants and *LC* obtained from separate experiments are used as a part of the routine calculation of *MRGlc* with Equation 21-50. The use of average estimates of the rate constants over the normal range of *MRGlc* values causes little error at late times (i.e., imaging 40 minutes after injection) because they appear with terms in Equation 21-50 with negative exponentials that become small when *T* becomes large.

In words, Equation 21-50 can be expressed as

$$\begin{pmatrix} \text{Regional Glucose} \\ \text{Metabolic Rate} \end{pmatrix}$$

$$= \left(\frac{\text{Plasma Glucose Conc.}}{\text{Lumped Constant}}\right)$$

$$\times \left[\frac{\begin{pmatrix} \text{Total } ^{18}\text{F} \\ \text{in Region} \end{pmatrix} - \begin{pmatrix} \text{Free } ^{18}\text{FDG} \\ \text{in Region} \end{pmatrix}}{\begin{pmatrix} \text{Total Net } ^{18}\text{FDG} \\ \text{Transported to the Region} \end{pmatrix}}\right] \quad (21\text{-}51)$$

The total ^{18}F minus free ^{18}FDG equals the tissue concentration of the reaction product, FDG-6-PO$_4$.

Equation 21-50 requires knowledge of the typical kinetics of the transport and phosphorylation processes to make intelligent decisions about scan duration and imaging time. Typically, approximately 40 minutes are required for the tracer to reach a near steady state in tissue after a bolus intravenous injection of FDG. A PET image usually is obtained at this time. The ROI value over the tissue or organ of interest, together with the other model parameters described earlier, is then used to calculate local values of *MRGlc*. Figure 21-17 illustrates a *parametric image* of *MRGlc* in the brain obtained using Equation 21-50 with data from an FDG PET image. A parametric image is one in which the parameter of interest (in this case *MRGlc*) is calculated on a pixel-by-pixel basis.

The FDG model contains several simplifications based on the approaches outlined in the preceding sections. The strategy of using an analog tracer effectively eliminates many alternative biochemical pathways for glucose metabolism and makes possible the simple three-compartment model. Additionally, the transport (between the first two compartments) and phosphorylation steps (between the second two compartments) represent combined steps of more complicated multistep processes. The exchange between substructures within each compartment is assumed to be rapid compared with exchange between compartments.

Even though blood flow has a major effect on tracer delivery, it is not included in the FDG model explicitly. The extraction of FDG normally is low enough (i.e., low value of $P \times S$) that the delivery of FDG has a low dependence on blood flow. In addition, the initial flow dependence on the delivery of FDG

0.6 µmol/min/g

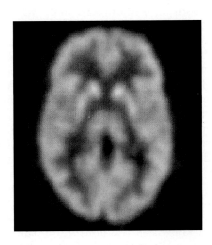

FIGURE 21-17 Parametric image of a transverse section through the brain showing *MRGlc* calculated on a pixel-by-pixel basis using Equation 21-50. (*Image courtesy Dr. Henry Huang, University of California–Los Angeles, Los Angeles, California.*)

0

progressively diminishes with time after injection.

In the tracer kinetic model, the FDG concentration in the vascular compartment shown in Figure 21-16 is measured by taking blood samples from the systemic circulation. Typically, it is assumed that the measured tissue data contain no significant counts from activity in the vascular space. In Equation 21-50 this is a good approximation at the typical imaging time of 40 minutes after injection. Models have been developed that add the vascular compartment to equations describing the tissue time-activity curve because in kinetic studies for measuring the rate constants, the amount of FDG in the tissue vascular pool is significant at early times after injection.

The deoxyglucose and FDG methods have been used to measure exogenous glucose use in many organs and tissues using autoradiography and PET, respectively. Examples include brain, heart, tumors, liver, kidney, and peripheral tissues.

6. Receptor Ligand Assays

Cell-surface *receptors* are important molecular targets that may be overexpressed or underexpressed in a range of important diseases states, for example in Parkinson's disease and many cancers. PET and SPECT can provide images of the distribution of these receptors by using tracers known as *radioligands* that are designed to bind specifically to the receptor of interest.[11,12] In most cases the binding is reversible, that is, over the time course of

an imaging study, the radioligand and receptor are likely to disassociate. The discussion that follows assumes this case.

A simple model for the interaction of a radioligand, L, with a receptor, R, is

$$L + R \underset{k_{\mathrm{off}}}{\overset{k_{\mathrm{on}}}{\rightleftharpoons}} RL \qquad (21\text{-}52)$$

where k_{on} and k_{off} are the rate of association and dissociation of L and R, respectively. At equilibrium, if the system is closed, there is no net change in the concentrations, and therefore

$$k_{\mathrm{on}}[L][R] = k_{\mathrm{off}}[RL] \qquad (21\text{-}53)$$

where [L], [R], and [RL] are the equilibrium concentrations of the unbound ligand, the receptor, and the bound receptor-ligand complex, respectively. The ratio $k_{\mathrm{off}}/k_{\mathrm{on}}$ is known as the *equilibrium dissociation constant* and is given the notation K_{d}. The total concentration of receptors is given by [R] + [RL] and is commonly denoted as B_{max}. Rearranging Equation 21-53 in terms of K_{d} and B_{max} yields

$$[RL] = \frac{B_{\mathrm{max}}[L]}{[L] + K_{\mathrm{d}}} \qquad (21\text{-}54)$$

It is possible to measure [RL] as a function of [L] in vitro and then fit the data to estimate both K_{d} and B_{max}. Thus quantitative information on the receptor concentration, and the strength of binding of the ligand to the receptor (commonly known as the *affinity*, and given by $1/K_{\mathrm{d}}$), can be obtained. In humans, such a study is impractical because it would require several injections of radioligand at different concentrations. As well, at higher concentrations of the ligand there would be concerns about possible pharmacologic effects resulting from the high levels of receptor occupancy. However, if a PET or SPECT study is carried out at tracer levels such that [RL] is small compared with [R] (i.e., the fraction of receptors occupied by the ligand is, for example, 10% or less), then $B_{max} \approx [R]$, and now from Equation 21-53, we see that

$$\frac{B_{max}}{K_d} = \frac{[RL]}{[L]} \qquad (21\text{-}55)$$

The ratio B_{max}/K_d is known as the *binding potential* (*BP*) and equals the ratio of bound radioligand to free radioligand concentration at equilibrium in this simple model. Although the situation in vivo following injection of a radioligand is quite complex, it turns out with appropriate modeling of dynamically acquired PET or SPECT data that it is possible to

estimate a parameter related to the in vitro *BP* as defined by Equation 21-55. Under the assumption that K_d does not change significantly with disease progression or treatment, images of *BP* will roughly reflect the relative concentration of receptors, B_{max}, in tissue. Such techniques have been widely applied to studies of receptor systems in the brain.

To see how the parameter *BP* is related to quantities that can be measured in a PET or SPECT scan, we start with a compartmental model as shown in Figure 21-18A. The injected radioligand is delivered via the flowing blood (plasma compartment) and is extracted with rate constant K_1 into the free compartment in the tissue. The rate constant K_1 represents the product of the blood flow and the unidirectional extraction fraction (see Section C.1) and has units of mL/min/g.*

---

*By convention, K_1 is denoted by a capital K because it has different units than the other rates constants, which are in units of inverse time.

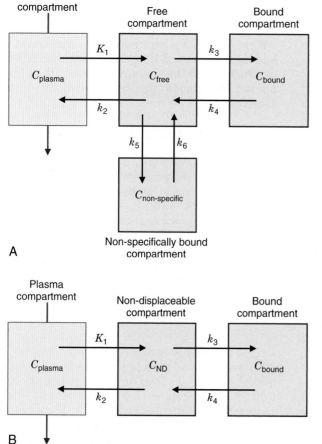

FIGURE 21-18 *A,* A general four-compartment model describing a radioligand that reversibly binds to a specific receptor. The rate constants for the exchange of the radioligand between compartments are denoted by K_1 through k_6, and the concentration of the radioligand in each compartment is denoted by *C. B,* A simplified version of the model in *A* in which the exchange between the "free" and "nonspecifically bound" compartments is assumed to be rapid (k_5 and k_6 >> k_3 and k_4) allowing these two compartments to be combined into a single compartment representing nondisplaceable radioligand. This reduces the number of model unknowns to four, increasing the robustness with which kinetic parameters can be estimated. This model is used in the analysis of many radioligand studies with PET and SPECT.

Once in the tissue, there are four possible fates for the radioligand. It can continue to exist in free form, it can bind to the receptor of interest, it can bind nonspecifically in the tissue to other proteins, or it can be transported back to the blood. Because any binding is reversible, there are rate constants both for binding of the ligand and for its dissociation. In sum, this model has four different compartments and six different rate constants.

Because of the limited temporal resolution and signal-to-noise characteristics of nuclear medicine images, it is not possible to robustly estimate six different kinetic rate constants from a dynamic PET or SPECT study. The model therefore is simplified by making the assumption that there is rapid exchange between the free and nonspecifically bound compartments, and that these two compartments can be combined together. This is a reasonable assumption if the radioligand has only low affinity binding to nonspecific targets such that it associates and dissociates rapidly (that is, k_5 and k_6 are large compared with k_3 and k_4). This combined compartment often is referred to as the *nondisplaceable compartment*. This results in the simplified kinetic model in Figure 21-18B, which consists of three compartments with four rate constants.

Based on conservation of mass, we can write simple differential equations for the concentration in the "nondisplaceable" and "specifically bound" compartments as

$$\frac{dC_{ND}(t)}{dt} = K_1 C_{plasma}(t) - k_2 C_{ND}(t) \\ - k_3 C_{ND}(t) + k_4 C_{bound}(t) \quad (21\text{-}56)$$

$$\frac{dC_{bound}(t)}{dt} = k_3 C_{ND}(t) - k_4 C_{bound}(t) \quad (21\text{-}57)$$

At equilibrium, there would be no net transfer of radioligand between compartments; therefore, from Equation 21-57,

$$k_3 C_{ND} = k_4 C_{bound} \Rightarrow \frac{C_{bound}}{C_{ND}} = \frac{k_3}{k_4} \quad (21\text{-}58)$$

C_{bound}/C_{ND} is the ratio of the specifically bound radioligand to the radioligand in the nondisplaceable tissue compartment at equilibrium. Thus it is closely related to the definition of the in vitro BP (Equation 21-55) and by convention[13] is called BP_{ND}. Examining Equation 21-56, at equilibrium, and using the relationship derived in Equation 21-58, one also obtains

$$K_1 C_{plasma}(t) + k_4 C_{bound}(t) = \\ \frac{k_2 k_4 C_{bound}(t)}{k_3} + k_4 C_{bound}(t) \\ \Rightarrow \frac{C_{bound}}{C_{plasma}} = \frac{K_1 k_3}{k_2 k_4} \quad (21\text{-}59)$$

This is the ratio of the specifically bound radioligand to the radioligand concentration in plasma at equilibrium. This is another form of BP and is given the symbol BP_P. Both definitions for BP can be found in the literature and only recently has consensus emerged on nomenclature.[13]

Equations 21-58 and 21-59 relate a parameter closely related to the in vitro equilibrium BP to kinetic rate constants that may be measured individually, or in combination, from a dynamic sequence of reconstructed PET or SPECT scans. There are several computational approaches for taking radioactivity values from such a dynamic sequence and converting them into estimates of the BP through the integration of Equations 21-56 and 21-57. The method used depends on the available data, for example, whether arterial blood samples were taken during the scan such that the $C_{plasma}(t)$ is known, and must also account for the fact that neither PET or SPECT can differentiate between radioactivity in C_{ND} and C_{bound}. Rather, the measurements are the sum of the radioligand in these two compartments as a function of time. Methods for computing BP from PET and SPECT data are reviewed in some detail in references 11 and 12. These types of approaches that integrate carefully designed radioligands, experimental protocols, and kinetic models are being used to study a number of important receptor systems, particularly in the brain (see Table 21-1). A representative example of such a study is shown in Figure 21-19.

F. SUMMARY

Tracer kinetic methods provide unique and accurate methods for measuring rates of physiologic, biochemical, and pharmacokinetic processes. This chapter has stressed some of the principles of tracer kinetic modeling in nuclear medicine and their relationship to measurements and descriptions of underlying physiologic processes. Many modeling approaches exist, and the complexity of the model that is used depends on the state of knowledge

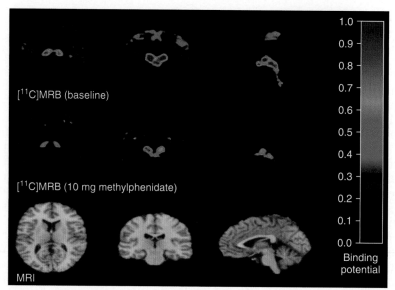

FIGURE 21-19   Images showing pixel-by-pixel calculations of binding potential (BP_{ND}) averaged over 11 healthy subjects for (S,S)-[^{11}C] methylreboxetine (MRB), a radioligand that selectively binds to the norepinephrine transporter (*top row*). This transporter is responsible for removing the neurotransmitters norepinephrine and dopamine from the synapses between neurons and is known to play an important role in several neurologic disorders, including attention deficit–hyperactivity disorder (ADHD). In this study, a decrease in binding potential was observed after administration of 10 mg of methylphenidate 75 minutes prior to radioligand injection (*middle row*). Methylphenidate is a drug that is used to treat ADHD. This study shows that methylphenidate binds to the norepinephrine transporter, thereby reducing the availability of receptor sites to which the radioligand may become bound. Sagittal, coronal, and transverse views are shown with corresponding anatomic slices from magnetic resonance imaging (*bottom row*). (*Adapted from Hannestad J, Gallezot J-D, Planeta-Wilson B, et al: Clinically relevant doses of methylphenidate significantly occupy norepinephrine transporters in humans in vivo. Biol Psych 68:854-860, 2010.*)

regarding the fate of the tracer in a biologic system and the quality of the image data that are used as input to the model. In nuclear medicine, it rarely is possible to extract more than three or four independent parameters in a model because of the finite temporal resolution and signal-to-noise ratio of the ROI data. In situations in which the fate of the tracer is complex (multiple metabolites, multiple modes of transport, and so forth), there can be considerable debate on the appropriate formulation of the model and the exact physiologic or functional meaning of the derived parameters. Once rigorous investigative studies are carried out to validate a tracer kinetic model, it often is possible to use simplified versions of the model that provide semiquantitative indices for use in the clinical setting, where it may not be possible to acquire dynamic sequences of images or obtain blood samples.

REFERENCES

1. Gambhir SS: Quantitative assay development for PET. In Phelps ME, editor: *PET: Molecular Imaging and Its Biological Applications*, New York, 2004, Springer-Verlag, pp 125-216.
2. Renkin EM: Transport of potassium-42 from blood to tissue in isolated mammalian skeletal muscles. *Am J Physiol* 197:1205-1210, 1959.
3. Crone C: Permeability of capillaries in various organs as determined by use of the indicator diffusion method. *Acta Physiol Scand* 58:292-305, 1963.
4. Lassen NA: *Tracer Kinetic Methods in Medical Physiology*, New York, 1979, Raven Press.
5. Kety SS, Schmidt CF: The nitrous oxide method for the quantitative determination of cerebral blood flow in man: Theory, procedure, and normal values. *J Clin Invest* 27:476-483, 1948.
6. Holden JE, Gatley SJ, Hichwa RD, et al: Regional cerebral blood flow using positron emission tomographic measurements of fluoromethane kinetics. *J Nucl Med* 22:1084-1088, 1981.
7. Huang S-C, Phelps ME, Hoffman EJ, Kuhl DE: A theoretical study of quantitative flow measurements with constant infusion of short-lived isotopes. *Phys Med Biol* 24:1151-1161, 1979.
8. Zieler K: Equations for measuring blood flow by external monitoring of radioisotopes. *Circ Res* 16:309-321, 1965.
9. Sokoloff L, Reivich M, Kennedy C, et al: The [C-14] deoxyglucose method for the measurement of local cerebral glucose utilization: Theory, procedure, and normal values in the conscious and anesthetized albino rat. *J Neurochem* 28:897-916, 1977.
10. Huang SC, Phelps ME, Hoffman EJ, et al: Noninvasive determination of local cerebral metabolic rate of glucose in man. *Am J Physiol* 238:E69-E82, 1980.

11. Slifstein M, Laruelle M: Models and methods for derivation of in vivo neuroreceptor parameters with PET and SPECT reversible radiotracers. *Nucl Med Biol* 28:595-608, 2001.
12. Ichise M, Meyer J, Yonekura Y: An introduction to PET and SPECT neuroreceptor quantification models. *J Nucl Med* 42:755-763, 2001.
13. Innis RB, Cunningham VJ, Delforge J, et al: Consensus nomenclature for in vivo imaging of reversibly binding radioligands. *J Cerebr Blood Flow Metabol* 27:1533-1539, 2007.

BIBLIOGRAPHY

Additional discussion of the theory, mathematical formulation, and application of compartmental modeling, including mathematical techniques in estimation theory for modeling, can be found in the following:

Cobelli C, Foster D, Toffolo G: *Tracer Kinetics in Biomedical Research: From Data to Model*, New York, 2001, Kluwer Academic Press.

Internal Radiation Dosimetry

Absorption of energy from ionizing radiation can cause damage to living tissues. This is used to advantage in radionuclide therapy, but it is a limitation for diagnostic applications because it is a potential hazard for the patient. In either case, it is necessary to analyze the energy distribution in body tissues quantitatively to ensure an accurate therapeutic prescription or to assess potential risks. The study of radiation effects on living organisms is the subject of *radiation biology* (or *radiobiology*) and is discussed in several excellent texts, some of which are listed at the end of this chapter.

One of the most important factors to be evaluated in the assessment of radiation effects on an organ is the amount of radiation energy deposited in that organ. Calculation of radiation energy deposited by internal radionuclides is the subject of *internal radiation dosimetry*. There are two general methods by which these calculations may be performed: the *classic method* and the *absorbed fraction method*. Although the classic method is somewhat simpler, and the results by the two methods are not greatly different, the absorbed fraction method (also known generally as the *MIRD method*, after the Medical Internal Radiation Dose Committee of the Society of Nuclear Medicine) is more versatile and gives more accurate results. Therefore it has gained wide acceptance as the standard method for performing internal dosimetry calculations. The procedures to be followed in using the absorbed fraction method are summarized in this chapter. Dosimetry calculations for external radiation sources as well as health physics aspects of radiation dosimetry are discussed in Chapter 23. Some radiation dose estimates for nuclear medicine procedures are summarized in Appendix E.

A. RADIATION DOSE AND EQUIVALENT DOSE: QUANTITIES AND UNITS

Radiation dose, D, refers to the quantity of radiation energy deposited in an absorber per gram of absorber material. This quantity applies to any kind of absorber material, including body tissues. The basic unit of radiation dose is the *gray*, abbreviated Gy*:

$$1\,\mathrm{Gy} = 1 \text{ joule energy deposited per kg absorber} \tag{22-1}$$

The traditional unit for absorbed dose is the *rad,* an acronym for radiation absorbed dose:

$$1\,\mathrm{rad} = 100 \text{ ergs energy deposited per g absorber} \tag{22-2}$$

Since 1 joule = 10^7 ergs, 1 Gy is equivalent to 100 rads or, alternatively, 1 rad = 10^{-2} Gy = 1 cGy. As is the case for units of activity, progress in the transition from traditional to SI units varies with geographic location, with SI units dominating practice in Europe, whereas traditional units still are commonplace in the United States. In this chapter, radiation doses are presented in grays, with values in rads also indicated in selected examples.

Equivalent dose, symbolically indicated by H_T, is a quantity that takes into account the relative biologic damage caused by radiation interacting with a particular tissue or organ. Tissue damage per gray of absorbed dose

---

*This unit is named after Harold Gray, a British medical physicist best known for his discovery of the "oxygen effect" in radiation therapy.

depends on the type and energy of the radiation, and how exactly the radiation deposits its energy in the tissue. For example, an α particle has a short range in tissue and deposits all of its energy in a very localized region. In contrast, γ rays and electrons deposit their energy over a wider area. Table 22-1 shows the radiation weighting factors, w_R, used to calculate equivalent dose for different types and energies of radiation. The SI unit of equivalent dose is the *sievert** (Sv). It is related to the average absorbed dose D in an organ or tissue, T, by

$$H_T = D_T \times w_R \qquad (22\text{-}3)$$

Equivalent dose replaces an older quantity known as the *dose equivalent*. The dose equivalent is based on the absorbed dose at a point in an organ (rather than an average across the whole organ) and is weighted by quality factors, Q, that are similar to w_R. The unit for dose equivalent also is the Sv.

The traditional unit for dose equivalent is the roentgen-equivalent man (rem). The conversion factor between traditional and SI units is

$$1 \text{ rem} = 10^{-2} \text{ Sv} = 1 \text{ cSv} = 10 \text{ mSv} \qquad (22\text{-}4)$$

---

*This unit is named after Rolf M. Sievert, best known for his development of elaborate mathematical models, including the Sievert integral, which for many years provided the basis for calculating radiation doses from implanted radium needles. He also constructed and performed many basic measurements with ionization chambers.

TABLE 22-1
WEIGHTING FACTORS FOR DIFFERENT TYPES OF RADIATION IN THE CALCULATION OF EQUIVALENT DOSE†

Type of Radiation	Radiation Weighting Factor, w_R
x rays	1
γ rays	1
Electrons, positrons	1
Neutrons	Continuous function of neutron energy
Protons >2 MeV	2
α particles, fission fragments, heavy ions	20

†Data from reference 1.

For radiations of interest in nuclear medicine (γ rays, x rays, electrons, and positrons) the radiation weighting factor is equal to 1. Therefore the equivalent dose or dose equivalent in Sv (or rem) is numerically equal to the absorbed dose in Gy (or rads).

B. CALCULATION OF RADIATION DOSE (MIRD METHOD)

1. Basic Procedure and Some Practical Problems

The absorbed fraction dosimetry method allows one to calculate the radiation dose delivered to a *target organ* from radioactivity contained in one or more *source organs* in the body (Fig. 22-1). The source and target may be the same organ, and, in fact, frequently the most important contributor to radiation dose is radioactivity contained within the target organ itself. Generally, organs other than the target organ are considered to be source organs if they contain concentrations of radioactivity that exceed the average concentration in the body.

The general procedure for calculating the radiation dose to a target organ from radioactivity in a source organ is a three-step process, as follows:

1. The amount of activity and time spent by the radioactivity in the source organ are determined. Obviously, the greater the activity and the longer the time that it is present, the greater is the radiation dose delivered by it.
2. The total amount of radiation energy emitted by the radioactivity in the source organ is calculated. This depends primarily on the energy of the radionuclide emissions and their frequency of emission (number per disintegration).
3. The fraction of energy emitted by the source organ that is absorbed by the target organ is determined. This depends on the type and energy of the emissions (absorption characteristics in body tissues) and on the anatomic relationships between source and target organs (size, shape, and distance between them).

Each of these steps involves certain difficulties. Step 2 involves physical characteristics of the radionuclide, which generally are known accurately. Step 3 involves patient anatomy, which can be quite different from one patient to the next. Step 1 is perhaps the most troublesome. Such data on radiopharmaceutical

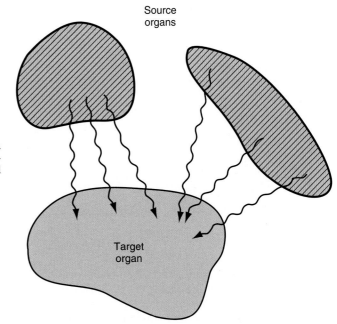

FIGURE 22-1 Absorbed dose delivered to a target organ from one or more source organs containing radioactivity is calculated by the absorbed fraction dosimetry method.

distribution as are available usually are obtained from studies on a relatively small number of human subjects or animals. There are variations in metabolism and distribution of radionuclides among human subjects, especially in different disease states. Also, the distribution of radioactivity within an organ may be inhomogeneous, leading to further uncertainties in the dose specification for that organ.

Because of these complications and variables, radiation dose calculations are made for anatomic models that incorporate "average" anatomic sizes and shapes. The radiation doses that are calculated are average values of D for the organs in this anatomic model. An exception is made when one is specifically interested in a *surface dose* to an organ from activity contained within that organ, for example, the dose to the bladder wall resulting from bladder contents. This is considered to have a value one-half the average dose to the organ or, in this case, the bladder contents.

In spite of the refined mathematical models used in the absorbed fraction model, the results obtained are only estimates of average values. Thus they should be used for guideline purposes only in evaluating the potential radiation effects on a patient.

2. Cumulated Activity, $\widetilde{A}$

The radiation dose delivered to a target organ depends on the amount of activity present in the source organ and on the length of time for

which the activity is present. The product of these two factors is the *cumulated activity* $\widetilde{A}$ in the source organ. The SI unit for cumulated activity is the becquerel • sec (Bq • sec). The corresponding traditional unit is the μCi • hr (1 μCi = 3.7×10^4 Bq; 1 hr = 3600 sec; therefore, 1 μCi • hr = $3.7 \times 10^4 \times 3600 = 1.332 \times 10^8$ Bq • sec = 1.332×10^2 MBq • sec). Cumulated activity is essentially a measure of the total number of radioactive disintegrations occurring during the time that radioactivity is present in the source organ. The radiation dose delivered by activity in a source organ is proportional to its cumulated activity.

Each radiotracer has its own unique spatial and temporal distribution in the body, as determined by radiotracer delivery, uptake, metabolism, clearance and excretion, and the physical decay of the radionuclide. The amount of activity contained in a source organ therefore generally changes with time. If the time-activity curve is known, the cumulated activity for a source organ is obtained by measuring the area under this curve (Fig. 22-2). Mathematically, if the time-activity curve is described by a function $A(t)$, then

$$\widetilde{A} \approx \int_0^\infty A(t)dt \qquad (22\text{-}5)$$

where it is assumed that activity is administered to the patient at time $t = 0$ and is measured to complete disappearance from the organ ($t = \infty$).

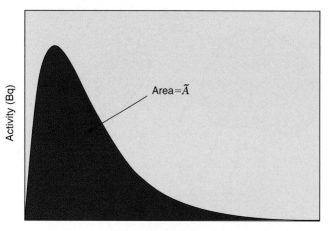

FIGURE 22-2 Hypothetical time-activity curve for radioactivity in a source organ. Cumulated activity $\widetilde{A}$ in Bq • sec is the area under the curve (equivalent to the integral in Equation 22-5).

To estimate the radiation dose received from a particular radiotracer, time-activity curves for all the major organs are required. These can be obtained from animal studies (which are then extrapolated with some uncertainty to the human), imaging studies in normal human subjects, prior knowledge of the tracer kinetics, or some combination of these approaches. Time-activity curves can be quite complex, and thus Equation 22-5 may be difficult to analyze. Frequently, however, certain assumptions can be made to simplify this calculation.

Situation 1: Uptake by the organ is "instantaneous" (i.e., very rapid with respect to the half-life of the radionuclide), and there is no biologic excretion. The time-activity curve then is described by ordinary radioactive decay (Equations 4-7 and 4-10):

$$A(t) = A_0 e^{-0.693t/T_p} \qquad (22\text{-}6)$$

where T_p is the physical half-life of the radionuclide and A_0 is the activity initially present in the organ. Thus

$$\widetilde{A} \approx A_0 \int_0^\infty e^{-0.693t/T_p}\, dt$$

$$= \frac{T_p A_0}{0.693} = 1.44 T_p A_0 \qquad (22\text{-}7)$$

The quantity $1.44 T_p$ is the average lifetime of the radionuclide (see Chapter 4, Section B.3). Thus the cumulated activity in a source organ, when eliminated by physical decay only, is the same as if activity were present at a constant level A_0 for a time equal to the average lifetime of the radionuclide (Fig. 22-3).

EXAMPLE 22-1

What is the cumulated activity in the liver for an injection of 100 MBq of a ^{99m}Tc-labeled sulfur colloid, assuming that 60% of the injected colloid is trapped by the liver and retained there indefinitely?

Answer

$$\widetilde{A} = 1.44 \times 100 \text{ MBq} \times 0.60 \times 6.0 \text{ hr}$$

$$= 518.4 \text{ MBq} \cdot \text{hr}$$

$$= 1.87 \times 10^6 \text{ MBq} \cdot \text{sec}$$

Situation 2: Uptake is instantaneous, and clearance is by biologic excretion only (no physical decay, or physical half-life very long in comparison with biologic excretion). In this situation, biologic excretion must be carefully analyzed. Frequently, it can be described by a set of exponential excretion components, with a fraction f_1 of the initial activity A_0 being excreted with a (biologic) half-life T_{b1}, a fraction f_2 with half-life T_{b2}, and so on (Fig. 22-4). The cumulated activity then is given by

$$\widetilde{A} \approx A_0 \int_0^\infty f_1 e^{-0.693t/T_{b1}}\, dt$$

$$+ A_0 \int_0^\infty f_2 e^{-0.693t/T_{b2}}\, dt + \ldots$$

$$= 1.44 T_{b1} f_1 A_0 + 1.44 T_{b2} f_2 A_0 + \ldots \qquad (22\text{-}8)$$

EXAMPLE 22-2

Suppose that 100 MBq of ^{99m}Tc-labeled micro-spheres are injected into a patient, with essentially instantaneous uptake of activity

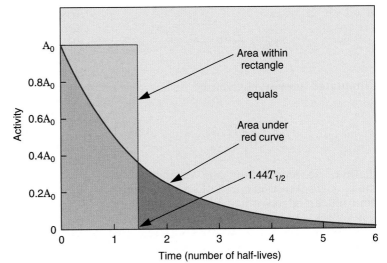

FIGURE 22-3 Illustration of relationship between $\tilde{A}$ and average lifetime $(1.44\ T_\mathrm{p})$ of a radionuclide for simple exponential decay.

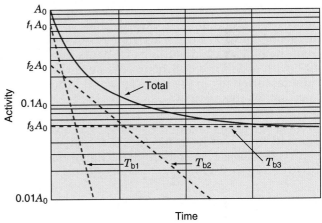

FIGURE 22-4 Illustration of a multicomponent exponential excretion curve. Fraction f_1 is excreted with biologic half-life $T_{\mathrm{b}1}$, f_2 with half-life $T_{\mathrm{b}2}$, f_3 with half-life $T_{\mathrm{b}3}$, and so on.

by the lungs. What is the cumulated activity in the lungs if 60% of the activity is excreted from the lungs with a biologic half-life of 15 minutes and 40% with a biologic half-life of 30 minutes?

Answer

Because ^{99m}Tc physical decay is much slower than the biologic excretion process, we may assume that no physical decay occurs during the time that activity is present in the lungs. Thus (Equation 22-8)

$$\tilde{A} = (1.44 \times 1/4\ \mathrm{hr} \times 0.60 \times 100\ \mathrm{MBq})$$
$$+ (1.44 \times 1/2\ \mathrm{hr} \times 0.40 \times 100\ \mathrm{MBq})$$
$$= (21.6 + 28.8)\ \mathrm{MBq \cdot hr}$$
$$= 50.4\ \mathrm{MBq \cdot hr}$$
$$= 1.81 \times 10^5\ \mathrm{MBq \cdot sec}$$

Situation 3: Uptake is instantaneous but clearance by both physical decay and biologic excretion are significant. In this case, if biologic excretion is described by a single-component exponential curve with biologic half-life T_b, and the physical half-life is T_p, then the total clearance is described by a single-component exponential curve with an *effective half-life T_e* given by*

$$\frac{1}{T_\mathrm{e}} = \frac{1}{T_\mathrm{p}} + \frac{1}{T_\mathrm{b}} \qquad (22\text{-}9)$$

---

*Equation 22-9 can be derived from Equations 4-2 and 4-9 by treating biologic excretion as the equivalent of a second pathway in a "branching" radioactive decay scheme.

or

$$T_e = \frac{T_p T_b}{(T_p + T_b)} \qquad (22\text{-}10)$$

Cumulated activity is given by

$$\widetilde{A} \approx 1.44 T_e A_0 \qquad (22\text{-}11)$$

If there is more than one component to the biologic excretion curve, then each component has an effective half-life given by Equation 22-9 for that component, and the cumulated activity is computed with effective half-lives replacing biologic half-lives in Equation 22-8.

EXAMPLE 22-3

Suppose in Example 22-2 that because of a metabolic defect 60% of the activity is excreted from the lungs with a half-life of 2 hours and 40% with a half-life of 3 hours. What is the cumulated activity in the lungs for a 100 MBq injection for this patient?

Answer
The effective half-lives for the two components of biologic excretion are (Equation 22-10)

$$T_{e1} = 6 \times 2/(6+2) = 1.5 \text{ hr}$$

$$T_{e2} = 6 \times 3/(6+3) = 2 \text{ hr}$$

Thus applying Equation 22-8, with T_e replacing T_b,

$$\widetilde{A} = (1.44 \times 1.5 \text{ hr} \times 0.60 \times 100 \text{ MBq})$$
$$+ (1.44 \times 2 \text{ hr} \times 0.40 \times 100 \text{ MBq})$$
$$= (129.6 + 115.2) \text{ MBq} \cdot \text{hr}$$
$$= 244.8 \text{ MBq} \cdot \text{hr}$$
$$= 8.81 \times 10^5 \text{ MBq} \cdot \text{sec}$$

Situation 4: Uptake is not instantaneous. The equations developed thus far will overestimate radiation doses when uptake by the source organ is not rapid in comparison with physical decay, that is, if a significant amount of physical decay occurs during the uptake process, before the activity reaches the source organ of interest. This situation arises with radionuclides that have a slow pattern of uptake in comparison with their physical half-life. Frequently, uptake can be described by an exponential equation of the form

$$A(t) = A_0 (1 - e^{-0.693t/T_u}) \qquad (22\text{-}12)$$

where T_u is the biologic uptake half-time. In this case, cumulated activity is given by

$$\widetilde{A} \approx 1.44 A_0 T_e (T_{ue} / T_u) \qquad (22\text{-}13)$$

where T_e is the effective excretion half-life (Equation 22-10) and T_{ue} is the effective uptake half-time

$$T_{ue} = \frac{T_u T_p}{T_u + T_p} \qquad (22\text{-}14)$$

EXAMPLE 22-4

A radioactive gas having a half-life of 20 seconds is injected in an intravenous solution. It appears in the lungs with an uptake half-time of 30 seconds and is excreted (by exhalation) with a biologic half-life of 10 seconds. What is the cumulated activity in the lungs for a 250-MBq injection?

Answer
The effective uptake half-time is (Equation 22-14)

$$T_{ue} = 20 \times 30/(20 + 30) = 12 \text{ sec}$$

and the effective excretion half-life is

$$T_e = 20 \times 10/(20 + 10) = 6.7 \text{ sec}$$

Thus, from Equation 22-13,

$$\widetilde{A} = 1.44 \times 250 \text{ MBq} \times 6.7 \text{ sec}$$
$$\times (12 \text{ sec}/30 \text{ sec})$$
$$= 964.8 \text{ MBq} \cdot \text{sec}$$

3. Equilibrium Absorbed Dose Constant, Δ

Given $\widetilde{A}$ for the source organ, the next step is to calculate the radiation energy emitted by this amount of cumulated activity. Energy emitted per unit of cumulated activity is given by the *equilibrium absorbed dose constant* Δ. The factor Δ must be calculated for each type of emission for the radionuclide. It is given by*

$$\Delta_i = 1.6 \times 10^{-13} N_i E_i \ (\text{Gy} \cdot \text{kg} / \text{Bq} \cdot \text{sec})$$
(22-15)

where E_i is the average energy (in MeV) of the ith emission and N_i is the relative frequency of that emission (number emitted per disintegration) by the radionuclide. In traditional units, the equilibrium absorbed dose constant is

$$\Delta_i = 2.13 N_i E_i \ (\text{rad} \cdot \text{g} / \mu\text{Ci} \cdot \text{hr}) \quad (22\text{-}16)$$

EXAMPLE 22-5

A certain radionuclide decays by emitting β particles in 100% of its disintegrations with $\overline{E}_\beta = 0.3$ MeV. This is followed immediately in 80% of its disintegrations by emission of a 0.2-MeV γ ray and in 20% by emission of a 0.195-MeV conversion electron and a 0.005-MeV characteristic x ray. What are the equilibrium absorbed dose constants for the emissions of this radionuclide?

Answer

$$\Delta_\beta = (1.6 \times 10^{-13}) \times 1.0 \times 0.30$$
$$= 4.80 \times 10^{-14} \ \text{Gy} \cdot \text{kg/Bq} \cdot \text{sec}$$
$$\Delta_\gamma = (1.6 \times 10^{-13}) \times 0.80 \times 0.20$$
$$= 2.56 \times 10^{-14} \ \text{Gy} \cdot \text{kg/Bq} \cdot \text{sec}$$
$$\Delta_e = (1.6 \times 10^{-13}) \times 0.20 \times 0.195$$
$$= 6.24 \times 10^{-15} \ \text{Gy} \cdot \text{kg/Bq} \cdot \text{sec}$$
$$\Delta_x = (1.6 \times 10^{-13}) \times 0.2 \times 0.005$$
$$= 1.60 \times 10^{-16} \ \text{Gy} \cdot \text{kg/Bq} \cdot \text{sec}$$

The product of cumulated activity $\widetilde{A}$ and equilibrium absorbed dose constant Δ_i is the radiation energy emitted by the ith emission, in Gy $\cdot$ kg, during the time that radioactivity is present in a source organ.

EXAMPLE 22-6

Assume that the radionuclide in Example 22-5 is used for the problem described in Example 22-4. What is the total amount of energy emitted from radioactivity contained in the lungs in Example 22-4?

*Essentially the energy emitted per nuclear disintegration: 1 MeV/dis = 1.6×10^{-13} Gy $\cdot$ kg/Bq $\cdot$ sec.

Answer

The total energy emitted per Bq $\cdot$ sec is the sum of the equilibrium absorbed dose constants for the β, γ, conversion electron and x-ray emissions:

$$\Delta = \Delta_\beta + \Delta_\gamma + \Delta_e + \Delta_x$$
$$= 8.0 \times 10^{-14} \ \text{Gy} \cdot \text{kg/Bq} \cdot \text{sec}$$
$$= 8.0 \times 10^{-8} \ \text{Gy} \cdot \text{kg/MBq} \cdot \text{sec}$$

The cumulated activity is 9.65×10^2 MBq $\cdot$ sec. Using these values and Equation 22-1, the total energy emitted is

$$\widetilde{A} \times \Delta = 9.65 \times 10^2 \ \text{MBq} \cdot \text{sec} \times 8.0$$
$$\times 10^{-8} \ \text{Gy} \cdot \text{kg/MBq} \cdot \text{sec}$$
$$= 7.72 \times 10^{-5} \ \text{Gy} \cdot \text{kg}$$
$$= 7.72 \times 10^{-5} \ \text{joules}$$

Values of Δ are presented in Appendix C for some of the radionuclides of interest in nuclear medicine. A full listing can be found in reference 2.

4. Absorbed Fraction, ϕ

The final step is to determine the fraction of the energy emitted by the source organ that is absorbed by the target organ. This is given by the *absorbed fraction* ϕ. The absorbed fraction depends on the amount of radiation energy reaching the target organ (tissue and distance attenuation between source and target organs) and on the volume and composition (e.g., lung, bone) of the target organ. Thus it depends on the type and energy of the emission and on the anatomic relationship of the source-target pair. In a dosimetry calculation, a value of ϕ must be determined for each type of emission from the radionuclide and for each source-target pair in the calculation. The notation $\phi_i(r_k \leftarrow r_h)$ is used to indicate absorbed fraction for energy delivered from a source organ (or region), r_h, to a target organ, r_k, for the ith emission of the radionuclide.

The total energy absorbed by a specific target organ thus is given by

Total energy absorbed (Gy $\cdot$ kg) =
$$\widetilde{A} \sum_i \phi_i(r_k \leftarrow r_h)\Delta_i$$
(22-17)

The summation Σ_i includes values of ϕ_i and Δ_i for all the emissions of the radionuclide and values of $\phi_i(r_k \leftarrow r_h)$ for the source-target pair. $\widetilde{A}$ is the cumulated activity in the source organ h. The energy absorbed by the target organ divided by the target organ mass m_t gives the *average absorbed dose* in grays to the target organ from activity in the source organ:

$$\bar{D}(r_k \leftarrow r_h) = \frac{\widetilde{A}}{m_t} \sum_i \phi_i(r_k \leftarrow r_h)\Delta_i$$

$$(22\text{-}18)$$

The total dose to the target organ then is obtained by summing the doses from all of the source organs, h, in the body.

Values of ϕ have been calculated for mathematical humanoid models incorporating organs and anatomic structures of "average" size and shape (Fig. 22-5). The model used for many years was that published by the MIRD committee of the Society of Nuclear Medicine.[3] Cristy and Eckerman[4] subsequently developed a series of models representing newborn, 1-year-old, 5-year-old, 10-year-old, 15-year-old, and adult individuals. Stabin and associates extended the model to women and pregnant women.[5] Organ masses for the adult male phantom developed by Cristy and Eckerman are given in Table 22-2. Most of the values for their adult male model are similar to the model originally developed by the MIRD committee; however, there are some significant differences as well, such as in the mass and values of ϕ for bone marrow. Consequently, the Cristy and Eckerman models now have replaced the older MIRD model.

Calculations of ϕ are complex, and the tables are quite lengthy for "penetrating" radiations (photons with energy $\gtrsim 10$ keV) because of the energy dependence of photon attenuation and absorption; however, the situation is simpler for nonpenetrating radiations (photons with energy ≤ 10 keV and electrons), for which it can be assumed that all of the emitted energy is "locally absorbed," that is, within the source organ itself. For these emissions, $\phi = 1$ when the target and the source are the same organ, $\phi = 0$ otherwise. In dosimetry calculations, it is useful to sum the equilibrium absorbed dose constants for the nonpenetrating radiations and treat them as a single parameter, Δ_{np}, because the absorbed fractions for all of these emissions are equal (unity when the source and target are the same organ, zero otherwise).

EXAMPLE 22-7

Compute the absorbed dose delivered to the lung by nonpenetrating radiations in the problem described by Examples 22-4 and 22-5.

Answer

The nonpenetrating radiations are the β particles ($\Delta_\beta = 4.80 \times 10^{-14}$ Gy $\cdot$ kg/Bq $\cdot$ sec), conversion electrons ($\Delta_e = 6.24 \times 10^{-15}$ Gy $\cdot$ kg/Bq $\cdot$ sec), and 5 keV characteristic x-rays ($\Delta_x = 1.60 \times 10^{-16}$ Gy $\cdot$ kg/Bq $\cdot$ sec). Thus $\Delta_{np} = 5.44 \times 10^{-14}$ Gy $\cdot$ kg/Bq $\cdot$ sec. Cumulated activity is $\widetilde{A} = 9.65 \times 10^8$ Bq $\cdot$ sec. Lung mass is 1 kg (see Table 22-2). Thus the average radiation dose delivered by these emissions to the lungs is

$$\bar{D} = (9.65 \times 10^8 \text{ Bq} \cdot \text{sec})$$
$$\times (5.44 \times 10^{-14} \text{ Gy} \cdot \text{kg/Bq} \cdot \text{sec})$$
$$\times (\phi = 1)/1 \text{ kg}$$
$$= 5.25 \times 10^{-5} \text{ Gy} \ (5.25 \text{ mrad})$$

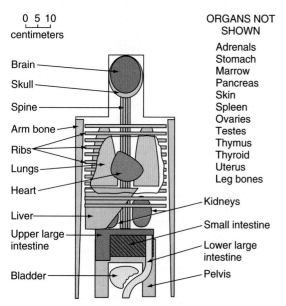

ANTERIOR VIEW OF THE PRINCIPAL ORGANS IN THE HEAD AND TRUNK OF THE PHANTOM

0 5 10
centimeters

Brain
Skull
Spine
Arm bone
Ribs
Lungs
Heart
Liver
Upper large intestine
Bladder

ORGANS NOT SHOWN

Adrenals
Stomach
Marrow
Pancreas
Skin
Spleen
Ovaries
Testes
Thymus
Thyroid
Uterus
Leg bones

Kidneys
Small intestine
Lower large intestine
Pelvis

FIGURE 22-5 Representation of an "average man" used for MIRD dose calculations and tables. (*Adapted with permission from Snyder WS, Fisher HL Jr, Ford MR, Warner GG: Estimates of absorbed fractions for monoenergetic photon sources uniformly distributed in various organs of a heterogenous phantom. J Nucl Med Suppl 3:9, 1969.*)

5. Specific Absorbed Fraction, Φ, and the Dose Reciprocity Theorem

The specific absorbed fraction is given by

$$\Phi = \frac{\phi}{m_t} \qquad (22\text{-}19)$$

TABLE 22-2
ORGAN MASSES FOR THE CRISTY AND ECKERMAN ADULT MALE PHANTOM

Organ	Mass (g)	Organ	Mass (g)
Adrenals	16.3	Lungs	1000
Brain	1420	Ovaries	8.71
Breasts (excluding skin)	351	Pancreas	94.3
Gallbladder contents	55.7	Skeleton	
Gallbladder wall	10.5	Active marrow	1120
Gastrointestinal Tract		Cortical bone	4000
Lower large intestine contents	143	Trabecular bone	1000
Lower large intestine wall	167	Skin	3010
Small intestine contents and wall	1100	Spleen	183
Stomach contents	260	Testes	39.1
Stomach wall	158	Thymus	20.9
Upper large intestine contents	232	Thyroid	20.7
Upper large intestine wall	220	Urinary bladder contents	211
Heart contents	454	Urinary bladder wall	47.6
Heart wall	316	Uterus	79.0
Kidneys	299	Remaining tissue	51,800
Liver	1910		

From Cristy M, Eckerman K: *Specific Absorbed Fractions of Energy at Various Ages From Internal Photon Sources (ORNL Report ORNL/TM-8381 V1–V7)*. Oak Ridge, TN, 1987, Oak Ridge National Laboratory.

It is the fraction of radiation emitted by the source organ that is absorbed *per gram* of target organ mass. The absorbed dose equation can be written using specific absorbed fractions as

$$\bar{D}(r_k \leftarrow r_h) = \tilde{A} \sum_i \Phi_i(r_k \leftarrow r_h)\Delta_i \quad (22\text{-}20)$$

The *dose reciprocity theorem* says that for a given organ pair the specific absorbed fraction is the same, regardless of which organ is the source and which is the target:

$$\Phi_i(r_k \leftarrow r_h) = \Phi_i(r_h \leftarrow r_k) \quad (22\text{-}21)$$

This simply says that the energy absorbed per gram is the same for radiation traveling from r_h to r_k as it is for radiation traveling from r_k to r_h, a fact that seems intuitively obvious.*

---

*Strictly speaking, the dose reciprocity theorem is precisely correct only when both the source and target materials are homogeneous absorbing materials. However, this requirement is not stringent and the theorem is sufficiently accurate (<1% error) for most applications in the human body. One situation in which the required conditions are not met is when red marrow is a source or target organ. This is evident in Tables 22-3 to 22-5, which are discussed in the next section.

The dose reciprocity theorem is useful when tables for ϕ are not available for all source-target organ pairs. If $\phi(r_k \leftarrow r_h)$ is known, then $\phi(r_h \leftarrow r_k)$ can be obtained from the dose reciprocity theorem. Rewriting Equation 22-21 in terms of ϕ:

$$\frac{\phi(r_h \leftarrow r_k)}{m_h} = \frac{\phi(r_k \leftarrow r_h)}{m_k} \quad (22\text{-}22)$$

$$\phi(r_h \leftarrow r_k) = \frac{m_h}{m_k} \times \phi(r_k \leftarrow r_h) \quad (22\text{-}23)$$

The specific absorbed fractions for a range of different phantoms are available in references 4 and 6.

6. Mean Dose per Cumulated Activity, *S*

Radiation dose calculations for penetrating radiations can be quite tedious, especially when there are multiple emissions to consider. The problem has been simplified by the introduction of *S* (sometimes also known as the dose factor, *DF*), the mean dose per unit cumulated activity:

$$S(r_k \leftarrow r_h) = \frac{1}{m_k} \sum_i \phi_i (r_k \leftarrow r_h) \Delta_i$$

$$= \sum_i \Phi_i (r_k \leftarrow r_h) \Delta_i$$

(22-24)

The quantity S has units of Gy/Bq · sec. It has been calculated for different source-target organ pairs for a wide variety of radionuclides of interest in nuclear medicine.[7] Tables 22-3 to 22-5 present values of S for ^{99m}Tc, ^{131}I, and ^{18}F, respectively. Given the values of S and cumulated activity, $\tilde{A}$, the average dose to an organ is given by

$$\bar{D}(r_k \leftarrow r_h) = \tilde{A} \times S(r_k \leftarrow r_h) \quad (22-25)$$

EXAMPLE 22-8

Calculate the radiation dose to the liver (LI) to an average adult male for an injection of 100 MBq of ^{99m}Tc sulfur colloid. Assume that 60% of the activity is trapped by the liver, 30% by the spleen (SP), and 10% by red bone marrow (RM), with instantaneous uptake and no biologic excretion.

Answer
The cumulated activities for the three source organs are (Equation 22-5)

$$\tilde{A}_{LI} = 1.44 \times 6.0 \text{ hr} \times 0.60 \times 100 \text{ MBq}$$
$$= 518.4 \text{ MBq} \cdot \text{hr}$$
$$= 1.87 \times 10^6 \text{ MBq} \cdot \text{sec}$$

$$\tilde{A}_{SP} = 1.44 \times 6.0 \text{ hr} \times 0.30 \times 100 \text{ MBq}$$
$$= 259.2 \text{ MBq} \cdot \text{hr}$$
$$= 9.33 \times 10^5 \text{ MBq} \cdot \text{sec}$$

$$\tilde{A}_{RM} = 1.44 \times 6.0 \text{ hr} \times 0.10 \times 100 \text{ MBq}$$
$$= 86.4 \text{ MBq} \cdot \text{hr}$$
$$= 3.11 \times 10^5 \text{ MBq} \cdot \text{sec}$$

The values of S for ^{99m}Tc are (see Table 22-3)

$$S(LI \leftarrow LI) = 3.16 \times 10^{-6} \text{ mGy/MBq} \cdot \text{sec}$$

$$S(LI \leftarrow SP) = 7.22 \times 10^{-8} \text{ mGy/MBq} \cdot \text{sec}$$

$$S(LI \leftarrow RM) = 8.96 \times 10^{-8} \text{ mGy/MBq} \cdot \text{sec}$$

Therefore, the absorbed doses are

$$\bar{D}(LI \leftarrow LI) = (1.87 \times 10^6 \text{ MBq} \cdot \text{sec})$$
$$\times (3.16 \times 10^{-6} \text{ mGy/MBq} \cdot \text{sec})$$
$$= 5.91 \text{ mGy}$$

$$\bar{D}(LI \leftarrow SP) = (9.33 \times 10^5 \text{ MBq} \cdot \text{sec})$$
$$\times (7.22 \times 10^{-8} \text{ mGy/MBq} \cdot \text{sec})$$
$$= 6.74 \times 10^{-2} \text{ mGy}$$

$$\bar{D}(LI \leftarrow RM) = (3.11 \times 10^5 \text{ MBq} \cdot \text{sec})$$
$$\times (8.96 \times 10^{-8} \text{ mGy/MBq} \cdot \text{sec})$$
$$= 2.79 \times 10^{-2} \text{ mGy}$$

The average total dose to the liver is therefore

$$\bar{D} = 5.91 + 6.74 \times 10^{-2} + 2.79 \times 10^{-2} \text{ mGy}$$
$$= 6.01 \text{ mGy} (\sim 0.6 \text{ rads})$$

Example 22-8 demonstrates that most of the dose delivered to an organ that concentrates the radionuclide arises from the radioactivity in the target organ itself [$\bar{D}(LI \leftarrow LI)$].

EXAMPLE 22-9

An adult male patient is to be treated with ^{131}I for hyperthyroidism. It is determined by prior studies with a tracer dose of ^{131}I that the patient's thyroidal iodine uptake is 60% and the biologic half-life of iodine in the thyroid gland is 2 days. Assuming instantaneous uptake ($T_u \ll T_p = 8$ days), what is the dose to the thyroid (THY) from radioactivity contained in the thyroid for this patient, per MBq ^{131}I?

Answer
The effective half-life of ^{131}I in the thyroid for this patient is (Equation 22-10)

$$T_e = 8 \times 2/(8 + 2) = 16/10 \text{ days}$$
$$= 1.38 \times 10^5 \text{ sec}$$

Therefore the cumulated activity per MBq administered is (Equation 22-11)

$$\tilde{A} = 1.44 \times (1.38 \times 10^5 \text{ sec})$$
$$\times 0.60 \times 1 \text{ MBq}$$
$$= 1.19 \times 10^5 \text{ MBq} \cdot \text{sec/MBq administered}$$

The dose per MBq · sec is (see Table 22-4):

$$S(THY \leftarrow THY) = 1.57 \times 10^{-3} \text{ mGy/MBq} \cdot \text{sec}$$

Thus the average absorbed dose for the thyroid is

$$\bar{D}\,(\text{THY} \leftarrow \text{THY})$$

$$= (1.19 \times 10^5 \text{ MBq} \cdot \text{sec/MBq administered})$$

$$\times \,(1.57 \times 10^{-3} \text{ mGy/MBq} \cdot \text{sec})$$

$$= 187 \text{ mGy/MBq administered}$$

$$(\text{or } 692 \text{ rads/mCi})$$

One could include the radiation dose to the thyroid from activity in other organs in the calculation performed in Example 22-9; however, inspection of Table 22-4 reveals that in comparison to the thyroid as the source organ, other organs have much smaller values of S (by approximately a factor of 500). This, plus the fact that other organs concentrate much less of the activity than the thyroid, eliminates the need to consider them as source organs in this calculation.

Examples 22-8 and 22-9 represent simplified situations in which only a few organs are involved and where the cumulated activities of the organs are relatively easy to estimate. In many cases, the calculations are more involved, with complex time-activity curves and more widespread distribution of the radiopharmaceutical among different organs. To facilitate dosimetry calculations, a software program[8] has been developed to calculate the absorbed dose to major organs from commonly employed radionuclides using the Cristy and Eckerman[4] and Stabin[5] phantom models of human anatomy. This greatly simplifies dose calculations, although it still is necessary to provide the cumulated activity data for each organ for the radiopharmaceutical of interest. Estimated radiation doses for a large number of commonly used radiopharmaceuticals are available from the Oak Ridge Institute for Science and Education.[9] By way of example, radiation dose estimates for ^{18}F-fluorodeoxyglucose (FDG) PET studies based on the Cristy and Eckerman adult male phantom are reproduced in Table 22-6.

7. Whole-Body Dose and Effective Dose

The complete output of a dose calculation is an estimate of the radiation dose for all the major organs in the body. This provides a large amount of information that is difficult to assimilate into a perception of the risk of a specific radiopharmaceutical study, or for comparison of the dose from a nuclear medicine procedure with that from other medical procedures that use radiation sources. For these reasons, it would be convenient to condense radiation dose estimates such as those in Table 22-6 into a single number. There are two different approaches to doing this: *whole-body* (or *total-body*) *dose* and *effective dose*.

The whole-body or total-body dose is the *total energy deposited in the body* divided by the *total mass of the body*, or in terms of the S factor for the total body (TB):

$$\bar{D}(\text{TB} \leftarrow \text{TB}) = \tilde{A} \times S(\text{TB} \leftarrow \text{TB})$$

$$(22\text{-}26)$$

This parameter was used for many years as the standard for evaluating the risks of different nuclear medicine procedures. However, the whole-body dose does not take into account the nonuniformity of dose distribution among the organs in the body, and its validity for comparing the risks of different nuclear medicine procedures is therefore questionable.

The effective dose, E, was introduced by the International Commission on Radiological Protection (ICRP)[1,10] as an attempt to characterize a nonuniform internal dose by a single number. This quantity was intended primarily for estimating radiation risks and doses received by radiation workers, although its extension to clinical nuclear medicine studies has been supported by the ICRP. The effective dose represents the whole-body dose that would result in the same overall risk as the nonuniform dose distribution actually delivered. This is achieved by assigning different *weighting factors* to the doses delivered to individual organs. The most recent recommended values for the tissue weighting factors, w_T, are shown in Table 22-7. The effective dose, which has units of sieverts, is calculated from

$$E = \sum_T w_T \times D_T \times w_R = \sum_T w_T \times H_T$$

$$(22\text{-}27)$$

where D_T is the average absorbed dose in organ T, w_T is the tissue weighting factor for organ T, and the summation is over all the organs listed in Table 22-7. H_T is the equivalent dose defined in Section A. As noted in Section A, $w_R = 1$ for all radiations used in diagnostic nuclear medicine procedures. An older quantity, *effective dose equivalent* (H_E), which uses slightly different tissue weighting factors, may be encountered in publications and in regulations established prior to 1991.

Text continued on page 424

TABLE 22-3

S VALUES (mGy/MBq • sec) FOR Tc-99m IN THE ADULT MALE PHANTOM*

Target Organs	Source Organs													
	Adrenals	Brain	Breasts	Gallbladder Contents	Lower Large Intestine	Small Intestine	Stomach	Upper Large Intestine	Heart Contents	Heart Wall	Kidneys	Liver	Lungs	Muscle
Adrenals	1.80E-04	4.18E-10	5.05E-08	3.13E-07	2.25E-08	7.46E-08	2.73E-07	9.58E-08	2.53E-07	2.85E-07	7.24E-07	4.35E-07	2.33E-07	1.12E-07
Brain	4.18E-10	4.23E-06	3.17E-09	1.49E-10	1.57E-11	3.91E-11	4.27E-10	4.68E-11	3.14E-09	2.54E-09	1.58E-10	8.16E-10	7.63E-09	2.21E-08
Breasts	5.05E-08	3.17E-09	1.14E-05	3.33E-08	2.28E-09	7.35E-09	5.73E-08	8.00E-09	2.41E-07	2.61E-07	1.99E-08	6.82E-08	2.33E-07	4.25E-08
Gallbladder wall	3.57E-07	1.54E-10	3.41E-08	3.37E-05	6.49E-08	4.38E-07	3.05E-07	7.53E-07	1.03E-07	1.22E-07	4.09E-07	8.70E-07	7.46E-08	1.19E-07
Lower large intestine wall	1.98E-08	1.32E-11	2.42E-09	5.93E-08	1.23E-05	5.92E-07	9.10E-08	2.14E-07	4.06E-09	4.90E-09	5.50E-08	1.44E-08	3.29E-09	1.34E-07
Small intestine	7.46E-08	3.91E-11	7.35E-09	4.58E-07	7.16E-07	4.22E-06	2.08E-07	1.25E-06	1.57E-08	2.06E-08	2.13E-07	1.16E-07	1.35E-08	1.12E-07
Stomach wall	2.85E-07	2.52E-10	5.93E-08	2.93E-07	1.24E-07	2.13E-07	8.53E-06	2.86E-07	1.66E-07	2.65E-07	2.53E-07	1.48E-07	1.19E-07	1.09E-07
Upper large intestine wall	9.41E-08	4.76E-11	7.51E-09	7.78E-07	3.10E-07	1.36E-06	2.65E-07	8.37E-06	2.12E-08	2.65E-08	2.12E-07	1.88E-07	1.81E-08	1.10E-07
Heart wall	2.85E-07	2.54E-09	2.61E-07	1.04E-07	5.42E-09	2.06E-08	2.33E-07	2.97E-08	5.48E-06	1.19E-05	8.22E-08	2.33E-07	4.40E-07	9.20E-08
Kidneys	7.24E-07	1.58E-10	1.99E-08	3.89E-07	7.10E-08	2.13E-07	2.73E-07	2.12E-07	6.45E-08	8.22E-08	1.32E-05	2.93E-07	6.66E-08	9.79E-08
Liver	4.35E-07	8.16E-10	6.82E-08	8.20E-07	1.80E-08	1.16E-07	1.47E-07	1.87E-07	2.13E-07	2.33E-07	2.93E-07	3.16E-06	1.97E-07	7.52E-08
Lungs	2.33E-07	7.63E-09	2.33E-07	7.09E-08	4.50E-09	1.35E-08	1.10E-07	1.77E-08	4.59E-07	4.40E-07	6.66E-08	2.09E-07	3.57E-06	9.36E-08
Muscle	1.12E-07	2.21E-08	4.25E-08	1.14E-07	1.23E-07	1.12E-07	9.96E-08	1.07E-07	8.83E-08	9.20E-08	9.79E-08	7.52E-08	9.34E-08	1.93E-07
Ovaries	3.14E-08	1.52E-11	2.61E-09	1.11E-07	1.26E-06	9.23E-07	5.85E-08	7.71E-07	4.55E-09	6.15E-09	7.02E-08	3.81E-08	5.39E-09	1.44E-07
Pancreas	1.09E-06	4.15E-10	6.22E-08	6.75E-07	5.21E-08	1.42E-07	1.23E-06	1.62E-07	2.65E-07	3.57E-07	4.97E-07	3.86E-07	1.74E-07	1.24E-07
Red marrow	2.53E-07	1.01E-07	5.52E-08	1.02E-07	2.01E-07	1.79E-07	7.50E-08	1.43E-07	1.11E-07	1.11E-07	1.71E-07	8.32E-08	1.11E-07	9.07E-08
Bone surfaces	2.67E-07	2.99E-07	7.76E-08	1.14E-07	1.82E-07	1.49E-07	1.03E-07	1.27E-07	1.60E-07	1.60E-07	1.62E-07	1.24E-07	1.66E-07	1.84E-07
Skin	3.41E-08	3.97E-08	7.63E-08	3.09E-08	3.62E-08	3.01E-08	3.41E-08	3.09E-08	3.41E-08	3.70E-08	3.79E-08	3.62E-08	4.02E-08	5.72E-08
Spleen	4.58E-07	5.19E-10	4.37E-08	1.33E-07	6.53E-08	1.01E-07	7.83E-07	1.05E-07	1.24E-07	1.67E-07	6.63E-07	7.22E-08	1.64E-07	1.03E-07
Testes	1.54E-09	1.46E-12	2.29E-07	6.91E-09	1.40E-07	2.61E-08	2.90E-09	1.92E-08	5.16E-10	6.16E-10	3.10E-09	1.57E-09	3.67E-10	9.89E-08
Thymus	5.66E-08	6.88E-09	2.29E-07	1.43E-08	2.04E-09	4.66E-09	3.65E-08	5.43E-09	8.87E-07	7.35E-07	1.73E-08	5.93E-08	2.85E-07	1.06E-07
Thyroid	8.11E-09	1.35E-07	3.01E-08	2.45E-08	2.48E-10	4.87E-10	2.62E-09	7.69E-10	5.17E-08	4.33E-08	2.95E-09	8.64E-09	8.82E-08	1.16E-07
Urinary bladder wall	7.55E-09	6.02E-12	1.33E-09	4.22E-08	4.98E-07	2.12E-07	1.73E-08	1.61E-07	2.22E-09	2.17E-09	1.87E-08	1.16E-08	1.33E-09	1.40E-07
Uterus	1.89E-08	1.31E-11	2.62E-09	1.16E-07	5.17E-07	8.37E-07	5.05E-08	3.97E-07	4.87E-09	5.47E-09	6.42E-08	3.29E-08	4.10E-09	1.43E-07
Total body	1.72E-07	1.25E-07	1.03E-07	1.36E-07	1.49E-07	1.59E-07	1.17E-07	1.41E-07	1.17E-07	1.65E-07	1.58E-07	1.59E-07	1.44E-07	1.33E-07

TABLE 22-3 CONTINUED

Target Organs					Source Organs									
	Ovaries	Pancreas	Red Marrow	Cortical Bone Surfaces	Trabecular Bone Surfaces	Cortical Bone Volume	Trabecular Bone Volume	Spleen	Testes	Thymus	Thyroid	Urinary Bladder Contents	Uterus	Total Body
Adrenals	3.14E-08	1.09E-06	2.41E-07	1.07E-07	1.07E-07	1.07E-07	1.07E-07	4.58E-07	1.54E-09	5.66E-08	8.11E-09	8.41E-09	1.89E-08	1.67E-07
Brain	1.52E-11	4.15E-10	8.08E-08	1.17E-07	1.17E-07	1.17E-07	1.17E-07	5.19E-10	1.46E-12	6.88E-09	1.35E-07	5.94E-12	1.31E-11	1.21E-07
Breasts	2.61E-09	6.22E-08	5.12E-08	3.10E-08	3.10E-08	3.10E-08	3.10E-08	4.37E-08	0.00E+00	2.29E-07	3.01E-08	1.30E-09	2.62E-09	1.00E-07
Gallbladder wall	9.91E-08	8.14E-07	1.15E-07	4.41E-08	4.41E-08	4.41E-08	4.41E-08	1.35E-07	6.71E-09	2.65E-08	2.64E-09	3.45E-08	1.15E-07	1.77E-07
Lower large intestine wall	1.12E-06	4.17E-08	1.98E-07	7.25E-08	7.25E-08	7.25E-08	7.25E-08	4.65E-08	1.96E-07	1.61E-09	2.06E-10	5.78E-07	4.97E-07	1.73E-07
Small intestine	9.23E-07	1.42E-07	1.86E-07	5.74E-08	5.74E-08	5.74E-08	5.74E-08	1.01E-07	2.61E-08	4.66E-09	4.87E-10	2.24E-07	8.37E-07	1.77E-07
Stomach wall	5.85E-08	1.26E-06	8.23E-08	3.89E-08	3.89E-08	3.89E-08	3.89E-08	7.75E-07	4.43E-09	3.61E-08	3.71E-09	2.10E-08	5.53E-08	1.58E-07
Upper large intestine wall	8.29E-07	1.69E-07	1.55E-07	5.02E-08	5.02E-08	5.02E-08	5.02E-08	1.06E-07	1.78E-08	5.31E-09	7.69E-10	1.60E-07	4.17E-07	1.72E-07
Heart wall	6.15E-09	3.57E-07	1.09E-07	5.74E-08	5.74E-08	5.74E-08	5.74E-08	1.67E-07	6.16E-10	7.35E-07	4.33E-08	2.22E-09	5.47E-08	1.61E-07
Kidneys	7.02E-08	4.97E-07	1.70E-07	6.22E-08	6.22E-08	6.22E-08	6.22E-08	6.63E-07	3.10E-09	1.73E-08	2.95E-09	2.00E-08	6.42E-08	1.54E-07
Liver	3.81E-08	3.86E-07	8.96E-08	4.82E-08	4.82E-08	4.82E-08	4.82E-08	7.22E-08	1.57E-09	5.93E-08	8.64E-09	1.17E-08	3.29E-08	1.55E-07
Lungs	5.39E-09	1.76E-07	1.09E-07	6.67E-08	6.67E-08	6.67E-08	6.67E-08	1.65E-07	3.67E-10	2.97E-07	8.82E-08	1.04E-09	4.10E-09	1.41E-07
Muscle	1.44E-07	1.24E-07	9.07E-08	7.45E-08	7.45E-08	7.45E-08	7.45E-08	1.03E-07	9.89E-08	1.06E-07	1.16E-07	1.30E-07	1.43E-07	1.31E-07
Ovaries	3.22E-04	3.65E-08	2.13E-07	6.54E-08	6.54E-08	6.54E-08	6.54E-08	3.85E-08	0.00E+00	1.94E-09	2.29E-10	5.41E-07	1.57E-06	1.81E-07
Pancreas	3.65E-08	3.73E-05	1.48E-07	6.54E-08	6.54E-08	6.54E-08	6.54E-08	1.28E-06	2.58E-09	6.13E-08	7.28E-09	1.38E-08	3.73E-08	1.79E-07
Red marrow	2.13E-07	1.40E-07	1.79E-06	2.01E-07	6.38E-07	2.01E-07	3.88E-07	8.43E-08	2.69E-08	8.25E-08	7.94E-08	8.02E-08	1.39E-07	1.31E-07
Bone surfaces	1.66E-07	1.67E-07	1.01E-06	2.73E-06	3.15E-06	7.26E-07	1.21E-06	1.26E-07	1.02E-07	1.23E-07	1.97E-07	1.05E-07	1.30E-07	2.59E-07
Skin	3.09E-08	3.01E-08	4.20E-08	4.68E-08	4.68E-08	4.68E-08	4.68E-08	3.49E-08	1.03E-07	4.39E-08	4.38E-08	3.90E-08	3.01E-08	9.14E-08
Spleen	3.85E-08	1.28E-06	9.19E-08	4.94E-08	4.94E-08	4.94E-08	4.94E-08	2.24E-05	2.17E-09	3.89E-08	7.83E-09	8.40E-09	2.57E-08	1.54E-07
Testes	0.00E+00	2.58E-09	3.09E-08	4.09E-08	4.09E-08	4.09E-08	4.09E-08	2.17E-09	8.64E-05	1.91E-10	2.31E-11	3.73E-07	0.00E+00	1.28E-07
Thymus	1.94E-09	6.13E-08	8.45E-08	4.83E-08	4.83E-08	4.83E-08	4.83E-08	3.89E-08	1.91E-10	1.50E-04	1.62E-07	8.06E-10	1.81E-09	1.43E-07
Thyroid	2.29E-10	7.28E-09	7.54E-08	7.67E-08	7.67E-08	7.67E-08	7.67E-08	7.83E-09	2.31E-11	1.62E-07	1.49E-04	9.56E-11	2.14E-10	1.45E-07
Urinary bladder wall	5.49E-07	1.38E-08	9.18E-08	4.01E-08	4.01E-08	4.01E-08	4.01E-08	8.04E-09	3.85E-07	8.14E-10	9.65E-11	1.10E-05	1.28E-06	1.67E-07
Uterus	1.57E-06	3.73E-08	1.54E-07	4.89E-08	4.89E-08	4.89E-08	4.89E-08	2.57E-08	0.00E+00	1.81E-09	2.14E-10	1.24E-06	4.70E-05	1.83E-07
Total body	1.87E-07	1.85E-07	1.52E-07	1.40E-07	1.40E-07	1.40E-07	1.40E-07	1.59E-07	1.32E-07	1.48E-07	1.49E-07	1.18E-07	1.88E-07	1.39E-07

*Data from References 6 and 7.

TABLE 22-4
S VALUES (mGy/MBq · sec) FOR I-131 IN THE ADULT MALE PHANTOM*

Target Organs	Adrenals	Brain	Breasts	Gallbladder Contents	Lower Large Intestine	Small Intestine	Stomach	Upper Large Intestine	Heart Contents	Heart Wall	Kidneys	Liver	Lungs	Muscle
Adrenals	1.97E-03	3.40E-09	1.65E-07	8.41E-07	7.95E-08	2.20E-07	7.93E-07	2.54E-07	6.95E-07	7.91E-07	2.05E-06	1.21E-06	6.71E-07	3.31E-07
Brain	3.40E-09	2.91E-05	1.43E-08	1.88E-09	2.62E-10	5.20E-10	3.50E-09	6.17E-10	1.68E-08	1.56E-08	1.47E-09	5.53E-09	3.34E-08	7.10E-08
Breasts	1.65E-07	1.43E-08	1.01E-04	1.09E-07	1.40E-08	3.33E-08	1.83E-07	3.85E-08	7.32E-07	7.97E-07	7.53E-08	2.18E-07	6.73E-07	1.36E-07
Gallbladder wall	9.04E-07	1.51E-09	1.15E-07	3.12E-04	1.79E-07	1.18E-06	8.17E-07	1.99E-06	2.79E-07	3.49E-07	1.09E-06	2.35E-06	2.16E-07	3.40E-07
Lower large intestine wall	7.68E-08	2.21E-10	1.42E-08	1.56E-07	1.19E-04	1.62E-06	2.52E-07	6.49E-07	1.91E-08	2.40E-08	1.67E-07	5.21E-08	1.65E-08	3.79E-07
Small intestine	2.20E-07	5.20E-10	3.33E-08	1.21E-06	1.94E-06	4.01E-05	5.41E-07	3.45E-06	5.72E-08	7.44E-08	5.93E-07	3.28E-07	5.05E-08	3.18E-07
Stomach wall	7.22E-07	2.68E-09	1.93E-07	7.80E-07	3.66E-07	5.96E-07	7.04E-05	7.97E-07	4.61E-07	7.00E-07	6.98E-07	4.24E-07	3.26E-07	3.13E-07
Upper large intestine wall	2.58E-07	6.23E-10	3.31E-08	2.07E-06	8.60E-07	3.80E-06	7.01E-07	7.57E-05	7.68E-08	9.14E-08	5.82E-07	5.16E-07	6.40E-08	3.11E-07
Heart wall	7.91E-07	1.56E-08	7.97E-07	2.88E-07	2.60E-08	7.44E-08	6.71E-07	9.44E-08	4.20E-05	1.09E-04	2.29E-07	6.37E-07	1.19E-06	2.64E-07
Kidneys	2.05E-06	1.47E-09	7.53E-08	1.03E-06	2.22E-07	5.93E-07	7.22E-07	5.85E-07	2.01E-07	2.29E-07	1.18E-04	8.19E-07	1.97E-07	2.89E-07
Liver	1.21E-06	5.53E-09	2.18E-07	2.20E-06	6.41E-08	3.28E-07	4.11E-07	5.29E-07	5.81E-07	6.37E-07	8.19E-07	2.15E-05	5.48E-07	2.21E-07
Lungs	6.71E-07	3.35E-08	6.74E-07	1.94E-07	2.19E-08	5.05E-08	3.07E-07	6.16E-08	1.25E-06	1.19E-06	1.97E-07	5.49E-07	3.40E-05	2.67E-07
Muscle	3.31E-07	7.10E-08	1.36E-07	3.27E-07	3.48E-07	3.18E-07	2.85E-07	3.03E-07	2.50E-07	2.64E-07	2.89E-07	2.21E-07	2.67E-07	1.42E-06
Ovaries	1.13E-07	2.65E-10	1.45E-08	2.99E-07	3.52E-06	2.47E-06	1.81E-07	2.08E-06	2.59E-08	3.16E-08	2.17E-07	1.20E-07	2.67E-08	4.07E-07
Pancreas	2.88E-06	4.18E-09	1.97E-07	1.83E-06	1.41E-07	3.98E-07	3.32E-06	4.46E-07	6.94E-07	9.41E-07	1.41E-06	1.03E-06	4.76E-07	3.53E-07
Red marrow	7.32E-07	2.77E-07	1.83E-07	3.07E-07	6.10E-07	5.14E-07	2.38E-07	4.27E-07	3.36E-07	3.36E-07	5.18E-07	2.63E-07	3.37E-07	2.75E-07
Bone surfaces	4.79E-07	5.34E-07	1.50E-07	1.90E-07	3.09E-07	2.46E-07	1.77E-07	2.14E-07	2.70E-07	2.70E-07	2.83E-07	2.17E-07	2.96E-07	3.28E-07
Skin	1.15E-07	1.40E-07	2.49E-07	1.02E-07	1.14E-07	9.94E-08	1.15E-07	1.02E-07	1.15E-07	1.20E-07	1.34E-07	1.18E-07	1.30E-07	1.89E-07
Spleen	1.25E-06	5.74E-09	1.51E-07	3.38E-07	1.77E-07	2.93E-07	2.09E-06	2.78E-07	3.26E-07	4.39E-07	1.87E-06	2.16E-07	4.54E-07	3.01E-07
Testes	1.02E-08	4.79E-11	0.00E+00	3.33E-08	4.43E-07	9.10E-08	1.86E-08	7.48E-08	4.30E-09	4.95E-09	1.91E-08	1.03E-08	3.16E-09	3.02E-07
Thymus	1.86E-07	3.52E-08	8.03E-07	5.65E-08	1.20E-08	2.20E-08	1.34E-07	2.73E-08	2.41E-06	1.98E-06	6.83E-08	1.77E-07	7.95E-07	3.11E-07
Thyroid	3.16E-08	4.21E-07	1.03E-07	1.50E-08	2.38E-09	3.04E-09	1.57E-08	5.97E-09	1.43E-07	1.36E-07	1.93E-08	3.51E-08	2.54E-07	3.38E-07
Urinary bladder wall	3.53E-08	1.34E-10	9.12E-09	1.60E-07	1.30E-06	5.76E-07	7.23E-08	4.55E-07	1.30E-08	9.31E-09	7.47E-08	5.03E-08	8.80E-09	3.97E-07
Uterus	8.02E-08	2.41E-10	1.70E-08	3.31E-07	1.39E-06	2.18E-06	1.51E-07	1.03E-06	2.53E-08	2.96E-08	2.00E-07	1.04E-07	2.03E-08	4.04E-07
Total body	8.01E-07	6.71E-07	6.33E-07	4.02E-07	6.01E-07	6.98E-07	4.26E-07	5.47E-07	4.33E-07	7.86E-07	7.71E-07	7.71E-07	7.22E-07	7.11E-07

TABLE 22-4 CONTINUED

Target Organs						Source Organs								
	Ovaries	Pancreas	Red Marrow	Cortical Bone Surfaces	Trabecular Bone Surfaces	Cortical Bone Volume	Trabecular Bone Volume	Spleen	Testes	Thymus	Thyroid	Urinary Bladder Contents	Uterus	Total Body
Adrenals	1.13E-07	2.88E-06	7.32E-07	3.31E-07	3.31E-07	3.31E-07	3.31E-07	1.25E-06	1.02E-08	1.86E-07	3.16E-08	4.16E-08	8.02E-08	8.00E-07
Brain	2.65E-10	4.18E-09	2.60E-07	3.74E-07	3.74E-07	3.74E-07	3.74E-07	5.74E-09	4.79E-11	3.52E-08	4.21E-07	1.33E-10	2.41E-10	6.71E-07
Breasts	1.45E-08	1.97E-07	1.78E-07	1.09E-07	1.09E-07	1.09E-07	1.09E-07	1.51E-07	0.00E+00	8.03E-07	1.03E-07	7.58E-09	1.70E-08	6.32E-07
Gallbladder wall	2.97E-07	2.07E-06	3.36E-07	1.34E-07	1.34E-07	1.34E-07	1.34E-07	3.46E-07	3.23E-08	9.17E-09	1.53E-08	1.20E-07	3.32E-07	8.28E-07
Lower large intestine wall	3.20E-06	1.30E-07	5.80E-07	2.23E-07	2.23E-07	2.23E-07	2.23E-07	1.34E-07	5.69E-07	9.83E-09	2.00E-09	1.54E-06	1.29E-06	8.16E-07
Small intestine	2.47E-06	3.98E-07	5.21E-07	1.67E-07	1.67E-07	1.67E-07	1.67E-07	2.93E-07	9.10E-08	2.20E-08	3.04E-09	5.77E-07	2.18E-06	8.28E-07
Stomach wall	1.86E-07	3.45E-06	2.46E-07	1.22E-07	1.22E-07	1.22E-07	1.22E-07	2.03E-06	2.27E-08	1.29E-07	2.17E-08	7.34E-08	1.73E-07	7.86E-07
Upper large intestine wall	2.23E-06	4.37E-07	4.42E-07	1.49E-07	1.49E-07	1.49E-07	1.49E-07	2.84E-07	6.84E-08	2.99E-08	5.97E-09	4.36E-07	1.09E-06	8.16E-07
Heart wall	3.16E-08	9.41E-07	3.23E-07	1.74E-07	1.74E-07	1.74E-07	1.74E-07	4.39E-07	4.95E-07	1.98E-06	1.36E-07	1.03E-08	2.96E-08	7.83E-07
Kidneys	2.17E-07	1.41E-06	5.16E-07	1.93E-07	1.93E-07	1.93E-07	1.93E-07	1.87E-06	1.91E-08	6.83E-08	1.93E-08	8.05E-08	2.00E-07	7.70E-07
Liver	1.20E-07	1.03E-06	2.68E-07	1.50E-07	1.50E-07	1.50E-07	1.50E-07	2.16E-07	1.03E-08	1.77E-07	3.51E-08	4.82E-08	1.04E-07	7.71E-07
Lungs	2.67E-08	4.76E-07	3.35E-07	2.06E-07	2.06E-07	2.06E-07	2.06E-07	4.55E-07	3.16E-09	7.96E-07	2.54E-07	6.19E-09	2.03E-08	7.22E-07
Muscle	4.07E-07	3.53E-07	2.75E-07	2.30E-07	2.30E-07	2.30E-07	2.30E-07	3.01E-07	3.02E-07	3.11E-07	3.38E-07	3.70E-07	4.04E-07	7.11E-07
Ovaries	3.63E-03	1.22E-07	5.99E-07	1.91E-07	1.91E-07	1.91E-07	1.91E-07	1.13E-07	0.00E+00	1.20E-08	2.36E-09	1.39E-06	3.99E-06	8.36E-07
Pancreas	1.22E-07	3.59E-04	4.44E-07	1.98E-07	1.98E-07	1.98E-07	1.98E-07	3.61E-06	1.52E-08	1.75E-07	3.22E-08	5.46E-08	1.24E-07	8.31E-07
Red marrow	5.99E-07	4.36E-07	1.55E-05	6.11E-07	6.91E-06	6.11E-07	5.08E-06	2.71E-07	9.62E-08	2.55E-07	2.38E-07	2.40E-07	4.07E-07	6.10E-07
Bone surfaces	2.79E-07	2.91E-07	8.20E-06	1.42E-05	1.83E-05	5.62E-06	1.05E-05	2.18E-07	1.86E-07	2.19E-07	3.42E-07	1.78E-07	2.13E-07	9.26E-07
Skin	1.02E-07	9.94E-08	1.46E-07	1.71E-07	1.71E-07	1.71E-07	1.71E-07	1.22E-07	3.29E-07	1.52E-07	1.45E-07	1.27E-07	9.94E-08	6.07E-07
Spleen	1.13E-07	3.61E-06	2.76E-07	1.50E-07	1.50E-07	1.50E-07	1.50E-07	1.95E-04	1.31E-08	1.05E-07	3.12E-08	4.30E-08	9.33E-08	7.71E-07
Testes	0.00E+00	1.52E-08	1.01E-07	1.32E-07	1.32E-07	1.32E-07	1.32E-07	1.31E-08	8.53E-04	2.02E-09	4.14E-10	1.03E-06	0.00E+00	7.10E-07
Thymus	1.20E-08	1.75E-07	2.57E-07	1.50E-07	1.50E-07	1.50E-07	1.50E-07	1.05E-07	2.02E-07	1.57E-03	4.42E-07	6.07E-09	1.14E-07	7.40E-07
Thyroid	2.36E-09	3.22E-08	2.33E-07	2.37E-07	2.37E-07	2.37E-07	2.37E-07	3.12E-08	4.14E-10	4.42E-07	1.57E-03	1.19E-09	2.22E-09	7.40E-07
Urinary bladder wall	1.45E-06	5.80E-08	2.56E-07	1.22E-07	1.22E-07	1.22E-07	1.22E-07	4.04E-08	1.03E-06	6.06E-09	1.19E-09	8.85E-05	3.45E-06	8.07E-07
Uterus	3.99E-06	1.24E-07	4.26E-07	1.48E-07	1.48E-07	1.48E-07	1.48E-07	9.33E-08	0.00E+00	1.14E-08	2.22E-09	3.28E-06	4.35E-04	8.42E-07
Total body	8.37E-07	8.31E-07	7.47E-07	7.14E-07	7.14E-07	7.14E-07	7.14E-07	7.71E-07	7.11E-07	7.41E-07	7.41E-07	3.65E-07	8.43E-07	7.16E-07

*Data from References 6 and 7.

TABLE 22-5
S VALUES (mGy/MBq • sec) FOR F-18 IN THE ADULT MALE PHANTOM*

Target Organs	Source Organs													
	Adrenals	Brain	Breasts	Gallbladder Contents	Lower Large Intestine	Small Intestine	Stomach	Upper Large Intestine	Heart Contents	Heart Wall	Kidneys	Liver	Lungs	Muscle
Adrenals	2.59E-03	1.13E-08	4.60E-07	2.06E-06	2.06E-07	5.70E-07	2.06E-06	6.48E-07	1.74E-06	2.05E-06	5.22E-06	3.01E-06	1.74E-06	8.54E-07
Brain	1.13E-08	4.62E-05	3.99E-08	6.70E-09	9.01E-10	1.78E-09	1.06E-08	2.11E-09	4.95E-08	4.78E-08	4.84E-09	1.76E-08	9.70E-08	1.90E-07
Breasts	4.60E-07	3.99E-08	1.46E-04	3.01E-07	4.30E-08	9.55E-08	4.92E-07	1.13E-07	1.90E-06	2.06E-06	2.07E-07	5.85E-07	1.74E-06	3.64E-07
Gallbladder wall	2.21E-06	5.00E-09	3.17E-07	4.38E-04	4.59E-07	3.00E-06	2.06E-06	4.89E-06	6.98E-07	9.32E-07	2.84E-06	5.85E-06	5.54E-07	8.70E-07
Lower large intestine wall	2.06E-07	7.54E-10	4.30E-08	3.97E-07	1.64E-04	4.11E-06	6.64E-07	1.74E-06	5.57E-08	7.00E-08	4.28E-07	1.40E-07	4.93E-08	9.65E-07
Small intestine	5.70E-07	1.78E-09	9.55E-08	3.00E-06	4.74E-06	5.49E-05	1.34E-06	8.69E-06	1.59E-07	2.06E-07	1.50E-06	8.38E-07	1.40E-07	8.07E-07
Stomach wall	1.74E-06	9.18E-09	5.23E-07	1.90E-06	9.34E-07	1.52E-06	1.03E-04	2.06E-06	1.16E-06	1.74E-06	1.74E-06	1.11E-06	8.23E-07	8.07E-07
Upper large intestine wall	6.81E-07	2.11E-09	9.54E-08	5.21E-06	2.21E-06	9.64E-06	1.74E-06	1.07E-04	2.06E-07	2.38E-07	1.47E-06	1.31E-06	1.75E-07	7.91E-07
Heart wall	2.05E-06	4.78E-08	2.06E-06	7.29E-07	7.64E-08	2.06E-07	1.74E-06	2.38E-07	6.32E-05	1.53E-04	5.86E-07	1.58E-06	3.00E-06	6.80E-07
Kidneys	5.22E-06	4.84E-09	2.07E-07	2.68E-06	5.85E-07	1.50E-06	1.74E-06	1.47E-06	5.39E-07	5.86E-07	1.68E-04	2.06E-06	5.08E-07	7.44E-07
Liver	3.01E-06	1.76E-08	5.85E-07	5.53E-06	1.75E-07	8.38E-07	1.04E-06	1.36E-06	1.45E-06	1.58E-06	2.06E-06	3.40E-05	1.38E-06	5.70E-07
Lungs	1.74E-06	9.70E-08	1.74E-06	4.91E-07	6.36E-08	1.40E-07	7.92E-07	1.74E-07	3.16E-06	3.00E-06	5.08E-07	1.38E-06	4.69E-05	6.80E-07
Muscle	8.54E-07	1.90E-07	3.64E-07	8.38E-07	8.86E-07	8.07E-07	7.28E-07	7.75E-07	6.33E-07	6.80E-07	7.44E-07	5.70E-07	6.80E-07	2.21E-06
Ovaries	2.69E-07	9.16E-10	4.47E-08	7.75E-07	9.02E-06	6.17E-06	4.90E-07	5.22E-06	7.81E-08	9.21E-08	5.70E-07	3.17E-07	7.62E-08	1.04E-06
Pancreas	7.11E-06	1.39E-08	5.23E-07	4.73E-06	3.65E-07	1.03E-06	8.22E-06	1.12E-06	1.74E-06	2.37E-06	3.64E-06	2.53E-06	1.23E-06	9.02E-07
Red marrow	1.90E-06	6.96E-07	4.91E-07	7.91E-07	1.58E-06	1.31E-06	6.32E-07	1.12E-06	8.85E-07	8.85E-07	1.34E-06	6.96E-07	8.85E-07	7.12E-07
Bone surfaces	1.06E-06	1.20E-06	3.49E-07	3.95E-07	6.65E-07	5.22E-07	3.79E-07	4.58E-07	5.85E-07	5.85E-07	6.17E-07	4.74E-07	6.64E-07	7.28E-07
Skin	3.17E-07	3.81E-07	6.65E-07	2.70E-07	3.01E-07	2.70E-07	3.17E-07	2.70E-07	3.17E-07	3.17E-07	3.65E-07	3.18E-07	3.48E-07	5.07E-07
Spleen	3.16E-06	1.91E-08	4.12E-07	8.25E-07	4.59E-07	7.59E-07	5.21E-06	6.96E-07	8.07E-07	1.11E-06	4.75E-06	5.54E-07	1.14E-06	7.75E-07
Testes	3.19E-08	1.66E-10	0.00E+00	9.69E-08	1.18E-06	2.54E-07	5.73E-08	2.06E-07	1.39E-08	1.59E-08	5.88E-08	3.20E-08	1.03E-08	7.91E-07
Thymus	4.90E-07	1.03E-07	2.21E-06	1.59E-07	3.66E-08	6.20E-08	3.65E-07	7.94E-08	6.02E-06	4.91E-06	1.90E-07	4.60E-07	2.06E-06	8.07E-07
Thyroid	8.92E-08	1.11E-06	2.86E-07	4.62E-08	7.87E-09	9.46E-09	4.78E-08	1.92E-08	3.64E-07	3.64E-07	5.87E-08	9.67E-08	6.65E-07	8.70E-07
Urinary bladder wall	1.03E-07	4.63E-10	2.87E-08	4.43E-07	3.16E-06	1.44E-06	2.06E-07	1.17E-06	3.98E-08	2.71E-08	2.06E-07	1.43E-07	2.71E-08	1.01E-06
Uterus	2.21E-07	8.35E-10	5.26E-08	8.54E-07	3.47E-06	5.38E-06	3.97E-07	2.53E-06	7.48E-08	8.74E-08	5.23E-07	2.70E-07	5.88E-08	1.03E-06
Total body	1.49E-06	1.17E-06	1.10E-06	9.67E-07	1.21E-06	1.34E-06	9.19E-07	1.14E-06	9.10E-07	1.46E-06	1.43E-06	1.43E-06	1.30E-06	1.28E-06

TABLE 22-5 CONTINUED

Target Organs	Source Organs													
	Ovaries	Pancreas	Red Marrow	Cortical Bone Surface	Trabecular Bone Surface	Cortical Bone Volume	Trabecular Bone Volume	Spleen	Testes	Thymus	Thyroid	Urinary Bladder Contents	Uterus	Total Body
Adrenals	2.69E-07	7.11E-06	1.90E-06	8.70E-07	8.70E-07	8.70E-07	8.70E-07	3.16E-06	3.19E-08	4.90E-07	8.92E-08	1.19E-07	2.21E-07	1.49E-06
Brain	9.16E-10	1.39E-08	6.96E-07	9.97E-07	9.97E-07	9.97E-07	9.97E-07	1.91E-08	1.66E-10	1.03E-07	1.11E-06	4.63E-10	8.35E-10	1.17E-06
Breasts	4.47E-08	5.23E-07	4.91E-07	3.01E-07	3.01E-07	3.01E-07	3.01E-07	4.12E-07	0.00E+00	2.21E-06	2.86E-07	2.23E-08	5.26E-08	1.10E-06
Gallbladder wall	7.59E-07	5.21E-06	8.68E-07	3.48E-07	3.48E-07	3.48E-07	3.48E-07	8.71E-07	9.37E-08	2.38E-07	4.62E-08	3.17E-07	8.40E-07	1.57E-06
Lower large intestine wall	8.22E-06	3.32E-07	1.49E-06	5.85E-07	5.85E-07	5.85E-07	5.85E-07	3.48E-07	1.46E-06	3.03E-08	6.59E-09	3.79E-06	3.16E-06	1.54E-06
Small intestine	6.17E-06	1.03E-06	1.31E-06	4.27E-07	4.27E-07	4.27E-07	4.27E-07	7.59E-07	2.54E-07	6.20E-08	9.46E-09	1.44E-06	5.38E-06	1.57E-06
Stomach wall	5.07E-07	8.69E-06	6.33E-07	3.16E-07	3.16E-07	3.16E-07	3.16E-07	5.06E-06	6.52E-08	3.65E-07	6.36E-08	1.90E-07	4.60E-07	1.47E-06
Upper large intestine wall	5.54E-06	1.09E-06	1.11E-06	3.80E-07	3.80E-07	3.80E-07	3.80E-07	6.97E-07	1.90E-07	9.06E-08	1.92E-08	1.09E-06	2.69E-06	1.54E-06
Heart wall	9.21E-08	2.37E-06	8.38E-07	4.58E-07	4.58E-07	4.58E-07	4.58E-07	1.11E-06	1.59E-08	4.91E-06	3.64E-07	3.03E-08	8.74E-08	1.46E-06
Kidneys	5.70E-07	3.64E-06	1.34E-06	5.06E-07	5.06E-07	5.06E-07	5.06E-07	4.75E-06	5.88E-08	1.90E-07	5.87E-08	2.22E-07	5.23E-07	1.43E-06
Liver	3.17E-07	2.53E-06	6.96E-07	3.95E-07	3.95E-07	3.95E-07	3.95E-07	5.54E-07	3.20E-08	4.60E-07	9.67E-08	1.35E-07	2.70E-07	1.43E-06
Lungs	7.62E-08	1.23E-06	8.85E-07	5.53E-07	5.53E-07	5.53E-07	5.53E-07	1.14E-06	1.03E-08	2.06E-06	6.65E-07	1.92E-08	5.88E-08	1.30E-06
Muscle	1.04E-06	9.02E-07	7.12E-07	6.01E-07	6.01E-07	6.01E-07	6.01E-07	7.75E-07	7.91E-07	8.07E-07	8.70E-07	9.49E-07	1.03E-06	1.28E-06
Ovaries	4.72E-03	3.17E-07	1.53E-06	4.90E-07	4.90E-07	4.90E-07	4.90E-07	2.69E-07	0.00E+00	3.67E-08	7.87E-09	3.47E-06	9.96E-06	1.58E-06
Pancreas	3.17E-07	4.97E-04	1.15E-06	5.22E-07	5.22E-07	5.22E-07	5.22E-07	9.16E-06	4.62E-08	4.42E-07	8.90E-08	1.49E-07	3.17E-07	1.57E-06
Red marrow	1.53E-06	1.15E-06	2.04E-05	1.58E-06	1.04E-05	1.58E-06	8.44E-06	7.28E-07	2.69E-07	6.65E-07	6.17E-07	6.17E-07	1.06E-06	1.21E-06
Bone surfaces	5.85E-07	6.32E-07	1.14E-05	1.59E-05	1.98E-05	8.73E-06	1.31E-05	4.75E-07	4.12E-07	4.75E-07	7.59E-07	3.64E-07	4.42E-07	1.59E-06
Skin	2.70E-07	2.70E-07	3.96E-07	4.75E-07	4.75E-07	4.75E-07	4.75E-07	3.33E-07	8.71E-07	4.12E-07	3.81E-07	3.33E-07	2.70E-07	1.03E-06
Spleen	2.69E-07	9.16E-06	7.28E-07	3.96E-07	3.96E-07	3.96E-07	3.96E-07	2.81E-04	3.99E-08	2.71E-07	8.72E-08	1.27E-07	2.54E-07	1.43E-06
Testes	0.00E+00	4.62E-08	2.69E-07	3.48E-07	3.48E-07	3.48E-07	3.48E-07	3.99E-08	1.17E-03	6.76E-09	1.45E-09	2.53E-06	0.00E+00	1.28E-06
Thymus	3.67E-08	4.42E-07	6.65E-07	3.96E-07	3.96E-07	3.96E-07	3.96E-07	2.71E-07	6.76E-09	2.11E-03	1.11E-06	1.92E-08	3.51E-08	1.35E-06
Thyroid	7.87E-09	8.90E-08	6.17E-07	6.33E-07	6.33E-07	6.33E-07	6.33E-07	8.72E-08	1.45E-09	1.11E-06	2.11E-03	4.04E-09	7.39E-09	1.35E-06
Urinary bladder wall	3.63E-06	1.58E-07	6.47E-07	3.16E-07	3.16E-07	3.16E-07	3.16E-07	1.19E-07	2.53E-06	1.92E-08	4.04E-09	1.31E-04	8.69E-06	1.52E-06
Uterus	9.96E-06	3.17E-07	1.06E-06	3.79E-07	3.79E-07	3.79E-07	3.79E-07	2.54E-07	0.00E+00	3.51E-08	7.39E-09	8.21E-06	6.07E-04	1.60E-06
Total body	1.58E-06	1.57E-06	1.36E-06	1.28E-06	1.28E-06	1.28E-06	1.28E-06	1.43E-06	1.28E-06	1.35E-06	1.35E-06	8.66E-07	1.60E-06	1.28E-06

*Data from References 6 and 7.

EXAMPLE 22-10

Estimate the effective dose E for an adult male following the injection of 250 MBq of ^{18}F-FDG. Assume any organs or tissues not shown in Table 22-6 have a radiation dose of 1.5×10^{-2} mGy/MBq (approximately the average of those that are listed).

Answer

Using the available dose values in Table 22-6 and the tissue weighting factors in Table 22-7, and assuming that $w_R = 1$, then the effective dose is given by:

$$E = 0.12 \times (1.3 \times 10^{-2} \text{ mGy/MBq})$$
$$\times \, 250 \text{ MBq (red marrow)}$$
$$+ \, 0.12 \times (3 \times 10^{-2} \text{ mGy/MBq})$$
$$\times \, 250 \text{ MBq (colon, as sum of upper}$$
$$\text{and lower large intestine walls)}$$
$$+ \, 0.12 \times (1.7 \times 10^{-2} \text{ mGy/MBq})$$
$$\times \, 250 \text{ MBq (lung)}$$
$$+ \, 0.12 \times (1.3 \times 10^{-2} \text{ mGy/MBq})$$
$$\times \, 250 \text{ MBq (stomach)}$$
$$+ \, 0.12 \times (9.2 \times 10^{-3} \text{ mGy/MBq})$$
$$\times \, 250 \text{ MBq (breast)}$$
$$+ \, 0.08 \times (1.3 \times 10^{-2} \text{ mGy/MBq})$$
$$\times \, 250 \text{ MBq (gonads)}$$
$$+ \, 0.04 \times (1.9 \times 10^{-1} \text{ mGy/MBq})$$
$$\times \, 250 \text{ MBq (bladder)}$$
$$+ \, 0.04 \times (1.5 \times 10^{-2} \text{ mGy/MBq})$$
$$\times \, 250 \text{ MBq (esophagus)}$$
$$+ \, 0.04 \times (1.6 \times 10^{-2} \text{ mGy/MBq})$$
$$\times \, 250 \text{ MBq (liver)}$$
$$+ \, 0.04 \times (1.0 \times 10^{-2} \text{ mGy/MBq})$$
$$\times \, 250 \text{ MBq (thyroid)}$$
$$+ \, 0.01 \times (1.2 \times 10^{-2} \text{ mGy/MBq})$$
$$\times \, 250 \text{ MBq (bone surfaces)}$$
$$+ \, 0.01 \times (1.9 \times 10^{-2} \text{ mGy/MBq})$$
$$\times \, 250 \text{ MBq (brain)}$$
$$+ \, 0.01 \times (1.5 \times 10^{-2} \text{ mGy/MBq})$$
$$\times \, 250 \text{ MBq (salivary glands)}$$
$$+ \, 0.01 \times (8.4 \times 10^{-3} \text{ mGy/MBq})$$
$$\times \, 250 \text{ MBq (skin)}$$
$$+ \, 0.00923 \times (2.67 \times 10^{-1} \text{ mGy/MBq})$$
$$\times \, 250 \text{ MBq (sum of 13 listed}$$
$$\text{remainder tissues)}$$
$$= 5.8 \text{ mSv (or 0.58 rem)}$$

Although effective dose is regarded as a better indicator of overall radiation risk than whole-body dose for radiation protection purposes, there still is debate about its relevance, and care should be taken in its use and interpretation. In particular, effective dose is not recommended for use in radionuclide therapy applications, nor should it be used to evaluate the risk from a radionuclide study to a specific individual. This is because calculations of E are based on an "average" human, whereas the actual dose can vary considerably with body shape and size, as well as the specific distribution of the radionuclide in the individual. This, along with some other general limitations of internal radiation dose estimates, is discussed in the next section.

8. Limitations of the MIRD Method

There are a number of important limitations in the MIRD approach for calculating radiation dose. Although Equation 22-18 is fundamentally correct, the values of ϕ are currently based on simplistic models of human anatomy that assume specific relationships in the shape, size, and location of various organs (see Fig. 22-5). More realistic models of the human body, based on medical imaging data and advanced computer modeling, are currently under development for dosimetry purposes. The MIRD formulation also implicitly assumes that activity is distributed uniformly within each organ and, furthermore, that energy is uniformly deposited throughout the organ. The assumption can cause a significant error in the calculated dose from nonpenetrating radiation (e.g., Auger electrons) when the activity is taken up in specific regions or cell types within an organ. Local radionuclide concentrations and, hence, the absorbed dose can be much higher than organ average calculations might suggest.

Calculation of cumulated activity, $\widetilde{A}$, also is problematic. Initially, with a new radiopharmaceutical, this must be determined from animal studies. There can be significant differences between the kinetics of a tracer in an animal model and in the human. Once a radiopharmaceutical is approved for human use, it is possible to obtain human data to estimate $\widetilde{A}$. However, values for healthy subjects may differ widely from those for patients and from one patient to the next because of pathophysiologic effects on uptake, clearance, and excretion of the radiopharmaceutical.

Despite these limitations, the MIRD method is a useful tool for comparing the average dose to various organs in patients for

TABLE 22-6
RADIATION DOSE ESTIMATES FOR ¹⁸F-FLUORODEOXYGLUCOSE IN AN ADULT SUBJECT*

Organ Dose	mGy/MBq Administered	Organ Dose	mGy/MBq Administered
Adrenals	1.3×10^{-2}	Muscle	1.1×10^{-2}
Brain	1.9×10^{-2}	Ovaries	1.7×10^{-2}
Breasts	9.2×10^{-3}	Pancreas	2.6×10^{-2}
Gallbladder wall	1.4×10^{-2}	Red marrow	1.3×10^{-2}
Lower large intestine wall	1.7×10^{-2}	Bone surfaces	1.2×10^{-2}
Small intestine	1.4×10^{-2}	Skin	8.4×10^{-3}
Stomach	1.3×10^{-2}	Spleen	3.7×10^{-2}
Upper large intestine wall	1.3×10^{-2}	Testes	1.3×10^{-2}
Heart wall	6.0×10^{-2}	Thymus	1.2×10^{-2}
Kidneys	2.0×10^{-2}	Thyroid	1.0×10^{-2}
Liver	1.6×10^{-2}	Urinary bladder wall	1.9×10^{-1}
Lungs	1.7×10^{-2}	Uterus	2.3×10^{-2}

*Data from reference 9.

TABLE 22-7
TISSUE WEIGHTING FACTORS USED FOR CALCULATION OF EFFECTIVE DOSE (*E*)*

Organ	w_T
Red marrow, colon, lungs, stomach, breast	0.12 each
Gonads	0.08
Bladder, liver, esophagus, thyroid	0.04 each
Brain, skin, salivary glands, bone surfaces	0.01 each
Remainder tissues (adrenals, extrathoracic region, gallbladder, heart, kidneys, lymphatic nodes, muscle, oral mucosa, pancreas, prostate [males], small intestine, spleen, thymus, uterus/cervix [females])	0.12 total (0.00923 each)

*From reference 1.

a wide variety of nuclear medicine procedures. It also is an essential tool in the approval process for new radiopharmaceuticals. In circumstances in which the assumptions on which the MIRD approach is based are unacceptable, more complex and involved methods can be used. For example, microdosimetric (cellular level) calculations should be done for radiopharmaceuticals that have very nonuniform uptake in radiosensitive organs. As well, radiation dose estimates for therapeutic

applications should incorporate data acquired on the specific patient, as described in Example 22-9.

REFERENCES

Society of Nuclear Medicine (MIRD) and ICRP publications, along with data available from references 6 and 9, provide the basic data for calculating absorbed doses:

1. International Commission on Radiological Protection: Recommendations of the International Commission on Radiological Protection. *ICRP Publication 103; Ann ICRP* 37(2-4), 2007.
2. Eckerman KF, Endo A: *MIRD: Radionuclide Data and Decay Schemes*, New York, 2008, Society of Nuclear Medicine.
3. Snyder W, Ford M, Warner G: *Estimates of Specific Absorbed Fractions for Photon Sources Uniformly Distributed in Various Organs of a Heterogeneous Phantom [MIRD Pamphlet No. 5 (revised)]*, New York, 1978, Society of Nuclear Medicine.
4. Cristy M, Eckerman K: *Specific Absorbed Fractions of Energy at Various Ages from Internal Photon Sources (ORNL Report ORNL/TM-8381 V1-V7)*, Oak Ridge, TN, 1987, Oak Ridge National Laboratory.
5. Stabin, M, Watson E, Cristy M, et al: *Mathematical Models of the Adult Female at Various Stages of Pregnancy (ORNL Report ORNL/TM-12907)*, Oak Ridge, TN, 1995, Oak Ridge National Laboratory.
6. Radiation Dose Assessment Resource (RADAR): http://www.doseinfo-radar.com [accessed December 17, 2011].
7. Stabin MG, Siegel JA: Physical models and dose factors for use in internal dose assessment. *Health Phys* 85: 294-310, 2003.
8. Stabin MG, Sparks RB, Crowe E: OLINDA/EXM: The second generation personal computer software for

internal dose assessment in nuclear medicine. *J Nucl Med* 46: 1023-1027, 2005.

9. Stabin MG, Stubbs JB, Toohey RE: *Radiation Dose Estimates for Radiopharmaceuticals (ORNL Report NUREG/CR-6345)*, Oak Ridge, TN, 1996, Oak Ridge Institute for Science and Education. Also available electronically at http://orise.orau.gov/files/reacts/dosetables.pdf [accessed December 17, 2011].

10. International Commission on Radiological Protection: 1990 Recommendations of the International Commission on Radiological Protection (ICRP Publication No. 60), New York, 1991, Pergamon Press.

BIBLIOGRAPHY

An excellent overview of the topics in this chapter is provided by the following detailed text:

Stabin MG: *Fundamentals of Nuclear Medicine Dosimetry*, New York, 2008, Springer.

A comprehensive collection of materials and data can be found at the Radiation Dose Assessment Resource website:

http://www.doseinfo-radar.com [accessed December 17, 2011].

A general guide for performing internal dosimetry calculations with the MIRD approach is the following:

Loevinger R, Budinger T, Watson E: *MIRD Primer for Absorbed Dose Calculations*, New York, 1991, Society of Nuclear Medicine.

The following book is useful for MIRD calculations at the cellular level:

Goddu SM, Howell RW, Bouchet LG, et al: *MIRD Cellular S Values*, New York, 1997, Society of Nuclear Medicine.

Recommended textbooks on basic radiation biology are the following:

Hall EJ: *Radiobiology for the Radiologist*, ed 7, New York, 2011, Lippincott, Williams & Wilkins.

Forshier S: *Essentials of Radiation Biology and Protection*, ed 2, Clifton Park, NY, 2008, Delmar Cengage Learning.

Radiation Safety and Health Physics

Chapter 22 dealt with the radiation dose received by patients undergoing nuclear medicine procedures. This chapter deals primarily with the exposure of personnel who work in nuclear medicine clinics and research laboratories and who are exposed to radiation in their normal working environment. Stored radioactive materials, handling of calibration sources, preparation of radioactive materials for patients and phantoms, and proximity to patients or phantoms to whom these preparations have been administered all are potential sources of radiation exposure. An additional problem is the potential for radiation exposure to nonlaboratory personnel, such as patient relatives, attending nursing staff, and even passers-by in the hallways adjacent to the laboratory.

The quantities of radioactive material used and radiation levels encountered in a nuclear medicine laboratory generally are well below what is necessary to cause any type of "radiation sickness." Of more concern are the long-term effects that may possibly result from chronic exposures to even low levels of radiation. The most important of these effects are genetic damage to cells (mutagenesis), damage to chromosomes (clastogenesis), and carcinogenesis.

Presently, our understanding of the effects of chronic exposure to low levels of radiation is far from complete. Radiation protection regulations and guidelines currently are based on a *linear nonthreshold (LNT) model,* which assumes that there is no "threshold dose" for these long-term effects and that the risk increases linearly with radiation dose.[1] There also are experiments, data, and proposed radiation injury models that are inconsistent with the LNT model.[2,3] Some scientists argue that studies involving low levels of radiation suggest that low doses actually

have a beneficial effect on health, resulting from stimulation of the immune system.[4] This effect is known as *radiation hormesis.* A vigorous debate about the biologic consequences of low levels of ionizing radiation, the relevance of absorbed dose estimates in assessing health risks, and the effect of these findings on regulations pertaining to radiation exposure is likely to continue for some years to come. Whatever the outcome of this debate, and even though the risks to personnel occupationally exposed to ionizing radiation in the nuclear medicine environment clearly are small (based on decades of historic data), common sense dictates that radiation exposures in and around the nuclear medicine laboratory be kept as low as is reasonably achievable.

When considering possible health effects to nuclear medicine patients or occupationally exposed personnel, it also is important to place the dose received in perspective by considering the radiation dose received by all of us from natural background sources. These sources include naturally occurring radionuclides in the body (e.g., ^{40}K), cosmic radiation, and radionuclides that occur naturally in the environment. Effective doses to individuals per year from these natural sources average approximately 2.4 mSv (typical range 1-13 mSv).[5] As shown in Example 22-10, a 250-MBq injection of ^{18}F-fluorodeoxyglucose leads to an effective dose of roughly 5.8 mSv (equal to the dose that would be received in approximately 1.7 years from nature). The average effective dose to the extremities of nuclear medicine technical personnel is on the order of 4 mSv per year.[6]

The analysis of problems in the handling of radiation sources and the development of safe handling practices are the general concerns of the broad field of *health physics.* The practices that are prescribed by this analysis are sometimes expressed formally as regulations and sometimes as "common sense" recommendations. In this chapter we primarily discuss aspects of health physics and radiation safety practices as they apply to the nuclear medicine laboratory. However, a further responsibility arises because nuclear medicine scientists and practitioners often are among the first people contacted (e.g., by the media) for information on public-health radiation issues. Therefore it is wise to know where reliable sources of information can be found. A number of international organizations such as the International Commission on Radiological Protection (ICRP), the United Nations Scientific Committee on the Effects of Atomic Radiation and the International Atomic Energy Agency provide useful reports and literature. Selected references and websites are provided at the end of this chapter.

A. QUANTITIES AND UNITS

1. Dose-Modifying Factors

For health physics purposes, specification of the radiation absorbed dose in grays (see Chapter 22, Section A) is inadequate for a complete and accurate assessment of potential radiation hazards. Although the relative risk of potential injury increases with increasing absorbed dose values, several other *dose-modifying factors* also must be taken into account.

1. *The part of the body exposed.* Total-body exposure carries a greater risk than partial-body exposure. Exposure of major organs in the trunk of the body is more serious than exposure to the extremities. The active blood-forming organs, the gonads, and the lens of the eye are especially sensitive to radiation damage. A superficial dose to the skin (e.g., from an external source of β particles) is less hazardous than the same dose delivered to greater depths (e.g., from an external source of γ rays or from internally deposited radioactivity).

2. *The time span over which the radiation dose is delivered.* A given number of grays delivered over a short period (e.g., minutes or hours) has a greater potential for damage than the same dose delivered over a long period (e.g., months or years).

3. *The age of the exposed individual.* Children are more susceptible to injurious

radiation effects than are adults. The developing embryo and fetus are especially sensitive.

4. *The type of radiation involved.* In general, densely ionizing radiation [i.e., high-linear energy transfer radiation (see Chapter 6, Section A.4)] such as α particles, fission fragments, and other nuclear particles, are more damaging per gray of absorbed dose than is less densely ionizing radiation, such as β particles and γ rays.

The dose-modifying factors in this list are taken into account in preparing regulations and making recommendations for handling of radioactive materials. For example, regulations specify different dose limits for different parts of the body, for different time periods, and for different age groups. To account for the differing hazards of different types of radiation, the *equivalent dose*, defined previously (see Chapter 22, Section A), is used. For most of the radiation encountered in nuclear medicine, the equivalent dose in sieverts (or rems) is numerically equal to the absorbed dose in grays (or rads), although it must be emphasized that equivalent dose and absorbed dose are not the same quantity and have different units. To account for differing hazards for different organs and tissue types, the equivalent dose is modified by organ-specific weighting factors to compute the *effective dose* (see Chapter 22, Section B.7) to an individual.

In some older texts, and in current United States federal regulations (see Section B), the related quantities *dose equivalent* (in place of equivalent dose) and *effective dose equivalent* (in place of effective dose) may be encountered. The conceptual difference is that equivalent dose is based on the average absorbed dose in a specific tissue or an organ, whereas dose equivalent is based on the absorbed dose at a point in tissue. There also are differences in the scaling factors used to convert the absorbed dose into these quantities. These quantities are summarized in Table 23-1. Broadly speaking, for nuclear medicine applications, equivalent dose and dose equivalent, as well as effective dose and effective dose equivalent, have similar numerical values.

2. Exposure and Air Kerma

For the purpose of describing radiation *levels* in a radiation environment, an additional quantity—*exposure*—has traditionally been used. Exposure refers to the amount of

TABLE 23-1
QUANTITIES USED IN HEALTH PHYSICS

Quantity	Symbol	Units	Definition	Comment
Equivalent dose	H_T	Sv	Average absorbed dose across a tissue or organ T with weighting factors that depend on the type and energy of radiation. See Chapter 22, Section A.	Replaces dose equivalent. See ICRP Publication 60 and updated radiation weighting factors in ICRP Publication 103.
Effective dose	E	Sv	Measure of absorbed dose to whole body based on multiplying equivalent dose by organ-specific weighting factors. See Chapter 22, Section B.7.	See ICRP Publication 60 (1991) and updated tissue weighting factors in ICRP Publication 103 (2007).
Dose equivalent	H	Sv	Absorbed dose at a point in an organ, with quality factors that depend on the type of radiation. See ICRP Publication 51.	Replaced by equivalent dose in ICRP Publication 60 but still used in U.S. Federal regulations in 2012.
Effective dose equivalent	H_E	Sv	Introduced in ICRP Publication 26 (1977) as a measure of effective radiation dose to the whole body. Is based on dose equivalent values multiplied by tissue weighting factors.	Replaced by effective dose in ICRP Publication 60 (1991) but still used in U.S. Federal regulations in 2012.
Exposure	X	C/kg	Amount of charge liberated per kg of air by a γ-ray or x-ray source.	Traditional units were the Roentgen (R) in which $1R = 2.58 \times 10^{-4}$ C/kg. Exposure replaced by air kerma.
Air kerma	K	Gy	Amount of kinetic energy released per kg of air by uncharged ionizing radiation (photons and neutrons).	For radionuclides used in nuclear medicine, the conversion between air kerma and exposure is $K(Gy) \approx X(C/kg) \times 33.7$.

References:
ICRP Publication 26: *Ann ICRP* 1: 3, 1977.
ICRP Publication 51: *Ann ICRP* 17: 2-3, 1987.
ICRP Publication 60: *Ann ICRP* 21: 1-3, 1991.
ICRP Publication 103: *Ann ICRP* 37: 2-4, 2007.
ICRP, International Commission on Radiological Protection.

ionization of air caused by a γ-ray or x-ray source. The traditional unit of exposure is the *roentgen* (R), with subunits of milliroentgens (1 mR = 10^{-3} R), microroentgens (1 µR = 10^{-6} R), and so on. An exposure of 1 R implies ionization liberating an amount of charge equal to 2.58×10^{-4} coulombs/kg of air, or approximately 2×10^9 ionizations per cc of dry air at standard temperature and pressure. An *exposure rate* of 1 R/min implies that this amount of ionization is produced during 1 minute. The SI unit for exposure is the coulomb/kg, with no special name. Thus 1 coulomb/kg $\approx$ 3876 R and 1 R = 2.58×10^{-4} coulombs/kg.

The use of the SI units for exposure is cumbersome, and therefore in the transition to SI units, exposure is being replaced by a related quantity known as *air kerma*. *Kerma* stands for *k*inetic *e*nergy *r*eleased in *m*edi*a*. *Exposure* refers to the ionization charge produced in air, whereas *air kerma* refers to the amount of kinetic energy released in air (Table 23-1). More precisely, the air kerma is the sum of the kinetic energy of all charged particles produced by interactions from a source of x rays or γ rays (through Compton scatter, photoelectric absorption, or pair production) per kg of air. The units of air kerma are grays (J/kg), the same as for absorbed dose. If all of the photon energy transferred to charged particles is deposited locally (in air, bremsstrahlung production is negligible, so this is a reasonable assumption), then the absorbed dose in air has the same value as the air kerma. Using the fact that 33.7 eV of energy is required to produce an ion pair in air (see Table 7-1), and assuming bremsstrahlung losses can be ignored, the relationship between exposure, X, and air kerma, K, can be calculated as:

$$K(\text{Gy}) \approx X(\text{C/kg}) \times 33.7 \qquad (23\text{-}1)$$

The conversion between traditional units of exposure and air kerma is given by:

$$K(\text{Gy}) \approx X(\text{R}) \times 0.00869 \qquad (23\text{-}2)$$

Exposure and air kerma are useful quantities because they can be measured using ionization chambers, which are basically ionization-measurement devices (Chapter 7, Section A.2). Specific instruments used for health physics measurements are described in Section E.

If the air kerma in Gy is known at a certain location, the absorbed dose in Gy that would be delivered to a person at that location can be estimated by means of a scaling factor, f. This factor is defined as the ratio of the absorbed dose in the medium of interest, D_{med}, to the absorbed dose in air, D_{air}:

$$f = D_{\text{med}}/D_{\text{air}} \approx D_{\text{med}}/K \qquad (23\text{-}3)$$

The factor f depends on the mass attenuation coefficients (Chapter 6, Section D.1) of the medium of interest and of air and is energy dependent. Figure 23-1 shows the value of f as a function of energy for bone and for soft tissues. For soft tissues, $f \approx 1.1$. The value is close to unity because the mass attenuation properties of soft tissues and air are similar. For low-energy photons ($E \leq 100$ keV), the value of f for bone is greater than unity. Because of photoelectric absorption by the heavier elements in bone (Ca and P), energy absorption in bone is greater than energy absorption by air at these energies; however, for most of the γ-ray energies commonly employed in nuclear medicine, the value of f for bone also is close to 1.

Thus for practical purposes air kerma (in grays) is approximately equal numerically to the absorbed dose in grays that would be received by an individual at that location, and in turn, as described in Section A.1, the absorbed dose in grays is numerically equal to the equivalent dose in sieverts. Because of their approximate numerical equivalence, grays and sieverts, or in traditional units, roentgens, rads and rems, are sometimes (mis)used as approximately interchangeable quantities; however, one should be aware that

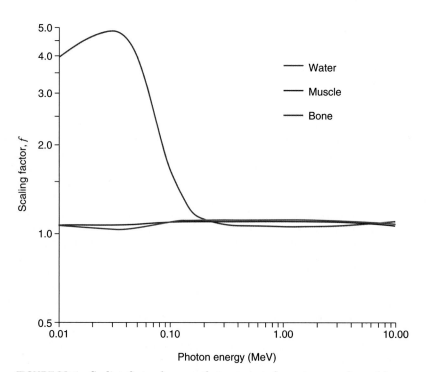

FIGURE 23-1 Scaling factor f versus photon energy for water, muscle, and bone.

they represent distinctly different physical quantities.

B. REGULATIONS PERTAINING TO THE USE OF RADIONUCLIDES

Regulations for the transport, handling, and exposure to ionizing radiation vary from country to country. The discussion in this section limits itself to the regulations in place in the United States in 2012 and uses selected regulations to highlight important regulatory concepts. A complete discussion of the many regulations involved is beyond the scope of this chapter. Also, the regulations are under constant review and subject to periodic changes. Therefore the regulations presented in this section should not be used to determine compliance without checking that they are still current. Further information may be obtained in the references at the end of the chapter or from institutional health physicists.

1. Nuclear Regulatory Commission Licensing and Regulations

The use and distribution of radioactive materials in the United States are under the primary control of the Nuclear Regulatory Commission (NRC). The NRC issues licenses to individuals and to institutions to possess and use radioactive materials. In addition to medical uses, industrial, research, educational, and other uses of radioactive materials also require NRC licensing. In some states, the NRC has entered into an agreement to transfer its regulatory and licensing functions to a radiation control agency within the state. Such states are called *agreement states.*

Medical licenses generally fall into one of two categories: specific licenses of limited scope or specific licenses of broad scope. Limited-scope licenses are for limited kinds and quantities of radionuclides, which are listed specifically in the license. They may be issued to individual physicians (e.g., in private offices) or to institutions (e.g., hospitals). Licenses issued to institutions also list the of individuals authorized to practice under the license.

Broad-scope licenses are issued to larger institutions that require greater licensing flexibility (e.g., basic research as well as medical uses in a university setting). Broad-scope licenses generally cover more radionuclides and greater quantities than do limited-scope licenses. The NRC permits the institutional radiation safety committee to authorize individuals to use radionuclides under the license rather than requiring them to be listed specifically on the license.

The NRC also issues regulations that must be observed by licensees in the use of radioactive materials. These regulations are published in Title 10 of the Code of Federal Regulations (CFR). Two of the more relevant sections of these regulations for nuclear medicine are Part 20 (10CFR20), covering radiation protection, and Part 35 (10CFR35), covering medical uses. The NRC regulations are based primarily on the recommendations of two advisory bodies, the ICRP and the National Council on Radiation Protection and Measurement (NCRP), as discussed in Section B.7. The NRC also periodically issues regulatory guides to assist licensees in the interpretation and implementation of its regulations.

In addition to the NRC, several other government agencies are involved in the regulation of radioactive materials, such as the U.S. Department of Transportation (shipping regulations) and the U.S. Food and Drug Administration (pharmaceutical aspects).

2. Restricted and Unrestricted Areas

The NRC regulations prescribe different maximum radiation limits for restricted and unrestricted areas. A restricted area is one "… access to which is controlled by the licensee for the purposes of protection of individuals from exposure to radiation and radioactive materials." Normally, restricted areas are not accessible to the general public, and generally they are occupied only by individuals whose employment responsibilities require them to work with radioactive materials and other radiation sources. Such individuals (e.g., nuclear medicine physicians, technicians, and radiochemists) are said to be *occupationally exposed.* Administrative staff, janitorial personnel, and facilities maintenance personnel generally are not included in this category. Restricted areas must be clearly marked with radiation warning signs.

3. Dose Limits

The *dose limits* specified in 10CFR20 are based on the general recommendations by the ICRP and NCRP (Section B.7) that an individual's total effective dose (see Chapter 22, Section B.7) should not exceed 50 mSv (5 rem) per year. Furthermore, 10CFR20 requires that the deep-dose equivalent (dose equivalent at a depth of 1 cm in tissue) to any individual organ or tissue (excluding the lens of the eye) should not exceed 500 mSv (50 rem) per year. The limit for shallow-dose equivalent (dose equivalent at a depth of 0.007 cm in tissue) to the skin and extremities also is 500 mSv (50 rem) per year. The most restrictive limit is to the lens of the eye, which has an annual limit of 150 mSv (15 rem). The annual occupational dose limits for minors (<18 years of age) are 10% of the annual dose limits specified for adult workers. The dose equivalent to an embryo or fetus should not exceed 5 mSv (0.5 rem).

These dose limits, which apply to occupationally exposed personnel, are called *occupational dose limits*. Occupational dose limits do not include radiation doses received by the occupationally exposed individual while that individual is undergoing a medical examination, nor do they include any radiation dose from natural radiation sources, such as cosmic rays and naturally occurring radioactivity in the environment.

Note that the regulations require the licensee to control radiation doses not only from licensed materials but from "other sources in the licensee's possession" as well (e.g., nonlicensed radioactive materials or an x-ray generator). Thus a licensee would be in violation of the regulations if the occupation limits were exceeded even if most of the radiation dose were caused by nonlicensed sources.

For individual members of the public, the annual effective dose equivalent limits are 1 mSv (0.1 rem). Radiation levels in *unrestricted areas* should deliver a radiation dose of less than 0.5 µSv/hr (0.05 mrem/hr), assuming continuous occupation of the area. Transient radiation levels of up to 20 µSv/hr (2 mrem/hr) are permitted.

4. Concentrations for Airborne Radioactivity in Restricted Areas

A particular problem in nuclear medicine laboratories is the potential for leakage or escape of radioactive gases (e.g., ^{133}Xe used in pulmonary function studies) or volatile radioactive material (e.g., concentrated ^{131}I solutions). The NRC regulations specify the concentrations for airborne radioactive materials that would result in the annual dose limits described in Section 3. These calculations assume that the workers are chronically exposed to these concentrations during a 2000-hour working year and that 2×10^4 mL of air is breathed per minute. Concentration

limits are shown in Table 23-2 for radionuclides that are used in nuclear medicine.

5. Environmental Concentrations and Concentrations for Sewage Disposal

The NRC regulations also specify the environmental concentrations of radioactivity in air and water and the concentration of radionuclides disposed of into sewage water, which would lead to the annual dose limits described in Section B.3 for the general public. Radioactive concentrations in sewage are of concern because they may eventually reach public water supplies. These limits assume continuous inhalation or ingestion by the general public over a period of 1 year, and they further assume that the average person breathes 2×10^4 mL of air per minute and has an annual water intake of 7.3×10^5 mL. Sewer water is assumed to be diluted by a factor of 10 before it is ingested. The methods for calculating these concentrations are described in 10CFR20. Table 23-3 shows these concentrations for several radionuclides of interest.

6. Record-Keeping Requirements

The NRC regulations require that rather extensive records be kept by the licensee. These include, among others, personnel dosimetry and radiation survey records (Section E), wipe testing records for sealed sources, summaries of quality control checks on radiation monitoring equipment, inventory

TABLE 23-2

CONCENTRATION OF AIRBORNE RADIOACTIVITY THAT WOULD RESULT IN THE ANNUAL DOSE LIMITS DESCRIBED IN SECTION B.3 FOR OCCUPATIONALLY EXPOSED PERSONNEL

Radionuclide	Air Concentration	
	µCi/mL	kBq/mL
^{3}H	2×10^{-5}	0.74
^{11}C	2×10^{-4}	7.4
^{14}C	1×10^{-6}	3.7×10^{-2}
^{18}F	3×10^{-5}	1.1
^{99m}Tc	6×10^{-5}	2.2
^{125}I	3×10^{-8}	1.1×10^{-3}
^{131}I	2×10^{-8}	7.4×10^{-4}
^{133}Xe	1×10^{-4}	3.7

Data from 10CFR20, Appendix B, Table 1.

TABLE 23-3

ENVIRONMENTAL CONCENTRATIONS (AIRBORNE AND WATER) AND SEWAGE CONCENTRATIONS THAT WOULD RESULT IN THE ANNUAL DOSE LIMITS DESCRIBED IN SECTION B.3 FOR THE GENERAL PUBLIC IF CONTINUOUSLY INHALED OR INGESTED

| | Environmental Concentrations | | | | Sewage Concentration | |
| | Air | | Water | | | |
Radionuclide	μCi/mL	kBq/mL	μCi/mL	kBq/mL	μCi/mL	kBq/mL
^{3}H	1×10^{-7}	3.7×10^{-3}	1×10^{-3}	3.70×10^1	1×10^{-2}	3.70×10^2
^{11}C	6×10^{-7}	2.2×10^{-2}	6×10^{-3}	2.22×10^2	6×10^{-2}	2.22×10^3
^{14}C	3×10^{-9}	1.1×10^{-4}	3×10^{-5}	1.11×10^0	3×10^{-4}	1.11×10^1
^{18}F	1×10^{-7}	3.7×10^{-3}	7×10^{-4}	2.59×10^1	7×10^{-3}	2.59×10^2
^{99m}Tc	2×10^{-7}	7.4×10^{-3}	1×10^{-3}	3.70×10^1	1×10^{-2}	3.70×10^2
^{125}I	3×10^{-10}	1.1×10^{-5}	2×10^{-6}	7.40×10^{-2}	2×10^{-5}	7.40×10^{-1}
^{131}I	2×10^{-10}	7.4×10^{-6}	1×10^{-6}	3.70×10^{-2}	1×10^{-5}	3.70×10^{-1}
^{133}Xe	5×10^{-7}	1.9×10^{-2}	—	—	—	—

Data from 10CFR20, Appendix B, Tables 2 and 3.

and disposal records, minutes of radiation safety committee meetings, and records of training in radiation safety of laboratory personnel. Maintenance of proper records is one of the major activities of an NRC licensee.

7. Recommendations of Advisory Bodies

The NRC regulatory limits are based on recommended radiation dose limits published by various advisory bodies. These bodies include the NCRP, a U.S. organization, and two international groups, the ICRP and the International Commission on Radiological Units (ICRU). The last group is concerned mostly with definitions of radiologic units. Recommendations from these groups do not carry the force of law; however, there is a tendency of regulatory agencies such as the NRC to convert them into law. Therefore, it is worthwhile to keep abreast of their recommendations.

Some titles of NCRP and ICRP reports that are applicable to nuclear medicine are listed in the references at the end of the chapter. Table 23-4 lists the dose limits currently recommended by the NCRP and the ICRP. Note that their coverage is somewhat broader than those appearing in the NRC regulations. Note also the restrictive limits placed on pregnant women with respect to the fetus. A pregnant, occupationally exposed woman may require special work restrictions to ensure that this dose limit for the fetus is not exceeded during her pregnancy.

C. SAFE HANDLING OF RADIOACTIVE MATERIALS

1. The ALARA Concept

Radiation dose limits and other restrictions specified in NRC regulations are legal limits that must not be exceeded at any time by an NRC licensee; however, they should not be considered as thresholds below which exposure to radiation is of no concern. Presently, although the radiation hazards associated with the limits specified in the regulations are very small, they are not assumed to be totally risk free, and any reasonable technique for reducing radiation dose may have potential benefits in the long run.

Recognizing this, the NCRP as early as 1954, and more recently the NRC regulations, have recommended as an operating philosophy that the objective of radiation safety practices should be not simply to keep radiation doses within legal limits but to keep them "*as low as reasonably achievable*" (ALARA). In its regulations, the NRC has defined ALARA to mean "as low as reasonably achievable taking into account the state of technology and economics of improvement in relation to benefits to the public health and safety, and other societal and socioeconomic considerations, and in relation to the use of atomic energy in the public interest." NRC Regulatory Guides 8.10 ("Operating Philosophy for Maintaining Occupational Exposures as Low

TABLE 23-4

DOSE LIMITS RECOMMENDED BY THE INTERNATIONAL COMMISSION ON RADIOLOGICAL PROTECTION AND THE NATIONAL COUNCIL ON RADIATION PROTECTION AND MEASUREMENT

	NCRP* (1993)	ICRP† (2007)
Occupational Exposure		
Effective dose: Annual	50 mSv	20 mSv/year averaged over 5 years, no more than 50 mSv in any one year
Effective dose: Cumulative	10 mSv × age (years)	100 mSv in any 5-year period
Equivalent dose: Annual	150 mSv to lens of eye 500 mSv to skin, hands, and feet	150 mSv to lens of eye 500 mSv to skin, hands, and feet
General Public Exposure		
Effective dose: Annual	1 mSv if continuous 5 mSv if infrequent	1 mSv; higher if needed as long as average over 5 years does not exceed 1 mSv
Equivalent dose: Annual	50 mSv to skin, hands, and feet	15 mSv to lens of eye 50 mSv to skin
Embryo-Fetus		
Equivalent dose	0.5 mSv per month once pregnancy declared	Same as general public exposure, even for occupational exposure

ICRP, International Commission on Radiological Protection; NCRP, National Council on Radiation Protection and Measurement.

*Limitation of Exposure to Ionizing Radiation (NCRP Report No. 116), 1993.

†2007 Recommendations of the International Commission on Radiological Protection (ICRP Publication No. 103). Annals of the ICRP, 2-4, 2007.

as Reasonably Achievable") and 8.18 ("Information Relevant to Ensuring that Occupational Radiation Exposures at Medical Institutions will be as Low as Reasonably Achievable") provide practical advice on implementation of ALARA principles.

The concept of ALARA has long been the operational objective of radiation safety practices in well-run nuclear medicine laboratories, and they have now taken on regulatory force. ALARA principles can be applied to the handling of radiation sources, to storage and shielding techniques, and to the design and layout of the laboratory. Some of the basic techniques for keeping radiation doses "ALARA" are discussed later.

2. Reduction of Radiation Doses from External Sources

Types of Sources. External sources are those that deliver a radiation dose from outside the body. The principal sources are γ-ray- and x-ray-emitting radionuclides in patients, syringes, vials, waste disposal areas, and so forth. Unshielded emitters emitting particles of sufficient energy to travel some distance in air (e.g., ^{32}P, but not ^{14}C) also constitute an external hazard, although β particles generally deliver only a superficial radiation dose to the skin.

Air Kerma Rate Constant. The air kerma caused by γ-ray and x-ray emitters can be estimated from the air kerma rate constant, Γ. This constant has a specific value for each radionuclide and is defined as the air kerma caused by γ-ray and x-ray emissions, in mGy per hour, at a distance of 1 m from an unshielded 1-GBq point source of that radionuclide. The units for Γ are mGy · m²/Gbq · hr. Calculation of the air kerma rate constant is based on the number of γ-ray and x-ray emissions from the radionuclide (number per disintegration) and their energies, and on the absorption coefficient of air at these energies. Values of Γ for some radionuclides used in nuclear medicine are summarized in Table 23-5. For practical health physics purposes, the calculation of Γ should only include γ and x rays above a certain minimum energy value because photons of a lower energy have such low penetrating power (e.g., through the walls of a syringe or vial) that they pose a negligible external hazard. The minimum energy used to compute the values in Table 23-5 is 20 keV.

To estimate the air kerma rate KR(mGy/hr) at a distance d(m) from an activity A(GBq) of a radionuclide having an air kerma rate constant Γ(mGy · m²/GBq · hr), the following equation is used:

$$KR = A\Gamma/d^2 \qquad (23\text{-}4)$$

TABLE 23-5

γ RAY AIR KERMA RATE CONSTANTS FOR SEVERAL RADIONUCLIDES OF INTEREST IN NUCLEAR MEDICINE

Radionuclide	Γ (mGy · m²/GBq · hr)	Radionuclide	Γ (mGy · m²/GBq · hr)
^{11}C	0.1393	^{99m}Tc	0.0141
^{13}N	0.1394	^{111}In	0.0831
^{15}O	0.1395	^{123}I	0.0361
^{18}F	0.1351	^{125}I	0.0377
^{57}Co	0.0141	^{131}I	0.0522
^{60}Co	0.3090	^{133}Xe	0.0143
^{67}Ga	0.0195	^{137}Cs/^{137}Ba	0.0821
^{68}Ga	0.1290	^{201}Tl	0.0102
^{99}Mo/^{99m}Tc (at equilibrium)	0.0336		

Data from Ninkovic MM, Raicevic JJ, Adrovic F: Air kerma rate constants for gamma emitters used most often in practice. *Radiat Prot Dos* 115:247-250, 2005.

The appearance of d^2 in the denominator of Equation 23-4 is an expression of the *inverse-square law* (see Chapter 11, Section A.2). Because γ rays and x rays are emitted isotropically (e.g., with no preferred direction) radiation intensity and dose levels decrease as the square of the distance from the source.

EXAMPLE 23-1

Calculate the air kerma rates at 10-cm and 300-cm distances from a syringe containing 1 GBq of ^{99m}Tc.

Answer
The air kerma rate constant Γ is 0.0141 mGy · m²/GBq · hr (see Table 23-5). Therefore from Equation 23-4, at 10 cm

$$KR = 1\,\text{GBq} \times 0.0141\,\text{mGy} \cdot \text{m}^2/\text{GBq} \cdot \text{hr} \div (0.1^2)\,\text{m}^2$$
$$= 1.41\,\text{mGy/hr}$$

and at 300 cm,

$$KR = 1\,\text{GBq} \times 0.0141\,\text{mGy} \cdot \text{m}^2/\text{GBq} \cdot \text{hr} \div (3^2)\,\text{m}^2$$
$$= 1.57 \times 10^{-3}\,\text{mGy/hr}$$
$$= 1.57\,\mu\text{Gy/hr}$$

The strong effect of distance on radiation dose equivalent rate is illustrated by Example 23-1.

In some texts, exposure rates, XR (roentgens/hr), may be encountered instead of air kerma rates. The relationship between the two is given by

$$XR\,(\text{R/hr}) \approx KR\,(\text{mGy/hr}) \times 0.115 \quad (23\text{-}5)$$

The equivalent dose rate to a particular tissue or organ and the effective dose rate to an individual can be estimated from the air kerma rates using the appropriate radiation weighting factors and tissue weighting factors (see Chapter 22, Sections A and B.7), but strongly depend on patient geometry and the direction from which the radiation is incident on the individual. Some publications estimate dose equivalent rates in mSv/hr using simplified tissue models. These values are numerically higher than the air kerma rates as they include the conversion from air kerma to tissue absorbed dose (a factor of ~1.11 in soft tissue), and also are increased by the contributions of Compton-scattered photons within the body.

Equation 23-4 is accurate for distances that are large in comparison to the physical size of the source; however, it is not valid at very small distances. For example, it predicts that the air kerma rate, and therefore the equivalent dose, becomes infinite as d approaches zero. A practical situation in which this problem arises is in the estimation of equivalent dose rates on contact with the source, for example, equivalent dose rates to the hand while handling syringes and vials. These equivalent dose rates have been determined experimentally for ^{99m}Tc. They range from about 0.14 mSv/MBq • hr on the surface of larger syringes (10 to 20 mL) up to approximately 0.7 mSv/MBq • hr for smaller syringes (1 to 2 mL).[7] Equivalent dose rates to the hands on contact with syringes containing ~1000 MBq of ^{99m}Tc can be in the range of several mSv/min, an obvious matter of concern in operations requiring handling of these sources. The use of syringe shielding therefore is indicated and can significantly reduce the radiation dose.

Time, Distance, Shielding (TDS) Rules. The basic principles for reducing radiation doses from external sources are described by the "TDS" rules, for time, distance, and shielding:

1. Decrease the *time* of exposure.
2. Increase the *distance* from the source.
3. Use *shielding* when practical and effective.

Time of exposure is decreased by working with or in the vicinity of radiation sources as rapidly as possible, consistent with good technique. Personnel should spend as little time as possible in "hot labs" and other high-level radiation areas. In particular, these areas should not be used for visiting, discussing problems unrelated to activities in the area, and so on. Laboratory monitors should be used in these areas to warn personnel when high-level radiation sources are present.

As shown by Example 23-1, *distance* can have a marked effect on radiation levels. Increasing distance always has a dose-reduction effect. Direct contact with radiation sources should be avoided by any available means, such as by using tongs to handle vials. Patient study areas (e.g., imaging rooms) should be arranged to permit the technician to operate instrumentation at reasonable distances (e.g., $\geq$2 m) from the patient. Separate waiting areas should be provided for patients who have been injected with radioactivity and for relatives, orderlies, and patients not requiring radioactive injections. Reception areas should not be used as waiting areas for radioactive patients. Storage areas for generators, radioactive trash, and other high-level sources should be remote from regularly occupied areas of the laboratory. Special attention should be given to their location in relation to unrestricted areas. (They also should be remote from imaging rooms and counting rooms to minimize instrument background levels.)

Examples of effective use of *shielding* are lead pigs for storage of vials and generators, lead-lined syringe holders, lead aprons, lead bricks for lining storage areas, and lead-lined drawing stations (Fig. 23-2). Leaded glass provides comfortable viewing and radiation protection simultaneously, especially

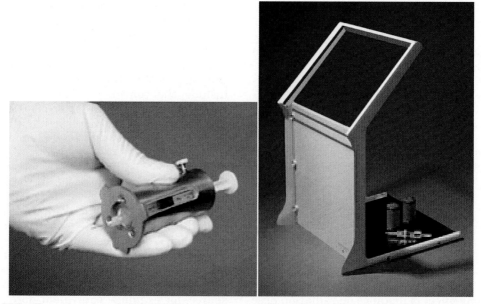

FIGURE 23-2 Examples of protective shielding devices used in nuclear medicine. *Left,* Shielded syringe holder designed for positron-emitting radionuclides. *Right,* Lead-lined drawing station for preparing and handling low-energy gamma-emitting radioactive materials ($\leq$150 keV) such as ^{99m}Tc. Lead-lined glass provides a good view of the work area. (*Photographs courtesy Biodex Medical Systems, Inc., Shirley, NJ.*)

for low-energy γ-ray and x-ray emitters (≤200 keV). Dose calibrators should be enclosed in a shielded area, using lead sheets or bricks, to avoid unnecessary exposure during measurement of radiopharmaceutical activity.

Table 6-4 lists tenth-value thicknesses of lead for several γ-ray and x-ray emitters. Small thicknesses of lead (≤1 mm) provide effective shielding for low-energy emitters (e.g., ^{133}Xe and ^{99m}Tc). Lead-lined aprons, which usually contain 0.25-mm- or 0.5-mm -equivalent lead thicknesses, provide a modest amount of radiation protection, but probably not enough to warrant their routine use in the nuclear medicine laboratory; however, they may be useful for specific applications, such as during handling of large quantities of ^{133}Xe ($E_\gamma = 81$ keV) or during elution of a ^{99m}Tc generator ($E_\gamma = 140$ keV). Greater thickness (>1 cm) is required for higher-energy γ-ray emitters, such as ^{131}I ($E_\gamma = 364$ keV); however, lead is still an effective shielding material at these energies. Concrete and similar materials find limited use for general purpose shielding in nuclear medicine.

Shielding is very effective for β emitters. A few millimeters of almost any solid material will stop even the most energetic β particles (see Fig. 6-10 and Table 6-1). In this case, however, low-Z materials (e.g., plastic, ordinary glass) are preferred over high-Z materials (e.g., leaded glass) to minimize bremsstrahlung production (see Equation 6-1). A good shielding arrangement for a high-energy β emitter, such as ^{32}P, is to use a plastic or glass container for the radioactive material to stop the β particles and then to place this in a lead container to absorb the bremsstrahlung radiation (see Fig. 6-3). A similar approach can be employed with positron emitters; however, the thickness of the lead must be substantial (tenth-value thickness for 511-keV photons is 13.4 mm) to provide effective shielding against the annihilation radiation.

3. Reduction of Radiation Doses from Internal Sources

Types of Sources. Nearly all nuclear medicine personnel are required at one time or another to work with radioactive sources in open or poorly sealed containers. There is always the possibility that in these operations some of the radioactive material will find its way into the body, where it delivers a radiation dose as an internal radiation source, or back to offices or other areas accessible to nonradiation

workers. Patient sweat or excreta, linens used on imaging tables, spillage occurring during transfer of activity between containers and syringes, radioactive trash, and radioactive gases released during pulmonary function tests are examples of potential sources.

A radioactive material that has been accidentally or carelessly ingested is an "uncontrolled source"; once it is inside the body, there is very little that can be done to reduce the radiation dose that it will deliver. (Techniques developed by the weapons and reactor industries for speeding the elimination of radioactive materials from the body generally are slow to act and thus are useful only for very long-lived radionuclides, and impractical for nuclear medicine.) The cardinal rule for keeping radiation doses from internal sources ALARA is to prevent the entry of the radioactive material into the body in the first place. To a certain extent, this is a matter of careful design of laboratory facilities, but equally as important, it is a matter of developing good laboratory work habits.

Some basic rules for avoiding internal radiation doses are the following:

1. No eating, drinking, smoking, or applying of cosmetics should occur in areas where open sources may be present (e.g., hot labs and patient study areas).
2. Lab coats and gloves should be worn when handling radioactive sources. Gloves should be handled so as to avoid contamination of their inside surfaces. Lab coats, aprons, and other protective clothing should stay in the laboratory (i.e., they should not worn outside the lab or taken home).
3. No foodstuffs or drinks should be stored where radioactive sources are kept, such as in laboratory refrigerators.
4. Pipetting should never be done by mouth.
5. Personnel should wash their hands after working with radioactive sources (a sink should be available in the laboratory), and they should be checked for contamination on a laboratory radiation monitor (Section E.1). Hands should also be monitored before going to lunch or on breaks and before leaving at the end of the day.
6. Work should be performed on absorbent pads to catch spills and prevent spattering of liquids.
7. Work with radioactive gases or other volatile materials (e.g., concentrated iodine solutions) should be performed

in a ventilated fume hood. These materials also should be stored in a hood.

8. Work areas should be kept tidy. Radioactive trash, contaminated pads, and so forth should be disposed of promptly.

9. Radioactive storage areas (e.g., hot labs) should not be used to store other materials, such as office supplies or linens.

10. Needless contamination of light switches, doorknobs, and other items that could result in unsuspected contamination to personnel should be avoided.

11. Containers with sharp or broken edges should not be used for radioactive materials.

12. Radioactive materials should be stored when they are not in use.

Studies with radioactive gases, such as ^{133}Xe, require special attention because of the potential for escape of radioactivity into the laboratory and beyond. Optimally, the laboratory ventilation system should be designed to maintain the laboratory under negative pressure relative to its surroundings and should be separate from other ventilation systems to prevent spread of airborne activity into other areas of the hospital. A gas-trapping system should be used to collect gases exhaled by the patient.

Most of the rules listed earlier in this section are of the common-sense variety and perhaps seem obvious; however, it is surprising how often they are violated through forgetfulness or indifference. This may explain the surprisingly high incidence of internal radionuclides found in some studies of nuclear medicine laboratory personnel (e.g., >70% incidence of radioactive iodine in thyroid glands).[6] Clearly, adherence to proper laboratory work rules is fundamental to the ALARA concept.

4. Laboratory Design

The principles of ALARA are enhanced by careful attention to laboratory design. Several design aspects have been mentioned already in relation to other problems, such as negative relative air pressure in laboratories employing volatile or gaseous radioactive materials and availability of a fume hood with its own exhaust system for storage of these materials. Some additional principles to be considered in laboratory design are the following:

1. Hot labs and radioactive storage areas should be located away from other busy work areas, public corridors, secretarial offices, and so on, and away from imaging and low-level counting rooms.

2. Work surfaces and floors should be constructed using smooth, nonabsorbent materials free from cracks and crevices.

3. Workbenches should be sufficiently sturdy to support lead shielding.

4. Washbasins and sinks should be conveniently available where unsealed radioactive materials are handled. It is desirable that sinks in hot labs have foot- or elbow-operated controls.

5. The laboratory design should permit separate storage of glassware and work tools (e.g., tongs, stirring devices) not used with radioactive materials to prevent needless contamination or mixture with similar items used with radioactive preparations.

5. Procedures for Handling Spills

Accidental spills of radioactive materials are infrequent occurrences in well-run nuclear medicine laboratories. Also, the quantities of radioactivity used in nuclear medicine do not create "life-threatening" hazards. Nevertheless, radioactive spills should not be treated as events completely without hazard, and laboratory personnel should be aware of the appropriate procedures to follow when spills do occur.

The steps to follow in dealing with a radioactive spill are (1) to *inform,* (2) to *contain,* and (3) to *decontaminate.*

1. Individuals in the immediate work area should be informed that a spill has occurred so they can avoid contamination if possible. Individuals outside the immediate area should be warned so they do not enter it. The radiation officer should be informed so that he or she may begin supervising further action as soon as possible.

2. By whatever means are reasonably possible, but without risking further hazards to themselves, laboratory personnel should attempt to contain the spill to prevent further spread of contamination. A flask that has been tipped over should be uprighted. Absorbent pads should be thrown over a liquid spill. Doors should be closed to prevent the escape of airborne radioactivity (e.g., gases, powders). The spill area should be closed off to prevent entry, especially by persons who might not be aware of the spill. Personnel monitoring for contamination should be started as soon as possible, so that

contaminated and uncontaminated persons can be segregated. To prevent the further spread of radioactivity, contaminated individuals should not be allowed to leave the area until they are decontaminated, and uncontaminated individuals (with the exception of appropriately protected emergency personnel and other designated personnel involved with the cleanup) should not be allowed to enter the spill area. Contamination monitoring should be done using a sensitive radiation monitoring instrument appropriate for the type of radioactivity involved. It is advisable that each laboratory have on hand a thin-window Geiger-Müller (GM) counter survey meter (Section E.1) for handling such situations.

3. Personnel decontamination procedures should receive first priority, followed by decontamination of work areas, and so on. Personnel involved in decontamination procedures should wear protective clothing to avoid becoming contaminated themselves in the process. Contaminated skin should be flushed thoroughly with water. Special attention should be given to open wounds and contamination around the eyes, nose, and mouth. Contaminated clothing should be removed and placed in plastic bags for storage. After major localized areas of personnel contamination have been attended to, a shower bath may be required to remove more widely distributed contamination.

Decontamination of laboratory and work areas should not be attempted except under the supervision of the radiation safety officer or radiation health physicist. If the work surfaces and floors are constructed from a non-absorbent material, soap and water is generally all that is needed for decontamination. Contaminated areas should be cleaned "from outside in" to minimize the spread of contamination. Porous or cracked surfaces may create difficult problems. If complete decontamination is not possible, it may be necessary to cover and shield the affected surfaces or perhaps even to remove and replace them.

D. DISPOSAL OF RADIOACTIVE WASTE

There are three general techniques for disposing of radioactive wastes.

1. *Dilute and disperse.* Small quantities of radioactive materials may be released into the environment—for example, radioactive gases into the ventilation system or liquid wastes into the sink—provided that the concentrations do not exceed the values specified in 10CFR20 (see Table 23-3). In keeping with the ALARA concept, however, this technique should not be used if reasonable alternatives are available (e.g., steps 2 and 3).

2. *Store and decay.* For materials having reasonably short half-lives (e.g., a few weeks or less), and if suitable storage space is available, this may be an economical and effective disposal technique. After a decay period of 10 half-lives has elapsed, only 0.1% of the initial activity remains. It is advisable to separate waste materials into two categories: those having half-lives shorter than 3 days and those having half-lives longer than 3 days, so that long-term accumulation of large volumes of waste material can be avoided. Disposal by decay of materials with half-lives longer than approximately 1 month is frequently impractical because of the long storage period required.

3. *Concentrate and bury.* This is frequently the only effective means of disposal of long-lived radioactivity, particularly if storage space is limited. A number of commercial companies provide this type of disposal service.

E. RADIATION MONITORING

1. Survey Meters and Laboratory Monitors

Survey meters are used to monitor radiation levels in and near laboratories where radioactive materials or other radiation sources are present. Generally, they are battery operated and portable. The radiation detector is usually an ionization chamber or a GM tube (see Figs. 7-3 and 7-11).

Ionization chamber types are calibrated to read exposure levels. The full-scale reading and range on the meter display is switch selectable, typically from 0 to 3 mR/hr up to 0 to 300 mR/hr. Many systems now also provide readings in units of air kerma. Some types have very thin mica or aluminum entrance windows for the ionization chamber and can be used to detect β particles as well as x rays or γ rays. Ionization chamber survey meters give reasonably accurate estimates of exposure rates (±10%) over most of the nuclear medicine energy range. Most ionization chamber survey meters do not have sealed chambers. Thus for greatest accuracy their readings should be corrected for ambient temperature and pressure variations (see Chapter 7, Section A.2, Equation 7-1). These corrections are small at sea-level pressures and room temperatures; however, the pressure correction factor may be significant at higher elevations (~20% at 1600 m).

The accuracy of an ionization chamber survey meter should be checked periodically (e.g., annually, or following major repairs) using a radiation source producing a known radiation exposure level. Sealed sources used in radiation therapy departments are useful for this purpose.

GM tube types of survey meters are more sensitive than ionization chamber types because they respond to individual ionizing radiation events. Most of these instruments have meters that display event counting rates (cpm). Some types with thin mica or aluminum entrance windows are suitable for detecting α and β particles as well as γ rays and x rays. Because of their relatively high sensitivity, GM-type survey meters are most useful for detecting small quantities of radioactivity from minor spills, in waste receptacles, and so on.

Laboratory monitors are very similar to survey meters, but they are designed to be used at a fixed location rather than as portable units. They are operated continuously; thus they are generally plugged into the wall rather than battery operated. Most have GM tube detectors and produce an audible clicking noise when radiation is detected in addition to having a meter display of counting rate. A laboratory monitor should be used in any area where large quantities of radioactivity are handled (e.g., in a radiopharmacy laboratory) to warn of the presence of high radiation levels. They also are useful for monitoring hands after operations requiring the handling of radioactivity.

2. Personnel Dosimeters

Personnel dosimeters are devices worn by laboratory personnel to monitor radiation doses from external sources. There are two general types: *dosimeter badges,* which are used to measure cumulative doses over periods of weeks or months, and *pocket dosimeters,* which are generally used for monitoring over a shorter term.

Dosimeter badges monitor radiation doses using either a small piece of x-ray film, or much more commonly, thermoluminescent dosimeter (TLD) chips (Fig. 23-3). TLDs generally use small "chips" of LiF, a material that gives off light when heated after it has been exposed to ionizing radiation. The amount of light given off is measured using a photomultiplier tube while the chip is heated in an oven inside a light-tight enclosure. The amount of light given off is used to estimate the radiation dose received. There also are dosimeter badges based on optically stimulated (rather than heat-stimulated) luminescence of materials such as Al_2O_3.

Dosimeter badge services are provided by a number of commercial suppliers. New badges are supplied at regular (e.g., monthly) intervals, and readings for the preceding period are reported back to the user, typically

FIGURE 23-3 Examples of personnel dosimeter badges. *Left,* Thermoluminescent dosimeter (TLD) badge, which usually contains several TLD chips, with one chip being exposed through a thin Mylar window in the badge to permit measurement of low-energy beta radiation. *Right,* Photograph of ring dosimeters that contain a single lithium fluoride TLD chip and are useful for measuring the dose to the skin and hands from the handling of radionuclides. Bar codes are used on most personnel monitors to permit easy identification and data logging when reading out the TLD chips. (*Courtesy Mirion Technologies, Irvine, CA.*)

within about a month. The reports provided by most companies are satisfactory for NRC record-keeping purposes.

Pocket dosimeters were described in Chapter 7, Section A.2 (Fig. 7-5). They are essentially ionization chamber devices that provide an immediate readout of radiation doses and thus are especially useful for measuring over short periods or when a rapid indication of results is needed, such as during complicated radiopharmaceutical preparation procedures.

3. Wipe Testing

Wipe testing is used to detect small amounts of radioactive contamination on bench-top surfaces, on the outside of shipping packages, and so on, or to detect small amounts of radioactive leakage from sealed radioactive sources. The surface is wiped with an alcohol-soaked patch of gauze or cotton-tipped swab, which is then counted in a well counter (for γ-emitting nuclides) or a liquid scintillation counter (for β emitters). Contamination below the kBq level can be detected by wipe testing. NRC regulations require periodic wipe testing of work areas and maintaining records of these tests.

REFERENCES

1. Evaluation of the Linear Nonthreshold Dose-Response Model for Ionizing Radiation (NCRP Report No. 136). Bethesda, MD, NCRP, 2001.
2. Rossi HH: Sensible radiation protection. *Health Physics* 70:394-395, 1996.
3. Simmons JA, Watt DE: *Radiation Protection Dosimetry: A Radical Reappraisal*. Madison, WI, 1999, Medical Physics Publishing Corp.
4. Kondo S: *Health Effects of Low-Level Radiation*. Madison, WI, 1993, Medical Physics Publishing Corp.
5. Source and Effects of Ionizing Radiation (United Nations Scientific Committee on the Effects of Atomic Radiation, 2008 Report to the General Assembly), New York, 2010, United Nations. Available at http://www.unscear.org/unscear/en/publications.html [accessed October 14, 2011].
6. Sources and Magnitude of Occupational and Public Exposures from Nuclear Medicine Procedures (NCRP Report No. 124). Bethesda, MD, NCRP, 1996.
7. Anger RT: Radiation protection in nuclear medicine. In *The Physics of Nuclear Medicine*, Chicago, 1977, American Association of Physicists in Medicine.

BIBLIOGRAPHY

A detailed discussion of health physics and radiation protection techniques can be found in the following textbooks:

Shapiro J: *Radiation Protection: A Guide for Scientists, Regulators, and Physicians*, 4th ed. Cambridge, MA, 2002, Harvard University Press.

Turner JE: *Atoms, Radiation, and Radiation Protection*, 2nd ed. New York, 1995, Wiley.

Regulatory documents can be found on the website of the Nuclear Regulatory Commission at http://www.nrc.gov [accessed October 14, 2011]. Relevant documents on this website include the following:

NRC Regulations:
10CFR20: http://www.nrc.gov/reading-rm/doc-collections/cfr/part020/index.html
10CFR35: http://www.nrc.gov/reading-rm/doc-collections/cfr/part035/index.html
NRC Regulatory Guides (available at http://www.nrc.gov/reading-rm/doc-collections/reg-guides/occupational-health/rg/):
8.10: Operating Philosophy for Maintaining Exposures as Low as Reasonably Achievable
8.18: Information Relevant to Ensuring that Occupational Radiation Exposures at Medical Institutions will be As Low As Reasonably Achievable
8.36: Radiation Dose to the Embryo/Fetus
8.39: Release of Patients Administered Radioactive Materials

The National Council on Radiation Protection and Measurement website is at http://www.ncrp.com [accessed October 14, 2011]. Important NCRP publications in addition to those listed in the references are the following:

Uncertainties in the Measurement and Dosimetry of External Radiation (NCRP Report No. 158), 2007.
Operational Radiation Safety Training (NCRP Report No. 134), 2000.
Radionuclide Exposure of the Embryo/Fetus (NCRP Report No. 128), 1998.
Operational Radiation Safety Program (NCRP Report No. 127), 1998.
Limitation of Exposure to Ionizing Radiation (NCRP Report No. 116), 1993.
Implementation of the Principle of As Low as Reasonably Achievable (ALARA) for Medical and Dental Personnel (NCRP Report No. 107), 1990.
Radiation Protection for Medical and Allied Health Personnel (NCRP Report No. 105), 1989.
Protection in Nuclear Medicine and Ultrasound Diagnostic Procedures in Children (NCRP Report No. 73), 1983.
Management of Persons Accidentally Contaminated with Radionuclides (NCRP Report No. 65), 1980.
Safe Handling of Radionuclides (NCRP Report No. 30), 1964.

The International Atomic Energy Agency website is at www.iaea.org [accessed October 14, 2011] and has several publications relevant to nuclear medicine that can be downloaded from the website: For example:

Nuclear Medicine Resources Manual, 2006.
Cyclotron Produced Radionuclides: Guidelines for Setting Up a Facility, Technical Reports Series No. 471, 2009.
Quality Assurance for Radioactivity Measurement in Nuclear Medicine, Technical Reports Series No. 454, 2006.
Applying Radiation Safety Standards in Nuclear Medicine, Safety Reports Series No. 40, 2005.

International bodies providing information or recommendations regarding radiation dose:

The International Commission on Radiological Protection website is at http://www.icrp.org [accessed October 14, 2011). An important ICRP publication with recommendations regarding dose limits is the following:

Recommendations of the International Commission on Radiological Protection (ICRP Publication No. 103), Annals of the ICRP, Vol. 37, 2-4, 2007.

The United Nations Scientific Committee on the Effects of Atomic Radiation website is at www.unscear.org/ [accessed October 14, 2011). Two comprehensive publications of interest are:

UNSCEAR 2008 Report: "Sources and effects of ionizing radiation."
UNSCEAR 2006 Report: "Effects of ionizing radiation."

Unit Conversions

Quantity	SI Units	Traditional Units
Activity	1 becquerel (Bq)	$= 2.703 \times 10^{-11}$ curies (Ci)
	3.7×10^{10} Bq	$= 1$ Ci
	1 MBq	$= 27.03$ μCi
	37 MBq	$= 1$ mCi
Absorbed Dose	1 gray (Gy)	$= 100$ rads
	1×10^{-2} Gy	$= 1$ rad
	1 mGy	$= 0.1$ rad
Equivalent Dose, Effective Dose	1 sievert (Sv)	$= 100$ rems
	1×10^{-2} Sv	$= 1$ rem
	1 mSv	$= 0.1$ rem
Exposure	1 C/kg air	$= 3876$ roentgen (R)
	2.58×10^{-4} C/kg air	$= 1$ R
Energy	1 joule (J)	$= 6.242 \times 10^{18}$ electron volts (eV)
	1.602×10^{-19} J	$= 1$ eV
Mass*	1 kilogram (kg)	$= 6.02214 \times 10^{26}$ unified atomic mass units (u)
	1.66054×10^{-27} kg	$= 1$ u
Pressure	1 pascal (Pa)	$= 7.501 \times 10^{-3}$ mm Hg (torr)
	1.333×10^{2} pascals (Pa)	$= 1$ mm Hg (torr)
	1 Pa	$= 9.869 \times 10^{-6}$ atmospheres (atm)
	1.013×10^{5} Pa	$= 1$ atm
Area	1 square meter (m²)	$= 1 \times 10^{28}$ barns
	1×10^{-28} m²	$= 1$ barn
Temperature*	x kelvin (K)	$= x - 273.15$ degrees centigrade (°C)
	$x + 273.15$ K	$= x$ °C
Magnetic Flux Density	1 tesla (T)	$= 1 \times 10^{4}$ gauss (G)
	1×10^{-4} T	$= 1$ G

*Note that mass and temperature are the only SI base units in this table. All others are derived from these and/or the five other SI base units.

A useful source of further information on SI units and unit conversions is the National Institute of Standards and Technology website, http://physics.nist.gov/cuu/Units/index.html. Accessed 4 November 2011.

Properties of the Naturally Occurring Elements

Name	Symbol	Atomic Number	Atomic Weight* (^{12}C scale)	Density† (g/cm^3)	$K_B{}^{‡}$ (keV)
Actinium	Ac	89	(227)	10.0	106.756
Aluminum	Al	13	26.982	2.70	1.560
Antimony	Sb	51	121.760	6.68	30.491
Argon	Ar	18	39.948	1.63§	3.203
Arsenic	As	33	74.922	5.75	11.867
Astatine	At	85	(210)	—	95.730
Barium	Ba	56	137.327	3.62	37.440
Beryllium	Be	4	9.012	1.85	0.111
Bismuth	Bi	83	208.980	9.79	90.526
Boron	B	5	10.811	2.34	0.188
Bromine	Br	35	79.904	3.10	13.474
Cadmium	Cd	48	112.411	8.69	26.711
Calcium	Ca	20	40.078	1.54	4.038
Carbon	C	6	12.011	2.2 (graphite) 3.51 (diamond)	0.284
Cerium	Ce	58	140.116	6.77	40.443
Cesium	Cs	55	132.905	1.87	35.985
Chlorine	Cl	17	35.453	2.90§	2.822
Chromium	Cr	24	51.996	7.15	5.989
Cobalt	Co	27	58.933	8.86	7.709
Copper	Cu	29	63.546	8.96	8.979
Dysprosium	Dy	66	162.500	8.55	53.789
Erbium	Er	68	167.260	9.07	57.486
Europium	Eu	63	151.964	5.24	48.519
Fluorine	F	9	18.998	1.55§	0.685
Francium	Fr	87	(223)	—	101.147
Gadolinium	Gd	64	157.250	7.90	50.239
Gallium	Ga	31	69.723	5.91	10.367
Germanium	Ge	32	72.610	5.32	11.103

Continued

Name	Symbol	Atomic Number	Atomic Weight* (¹²C scale)	Density‡ (g/cm³)	K_B‡ (keV)
Gold	Au	79	196.967	19.3	80.725
Hafnium	Hf	72	178.490	13.3	65.351
Helium	He	2	4.003	0.164§	0.025
Holmium	Ho	67	162.930	8.80	55.618
Hydrogen	H	1	1.008	0.082§	0.014
Indium	In	49	114.818	7.31	27.940
Iodine	I	53	126.904	4.93	33.169
Iridium	Ir	77	192.217	22.6	76.111
Iron	Fe	26	55.845	7.87	7.112
Krypton	Kr	36	83.800	3.43§	14.326
Lanthanum	La	57	138.906	6.15	38.925
Lead	Pb	82	207.200	11.3	88.005
Lithium	Li	3	6.941	0.534	0.055
Lutetium	Lu	71	174.967	9.84	63.314
Magnesium	Mg	12	24.305	1.74	1.305
Manganese	Mn	25	54.938	7.3	6.539
Mercury	Hg	80	200.590	13.5	83.102
Molybdenum	Mo	42	95.940	10.2	20.000
Neodymium	Nd	60	144.240	7.01	43.569
Neon	Ne	10	20.180	0.825§	0.867
Nickel	Ni	28	58.693	8.90	8.333
Niobium	Nb	41	92.906	8.57	18.986
Nitrogen	N	7	14.007	1.15§	0.402
Osmium	Os	76	190.230	22.59	73.871
Oxygen	O	8	15.999	1.31§	0.532
Palladium	Pd	46	106.420	12.0	24.345
Phosphorus	P	15	30.974	1.82 (white) 2.16 (red) 2.69 (black)	2.146
Platinum	Pt	78	195.078	21.5	78.395
Polonium	Po	84	(209)	9.2	93.105
Potassium	K	19	39.098	0.89	3.607
Praseodymium	Pr	59	140.908	6.77	41.991
Promethium	Pm	61	(145)	7.26	45.184
Protactinium	Pa	91	231.036	15.4	112.601
Radium	Ra	88	(226)	5	103.922
Radon	Rn	86	(222)	9.074§	98.404
Rhenium	Re	75	186.207	20.8	71.676
Rhodium	Rh	45	102.906	12.4	23.220
Rubidium	Rb	37	85.468	1.53	15.200
Ruthenium	Ru	44	101.070	12.1	22.117
Samarium	Sm	62	150.360	7.52	46.834
Scandium	Sc	21	44.067	2.99	4.493

Name	Symbol	Atomic Number	Atomic Weight* (^{12}C scale)	Density† (g/cm^3)	$K_B{}^\ddagger$ (keV)
Selenium	Se	34	78.960	4.81 (gray) 4.39 (α form) 4.28 (vitreous)	12.658
Silicon	Si	14	28.086	2.33	1.839
Silver	Ag	47	107.868	10.5	25.514
Sodium	Na	11	22.990	0.97	1.072
Strontium	Sr	38	87.620	2.64	16.105
Sulfur	S	16	32.066	2.07 (rhombic) 2.00 (monoclinic)	2.472
Tantalum	Ta	73	180.948	16.4	67.416
Technetium	Tc	43	(98)	11	21.044
Tellurium	Te	52	127.600	6.23	31.814
Terbium	Tb	65	158.925	8.23	51.996
Thallium	Tl	81	204.383	11.8	85.530
Thorium	Th	90	232.038	11.7	109.651
Thulium	Tm	69	168.934	9.32	59.390
Tin	Sn	50	118.710	5.77 (gray) 7.29 (white)	29.200
Titanium	Ti	22	47.867	4.51	4.966
Tungsten	W	74	183.840	19.3	69.525
Uranium	U	92	238.029	19.1	115.606
Vanadium	V	23	50.942	6.0	5.465
Xenon	Xe	54	131.290	5.37§	34.561
Ytterbium	Yb	70	173.040	6.90	61.332
Yttrium	Y	39	88.906	4.47	17.038
Zinc	Zn	30	65.390	7.13	9.659
Zirconium	Zr	40	91.224	6.52	17.998

*Values averaged for the elements in their natural abundance. For nuclides with no stable isotopes, value in parentheses corresponds to mass number of most stable isotope. Values from http://physics.nist.gov/PhysRefData. Accessed 16 June 2011.

†Values averaged for the elements in their natural abundance. Values from The Handbook of Chemistry and Physics, 91st ed., CRC Press, 2010-2011.

‡K-shell binding energies. Values from http://www.nist.gov/pml/data/xraytrans/index.cfm. Accessed 23 November 2011.

§Densities for gases in g/liter at 0° C and pressure of 1 atm.

appendix
C

Decay Characteristics of Some Medically Important Radionuclides

The figures show nuclear decay scheme diagrams using the conventions described in Chapter 3. In the tables accompanying the decay diagrams, the first column is the type of radiation emitted, $y(i)$ is the frequency of the ith emission per nuclear decay in $(Bq \cdot sec)^{-1}$, $E(i)$ is the corresponding transition energy for the emission in MeV (given as the average energy for beta decay), and $y(i) \times E(i)$ is the average energy emitted per decay. [Figures from ICRP Publication No. 38, Radionuclide Transformations: Energy and Intensity of Emissions. In Annals of the ICRP (International Commission on Radiological Protection). Oxford, Pergamon Press, 1983.]

Legend for radiation listed in decay tables:

γ	gamma ray
β^-	beta-minus particle
β^+	beta-plus particle
$\gamma\pm$	annihilation photons
ce-K, ce-L, etc....	internal conversion electrons ejected from the K, L, etc.... shell (Chapter 3, Section E)
Auger-XXX	Auger electrons (see Chapter 2, Section C.3 for explanation of notation)
K_α, K_β etc, ... X ray	characteristic x rays (see Chapter 2, Table 2-1 for notation)
ΔE	residual low-energy radiation (mainly Auger processes) not easily described by individual discrete transitions

HYDROGEN-3

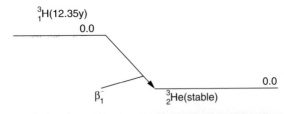

Half Life = 12.35 Years			
Decay Mode(s): β^-			
Radiation	$y(i)$ (Bq•s)$^{-1}$	$E(i)$ (MeV)	$y(i) \times E(i)$
β^- 1	1.00E 00	5.683E-03*	5.68E-03
LISTED β, ce AND Auger RADIATIONS			5.68E-03
LISTED RADIATIONS			5.68E-03

*AVERAGE ENERGY (MeV)

HELIUM-3 DAUGHTER IS STABLE.

CARBON-11

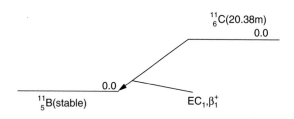

Half Life = 20.38 Minutes			
Decay Mode(s): EC, β⁺			
Radiation	*y(i)* (Bq•s)⁻¹	*E(i)* (MeV)	*y(i)* × *E(i)*
β⁺ 1	9.98E-01	3.855E-01*	3.85E-01
γ±	2.00E 00	5.110E-01	1.02E 00
Kα₁ X ray	1.62E-06	1.833E-04	2.97E-10
Kα₂ X ray	8.10E-07	1.833E-04	1.48E-10
LISTED X, γ AND γ± RADIATIONS			1.02E 00
LISTED β, ce AND Auger RADIATIONS			3.85E-01
LISTED RADIATIONS			1.40E 00

*AVERAGE ENERGY (MeV)
BORON-11 DAUGHTER IS STABLE.

NITROGEN-13

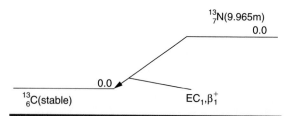

Half Life = 9.965 Minutes			
Decay Mode(s): EC, β⁺			
Radiation	*y(i)* (Bq•s)⁻¹	*E(i)* (MeV)	*y(i)* × *E(i)*
β⁺ 1	9.98E-01	4.918E-01*	4.91E-01
γ±	2.00E 00	5.110E-01	1.02E 00
Kα₁ X ray	2.38E-06	2.774E-04	6.59E-10
Kα₂ X ray	1.19E-06	2.774E-04	3.30E-10
Auger-KLL	1.80E-03	2.564E-04*	4.61E-07
LISTED X, γ AND γ± RADIATIONS			1.02E 00
LISTED β, ce AND Auger RADIATIONS			4.91E-01
LISTED RADIATIONS			1.51E 00

*AVERAGE ENERGY (MeV)
CARBON-13 DAUGHTER IS STABLE.

CARBON-14

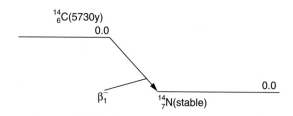

Half Life = 5730 Years			
Decay Mode(s): β⁻			
Radiation	*y(i)* (Bq•s)⁻¹	*E(i)* (MeV)	*y(i)* × *E(i)*
β⁻ 1	1.00E 00	4.945E-02*	4.95E-02
LISTED β, ce AND Auger RADIATIONS			4.95E-02
LISTED RADIATIONS			4.95E-02

*AVERAGE ENERGY (MeV)
NITROGEN-14 DAUGHTER IS STABLE.

OXYGEN-15

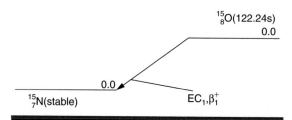

Half Life = 122.24 Seconds			
Decay Mode(s): EC, β⁺			
Radiation	*y(i)* (Bq•s)⁻¹	*E(i)* (MeV)	*y(i)* × *E(i)*
β⁺ 1	9.99E-01	7.353E-01*	7.34E-01
γ±	2.00E 00	5.110E-01	1.02E 00
Kα₁ X ray	2.65E-06	3.924E-04	1.04E-09
Kα₂ X ray	1.32E-06	3.924E-04	5.19E-10
Auger-KLL	1.13E-03	3.684E-04*	4.15E-07
LISTED X, γ AND γ± RADIATIONS			1.02E 00
LISTED β, ce AND Auger RADIATIONS			7.34E-01
LISTED RADIATIONS			1.76E 00

*AVERAGE ENERGY (MeV)
NITROGEN-15 DAUGHTER IS STABLE.

FLUORINE-18

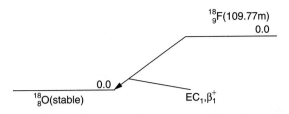

Half Life = 109.77 Minutes			
Decay Mode(s): EC, β⁺			
Radiation	$y(i)$ (Bq•s)⁻¹	$E(i)$ (MeV)	$y(i) \times E(i)$
β⁺ 1	9.67E-01	2.498E-01*	2.42E-01
γ±	1.93E 00	5.110E-01	9.86E-01
LISTED X, γ AND γ± RADIATIONS			1.02E 00
LISTED β, ce AND Auger RADIATIONS			2.50E-01
LISTED RADIATIONS			1.27E 00

*AVERAGE ENERGY (MeV)
OXYGEN-18 DAUGHTER IS STABLE.

SODIUM-22

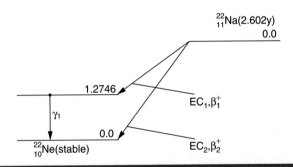

Half Life = 2.602 Years			
Decay Mode(s): EC, β⁺			
Radiation	$y(i)$ (Bq•s)⁻¹	$E(i)$ (MeV)	$y(i) \times E(i)$
β⁺ 1	8.98E-01	2.154E-01*	1.94E-01
β⁺ 2	6.00E-04	8.350E-01*	5.01E-04
γ±	1.80E 00	5.110E-01	9.19E-01
γ 1	9.99E-01	1.275E 00	1.27E 00
ce-K, γ 1	6.43E-06	1.274E 00	8.19E-06
ce-L₁, γ 1	3.77E-07	1.274E 00	4.81E-07
ce-L₂, γ 1	2.07E-10	1.275E 00	2.64E-10
ce-L₃, γ 1	3.40E-10	1.275E 00	4.33E-10
Kα₁ X ray	9.42E-04	8.486E-04	7.99E-07
Kα₂ X ray	4.72E-04	8.486E-04	4.01E-07
Kα₃ X ray	1.19E-12	8.219E-04	9.76E-16
Auger-KLL	9.96E-02	8.006E-04*	7.97E-05
LISTED X, γ AND γ± RADIATIONS			2.19E 00
LISTED β, ce AND Auger RADIATIONS			1.94E-01
LISTED RADIATIONS			2.39E 00

*AVERAGE ENERGY (MeV)
NEON-22 DAUGHTER IS STABLE.

PHOSPHORUS-32

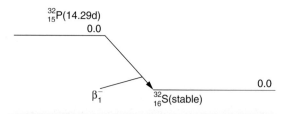

$^{32}_{15}$P(14.29d)
0.0

0.0

β^-_1 $^{32}_{16}$S(stable)

Half Life = 14.29 Days		
Decay Mode(s): β⁻		
Radiation $y(i)$ (Bq•s)⁻¹	$E(i)$ (MeV)	$y(i) \times E(i)$
β⁻ 1 1.00E 00	6.947E-01*	6.95E-01
LISTED β, ce AND Auger RADIATIONS		6.95E-01
LISTED RADIATIONS		6.95E-01

*AVERAGE ENERGY (MeV)
SULFUR-32 DAUGHTER IS STABLE.

SULFUR-35

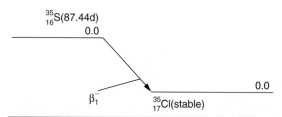

$^{35}_{16}$S(87.44d)
0.0

0.0

β^-_1 $^{35}_{17}$Cl(stable)

Half Life = 87.44 Days		
Decay Mode(s): β⁻		
Radiation $y(i)$ (Bq•s)⁻¹	$E(i)$ (MeV)	$y(i) \times E(i)$
β⁻ 1 1.00E 00	4.883E-02*	4.88E-02
LISTED β, ce AND Auger RADIATIONS		4.88E-02
LISTED RADIATIONS		4.88E-02

*AVERAGE ENERGY (MeV)
CHLORINE-35 DAUGHTER IS STABLE.

CHROMIUM-51

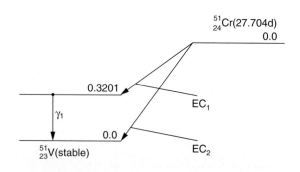

$^{51}_{24}$Cr(27.704d)
0.0

0.3201   EC₁

γ_1

0.0   EC₂

$^{51}_{23}$V(stable)

Half Life = 27.704 Days		
Decay Mode(s): EC		
Radiation	$y(i)$ (Bq•s)⁻¹	$E(i)$ (MeV)	$y(i) \times E(i)$
γ 1	9.83E-02	3.201E-01	3.15E-02
ce-K, γ 1	1.52E-04	3.146E-01	4.78E-05
ce-L₁, γ 1	1.38E-05	3.194E-01	4.41E-06
Kα₁ X-ray	1.33E-01	4.952E-03	6.59E-04
Kα₂ X-ray	6.70E-02	4.945E-03	3.31E-04
Kβ₁ X-ray	1.76E-02	5.427E-03	9.53E-05
Kβ₃ X-ray	8.88E-03	5.427E-03	4.82E-05
Auger-KLL	5.58E-01	4.339E-03*	2.42E-03
Auger-KLX	1.13E-01	4.876E-03*	5.49E-04
Auger-KXY	8.59E-03	5.386E-03*	4.63E-05
Auger-LMM	1.51E 00	4.859E-04*	7.34E-04
Auger-LMX	1.05E-02	5.183E-04*	5.45E-06
Auger-MXY	3.19E 00	1.603E-05*	5.12E-05
LISTED X γ AND γ± RADIATIONS			3.26E-02
OMITTED X γ AND γ± RADIATIONS**			5.89E-07
LISTED β, ce AND Auger RADIATIONS			3.86E-03
OMITTED β, ce AND Auger RADIATIONS**			8.33E-08
LISTED RADIATIONS			3.65E-02
OMITTED RADIATIONS**			6.72E-07

*AVERAGE ENERGY (MeV)
**EACH OMITTED TRANSITION CONTRIBUTES <0.100% TO Σ$y(i) \times E(i)$ IN ITS CATEGORY.
VANADIUM-51 DAUGHTER IS STABLE.

COBALT-57

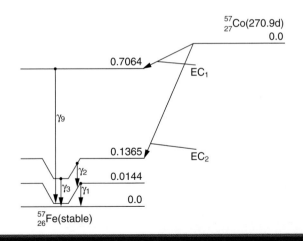

Half Life = 270.9 Days			
Decay Mode(s): EC			
Radiation	$y(i)$ $(Bq \cdot s)^{-1}$	$E(i)$ (MeV)	$y(i) \times E(i)$
γ 1	9.19E-02	1.441E-02	1.32E-03
ce-K, γ 1	7.13E-01	7.301E-03	5.20E-03
ce-L$_1$, γ 1	6.80E-02	1.357E-02	9.22E-04
ce-L$_2$, γ 1	4.20E-03	1.369E-02	5.75E-05
ce-L$_3$, γ 1	1.69E-03	1.370E-02	2.31E-05
γ 2	8.56E-01	1.221E-01	1.04E-01
ce-K, γ 2	1.84E-02	1.150E-01	2.12E-03
ce-L$_1$, γ 2	1.73E-03	1.212E-01	2.10E-04
γ 3	1.06E-01	1.365E-01	1.45E-02
ce-K, γ 3	1.43E-02	1.294E-01	1.84E-03
ce-L$_1$, γ 3	1.27E-03	1.356E-01	1.73E-04
γ 9	1.60E-03	6.920E-01	1.11E-03
Kα$_1$ X ray	3.34E-01	6.404E-03	2.14E-03
Kα$_2$ X ray	1.69E-01	6.391E-03	1.08E-03
Kβ$_1$ X ray	4.51E-02	7.058E-03	3.19E-04
Kβ$_3$ X ray	2.29E-02	7.058E-03	1.61E-04
Auger-KLL	8.54E-01	5.574E-03*	4.76E-03
Auger-KLX	2.04E-01	6.302E-03*	1.29E-03
Auger-KXY	1.79E-02	7.000E-03*	1.25E-04
Auger-LMM	2.43E 00	6.703E-04*	1.63E-03
Auger-LMX	1.54E-01	7.067E-04*	1.09E-04
Auger-MXY	5.33E 00	2.232E-05*	1.19E-04
LISTED X, γ AND γ± RADIATIONS			1.25E-01
OMITTED X, γ AND γ± RADIATIONS**			1.57E-04
LISTED β, ce AND Auger RADIATIONS			1.86E-02
OMITTED β, ce AND Auger RADIATIONS**			4.08E-05
LISTED RADIATIONS			1.44E-01
OMITTED RADIATIONS**			1.98E-04

*AVERAGE ENERGY (MeV)
**EACH OMITTED TRANSITION CONTRIBUTES <0.100% TO $\Sigma y(i) \times E(i)$ IN ITS CATEGORY.
IRON-57 DAUGHTER IS STABLE.

COBALT-60

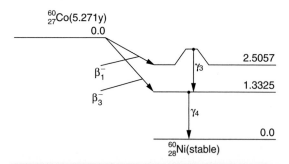

Half Life = 5.271 Years			
Decay Mode(s): β⁻			
Radiation	$y(i)$ (Bq•s)⁻¹	$E(i)$ (MeV)	$y(i) \times E(i)$
β⁻ 1	9.99E-01	9.577E-02*	9.57E-02
β⁻ 3	8.00E-04	6.258E-01*	5.01E-04
γ 3	9.99E-01	1.173E 00	1.17E 00
ce-K, γ 3	1.50E-04	1.165E 00	1.75E-04
γ 4	1.00E 00	1.332E 00	1.33E 00
ce-K, γ 4	1.14E-04	1.324E 00	1.50E-04
LISTED X, γ AND γ± RADIATIONS			2.50E 00
OMITTED X, γ AND γ± RADIATIONS**			1.14E-04
LISTED β, ce AND Auger RADIATIONS			9.65E-02
OMITTED β, ce AND Auger RADIATIONS**			4.73E-05
LISTED RADIATIONS			2.60E 00
OMITTED RADIATIONS**			1.61E-04

*AVERAGE ENERGY (MeV)
**EACH OMITTED TRANSITION CONTRIBUTES <0.100% TO Σ$y(i) \times E(i)$ IN ITS CATEGORY.
NICKEL-60 DAUGHTER IS STABLE.

COPPER-62

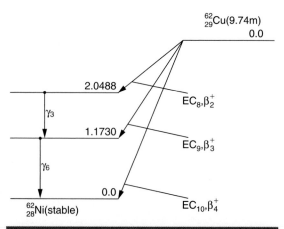

Half Life = 9.74 Minutes			
Decay Mode(s): EC, β⁺			
Radiation	$y(i)$ (Bq•s)⁻¹	$E(i)$ (MeV)	$y(i) \times E(i)$
β⁺ 4	9.76E-01	1.315E 00*	1.28E 00
γ±	1.96E 00	5.110E-01	1.00E 00
γ 3	1.47E-03	8.757E-01	1.29E-03
γ 6	3.35E-03	1.173E 00	3.93E-03
LISTED X, γ AND γ± RADIATIONS			1.00E 00
OMITTED X, γ AND γ± RADIATIONS**			1.92E-03
LISTED β, ce AND Auger RADIATIONS			1.28E 00
OMITTED β, ce AND Auger RADIATIONS**			1.48E-03
LISTED RADIATIONS			2.29E 00
OMITTED RADIATIONS**			3.40E-03

*AVERAGE ENERGY (MeV)
**EACH OMITTED TRANSITION CONTRIBUTES <0.100% TO Σ$y(i) \times E(i)$ IN ITS CATEGORY.
NICKEL-62 DAUGHTER IS STABLE.

COPPER-64

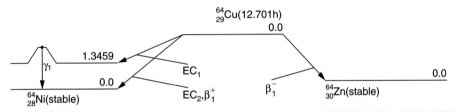

Half Life = 12.701 Hours			
Decay Mode(s): β⁻, EC, β⁺			
Radiation	**$y(i)$ (Bq•s)$^{-1}$**	**$E(i)$ (MeV)**	**$y(i) \times E(i)$**
β⁻ 1	3.72E-01	1.902E-01*	7.08E-02
β⁺ 1	1.79E-01	2.781E-01*	4.97E-02
γ±	3.58E-01	5.110E-01	1.83E-01
γ 1	4.90E-03	1.346E 00	6.59E-03
Kα₁ X ray	9.78E-02	7.478E-03	7.31E-04
Kα₂ X ray	4.97E-02	7.461E-03	3.71E-04
Auger-KLL	1.84E-01	6.489E-03*	1.20E-03
Auger-KLX	4.83E-02	7.356E-03*	3.55E-04
Auger-LMM	5.66E-01	8.103E-04*	4.59E-04
LISTED X, γ AND γ± RADIATIONS			1.90E-01
OMITTED X, γ AND γ± RADIATIONS**			1.68E-04
LISTED β, ce AND Auger RADIATIONS			1.23E-01
OMITTED β, ce AND Auger RADIATIONS**			1.06E-04
LISTED RADIATIONS			3.13E-01
OMITTED RADIATIONS**			2.74E-04

*AVERAGE ENERGY (MeV)

**EACH OMITTED TRANSITION CONTRIBUTES <0.100% TO $\Sigma y(i) \times E(i)$ IN ITS CATEGORY.

ZINC-64 DAUGHTER, YIELD 3.72E-01, IS STABLE.

NICKEL-64 DAUGHTER, YIELD 6.28E-01, IS STABLE.

GALLIUM-67

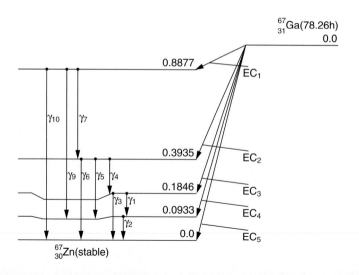

	Half Life = 78.26 Hours						
			Decay Mode(s): EC				
Radiation	y(i) (Bq•s)⁻¹	E(i) (MeV)	y(i) × E(i)	Radiation	y(i) (Bq•s)⁻¹	E(i) (MeV)	y(i) × E(i)
γ 1	3.07E-02	9.127E-02	2.80E-03	γ 10	1.45E-03	8.877E-01	1.29E-03
ce-K, γ 1	2.23E-03	8.161E-02	1.82E-04	Kα₁ X ray	3.28E-01	8.639E-03	2.83E-03
γ 2	3.83E-01	9.331E-02	3.57E-02	Kα₂ X ray	1.67E-01	8.616E-03	1.44E-03
ce-K, γ 2	2.87E-01	8.365E-02	2.40E-02	Kβ₁ X ray	4.49E-02	9.572E-03	4.30E-04
ce-L₁, γ 2	2.54E-02	9.212E-02	2.34E-03	Kβ₃ X ray	2.30E-02	9.572E-03	2.20E-04
ce-L₂, γ 2	3.98E-03	9.227E-02	3.67E-04	Auger-KLL	4.67E-01	7.466E-03*	3.49E-03
ce-L₃, γ 2	5.81E-03	9.229E-02	5.36E-04	Auger-KLX	1.33E-01	8.482E-03*	1.12E-03
ce-M, γ 2	5.17E-03	9.322E-02*	4.82E-04	Auger-KXY	1.31E-02	9.473E-03*	1.24E-04
γ 3	2.09E-01	1.846E-01	3.87E-02	Auger-LMM	1.55E 00	9.444E-04*	1.46E-03
ce-K, γ 3	4.07E-03	1.749E-01	7.11E-04	Auger-LMX	1.43E-01	1.020E-03*	1.46E-04
ce-L₁, γ 3	3.87E-04	1.834E-01	7.11E-05	Auger-MXY	3.49E 00	4.566E-05*	1.60E-04
γ 4	2.37E-02	2.090E-01	4.94E-03	LISTED X, γ AND γ± RADIATIONS			1.58E-01
ce-K, γ 4	1.90E-04	1.993E-01	3.79E-05	OMITTED X, γ AND γ± RADIATIONS**			8.52E-05
γ 5	1.68E-01	3.002E-01	5.04E-02	LISTED β, ce AND Auger RADIATIONS			3.54E-02
ce-K, γ 5	5.83E-04	2.906E-01	1.69E-04	OMITTED β, ce AND Auger RADIATIONS**			1.04E-04
γ 6	4.70E-02	3.935E-01	1.85E-02				
γ 7	6.86E-04	4.942E-01	3.39E-04	LISTED RADIATIONS			1.93E-01
γ 9	5.13E-04	7.944E-01	4.08E-04	OMITTED RADIATIONS**			1.89E-04

*AVERAGE ENERGY (MeV)
**EACH OMITTED TRANSITION CONTRIBUTES <0.100% TO Σy(i) × E(i) IN ITS CATEGORY.
ZINC-67 DAUGHTER IS STABLE.

GALLIUM-68

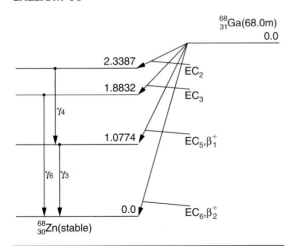

$^{68}_{31}$Ga(68.0m)
0.0

2.3387   EC$_2$
1.8832   EC$_3$
γ_4
1.0774   EC$_5$,β_1^+
γ_6 γ_3
0.0   EC$_6$,β_2^+
$^{68}_{30}$Zn(stable)

Half Life = 68 Minutes			
Decay Mode(s): EC, β^+			
Radiation	$y(i)$ **(Bq•s)$^{-1}$**	$E(i)$ **(MeV)**	$y(i) \times E(i)$
β^+ 1	1.08E-02	3.526E-01*	3.80E-03
β^+ 2	8.79E-01	8.358E-01*	7.35E-01
$\gamma\pm$	1.78E 00	5.110E-01	9.10E-01
γ 3	3.30E-02	1.077E 00	3.56E-02
γ 4	9.90E-04	1.261E 00	1.25E-03
γ 6	1.43E-03	1.883E 00	2.69E-03
LISTED X, γ AND $\gamma\pm$ RADIATIONS			9.49E-01
OMITTED X, γ AND $\gamma\pm$ RADIATIONS**			1.61E-03
LISTED β, ce AND Auger RADIATIONS			7.39E-01
OMITTED β, ce AND Auger RADIATIONS**			5.45E-04
LISTED RADIATIONS			1.69E 00
OMITTED RADIATIONS**			2.15E-03

*AVERAGE ENERGY (MeV)
**EACH OMITTED TRANSITION CONTRIBUTES <0.100% TO $\Sigma y(i) \times E(i)$ IN ITS CATEGORY.
ZINC-68 DAUGHTER IS STABLE.

GERMANIUM-68

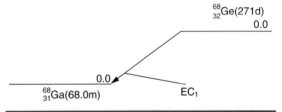

$^{68}_{32}$Ge(271d)
0.0

0.0
$^{68}_{31}$Ga(68.0m)   EC$_1$

Half Life = 271 Days			
Decay Mode(s): EC			
Radiation	$y(i)$ **(Bq•s)$^{-1}$**	$E(i)$ **(MeV)**	$y(i) \times E(i)$
Kα_1 X ray	2.55E-01	9.252E-03	2.36E-03
Kα_2 X ray	1.31E-01	9.225E-03	1.20E-03
Kα_3 X ray	3.93E-07	9.069E-03	3.56E-09
Kβ_1 X ray	3.59E-02	1.026E-02	3.68E-04
Kβ_3 X ray	1.83E-02	1.026E-02	1.88E-04
Kβ_5 X ray	6.43E-05	1.035E-02	6.66E-07
Lα X ray	4.55E-03	1.098E-03*	5.00E-06
Lβ X ray	1.91E-03	1.131E-03*	2.16E-06
Lη X ray	1.13E-04	9.842E-04	1.11E-07
Ll X ray	2.29E-04	9.573E-04	2.20E-07
Auger-KLL	3.19E-01	7.976E-03*	2.55E-03
Auger-KLX	9.39E-02	9.074E-03*	8.52E-04
Auger-KXY	9.49E-03	1.015E-02*	9.63E-05
Auger-LMM	1.11E 00	1.017E-03*	1.13E-03
Auger-LMX	1.18E-01	1.108E-03*	1.31E-04
Auger-MXY	2.53E 00	5.928E-05*	1.50E-04
LISTED X, γ AND $\gamma\pm$ RADIATIONS			4.13E-03
LISTED β, ce AND Auger RADIATIONS			4.90E-03
LISTED RADIATIONS			9.03E-03

*AVERAGE ENERGY (MeV)
GALLIUM-68 DAUGHTER IS RADIOACTIVE.

RUBIDIUM-82

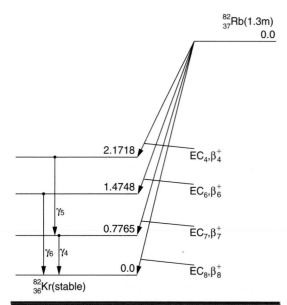

$^{82}_{37}$Rb(1.3m)

Half Life = 1.3 Minutes			
Decay Mode(s): EC, β⁺			
Radiation	**y(i) (Bq•s)⁻¹**	**E(i) (MeV)**	**y(i) × E(i)**
β⁺ 4	2.76E-03	5.174E-01*	1.43E-03
β⁺ 6	1.72E-03	8.325E-01*	1.44E-03
β⁺ 7	1.16E-01	1.157E 00*	1.34E-01
β⁺ 8	8.33E-01	1.523E 00*	1.27E 00
γ±	1.91E 00	5.110E-01	9.75E-01
γ 4	1.34E-01	7.765E-01	1.04E-01
γ 5	5.05E-03	1.395E 00	7.05E-03
γ 6	9.38E-04	1.475E 00	1.38E-03
LISTED X, γ AND γ± RADIATIONS			1.09E 00
OMITTED X, γ AND γ± RADIATIONS**			5.36E-03
LISTED β, ce AND Auger RADIATIONS			1.41E 00
OMITTED β, ce AND Auger RADIATIONS**			7.03E-04
LISTED RADIATIONS			2.49E 00
OMITTED RADIATIONS**			6.06E-03

*AVERAGE ENERGY (MeV)
**EACH OMITTED TRANSITION CONTRIBUTES <0.100% TO Σy(i) × E(i) IN ITS CATEGORY.
KRYPTON-82 DAUGHTER IS STABLE.

YTTRIUM-90

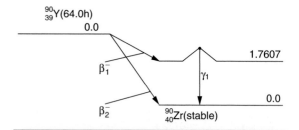

$^{90}_{39}$Y(64.0h)

Half Life = 64 Hours			
Decay Mode(s): β⁻			
Radiation	**y(i) (Bq•s)⁻¹**	**E(i) (MeV)**	**y(i) × E(i)**
β⁻ 2	1.00E 00	9.348E-01*	9.35E-01
Kα₁ X ray	5.79E-05	1.578E-02	9.13E-07
Kα₂ X ray	3.03E-05	1.569E-02	4.75E-07
Kβ₁ X ray	9.55E-06	1.767E-02	1.69E-07
Kβ₂ X ray	2.14E-06	1.797E-02	3.85E-08
Kβ₃ X ray	4.89E-06	1.765E-02	8.63E-08
Lα X ray	2.29E-06	2.042E-03*	4.67E-09
Lβ X ray	1.61E-06	2.130E-03*	3.43E-09
LISTED X, γ AND γ± RADIATIONS			1.69E-06
OMITTED X, γ AND γ± RADIATIONS**			1.25E-09
LISTED β, ce AND Auger RADIATIONS			9.35E-01
OMITTED β, ce AND Auger RADIATIONS**			3.10E-04
LISTED RADIATIONS			9.35E-01
OMITTED RADIATIONS**			3.10E-04

*AVERAGE ENERGY (MeV)
**EACH OMITTED TRANSITION CONTRIBUTES <0.100% TO Σy(i) × E(i) IN ITS CATEGORY.
ZIRCONIUM-90 DAUGHTER IS STABLE.

ZIRCONIUM-89

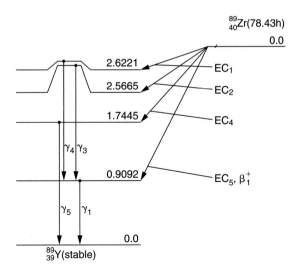

Half Life = 78.43 Hours			
Decay Mode(s): EC, β⁺			
Radiation	**y(i) (Bq•s)⁻¹**	**E(i) (MeV)**	**y(i) × E(i)**
β⁺ 1	2.26E-01	3.949E-01*	8.94E-02
γ±	4.53E-01	5.110E-01	2.31E-01
γ 1	9.99E-01	9.092E-01	9.08E-01
ce-K₂, γ 1	7.44E-03	8.922E-01	6.63E-03
ce-L₁, γ 1	8.26E-04	9.068E-01	7.49E-04
γ 3	9.99E-04	1.657E 00	1.66E-03
γ 4	7.69E-03	1.713E 00	1.32E-02
γ 5	1.30E-03	1.744E 00	2.26E-03
Kα₁ X ray	2.69E-01	1.496E-02	4.03E-03
Kα₂ X ray	1.40E-01	1.488E-02	2.09E-03
Auger-KLL	1.40E-01	1.262E-02*	1.76E-03
Auger-KLX	5.14E-02	1.453E-02*	7.47E-04
Auger-LMM	5.58E-01	1.664E-03*	9.28E-04
Auger-LMX	2.21E-01	1.960E-03*	4.34E-04
Auger-MXY	1.50E 00	2.250E 04	3.38E-04
LISTED X, γ AND γ± RADIATIONS			1.16E 00
OMITTED X, γ AND γ± RADIATIONS**			2.45E-03
LISTED β, ce AND Auger RADIATIONS			1.01E-01
OMITTED β, ce AND Auger RADIATIONS**			3.45E-04
LISTED RADIATIONS			1.26E 00
OMITTED RADIATIONS**			2.79E-03

Yttrium-89 DAUGHTER IS STABLE.
*AVERAGE ENERGY (MeV)
**EACH OMITTED TRANSITION CONTRIBUTES <0.100% TO Σy(i) × E(i) IN ITS CATEGORY.

MOLYBDENUM-99

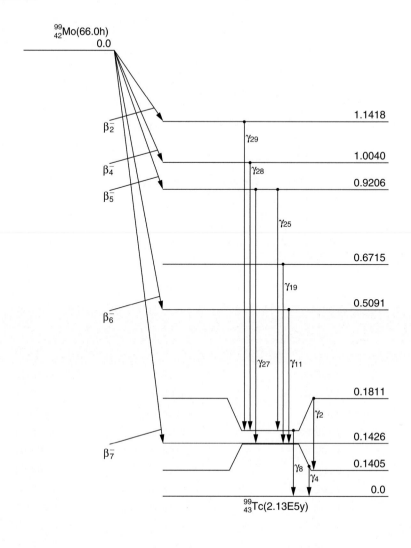

Half Life = 66 Hours			
Decay Mode(s): β⁻			
Radiation	$y(i)$ (Bq•s)$^{-1}$	$E(i)$ (MeV)	$y(i) \times E(i)$
β⁻ 5	1.66E-01	1.330E-01*	2.20E-02
β⁻ 6	1.17E-02	2.895E-01*	3.39E-03
β⁻ 7	8.20E-01	4.426E-01*	3.63E-01
γ 2	1.16E-02	4.059E-02	4.69E-04
ce-K, γ 2	3.77E-02	1.954E-02	7.38E-04
γ 4	4.95E-02	1.405E-01	6.95E-03
ce-K, γ 4	4.89E-03	1.194E-01	5.84E-04
γ 8	6.06E-02	1.811E-01	1.10E-02
ce-K, γ 8	7.62E-03	1.600E-01	1.22E-03
γ 11	1.19E-02	3.664E-01	4.37E-03
γ 19	5.45E-04	5.288E-01	2.88E-04
γ 24	2.60E-04	6.218E-01	1.61E-04
γ 25	1.22E-01	7.395E-01	9.02E-02
γ 27	4.32E-02	7.779E-01	3.36E-02
γ 28	1.33E-03	8.230E-01	1.09E-03
γ 29	9.76E-04	9.608E-01	9.37E-04
Kα₁ X ray	2.15E-02	1.837E-02	3.95E-04
Kα₂ X ray	1.13E-02	1.825E-02	2.06E-04
LISTED X, γ AND γ± RADIATIONS			1.50E-01
OMITTED X, γ AND γ± RADIATIONS**			4.87E-04
LISTED β, ce AND Auger RADIATIONS			3.91E-01
OMITTED β, ce AND Auger RADIATIONS**			1.33E-03
LISTED RADIATIONS			5.41E-01
OMITTED RADIATIONS**			1.82E-03

*AVERAGE ENERGY (MeV)
**EACH OMITTED TRANSITION CONTRIBUTES <0.100% TO $\Sigma y(i) \times E(i)$ IN ITS CATEGORY.
TECHNETIUM-99M DAUGHTER, YIELD 8.76E-01, IS RADIOACTIVE.
TECHNETIUM-99 DAUGHTER, YIELD 1.24E-01, IS RADIOACTIVE.

TECHNETIUM-99M

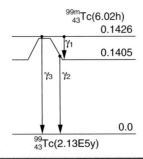

	Half Life = 6.02 Hours		
	Decay Mode(s): IT		
Radiation	$y(i)$ **(Bq•s)$^{-1}$**	$E(i)$ **(MeV)**	$y(i) \times E(i)$
ce-M, γ 1	9.14E-01	1.749E-03*	1.60E-03
ce-N$^+$, γ 1	7.57E-02	2.174E-03*	1.65E-04
γ 2	8.89E-01	1.405E-01	1.25E-01
ce-K, γ 2	8.79E-02	1.194E-01	1.05E-02
ce-L$_1$, γ 2	9.67E-03	1.374E-01	1.33E-03
ce-L$_2$, γ 2	6.10E-04	1.377E-01	8.40E-05
ce-L$_3$, γ 2	3.01E-04	1.378E-01	4.15E-05
ce-M, γ 2	1.92E-03	1.400E-01*	2.70E-04
ce-N$^+$, γ 2	3.71E-04	1.405E-01*	5.21E-05
ce-K, γ 3	6.91E-03	1.216E-01	8.41E-04
ce-L$_1$, γ 3	1.17E-03	1.396E-01	1.63E-04
ce-L$_2$, γ 3	2.43E-04	1.399E-01	3.39E-05
ce-L$_3$, γ 3	7.40E-04	1.400E-01	1.04E-04
ce-M, γ 3	4.19E-04	1.422E-01*	5.97E-05
Kα$_1$ X ray	4.03E-02	1.837E-02	7.39E-04
Kα$_2$ X ray	2.12E-02	1.825E-02	3.86E-04
Kβ$_1$ X ray	6.88E-03	2.062E-02	1.42E-04
Auger-KLL	1.45E-02	1.535E-02*	2.23E-04
Auger-KLX	5.76E-03	1.777E-02*	1.02E-04
Auger-LMM	7.10E-02	2.053E-03*	1.46E-04
Auger-LMX	3.05E-02	2.468E-03*	7.53E-05
Auger-MXY	1.11E 00	4.090E-04*	4.54E-04
LISTED X, γ AND γ± RADIATIONS			1.26E-01
OMITTED X, γ AND γ± RADIATIONS**			1.58E-04
LISTED β, ce AND Auger RADIATIONS			1.62E-02
OMITTED β, ce AND Auger RADIATIONS**			3.88E-05
LISTED RADIATIONS			1.42E-01
OMITTED RADIATIONS**			1.96E-04

*AVERAGE ENERGY (MeV)
**EACH OMITTED TRANSITION CONTRIBUTES <0.100% TO Σ$y(i) \times E(i)$ IN ITS CATEGORY.
TECHNETIUM-99 DAUGHTER IS RADIOACTIVE.

INDIUM-111

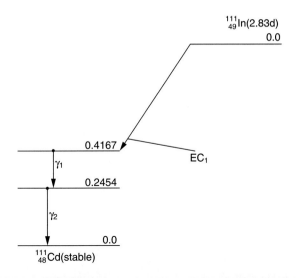

Half Life = 2.83 Days							
Decay Mode(s): EC							
Radiation	$y(i)$ (Bq•s)$^{-1}$	$E(i)$ (MeV)	$y(i) \times E(i)$	Radiation	$y(i)$ (Bq•s)$^{-1}$	$E(i)$ (MeV)	$y(i) \times E(i)$
γ 1	9.05E-01	1.713E-01	1.55E-01	Kβ$_2$ X ray	2.35E-02	2.664E-02	6.26E-04
ce-K, γ 1	8.27E-02	1.446E-01	1.19E-02	Kβ$_3$ X ray	4.14E-02	2.606E-02	1.08E-03
ce-L$_1$, γ 1	9.51E-03	1.673E-01	1.59E-03	Auger-KLL	1.06E-01	1.917E-02*	2.03E-03
ce-L$_2$, γ 1	5.32E-04	1.676E-01	8.91E-05	Auger-KLX	4.55E-02	2.232E-02*	1.02E-03
ce-M, γ 1	1.95E-03	1.707E-01*	3.33E-04	Auger-KXY	5.85E-03	2.544E-02*	1.49E-04
ce-N$^+$, γ 1	4.08E-04	1.713E-01*	6.99E-05	Auger-LMM	6.73E-01	2.590E-03*	1.74E-03
γ 2	9.40E-01	2.454E-01	2.31E-01	Auger-LMX	3.06E-01	3.187E-03*	9.75E-04
ce-K, γ 2	5.03E-02	2.187E-01	1.10E-02	Auger-LXY	3.86E-02	3.583E-03*	1.38E-04
ce-L$_1$, γ 2	5.15E-03	2.414E-01	1.24E-03	Auger-MXY	1.91E 00	5.104E-04*	9.75E-04
ce-L$_2$, γ 2	1.38E-03	2.417E-01	3.32E-04	LISTED X, γ AND γ± RADIATIONS			4.05E-01
ce-L$_3$, γ 2	1.32E-03	2.419E-01	3.19E-04	OMITTED X, γ AND γ± RADIATIONS**			2.00E-04
ce-M, γ 2	1.52E-03	2.448E-01*	3.71E-04	LISTED β, ce AND Auger RADIATIONS			3.44E-02
ce-N$^+$, γ 2	3.01E-04	2.454E-01*	7.39E-05	OMITTED β, ce AND Auger RADIATIONS**			3.14E-05
Kα$_1$ X ray	4.43E-01	2.317E-02	1.03E-02				
Kα$_2$ X ray	2.36E-01	2.298E-02	5.42E-03	LISTED RADIATIONS			4.40E-01
Kβ$_1$ X ray	8.07E-02	2.609E-02	2.10E-03	OMITTED RADIATIONS**			2.31E-04

*AVERAGE ENERGY (MeV)
**EACH OMITTED TRANSITION CONTRIBUTES <0.100% TO Σ$y(i) \times E(i)$ IN ITS CATEGORY.
CADMIUM-111 DAUGHTER IS STABLE.

IODINE-123

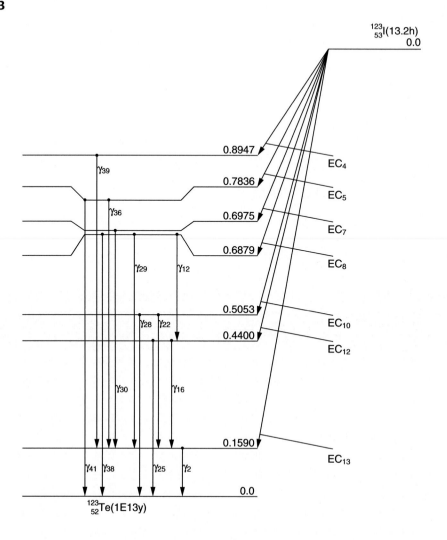

Half Life = 13.2 Hours			
Decay Mode(s): EC			
Radiation	**_y(i)_ (Bq•s)⁻¹**	**_E(i)_ (MeV)**	**_y(i)_ × _E(i)_**
γ 2	8.28E-01	1.590E-01	1.32E-01
ce-K, γ 2	1.35E-01	1.272E-01	1.72E-02
ce-L₁, γ 2	1.60E-02	1.540E-01	2.47E-03
ce-L₂, γ 2	1.09E-03	1.544E-01	1.69E-04
ce-L₃, γ 2	3.46E-04	1.546E-01	5.35E-05
ce-M, γ 2	3.46E-03	1.582E-01*	5.48E-04
ce-N⁺, γ 2	8.27E-04	1.590E-01*	1.32E-04
γ 12	7.07E-04	2.480E-01	1.75E-04
γ 16	7.86E-04	2.810E-01	2.21E-04
γ 22	1.25E-03	3.463E-01	4.33E-04
γ 25	4.25E-03	4.400E-01	1.87E-03
γ 28	3.14E-03	5.053E-01	1.59E-03
γ 29	1.38E-02	5.290E-01	7.31E-03
ce-K, γ 29	9.90E-05	4.971E-01	4.92E-05
γ 30	3.79E-03	5.385E-01	2.04E-03
γ 36	8.28E-04	6.246E-01	5.17E-04
γ 38	2.66E-04	6.879E-01	1.83E-04
γ 39	6.12E-04	7.358E-01	4.50E-04
γ 41	5.90E-04	7.836E-01	4.62E-04
Kα₁ X ray	4.58E-01	2.747E-02	1.26E-02
Kα₂ X ray	2.46E-01	2.720E-02	6.70E-03
Kβ₁ X ray	8.66E-02	3.100E-02	2.69E-03
Kβ₂ X ray	2.66E-02	3.171E-02	8.43E-04
Kβ₃ X ray	4.46E-02	3.094E-02	1.38E-03
Auger-KLL	8.15E-02	2.254E-02*	1.84E-03
Auger-KLX	3.69E-02	2.635E-02*	9.73E-04
Auger-KXY	4.92E-03	3.013E-02*	1.48E-04
Auger-LMM	6.06E-01	3.080E-03*	1.87E-03
Auger-LMX	3.11E-01	3.849E-03*	1.20E-03
Auger-LXY	4.40E-02	4.380E-03*	1.93E-04
Auger-MXY	1.80E 00	6.991E-04*	1.26E-03
LISTED X, γ AND γ± RADIATIONS			1.71E-01
OMITTED X, γ AND γ± RADIATIONS**			6.76E-04
LISTED β, ce AND Auger RADIATIONS			2.80E-02
OMITTED β, ce AND Auger RADIATIONS**			1.21E-04
LISTED RADIATIONS			1.99E-01
OMITTED RADIATIONS**			7.97E-04

*AVERAGE ENERGY (MeV)
**EACH OMITTED TRANSITION CONTRIBUTES <0.100% TO Σ_y(i) × E(i) IN ITS CATEGORY.
TELLURIUM-123M DAUGHTER, YIELD 5.00E-05, IS RADIOACTIVE.
TELLURIUM-123 DAUGHTER, YIELD 9.999E-01, IS RADIOACTIVE.

IODINE-124

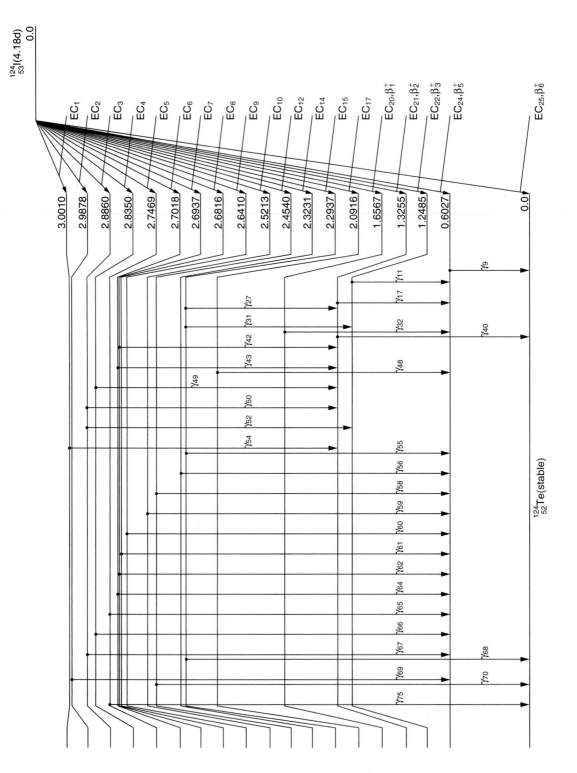

Half Life = 4.18 Days

Decay Mode(s): EC, β+

Radiation	$y(i)$ (Bq·s)$^{-1}$	$E(i)$ (MeV)	$y(i) \times E(i)$
β+ 2	2.92E-03	3.657E-01*	1.07E-03
β+ 5	1.12E-01	6.859E-01*	7.66E-02
β+ 6	1.12E-01	9.736E-01*	1.09E-01
γ±	4.53E-01	5.110E-01	2.31E-01
γ 9	6.11E-01	6.027E-01	3.68E-01
ce-K, γ 9	2.57E-03	5.709E-01	1.47E-03
γ 11	9.56E-03	6.458E-01	6.17E-03
γ 17	1.01E-01	7.228E-01	7.27E-02
ce-K, γ 17	3.42E-04	6.910E-01	2.36E-04
γ 27	4.20E-03	9.682E-01	4.06E-03
γ 31	4.30E-03	1.045E 00	4.49E-03
γ 32	1.20E-03	1.054E 00	1.26E-03
γ 40	1.45E-02	1.325E 00	1.92E-02
γ 42	2.90E-03	1.368E 00	3.97E-03
γ 43	1.69E-02	1.376E 00	2.32E-02
γ 48	1.80E-03	1.489E 00	2.68E-03
γ 49	3.03E-02	1.509E 00	4.57E-02
γ 50	1.70E-03	1.560E 00	2.65E-03
γ 52	2.00E-03	1.638E 00	3.27E-03
γ 54	1.10E-03	1.676E 00	1.84E-03
γ 55	1.06E-01	1.691E 00	1.79E-01
γ 56	1.70E-03	1.720E 00	2.92E-03
γ 58	2.10E-03	1.851E 00	3.89E-03
γ 59	1.60E-03	1.919E 00	3.07E-03
γ 60	3.40E-03	2.038E 00	6.93E-03

Radiation	$y(i)$ (Bq·s)$^{-1}$	$E(i)$ (MeV)	$y(i) \times E(i)$
γ 61	3.50E-03	2.079E 00	7.27E-03
γ 62	5.80E-03	2.091E 00	1.21E-02
γ 64	1.40E-03	2.099E 00	2.94E-03
γ 65	1.10E-03	2.144E 00	2.36E-03
γ 66	5.80E-03	2.232E 00	1.29E-02
γ 67	6.70E-03	2.283E 00	1.53E-02
γ 68	1.00E-03	2.294E 00	2.29E-03
γ 69	2.00E-03	2.385E 00	4.77E-03
γ 70	1.10E-03	2.454E 00	2.70E-03
γ 75	5.70E-03	2.747E 00	1.57E-02
Kα₁ X ray	3.08E-01	2.747E-02	8.45E-03
Kα₂ X ray	1.65E-01	2.720E-02	4.49E-03
Kβ X ray	5.81E-02	3.100E-02	1.80E-03
Auger-KLL	5.47E-02	2.254E-02*	1.23E-03
Auger-KLX	2.48E-02	2.635E-02*	6.53E-04
Auger-LMM	4.07E-01	3.080E-03*	1.25E-03
Auger-LMX	2.09E-01	3.849E-03*	8.03E-04
Auger-MXY	1.21E 00	6.991E-04*	8.44E-04
LISTED X, γ AND γ± RADIATIONS			1.08E 00
OMITTED X, γ AND γ± RADIATIONS**			1.82E-02
LISTED β, ce AND Auger RADIATIONS			1.93E-01
OMITTED β, ce AND Auger RADIATIONS**			7.71E-04
LISTED RADIATIONS			1.27E 00
OMITTED RADIATIONS**			1.90E-02

*AVERAGE ENERGY (MeV)
**EACH OMITTED TRANSITION CONTRIBUTES <0.100% TO $\Sigma y(i) \times E(i)$ IN ITS CATEGORY.

IODINE-125

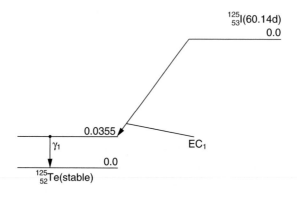

Half Life = 60.14 Days			
Decay Mode(s): EC			
Radiation	$y(i)$ $(Bq \cdot s)^{-1}$	$E(i)$ (MeV)	$y(i) \times E(i)$
$\gamma\,1$	6.67E-02	3.549E-02	2.37E-03
ce-K, $\gamma\,1$	8.03E-01	3.678E-03	2.95E-03
ce-L_1, $\gamma\,1$	9.52E-02	3.055E-02	2.91E-03
ce-L_2, $\gamma\,1$	7.64E-03	3.088E-02	2.36E-04
ce-L_3, $\gamma\,1$	1.91E-03	3.115E-02	5.96E-05
ce-M, $\gamma\,1$	2.09E-02	3.467E-02*	7.25E-04
ce-N^+, $\gamma\,1$	4.96E-03	3.549E-02*	1.76E-04
$K\alpha_1$ X ray	7.41E-01	2.747E-02	2.04E-02
$K\alpha_2$ X ray	3.98E-01	2.720E-02	1.08E-02
$K\beta_1$ X ray	1.40E-01	3.100E-02	4.34E-03
$K\beta_2$ X ray	4.30E-02	3.171E-02	1.36E-03
$K\beta_3$ X ray	7.20E-02	3.094E-02	2.23E-03
$K\beta_5$ X ray	1.44E-03	3.124E-02	4.51E-05
$L\alpha$ X ray	6.14E-02	3.768E-03*	2.31E-04
$L\beta$ X ray	5.93E-02	4.092E-03*	2.43E-04
Auger-KLL	1.32E-01	2.254E-02*	2.97E-03
Auger-KLX	5.97E-02	2.635E-02*	1.57E-03

IODINE-125—cont'd

Half Life = 60.14 Days			
Decay Mode(s): EC			
Radiation	**$y(i)$ (Bq•s)$^{-1}$**	**$E(i)$ (MeV)**	**$y(i) \times E(i)$**
Auger-KXY	7.95E-03	3.013E-02*	2.40E-04
Auger-LMM	1.01E 00	3.086E-03*	3.11E-03
Auger-LMX	5.17E-01	3.855E-03*	1.99E-03
Auger-LXY	7.33E-02	4.386E-03*	3.21E-04
Auger-MXY	2.99E 00	6.989E-04*	2.09E-03
ΔE	6.22E-01	5.577E-05*	3.47E-05
LISTED X, γ AND $\gamma\pm$ RADIATIONS			4.20E-02
OMITTED X, γ AND $\gamma\pm$ RADIATIONS**			4.58E-05
LISTED β, ce AND Auger RADIATIONS			1.94E-02
LISTED RADIATIONS			6.14E-02
OMITTED RADIATIONS**			4.58E-05

*AVERAGE ENERGY (MeV)
**EACH OMITTED TRANSITION CONTRIBUTES <0.100% TO $\Sigma y(i) \times E(i)$ IN ITS CATEGORY.
TELLURIUM-125 DAUGHTER IS STABLE.

IODINE-131

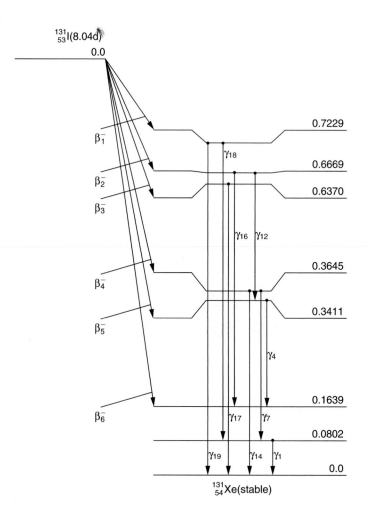

Half Life = 8.04 Days			
Decay Mode(s): β⁻			
Radiation	**$y(i)$ (Bq•s)⁻¹**	**$E(i)$ (MeV)**	**$y(i) \times E(i)$**
β⁻ 1	2.13E-02	6.935E-02*	1.48E-03
β⁻ 2	6.20E-03	8.693E-02*	5.39E-04
β⁻ 3	7.36E-02	9.660E-02*	7.11E-03
β⁻ 4	8.94E-01	1.915E-01*	1.71E-01
β⁻ 6	4.20E-03	2.832E-01*	1.19E-03
γ 1	2.62E-02	8.018E-02	2.10E-03
ce-K, γ 1	3.63E-02	4.562E-02	1.66E-03
ce-L₁, γ 1	4.30E-03	7.473E-02	3.21E-04
γ 4	2.65E-03	1.772E-01	4.70E-04
γ 7	6.06E-02	2.843E-01	1.72E-02
ce-K, γ 7	2.48E-03	2.497E-01	6.20E-04
γ 12	2.51E-03	3.258E-01	8.18E-04
γ 14	8.12E-01	3.645E-01	2.96E-01
ce-K, γ 14	1.55E-02	3.299E-01	5.10E-03
ce-L₁, γ 14	1.71E-03	3.590E-01	6.13E-04
γ 16	3.61E-03	5.030E-01	1.82E-03
γ 17	7.27E-02	6.370E-01	4.63E-02
γ 18	2.20E-03	6.427E-01	1.41E-03
γ 19	1.80E-02	7.229E-01	1.30E-02
Kα₁ X ray	2.59E-02	2.978E-02	7.72E-04
Kα₂ X ray	1.40E-02	2.946E-02	4.12E-04
LISTED X, γ AND γ± RADIATIONS			3.80E-01
OMITTED X, γ AND γ± RADIATIONS**			1.09E-03
LISTED β, ce AND Auger RADIATIONS			1.90E-01
OMITTED β, ce AND Auger RADIATIONS**			1.86E-03
LISTED RADIATIONS			5.70E-01
OMITTED RADIATIONS**			2.95E-03

*AVERAGE ENERGY (MeV)
**EACH OMITTED TRANSITION CONTRIBUTES <0.100% TO Σ$y(i) \times E(i)$ IN ITS CATEGORY.
XENON-131M DAUGHTER, YIELD 1.11E-02, IS RADIOACTIVE.
XENON-131 DAUGHTER, YIELD 9.889E-01, IS STABLE.

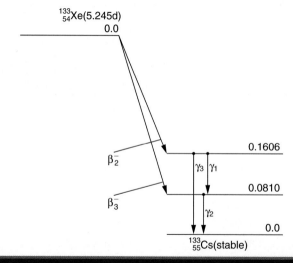

	Half Life = 5.245 Days		
	Decay Mode(s): β⁻		
Radiation	$y(i)$ (Bq•s)⁻¹	$E(i)$ (MeV)	$y(i) \times E(i)$
β⁻ 2	6.60E-03	7.502E-02*	4.95E-04
β⁻ 3	9.93E-01	1.005E-01*	9.98E-02
γ 1	2.11E-03	7.962E-02	1.68E-04
ce-K, γ 1	3.17E-03	4.364E-02	1.38E-04
γ 2	3.74E-01	8.100E-02	3.03E-02
ce-K, γ 2	5.35E-01	4.501E-02	2.41E-02
ce-L₁, γ 2	6.52E-02	7.528E-02	4.91E-03
ce-L₂, γ 2	4.91E-03	7.564E-02	3.72E-04
ce-M, γ 2	1.45E-02	8.000E-02*	1.16E-03
ce-N⁺, γ 2	3.80E-03	8.100E-02*	3.08E-04
γ 3	6.20E-04	1.606E-01	9.96E-05
Kα₁ X ray	2.53E-01	3.097E-02	7.85E-03
Kα₂ X ray	1.37E-01	3.063E-02	4.20E-03
Kβ₁ X ray	4.89E-02	3.499E-02	1.71E-03
Kβ₂ X ray	1.70E-02	3.584E-02	6.08E-04
Kβ₃ X ray	2.52E-02	3.492E-02	8.80E-04
Lα X ray	2.44E-02	4.285E-03*	1.04E-04
Lα X ray	2.32E-02	4.694E-03*	1.09E-04
Auger-KLL	3.69E-02	2.524E-02*	9.31E-04
Auger-KLX	1.73E-02	2.961E-02*	5.12E-04
Auger-LMM	3.03E-01	3.441E-03*	1.04E-03
Auger-LMX	1.72E-01	4.344E-03*	7.45E-04
Auger-MXY	9.33E-01	8.695E-04*	8.11E-04
LISTED X, γ AND γ± RADIATIONS			4.60E-02
OMITTED X, γ AND γ± RADIATIONS**			6.66E-05
LISTED β, ce AND Auger RADIATIONS			1.35E-01
OMITTED β, ce AND Auger RADIATIONS**			3.82E-04
LISTED RADIATIONS			1.81E-01
OMITTED RADIATIONS**			4.49E-04

*AVERAGE ENERGY (MeV)
**EACH OMITTED TRANSITION CONTRIBUTES <0.100% TO Σy(i) × E(i) IN ITS CATEGORY.
CESIUM-133 DAUGHTER IS STABLE.

CESIUM-137

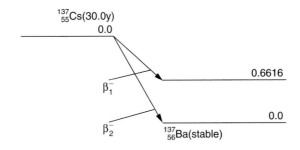

Half Life = 30 Years			
Decay Mode(s): β⁻			
Radiation	**$y(i)$ (Bq•s)⁻¹**	**$E(i)$ (MeV)**	**$y(i) \times E(i)$**
β⁻ 1	9.46E-01	1.734E-01*	1.64E-01
β⁻ 2	5.40E-02	4.246E-01*	2.29E-02
LISTED β, ce AND Auger RADIATIONS			1.87E-01
LISTED RADIATIONS			1.87E-01

*AVERAGE ENERGY (MeV)

BARIUM-137M DAUGHTER, YIELD 9.46E-01, IS RADIOACTIVE.

BARIUM-137 DAUGHTER, YIELD 5.40E-02, IS STABLE.

THALLIUM-201

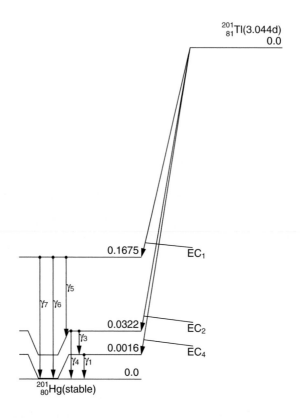

Half Life = 3.044 Days			
Decay Mode(s): EC			
Radiation	**$y(i)$ (Bq•s)$^{-1}$**	**$E(i)$ (MeV)**	**$y(i) \times E(i)$**
ce-N$^+$, γ 1	6.10E-01	1.570E-03*	9.58E-04
ce-L$_1$, γ 3	7.30E-02	1.576E-02	1.15E-03
ce-L$_2$, γ 3	7.55E-03	1.639E-02	1.24E-04
ce-M, γ 3	1.90E-02	2.775E-02*	5.27E-04
ce-N$^+$, γ 3	6.78E-03	3.060E-02*	2.07E-04
ce-L$_1$, γ 4	6.28E-02	1.735E-02	1.09E-03
ce-L$_2$, γ 4	6.53E-03	1.798E-02	1.17E-04
ce-M, γ 4	1.63E-02	2.934E-02*	4.80E-04
ce-N$^+$, γ 4	5.74E-03	3.219E-02*	1.85E-04
γ 5	2.65E-02	1.353E-01	3.59E-03
ce-K, γ 5	7.47E-02	5.224E-02	3.90E-03
ce-L$_1$, γ5	1.14E-02	1.205E-01	1.37E-03
ce-L$_2$, γ 5	1.20E-03	1.211E-01	1.45E-04
ce-M, γ 5	2.97E-03	1.325E-01*	3.93E-04
ce-N$^+$, γ 5	9.58E-04	1.353E-01*	1.30E-04
γ 6	1.60E-03	1.659E-01	2.65E-04
ce-K, γ 6	2.26E-03	8.278E-02	1.87E-04
ce-L$_1$, γ 6	3.42E-04	1.510E-01	5.16E-05

THALLIUM-201—cont'd

Half Life = 3.044 Days			
Decay Mode(s): EC			
Radiation	y(i) (Bq•s)$^{-1}$	E(i) (MeV)	y(i) × E(i)
γ 7	1.00E-01	1.674E-01	1.67E-02
ce-K, γ 7	1.54E-01	8.433E-02	1.30E-02
ce-L$_1$, γ 7	2.35E-02	1.526E-01	3.58E-03
ce-L$_2$, γ 7	2.48E-03	1.532E-01	3.80E-04
ce-L$_3$, γ 7	2.80E-04	1.551E-01	4.35E-05
ce-M, γ 7	6.10E-03	1.646E-01*	1.00E-03
ce-N$^+$, γ 7	1.96E-03	1.674E-01*	3.29E-04
Kα$_1$ X ray	4.62E-01	7.082E-02	3.27E-02
Kα$_2$ X ray	2.72E-01	6.889E-02	1.87E-02
Kβ$_1$ X ray	1.05E-01	8.026E-02	8.45E-03
Kβ$_2$ X ray	4.43E-02	8.258E-02	3.66E-03
Kβ$_3$ X ray	5.48E-02	7.982E-02	4.37E-03
Kβ$_5$ X ray	2.88E-03	8.077E-02	2.33E-04
Lα X ray	1.90E-01	9.980E-03*	1.90E-03
Lβ X ray	1.82E-01	1.185E-02*	2.15E-03
Lγ X ray	3.40E-02	1.397E-02*	4.75E-04
Auger-KLL	2.01E-02	5.526E-02*	1.11E-03
Auger-KLX	1.12E-02	6.652E-02*	7.45E-04
Auger-KXY	1.80E-03	7.733E-02*	1.39E-04
Auger-LMM	4.30E-01	7.753E-03*	3.34E-03
Auger-LMX	2.78E-01	1.022E-02*	2.84E-03
Auger-LXY	4.63E-02	1.214E-02*	5.62E-04
Auger-MXY	1.74E 00	2.673E-03*	4.66E-03
ΔE	1.05E 00	5.204E-04*	5.45E-04
LISTED X, γ AND γ± RADIATIONS			9.32E-02
OMITTED X, γ AND γ± RADIATIONS**			2.52E-04
LISTED β, ce AND Auger RADIATIONS			4.33E-02
OMITTED β, ce AND Auger RADIATIONS**			1.16E-04
LISTED RADIATIONS			1.37E-01
OMITTED RADIATIONS**			3.68E-04

*AVERAGE ENERGY (MeV)
**EACH OMITTED TRANSITION CONTRIBUTES <0.100% TO Σy(i) × E(i) IN ITS CATEGORY.
MERCURY-201 DAUGHTER IS STABLE.

appendix

D

Mass Attenuation Coefficients for Water, NaI(Tl), $Bi_4Ge_3O_{12}$, $Cd_{0.8}Zn_{0.2}Te$, and Lead

Photon Energy (MeV)	Attenuation Coefficient, μ (cm²/g)*				
	H_2O	NaI(Tl)	$Bi_4Ge_3O_{12}$	$Cd_{0.8}Zn_{0.2}Te$	Pb
ρ (g/cm³) →	1.00	3.67	7.13	5.81	11.34
0.02	0.721	20.7	65.8	21.4	84.0
0.0267	—	—	—	9.5	—
0.0267	—	—	—	25.8	—
0.03	0.329	6.71	22.6	19.1	28.9
0.0318	—	—	—	16.3	—
0.0318	—	—	—	33.2	—
0.0332	—	5.08	—	—	—
0.0332	—	29.87	—	—	—
0.04	0.240	18.35	10.5	18.2	13.4
0.05	0.208	10.18	5.76	10.0	7.39
0.06	0.192	6.23	3.53	6.10	4.53
0.08	0.176	2.86	1.66	2.78	2.11
0.088	—	—	—	—	1.65
0.088	—	—	—	—	7.42
0.0905	—	—	1.21	—	—
0.0905	—	—	4.92	—	—
0.1	0.165	1.58	3.82	1.52	5.34
0.15	0.148	0.566	1.39	0.544	1.91
0.2	0.136	0.302	0.696	0.290	0.936
0.3	0.118	0.153	0.294	0.148	0.373
0.4	0.106	0.110	0.179	0.107	0.215
0.5	0.0967	0.0904	0.131	0.0877	0.150

Photon Energy (MeV)	Attenuation Coefficient, μ (cm^2/g)*				
	H_2O	NaI(Tl)	$Bi_4Ge_3O_{12}$	$Cd_{0.8}Zn_{0.2}Te$	Pb
$\rho\ (g/cm^3) \rightarrow$	1.00	3.67	7.13	5.81	11.34
0.6	0.0894	0.0790	0.105	0.0769	0.117
0.8	0.0786	0.0657	0.0794	0.0641	0.0841
1.0	0.0707	0.0576	0.0661	0.0562	0.0680
1.022	0.0699	0.0569	0.0650	0.0556	0.0668
1.5	0.0575	0.0464	0.0506	0.0454	0.0509
2.0	0.0494	0.0412	0.0447	0.0404	0.0453
3.0	0.0397	0.0367	0.0402	0.0363	0.0420
4.0	0.0340	0.0351	0.0388	0.0350	0.0418
5.0	0.0303	0.0347	0.0387	0.0349	0.0426
6.0	0.0277	0.0348	0.0391	0.0352	0.0438
8.0	0.0243	0.0358	0.0406	0.0365	0.0467
10.0	0.0222	0.0372	0.0424	0.0382	0.0497

*Values without coherent scattering, obtained from reference 2 in Chapter 6.

H_2O, water; NaI(Tl), thallium-doped sodium iodide; $Bi_4Ge_3O_{12}$, bismuth germanate; $Cd_{0.8}Zn_{0.2}Te$, cadmium zinc telluride; Pb, lead.

E

Effective Dose Equivalent (mSv/MBq) and Radiation Absorbed Dose Estimates (mGy/MBq) to Adult Subjects from Selected Internally Administered Radiopharmaceuticals

Radiopharmaceutical	Route of Administration	Effective Dose Equivalent* (mSv/MBq)	Absorbed Dose for Selected Organs (mGy/MBq)†	
			Organ	Dose
^{13}N-ammonia	Intravenous	2.2×10^{-3}	Bladder wall	6.9×10^{-3}
			Brain	4.7×10^{-3}
			Liver	3.8×10^{-3}
			Gonads	1.7×10^{-3}
			Red marrow	1.8×10^{-3}
^{15}O-water	Intravenous	1.1×10^{-3}	Heart wall	2.2×10^{-3}
			Kidneys	1.9×10^{-3}
			Lungs	1.9×10^{-3}
			Gonads	6.7×10^{-4}
			Red marrow	9.0×10^{-4}
^{18}F-fluorodeoxyglucose (FDG)	Intravenous	3.0×10^{-2}	Bladder wall	1.9×10^{-1}
			Heart wall	6.0×10^{-2}
			Spleen	3.7×10^{-2}
			Gonads	1.7×10^{-2}
			Red marrow	1.3×10^{-2}
^{67}Ga-citrate	Intravenous	1.1×10^{-1}	Bone surfaces	3.2×10^{-1}
			LLI wall	2.6×10^{-1}
			ULI wall	1.5×10^{-1}
			Gonads	8.7×10^{-2}
			Red marrow	1.2×10^{-1}

Radiopharmaceutical	Route of Administration	Effective Dose Equivalent* (mSv/MBq)	Absorbed Dose for Selected Organs (mGy/MBq)†	
			Organ	*Dose*
^{99m}Tc-pertechnetate	Intravenous	1.1×10^{-2}	Bladder wall	3.6×10^{-2}
			ULI wall	2.8×10^{-2}
			LLI wall	2.7×10^{-2}
			Gonads	8.6×10^{-3}
			Red marrow	3.3×10^{-3}
^{99m}Tc-MDP	Intravenous	6.1×10^{-3}	Bone surfaces	3.5×10^{-2}
			Bladder wall	3.3×10^{-2}
			Kidneys	8.6×10^{-3}
			Gonads	3.3×10^{-3}
			Red marrow	5.4×10^{-3}
^{99m}Tc-sestamibi	Intravenous	1.5×10^{-2}	ULI wall	5.0×10^{-2}
			Bladder wall	3.7×10^{-2}
			LLI wall	3.7×10^{-2}
			Gonads	1.4×10^{-2}
			Red marrow	4.5×10^{-3}
^{99m}Tc-DTPA	Intravenous	8.2×10^{-3}	Bladder wall	7.7×10^{-2}
			Uterus	1.0×10^{-2}
			Kidneys	5.7×10^{-3}
			Gonads	5.5×10^{-3}
			Red marrow	2.2×10^{-3}
^{99m}Tc-mertiatide (MAG3)	Intravenous	1.2×10^{-2}	Bladder wall	1.4×10^{-1}
			Uterus	1.5×10^{-2}
			LLI wall	7.1×10^{-3}
			Gonads	6.6×10^{-3}
			Red marrow	1.1×10^{-3}
^{99m}Tc-exametazime (HMPAO)	Intravenous	1.4×10^{-2}	Gallbladder	5.1×10^{-2}
			Kidneys	3.5×10^{-2}
			Bladder wall	2.8×10^{-2}
			Gonads	7.0×10^{-3}
			Red marrow	3.5×10^{-3}
^{111}In-DTPA	Intravenous	4.1×10^{-2}	Bladder wall	4.3×10^{-1}
			Uterus	5.4×10^{-2}
			LLI wall	2.8×10^{-2}
			Gonads	2.5×10^{-2}
			Red marrow	8.2×10^{-3}
^{123}I-sodium iodide	Oral	1.2×10^{-1}	Thyroid	3.4×10^{0}
			Bladder wall	9.6×10^{-2}
			Stomach	5.5×10^{-2}
			Gonads	1.2×10^{-2}
			Red marrow	5.8×10^{-3}
^{131}I-sodium iodide	Oral	1.1×10^{1}	Thyroid	3.4×10^{2}
			Bladder wall	6.2×10^{-1}
			Stomach	3.6×10^{-1}
			Gonads	4.7×10^{-3}
			Red marrow	8.3×10^{-2}
^{201}Tl-thallium chloride	Intravenous	1.6×10^{-1}	Thyroid	6.2×10^{-1}
			Kidneys	4.6×10^{-1}
			Small intestine	4.5×10^{-1}
			Gonads	2.0×10^{-1}
			Red marrow	5.5×10^{-2}

Data from Stabin MG, Stubbs JB, Toohey RE: Radiation Dose Estimates for Radiopharmaceuticals, Oak Ridge Institute for Science and Education, NUREG/CR-6345, 1996. Available electronically at http://www.nrc.gov/reading-rm/doc-collections/nuregs/contract/cr6345/. Accessed 5 November 2011.

DTPA, diethylenetriaminepenta acetic acid; HMPAO, hexamethylpropyleneamine oxime; LLI, lower large intestine; MDP, methylene diphosphonate; ULI, upper large intestine.

*Effective dose equivalent has now been replaced by the quantity effective dose which uses somewhat different tissue weighting factors (see Chapter 22, Section B.7).

†Three organs with highest absorbed doses and the absorbed dose to red marrow and the gonads which are both radiosensitive. Values given for gonads are the higher of the absorbed dose estimates for the testes and ovaries.

The Fourier Transform

Fourier transforms (FTs) play an important role in tomographic reconstruction (see Chapter 16) and in computer implementation of convolutions (see Appendix G). As well, they are the basis for the modulation transfer function (MTF), one of the methods for evaluating spatial resolution of imaging systems (see Chapter 15) and many other methods for image analysis and image processing. This appendix is intended to provide an intuitive description of what the FT is, how it is calculated, and some of its properties. For simplicity, the analysis is presented for one-dimensional examples. The extension to higher dimensions is relatively straightforward. A detailed treatment is beyond the scope of this text. The interested reader is referred to the excellent texts by Bracewell for further information.[1,2]

A. THE FOURIER TRANSFORM: WHAT IT REPRESENTS

The FT is an alternative manner for representing a mathematical function or mathematical data. For example, suppose that the function $f(x)$ represents an image intensity profile. It can be shown that, so long as $f(x)$ has "reasonable properties," the profile can be represented as a sum of sine and cosine functions of different frequencies extending along the x-axis. The FT of $f(x)$, denoted as $F(k)$, represents the amplitudes of the sine and cosine functions for different *spatial frequencies, k.* Spatial frequency reflects how rapidly a sine or cosine function oscillates along the x-axis and has units of "cycles per distance," such as cycles per cm,* or cm^{-1}.

The concept of spatial frequency is illustrated in Figure F-1. Slow oscillations represent low spatial frequencies and rapid oscillations represent high frequencies. If $f(x)$ represents an image profile, the former would represent primarily the coarse structures, whereas the latter would represent fine details.

Thus $F(k)$ is a representation of the image profile in *k-space,* or *spatial-frequency space,* whereas $f(x)$ is a representation of the profile in *object space* (sometimes also called *distance space*). Either $f(x)$ or $F(k)$ is a valid representation of the image intensity profile and, as shown subsequently, either one can be derived from the other. Another notation for the FT is

$$F(k) = \mathscr{F}[f(x)] \qquad \text{(F-1)}$$

where the symbol $\mathscr{F}$ denotes the *operation* of computing an FT.

FTs can be computed for functions in other coordinate spaces as well. For example, in audio technology, signal intensity varies as a function of time, $t,$ and its FT describes the function in terms of temporal frequencies, $v,$ expressed as cycles per second, or Hz (sec^{-1}). Coordinate pairs that are represented by a function and its FT, such as x and $k,$ or t and $v,$ are referred to as *conjugate variables.*

B. CALCULATING FOURIER TRANSFORMS

FTs can be calculated for continuous functions, which have values for all values of $x,$ or for discrete functions, which have values only at discrete *sampling intervals* Δx. A discrete function may be a sampled version of a continuous function. Continuous FTs often are employed for theoretical modeling or analysis. However, because the signals of interest in nuclear medicine imaging always are sampled (or "digitized") in projection bins, image pixels,

---

*As noted in Chapter 15, the notation k is used in physics to denote "spatial radians per distance" and the notation $\bar{k}$, or "k-bar," is used to denote "cycles per distance." Mathematically, $\bar{k} = k/2\pi$, because there are 2π radians per cycle. However, for notational simplicity, we use k for "cycles per distance" in this text.

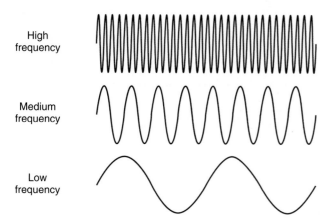

High frequency

Medium frequency

Low frequency

x

FIGURE F-1 Sinusoidal functions representing low, medium, and high spatial frequencies.

and so on, we will restrict our analysis to the *discrete Fourier transform* (DFT).

In practice, a function representing a signal, such as an image intensity profile $f(x)$, is sampled at a finite number of points, N, at equally spaced intervals Δx. The range of x over which the function is sampled, $N\Delta x$, is the field-of-view along x, denoted by FOV_x. For detectors with discrete detector elements, the sampling interval generally is the width of a detector element and the number of samples is the number of elements across the detector array. For gamma cameras, N and Δx (or FOV_x) can be chosen somewhat arbitrarily under computer control. In this case, FOV_x may or may not equal the physical field of view of the detector. It simply refers to the distance along the x-axis over which data are sampled, which may be smaller (or even larger) than the physical dimensions of the detector. Although N can be any integer, it usually is a power of 2 (64, 128, 256, etc.). Using a power of 2 allows one to use a very fast computational algorithm called the *fast Fourier transform* (FFT). FFTs are provided in many mathematical software packages. The basic algorithm is described in reference 3.

Consider the case in which $f(x)$ is sampled at N points over a range of x values from $x = 0$ to $(N-1)\Delta x$. The sampled data are represented as

$$f_j = f(j\Delta x) \quad j = 0, 1, 2, \ldots N-1 \quad \text{(F-2)}$$

Given N sampled points, the discrete FT of f is calculated from the following equation

$$F(m\Delta k) = F_m = (1/N)\sum_{j=0}^{N-1} f_j e^{-i2\pi mj/N}$$
$$m = 0, 1, 2, \ldots N-1 \quad \text{(F-3)}$$

where $i = \sqrt{-1}$ is a complex ("imaginary") number. Thus values for F are computed at N points at intervals Δk along the k-axis, ranging from $k = 0$ to $k_{\max} = (N-1)\Delta k$. The "zero-frequency" term actually represents a constant equal to the average value of $f(x)$ within the digitized field of view, FOV_x. We will designate the range of k-space over which the FT is computed, $N\Delta k$, as FOV_k (i.e., the "field of view" in k-space).

The complex exponential term in Equation F-3 represents *Euler's equation*, which states that

$$e^{i\theta} = \cos\theta + i\sin\theta \quad \text{(F-4)}$$

Thus in general, the values of F_m are complex numbers with "real" and "imaginary" parts. This does not imply that the FT is some sort of "imaginary" entity. Rather, the use of complex numbers in Equation F-3 provides a mathematically convenient way for computing the amplitudes of "cosine" functions versus "sine" functions in the representation of $f(x)$.

Conversely, if one is provided with a set of N values for F, they can be used to compute N values for f using the *inverse* discrete FT, described by

$$f(j\Delta x) = f_j = \sum_{\kappa=0}^{N-1} F_\kappa e^{i2\pi j\kappa/N} \quad \text{(F-5)}$$

This equation differs from Equation F-3 only in the sign of the exponent and the appearance of the $(1/N)$ term in the former. (Some alternative formulations for Equations F-3 and F-5 show different placements for N.) To distinguish between the two, Equation F-3 sometimes is referred to as the *forward* FT.

Finally, if the original data are properly sampled (e.g., if they meet the requirements of the sampling theorem discussed later), the inverse FT of the forward FT returns an exact replica of the original data. In other words,

$$\mathscr{F}^{-1}[\mathscr{F}[f(x)]] = f(x) \qquad \text{(F-6)}$$

C. SOME PROPERTIES OF FOURIER TRANSFORMS

Equations F-3 and F-5 are mathematically powerful, but not exactly transparent regarding how they operate. Therefore we will not attempt to analyze them in any detail. Rather, we will focus only on some properties of the discrete FT.

As noted previously, the FT of a function that is sampled at N points also has N values. It can be shown that the following relationships apply regarding data intervals and range in k-space:

$$\Delta k = 1/\text{FOV}_x \qquad \text{(F-7)}$$

$$\text{FOV}_k = 1/\Delta x \qquad \text{(F-8)}$$

Thus the *data interval in k-space* is inversely proportional to the *data range in object space,* and conversely, the *data range in k-space* is inversely proportional to the *data interval in object space.* The relationships between data ranges and intervals are completely symmetrical between object space and k-space. Figure F-2 illustrates these relationships.

Equations F-7 and F-8 have important practical ramifications. Specifically, a fundamental assumption underlying the discrete FT algorithm is that the sampled function $f(x)$ actually extends indefinitely along the x-axis, repeating itself at intervals $N\Delta x = \text{FOV}_x$. In other words, it assumes that the sampled data simply repeat themselves to infinity in both directions along the x-axis. Conversely, it assumes as well that $F(k)$ repeats itself at intervals $N\Delta k$, again to infinity, in both directions along the k-axis. This means, for example, that because of cyclic repetition, $F(N\Delta k) = F(0)$. This result can be understood by observing that $N\Delta k = \text{FOV}_k$ and, according to Equation F-8, $\text{FOV}_k = 1/\Delta x$.

Thus if a sinusoidal oscillation with spatial frequency $N\Delta k$ is present in the underlying sampled function, it would be sampled precisely once per cycle. Furthermore, with precise timing, it would be sampled always at the same point during the cycle. Thus the sampled data for a spatial frequency $k = N\Delta k$ would be indistinguishable from a "constant" signal level (i.e., $k = 0$). Similarly, the next higher frequency, $(N + 1)\Delta k$, would be sampled less than once per cycle and would be indistinguishable from a lower frequency, which turns out to be k_1. This effect, in which a higher

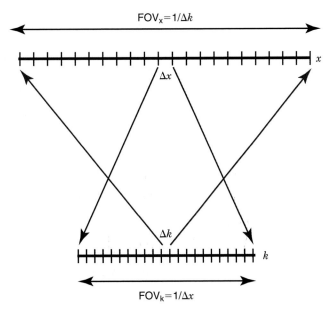

FIGURE F-2 Illustration of the relationships between sampling ranges and sampling intervals in object space and k-space.

frequency is falsely identified as a lower one, is called *aliasing.**

From this illustration, it is apparent that the highest spatial frequency that can be represented accurately in a discrete FT is determined by the sampling interval along the x-axis. However, there is yet another complication. Specifically, an image (or any object-space profile) can contain both positive and negative frequencies. For a cosine function, there is no difference, because $\cos(-x) = \cos(x)$. However, for a sine function, $\sin(-x) = -\sin(x)$, so the positive and negative oscillations are reversed. In terms of the level of "detail" contained in an image or profile, the difference between positive and negative frequencies is unimportant. The practical importance is that the use of "negative frequencies" allows one to obtain consistent representations for the FT of a function $f(x)$ that is shifted along the x-axis, but otherwise unchanged. This is the basis of the "shift theorem," which is discussed in advanced texts.[4]

The presence of negative frequencies means that the highest spatial frequency that can be computed by a discrete FT is not $N\Delta k$, but only half this value. This frequency, called the *Nyquist frequency,* can be derived from Equations F-6 and F-7 and usually is expressed as

$$k_{Nyquist} = 1/(2 \times \Delta x) \qquad \text{(F-9)}$$

This equation is the basis of the *sampling theorem,*[5] which states that, to accurately compute (or "recover") a specified spatial frequency requires a sampling rate of at least two samples per cycle at that frequency. Frequencies higher than the Nyquist frequency cannot be distinguished from lower frequencies. An example is illustrated in Figure F-3. Profiles (or signals) containing frequencies that are higher than the Nyquist frequency specified in Equation F-9 are said to be *undersampled.*

Undersampling and aliasing can lead to image distortions, errors in data analysis, and so forth. Figure F-4 illustrates their effects in k-space. In this hypothetical example, a sampling rate was used corresponding to $k_{Nyquist} = 8$ cm^{-1}; however, the

sampled function actually contained frequencies out to ± 10 cm^{-1}. Because of undersampling, frequencies between 8 and 10 cm^{-1} are "wrapped around" and overlap the lower end of the negative frequency portion of the spectrum. A similar wraparound of negative frequencies occurs at the high end of the spectrum. When the FT is calculated, the overlapped portions are simply added together in the *aliased regions* of the spectrum.

This effect and its consequences for emission CT reconstruction are discussed in Chapter 16, Section C.1. Similar effects occur in other applications of FTs [e.g., computations of MTFs (see Chapter 15, Section B.2)]. The sampling theorem also has practical ramifications for data acquisition with arrays of discrete detector elements. It can be shown that a linear array of detector elements, each of width d, with negligibly small spacings between elements, can detect spatial frequencies out to $k = 1/d$ and even beyond. (See Figure F-5C, which is essentially the MTF for an aperture of width $d = 1$.) However, with only one sample per distance d, the Nyquist frequency is $k_{Nyquist} = 1/(2d) = (1/2) \times (1/d)$ (i.e., one half the frequency that would be desired for $k_{max} = 1/d$). Thus spatial frequencies in the range $1/(2d) < k < 1/d$ are aliased back into lower frequency portions of the spectrum, where they can cause distortions of the recorded profile. One way to avoid this is to acquire a second set of data with the detector array "shifted" by one half the width of the individual detector elements and to insert these data points between those for the "unshifted" data. This reduces the sampling interval from d to $d/2$, thereby suppressing the effects of aliasing.

This analysis applies when the spatial resolution of the individual detector elements is essentially the width of the element. This would apply, for example, to collimated linear arrays operating in single-photon counting mode. The situation is yet more complicated in positron emission tomography (PET), for which the spatial resolution of a pair of detectors of width d can be as small as $d/2$ (see Chapter 18, Section A.3). In this case, interleaving is essential to achieve the required sampling frequency. Techniques for accomplishing this in PET imaging are described in Chapter 18, Section A.6.

It is important to realize that aliasing cannot be "undone" by postprocessing of undersampled data. If one starts with an undersampled data set, there is no way to

---

*Most readers will have seen examples of aliasing in motion pictures, in which undersampling in the frame rate leads to the appearance of false and even backward rotations of the spokes of wagon wheels, helicopter blades, and so forth.

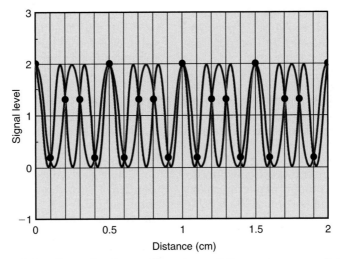

FIGURE F-3 Illustration of the effects of undersampling of a spatially varying source distribution. The *red curve* represents the actual signal, which has a spatial frequency of 6 cycles per cm (6 cm⁻¹). *Black dots* represent sampled points. The sampling distance is $\Delta x = 0.1$ cm, for which the Nyquist frequency is 5 cm⁻¹. As a result of undersampling, the sampled data for a spatial frequency of 6 cm⁻¹ cannot be distinguished from a signal corresponding to a spatial frequency of 4 cm⁻¹ (*green line*). The false representation of higher frequencies as lower frequencies is called *aliasing.*

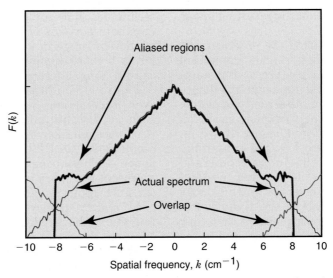

FIGURE F-4 Illustration of the effects of aliasing in k-space. The graphs represent hypothetical Fourier transforms (FTs). Data in object space were sampled at 16 points per cm (cm⁻¹) for computing the FT. The Nyquist frequency for this sampling rate is 8 cm⁻¹. However, the actual spatial-frequency spectrum (*blue line*) has significant frequency content extending out to 10 cm⁻¹. As a result of undersampling, the replicated ends of the actual spectrum (*orange line*) overlap and add to the calculated spectrum (*heavy red line*), creating an aliased region of the spectrum. The calculated spectrum does not accurately represent the true FT for the object.

"unwrap" the spectral overlap illustrated in Figure F-4.

Aliasing can be avoided in a number of ways, including the following:

1. The most direct way is to use a sampling interval along the x-axis (or along whatever object-space axis is being sampled) that is sufficiently small to meet the requirements of the sampling theorem for all frequencies present in the data. One way to achieve this is to acquire more data points along the sampled distance. However, for applications involving FFTs, this requires increasing the

number of samples (e.g., pixels) by a power of 2, which could be impractical for reasons of dataset size, computation time, and so on.

2. If increasing the number of samples is not practical, one could use the same number of samples but apply them over a smaller field-of-view, thereby decreasing the sampling interval Δx. This would increase the Nyquist frequency (Equation F-9); however, reducing FOV_x also has practical constraints. For example, the field-of-view must provide full coverage of the scanned portion of the body in computed tomography.

3. A third alternative is to deliberately "blur" the input profile (or image) before it is recorded by the detectors so as to suppress the potentially undersampled high-frequency components. In the context of image profiles, this means that the data projected onto the detector itself must be "blurred" (e.g., by using a coarser collimator). The obvious disadvantage of this approach is that the resulting image also is blurred.

If none of these options are practical, the effects of undersampling can be minimized (but not eliminated) by postprocessing steps. One approach for already undersampled data is to simply "lop off" or otherwise completely suppress portions of the frequency spectrum that might be affected by aliasing. However, as illustrated in Figure F-4, the potentially affected range is somewhat unpredictable and may be large. This also is a rather wasteful approach, because one ends up throwing away not only frequencies above the Nyquist frequency but any lower frequencies that they overlap after they are wrapped around in the spectrum.

D. SOME EXAMPLES OF FOURIER TRANSFORMS

Figure F-5 shows some one-dimensional examples of functions and their FTs. Note that the functions are represented as continuous functions. If DFTs were involved, the underlying functions would be the same, but the data would be represented by "points" on the curves, following the rules outlined in the preceding section.

In Figure F-5A, $f(x)$ is a constant, as would be all sampled values of it. Therefore the only

spatial frequency that has nonzero amplitude is $k = 0$, and that amplitude is equal to the value of the constant.

Figure F-5B shows a cosine wave of spatial frequency 1 cycle per cm and amplitude 1.0 superimposed on a constant value of 1 unit. In this case, $f(x) = 0.5 \cos (2\pi x) + 1$. The FT has an amplitude of 1 at $k = 0$, and amplitudes of 0.5 each at ± 1 cm^{-1}. This ambiguity arises from the fact that the FT cannot distinguish between positive versus negative cosine functions at these frequencies. (They are in fact identical.) Thus it assigns half of the observed amplitude to each value. (The FT of a sine function is similar, except that the value for $k = -1$ cm^{-1} is -0.5. This is because the FT cannot distinguish a positive sine wave of negative frequency from a negative sine wave of positive frequency.)

Figure F-5C shows a "boxcar" function (sometimes also called a "rect" function) for which $f(x) = 1$ when x is within the range $-0.5 \le x \le +0.5$ and $f(x) = 0$ outside this range. This function crops up often in signal-processing techniques that employ "windowing" filters to eliminate portions of the spectrum outside a windowed range in k-space. In this case, the FT is a more complicated function (specifically, a "sinc function").

Figure F-5C illustrates that FTs can exist even for functions with unusual properties, such as discontinuities. Also, although $f(x)$ itself does not have apparent "high-frequency oscillations," its FT has high-frequency components. These components are necessary to represent the sharp edges of the boxcar. Thus the MTF of an imaging system must have a good high-frequency response to faithfully reproduce sharp edges and fine details. In addition, the figure indicates that windowing filters with sharp edges in k-space can create "ringing artifacts" (sometimes called *Gibbs phenomenon*) in object space. These features, as well as illustrations of FTs of many other mathematical functions, are presented and discussed in reference 1.

Finally, Figure F-6 shows a brain image, an intensity profile for a strip across the image, and the FT of that profile. Note that most of the "information" is contained in the low-frequency portion of the spectrum. This does not mean that the high-frequency response (MTF) of the imaging system is unimportant. Indeed, because the high-frequency components are relatively weak to begin with, it is important that the imaging system be able to faithfully preserve them.

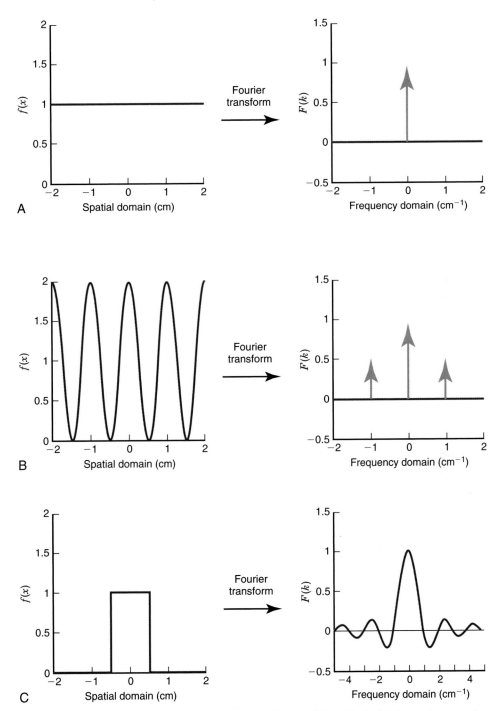

FIGURE F-5   Three functions and their Fourier transforms (FTs). A, $f(x) = 1.0$. Its FT has only one value, at $k = 0$. B, $f(x) = \cos(2\pi x) + 1$. Its FT has values at $k = 0$ and at $k = \pm 1$ cm^{-1}, corresponding to the spatial frequency of the cosine function. C, A boxcar function, $f(x) = 1$ when $-0.5 < x < 0.5$, $f(x) = 0$ otherwise. Its FT is a sinc function, $\mathrm{sinc}(x) = [\sin(\pi x)]/(\pi x)$. Note that this also is the modulation transfer function of an aperture or a detector element of unit width.

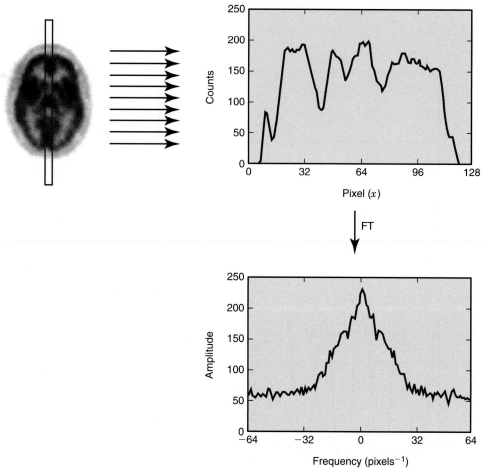

FIGURE F-6 Frequency spectrum for a profile recorded through a brain image. The box on the image indicates the sampling line for the profile (i.e., the direction x is vertical).

REFERENCES

1. Bracewell RN: *The Fourier Transform and Its Applications*, ed 3, New York, 2000, McGraw-Hill.
2. Bracewell RN: *Fourier Analysis and Imaging*, New York, 2004, Kluwer Academic Publishers.
3. Press WH, Teukolsky SA, Vetterling WT, Flannery BP: *Numerical Recipes in C*, ed 2, Cambridge, 1992, United Kingdom, Chapter 12.
4. Bracewell RN: *The Fourier Transform and Its Applications*, ed 3, New York, 2000, McGraw-Hill, pp 111-113.
5. Bracewell RN: *The Fourier Transform and Its Applications*, ed 3, New York, 2000, McGraw-Hill, Chapter 10.

Convolution

Convolution is a mathematical technique that is used in imaging to compute the effects of two (or more) simultaneously operating "blurring" functions. For example, the image projected by a collimator onto an Anger camera detector is blurred by the collimator holes (see Chapter 14, Sections C and D). Suppose that the collimator point-spread function is described by a one-dimensional function $f(x)$. That profile is projected onto the camera detector where it is blurred again by the intrinsic resolution of the detector (see Chapter 14, Section A.1). Suppose that the detector point-spread function is described by a function $g(x)$. Then the combined effects or the collimator and detector blurring are described by the convolution, $h(x)$, symbolized as

$$h(x) = f(x) * g(x) \qquad \text{(G-1)}$$

An alternative notation is

$$h(x) = f(x) \circledast g(x) \qquad \text{(G-2)}$$

The function $h(x)$ is thus the *system point-spread function* that accounts for both collimator and detector blurring. Mathematical convolutions find use in other fields as well, including statistics, electrical engineering, and general signal processing.

Mathematically, the convolution of two continuous functions is described by[1]

$$h(x) \int_{-\infty}^{\infty} f(u)g(x-u)du \qquad \text{(G-3)}$$

Equation G-3 is useful for theoretical development and analysis (e.g., for system design). However, most practical uses of convolution in the nuclear medicine laboratory involve discrete, rather than continuous functions (e.g., a projection profile for emission computed tomography is a discrete series of numbers). One representation for discrete

functions represented by a series of n values is

$$f = [a_0, a_1, \ldots, a_{n-1}] \qquad \text{(G-4)}$$

$$g = [b_0, b_1, \ldots, b_{n-1}] \qquad \text{(G-5)}$$

A more compact notation is $f(x_i)$. Here it is understood that $x_i = (i \times \Delta x)$, in which Δx is the sampling interval along the x-axis and $i = 0, 1, 2, \ldots, (n-1)$.

Using the notation of Equations G-4 and G-5, the convolution of two discrete functions is described by

$$h_j = \sum_{i=0}^{j} (a_i \times b_{j-i}) \qquad \text{(G-6)}$$

The subscript j runs from 0 to $(2n - 1)$; in other words, the convolution yields $(2n - 1)$ values for h.[†]

Analogous expressions exist for two-dimensional convolutions (i.e., functions of two variables, x and y). Convolutions frequently are performed for two-dimensional image processing (e.g., image smoothing—another form of "blurring"); however, the concepts embodied in Equations G-3 and G-6 are most easily demonstrated using one-dimensional graphical examples.

Consider first the convolution of two continuous functions. In Figure G-1A, the function $f(u)$ is plotted as a function of u. (Note that this is only a change of notation for mathematical representation. It is not a

---

[†]Note that some of the values for a and b appearing in Equation G–6 do not exist. Specifically, there are no values of a or b when their subscripts exceed $(n - 1)$, both of which occur in the equation as written. For computer implementation, the practical solution is to increase the length of the arrays for f and g by adding zeros to the list of values in Equations G–4 and G–5. This is sometimes called "zero padding." Additional values for h that are generated by this step are discarded.

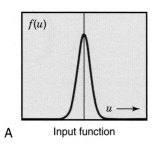

A   Input function

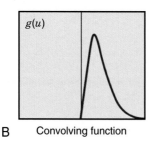

B   Convolving function

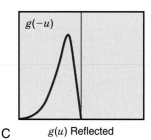

C   g(u) Reflected

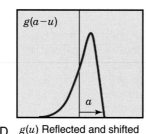

D   g(u) Reflected and shifted

FIGURE G-1 Illustration of the steps involved in determining the value of the convolution of two functions, $f(x)$ and $g(x)$, at $x = a$. *A* and *B*, the variable x is replaced by u. *C*, The convolving function, g, is reflected through the origin, $u = 0$. *D*, The reflected function is shifted by a distance, $x = a$, to the right. *E*, The function f and the shifted function, g, are overlapped. *F*, The product of f and the shifted function g is taken at each value of u. The area under the curve representing this product is $h(a)$, the value of the convolution at the value of $x = a$.

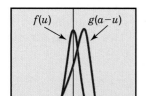

E   Functions superimposed

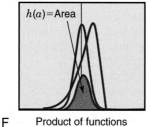

F   Product of functions

change of coordinate systems.) The convolving function, $g(u)$ is shown in Figure G-1B. Figure G-1C shows $g(-u)$, which is $g(u)$ reflected through the origin. In Figure G-1D, g is shifted by a distance a along the u-axis. This is $g(a - u)$, or $g(x - u)$ for $x = a$ in Equation G-3. According to Equation G-3, the integral of the product of f and the reflected and shifted function g is the value of the convolution function, h at $x = a$; in other words, $h(a)$. This is proportional to the shaded area in Figure G-1F. These steps (reflect, shift, take the product, and integrate) are repeated for all values of x (i.e., a in Figure G-1D) to obtain the full functional representation of $h(x)$.

Figure G-2 illustrates the convolution of two discrete functions, $f = [1,2,1]$ and $g = [0,2,1]$. As in Figure G-1, the convolving function, g, first is reflected through the origin (middle row, left). The product of the reflected

but unshifted $(a = 0)$ version of g and f is formed and summed across both functions. This is $h(0)$, which happens to have a value of zero in this example. The reflected and shifted version of g then is shifted along the horizontal axis and the process is repeated at unit increments of a. In each successive illustration for different values of a, the product is proportional to the shaded area. The process is repeated until g has passed completely over the function f and only values of zero are obtained (bottom row, right). As indicated earlier, this requires $(2n - 1)$ steps, yielding $(2n - 1)$ values for h.

Some important properties of convolutions, either continuous or discrete, are that they are commutative, distributive over addition, and associative. These properties are respectively described by the following equations.

$$f(x) * g(x) = g(x) * f(x) \qquad \text{(G-7)}$$

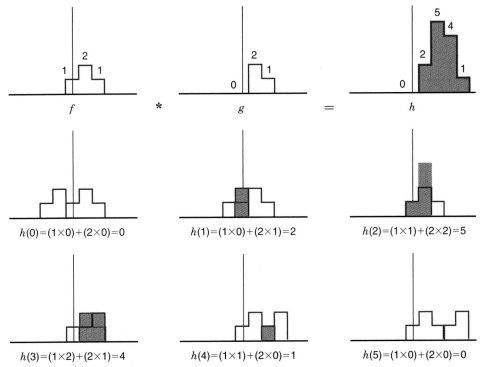

FIGURE G-2 *Top row,* The convolution of two discrete functions, $f(x) = [1,2,1]$ and $g(x) = [0,2,1]$, is given by $h(x) = [0,2,5,4,1]$. *Bottom rows,* The convolving function, $g(x)$, is reflected and then progressively shifted (*orange curves*) in unit increments of x across the stationary function, $f(x)$ (*blue curve*). The product of the overlapping functions is formed at each shift increment (*shaded areas*). The summation of the product (*shaded yellow areas*) is the value of the convolution, h, for the value of $x =$ to the shift distance. Where the reflected and shifted version of g does not overlap with f (e.g., $x = 0$ and $x = 5$), the convolution h is zero.

$$f(x) * [g(x) + h(x)] = [f(x) * g(x)] + [f(x) * h(x)]$$
(G-8)

$$f(x) * [g(x) * h(x)] = [f(x) * g(x)] * h(x) \quad \text{(G-9)}$$

Equation G-7 (commutative property) says that the order of the convolution can be reversed. Thus the result would be the same for a gamma camera image if the intrinsic blurring occurred before or after collimator blurring.

Equation G-8 (distributive property) says that the convolution of the sum is the same as the sum of the convolutions. Thus if two images are projected simultaneously onto the detector and then blurred, the same result is obtained as if the two images were projected separately, blurred, and then added.

Finally, Equation G-9 (associative property) says that the order of convolution does not affect the outcome. Thus if multiple blurring effects are present, the order in which they occur does not affect the outcome. This property is useful for image processing operations that employ convolutions. (Note that $h(x)$ in Equation G-9 refers to a third blurring

function, not the convolution of f and g as used in Equations G-1 to G-3.)

Convolutions have other useful properties. For example, Figure G-3 illustrates the convolution of two Gaussian functions. Gaussian functions often are used to approximate the blurring caused by collimators, intrinsic resolution of the detector, and so on. These functions are characterized by a mean value, μ, and variance, σ^2, and are of the form

$$f(x) = \frac{1}{\sqrt{2\pi\sigma^2}} e^{-(x-\mu)^2 / 2\sigma^2} \quad \text{(G-10)}$$

It can be shown that the convolution of two Gaussian functions, with means μ_1 and μ_2 and variances σ_1 and σ_2, is a third Gaussian function with mean and variance given by

$$\mu_3 = \mu_1 + \mu_2 \quad \text{(G-11)}$$

$$\sigma_3^2 = \sigma_1^2 + \sigma_2^2 \quad \text{(G-12)}$$

Equation G-12 applies as well to the full width at half-maximum (FWHM) for the Gaussian functions. Thus, if two Gaussian

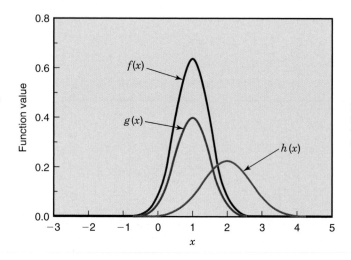

FIGURE G-3 The convolution of two Gaussian functions, $f(x)$ and $g(x)$, each having a mean value $\mu = 1.0$ and variance $\sigma^2 = 0.25$, is another Gaussian, $h(x)$, with $\mu = 2.0$ and variance $\sigma^2 = 0.5$.

functions, f and g, are characterized by FWHM(f) and FWHM(g), the full width at half-maximum of their convolution, h, is given by

$$\text{FWHM}(h) = \sqrt{\text{FWHM}(f)^2 + \text{FWHM}(g)^2}$$
(G-13)

This can be used to calculate the effect of combining the spatial resolutions of two components in an imaging system. For example, if a scintillation camera has an intrinsic resolution of 4 mm FWHM and a collimator resolution of 10 mm FWHM, and both can be reasonably approximated by a Gaussian shape, then their combined blurring effect is described by another Gaussian function with FWHM = $(10^2 + 4^2)^{1/2}$ = 10.8 mm. This would describe the system resolution for the detector/collimator combination (see Chapter 14, Section C.4). Note that blurring effects are not simply additive (i.e., 10 mm + 4 mm = 14 mm) and that the combined effect is dominated by the component that produces the greatest amount of blurring (the collimator, in this example).

One final useful property of convolution is given by the *convolution theorem*. If the Fourier transforms (FTs) of two functions $f(x)$ and $g(x)$ are $F(k)$ and $G(k)$, then the FT of their convolution has the properties that

$$\mathscr{F}[f(x) * g(x)] = F(k) \times G(k) \quad \text{(G-14)}$$

$$\mathscr{F}^{-1}[F(k) \times G(k)] = f(x) * g(x) \quad \text{(G-15)}$$

Here, $\mathscr{F}$ represents the operation of computing the FT and $\mathscr{F}^{-1}$ represents the inverse FT, as described in Appendix F. Thus convolution of two functions in the spatial domain is equivalent to point-by-point multiplication of their spectra in the spatial-frequency domain.

One practical application of the convolution theorem is that the modulation transfer function (MTF) of a system (see Chapter 15, Section B.2) is equal to the product of the MTFs of its individual components. This can be very helpful when the shapes of their point-spread functions are varied and non-Gaussian. Another practical application is that it provides a convenient alternative to Equation G-6 for computing the convolution of two functions. That equation requires many multiplications and data shifts, which can be very time consuming and tedious to implement in a computer. The convolution approach requires only the calculation of two FTs, a single point-by-point multiplication of these FTs, and the calculation of an inverse FT, all which can be performed much more rapidly by a computer than the tedious process described by Equation G-6. It should be noted that the convolution theorem can be extended to an arbitrary number of functions or dimensions.[2]

REFERENCES

1. Bracewell RN: *The Fourier Transform and Its Applications*, New York, 2000, McGraw-Hill, Chapter 3.
2. Bracewell RN: *Fourier Analysis and Imaging*. New York, 2004, Kluwer Academic Publishers, Chapter 5.

Index

Note: Page numbers followed by f indicate figures; those followed by t indicate tables.